NUMERICAL MATH
COM

Seri

G. H. GOLUB   A. GREENBAUM
A. M. STUART   E. SÜLI

NUMERICAL MATHEMATICS AND SCIENTIFIC COMPUTATION

Books in the series

Monographs marked with an asterix (*) appeared in the series 'Monographs in Numerical Analysis' which has been folded into, and is continued by, the current series.

Finite Elements and Fast Iterative Solvers: with Applications in Incompressible Fluid Dynamics

Howard C. Elman   David J. Silvester
Andrew J. Wathen

OXFORD

UNIVERSITY PRESS

*This book has been printed digitally and produced in a standard specification
in order to ensure its continuing availability*

OXFORD
UNIVERSITY PRESS

Great Clarendon Street, Oxford OX2 6DP

Oxford University Press is a department of the University of Oxford.
It furthers the University's objective of excellence in research, scholarship,
and education by publishing worldwide in

Oxford New York

Auckland Cape Town Dar es Salaam Hong Kong Karachi
Kuala Lumpur Madrid Melbourne Mexico City Nairobi
New Delhi Shanghai Taipei Toronto
With offices in
Argentina Austria Brazil Chile Czech Republic France Greece
Guatemala Hungary Italy Japan South Korea Poland Portugal
Singapore Switzerland Thailand Turkey Ukraine Vietnam

Oxford is a registered trade mark of Oxford University Press
in the UK and in certain other countries

Published in the United States
by Oxford University Press Inc., New York

© Oxford University Press 2005

ISBN 978-0-19-852868-5

To

Barbara, Alison and Gill

PREFACE

The subject of this book is the efficient solution of partial differential equations (PDEs) that arise when modelling incompressible fluid flow. The material is organized into four groups of two chapters each, covering the *Poisson equation* (Chapters 1 and 2); the *convection–diffusion equation* (Chapters 3 and 4); the *Stokes equations* (Chapters 5 and 6); and the *Navier–Stokes equations* (Chapters 7 and 8). These equations represent important models within the domain of computational fluid dynamics, but they also arise in many other settings. For each PDE model, there is a chapter concerned with finite element discretization, and a companion chapter concerned with efficient iterative solution of the algebraic equations obtained from discretization. As we will see, the first three models constitute subproblems of the last one, and the solution strategies that are developed for the Navier–Stokes equations build directly upon those of the preceding chapters. Three distinctive features of the book are the following.

1. **Scope**. We take the reader from the simplest elliptic PDE, namely the Poisson equation, to the nonlinear Navier–Stokes equations, the fundamental model of steady incompressible fluid flow.

2. **Practicality**. Our emphasis is on practical issues. A posteriori error estimation and robust and efficient iterative solution of linear systems of equations are discussed for each of the model problems. Approximation and linear system solution are not treated as separate topics; rather, state-of-the-art methods from numerical linear algebra (in particular, Krylov subspace methods and multilevel preconditioners) are placed into the context of solving particular discrete PDEs. Integration of material in this manner enables us to prove rigorous results on the convergence of the linear system solvers using properties of the discrete PDE operators. This approach is firmly rooted in the finite element multigrid literature, and we are hopeful that this book will play a bridge-building role between the PDE and linear algebra research communities.

3. **Style**. We take a contemporary view of scientific computing, and have attempted to couple rigorous mathematical analysis with computational examples. Thus, rather than being an end in itself, all of the analysis contained in the book is motivated by numerical experiments. Working in the "computational laboratory" is what ultimately drives our research and makes our scientific lives such fun. We have deliberately tried to develop theoretical results in a constructive style. Computational exercises reinforcing and complementing the analysis are included at the end of each chapter.

The material in the first four chapters has been taught to advanced undergraduate and first year graduate students at UMIST, the University of Manchester, the University of Maryland, and the University of Oxford over the past four years. Master's level courses in Mathematics, Computing or Engineering departments could be based on any of these chapters. The material in the final four chapters is presented at a research level; the algorithmic technology for the Navier–Stokes equations in the final chapter represents state-of-the-art research. The book is largely self-contained, although some mathematical details are left to more theoretical references. Having read our material first, students seem better able to delve into a more mathematical exposition (like that of Brenner & Scott [28], Brezzi & Fortin [33], Girault & Raviart [85] and Roos, Stynes & Tobiska [159]) and appreciate the technicalities.

Our intention is that the computational exercises should be done using `matlab`, see `http://www.mathworks.com`, specifically by running the Incompressible Flow Iterative Solution Software (IFISS) software that was developed to generate the numerical results in the text. Indeed, the software gives the reader the means to replicate all of the computational results herein (solution and error plots, solver convergence curves). The majority of IFISS is available in `matlab` source code, through the web sites

`http://www.manchester.ac.uk/ifiss`
`http://www.cs.umd.edu/~elman/ifiss.html`

or do a web search for 'IFISS'. We have used early versions of the IFISS software to support technical workshops that we have given in the last decade on fast solvers and modeling of incompressible flow. Instructors can use this material to design scientific computing courses based on the book. We are indebted to Alison Ramage of the University of Strathclyde, who contributed all the multigrid functions that are included the IFISS software. We are also grateful to Nick Watson who wrote some of the IFISS error estimation software.

We gratefully acknowledge the U. K. Engineering and Physical Sciences Research Council and U. S. National Science Foundation, and the University of Manchester Institute of Science and Technology, University of Maryland, and University of Oxford, for support of the research that led to the development of this book and for support for numerous long-term visits among the authors.

We wish to thank the following colleagues who have read preliminary versions of sections of the book: Pavel Bochev, Andrew Cliffe, Judy Ford, Phil Gresho, Max Gunzburger, Matthias Heil, Nick Higham, Seymour Parter, Alison Ramage, Arnold Reusken, Hans-Georg Roos, Valeria Simoncini, Alastair Spence, Gil Strang, Martin Stynes, Lutz Tobiska and Nick Trefethen. Their comments led to many improvements and their enthusiasm helped keep us sane in moments of despair. We especially wish to thank Endre Süli for his careful reading of several chapters and Catherine Powell for invaluable comments on the whole manuscript. In addition, the students in the University of Maryland course CMSC/AMSC 661 running in 2003 and 2004 served as involuntary guinea pigs in helping clarify the text and in improving the exercises in the first two chapters. Finally, we wish to thank Gene Golub and Cleve Moler. Cleve gave us `matlab` and changed our lives. The early contact among the three authors was facilitated by visits to Stanford University with invitations from Gene. Gene was instrumental in putting us in contact with Oxford University Press in the first instance, and more importantly, has been an inspiration to us throughout our academic careers.

CONTENTS

0

MODELS OF INCOMPRESSIBLE FLUID FLOW

To set the scene for subsequent chapters, the basic PDE models that are the focus of the book are derived in this chapter. Our presentation is rudimentary. Readers who would like to learn more about the physics underlying fluid flow modelling are advised to consult the books of Acheson [1] or Batchelor [12]. The fundamental principles are conservation of mass and conservation of momentum, or rather, that forces effect a change in momentum as described by Newton's famous Second Law of Motion.

Consider a fluid of density ρ moving in a region of three-dimensional space Ω. Suppose a particular small volume of fluid (imagine a dyed particle or bubble moving in the fluid) is at the position $\vec{x}$ with respect to some fixed coordinates at time t. If in a small interval of time δt this particle moves to position $\vec{x} + \delta \vec{x}$, then the velocity at position $\vec{x}$ and time t is

$$\vec{u} = (u_x, u_y, u_z) := \lim_{\delta t \to 0} \frac{\delta \vec{x}}{\delta t}.$$

Each of the velocity components u_x, u_y and u_z is a function of the coordinates x, y and z as well as the time t. Inside any particular fixed closed surface ∂D enclosing a volume $D \subset \Omega$, the total mass of fluid is $\int_D \rho \, d\Omega$, where $d\Omega = dxdydz$ is the increment of volume. The amount of fluid flowing out of D across ∂D is

$$\int_{\partial D} \rho \vec{u} \cdot \vec{n} \, dS,$$

where $\vec{n}$ is the unit normal vector to ∂D pointing outwards from D and dS is the increment of surface area. Therefore, since mass is conserved,

the rate of change of mass in D equals
the amount of fluid flowing into D across ∂D.

We express this mathematically as

$$\frac{d}{dt} \int_D \rho \, d\Omega = - \int_{\partial D} \rho \vec{u} \cdot \vec{n} \, dS.$$

Employing the Divergence theorem, which says that for a smooth enough vector field $\vec{v}$ and any region R with smooth enough boundary ∂R,

$$\int_{\partial R} \vec{v} \cdot \vec{n} \, dS = \int_R \nabla \cdot \vec{v} \, d\Omega,$$

1

and noting that D is fixed, yields

$$\frac{\mathrm{d}}{\mathrm{d}t}\int_D \rho\,\mathrm{d}\Omega + \int_{\partial D} \rho\vec{u}\cdot\vec{n}\,\mathrm{d}S = \int_D \frac{\partial\rho}{\partial t} + \nabla\cdot(\rho\vec{u})\,\mathrm{d}\Omega = 0,$$

where we have swapped the order of differentiation (with respect to time) and integration (with respect to space) in the first term. Since D is any volume, it follows that

$$\frac{\partial\rho}{\partial t} + \nabla\cdot(\rho\vec{u}) = 0 \quad\text{in } \Omega.$$

For an incompressible and homogeneous fluid the density is constant both with respect to time and the spatial coordinates. Hence

$$\nabla\cdot\vec{u} := \frac{\partial u_x}{\partial x} + \frac{\partial u_y}{\partial y} + \frac{\partial u_z}{\partial z} = 0 \quad\text{in } \Omega.$$

In order to express the conservation of momentum, we need to define the acceleration of the fluid. To this end, suppose that the velocity of the small (dyed) volume of fluid at time t is $\vec{u}$ as above and that at $t + \delta t$ it is $\vec{u} + \delta\vec{u}$. Since the velocity depends on both position and time, we explicitly write $\vec{u} = \vec{q}(\vec{x}, t)$ so that

$$\vec{u} + \delta\vec{u} = \vec{q}(\vec{x} + \delta\vec{x}, t + \delta t)$$

or, rearranging slightly,

$$\delta\vec{u} = \vec{q}(\vec{x} + \delta\vec{x}, t + \delta t) - \vec{q}(\vec{x}, t)$$
$$= \vec{q}(\vec{x} + \delta\vec{x}, t + \delta t) - \vec{q}(\vec{x}, t + \delta t) + \vec{q}(\vec{x}, t + \delta t) - \vec{q}(\vec{x}, t).$$

Then, using Taylor series we get

$$\vec{q}(\vec{x} + \delta\vec{x}, t + \delta t) - \vec{q}(\vec{x}, t + \delta t) = (\delta\vec{x}\cdot\nabla)\,\vec{q}(\vec{x}, t + \delta t) + \mathcal{O}(\|\delta\vec{x}\|^2)$$

and

$$\vec{q}(\vec{x}, t + \delta t) - \vec{q}(\vec{x}, t) = \delta t\frac{\partial}{\partial t}\vec{q}(\vec{x}, t) + \mathcal{O}(\delta t^2).$$

Dividing by δt, taking the limit as $\delta t \to 0$ and using the definition of $\vec{u} := \delta\vec{x}/\delta t$, we see that the acceleration is given by

$$\frac{\mathrm{d}\vec{u}}{\mathrm{d}t} = \lim_{\delta t\to 0}\frac{\delta\vec{u}}{\delta t} = \frac{\partial\vec{u}}{\partial t} + (\vec{u}\cdot\nabla)\vec{u}.$$

This derivative, the so-called *convective derivative*,

$$\frac{\mathrm{d}(\cdot)}{\mathrm{d}t} := \frac{\partial(\cdot)}{\partial t} + (\vec{u}\cdot\nabla)(\cdot),$$

expresses the rate of change of either a scalar quantity (or of each scalar component of a vector quantity) that is "following the fluid". Thus to summarize, the fluid acceleration is the convective derivative of the velocity. Note that the acceleration is nonlinear in $\vec{u}$.

For the volume D, which is fixed in space and time, the rate of change of momentum is the product of the mass and the acceleration, or

$$\int_D \rho \left(\frac{\partial \vec{u}}{\partial t} + (\vec{u} \cdot \nabla)\vec{u} \right) d\Omega.$$

An *ideal fluid* is an incompressible and homogeneous fluid that has no viscosity. For such a fluid, the only forces are due to the pressure, p, and any external *body force*, $\vec{f}$, such as gravity. Thus, the total force acting on the fluid contained in D are the pressure of the surrounding fluid, and the effect of the body force $\vec{f}$,

$$\int_{\partial D} p(-\vec{n}) \, dS + \int_D \rho \vec{f} \, d\Omega.$$

Taking $\vec{v} = p\vec{c}$ in the Divergence theorem with $\vec{c}$ being a constant vector pointing in an arbitrary direction gives

$$\int_D \nabla \cdot (p\vec{c}) \, d\Omega = \int_{\partial D} p\vec{c} \cdot \vec{n} \, dS.$$

However,

$$\nabla \cdot (p\vec{c}) = p \nabla \cdot \vec{c} + \vec{c} \cdot \nabla p = \vec{c} \cdot \nabla p$$

since $\vec{c}$ is constant. Moreover, $\vec{c}$ can be taken outside the integrals since it does not vary in D or on ∂D. This leads to

$$\vec{c} \cdot \left(\int_D \nabla p \, d\Omega - \int_{\partial D} p\vec{n} \, dS \right) = 0.$$

Since $\vec{c}$ points in an arbitrary direction it then follows that

$$\int_D \nabla p \, d\Omega = \int_{\partial D} p\vec{n} \, dS.$$

Newton's Second Law of Motion applied to the fluid in D requires that,

the rate of change of momentum of fluid in D equals
the sum of the external forces.

Expressed in mathematical terms, we have that

$$\int_D \rho \left(\frac{\partial \vec{u}}{\partial t} + (\vec{u} \cdot \nabla)\vec{u} \right) = \int_{\partial D} p(-\vec{n}) \, dS + \int_D \rho \vec{f} \, d\Omega.$$

This leads to

$$\int_D \rho \left(\frac{\partial \vec{u}}{\partial t} + (\vec{u} \cdot \nabla)\vec{u} \right) + \nabla p - \rho \vec{f} \, d\Omega = 0.$$

Then, since D is an arbitrary volume, we obtain the *Euler Equations* for an ideal incompressible fluid,

$$\frac{\partial \vec{u}}{\partial t} + (\vec{u} \cdot \nabla)\vec{u} = -\frac{1}{\rho}\nabla p + \vec{f} \quad \text{in } \Omega$$

$$\nabla \cdot \vec{u} = 0 \quad \text{in } \Omega.$$

Written in "longhand" and assuming that $\vec{f} = (g_x, g_y, g_z)^T$, the Euler system is as follows,

$$\frac{\partial u_x}{\partial t} + u_x \frac{\partial u_x}{\partial x} + u_y \frac{\partial u_x}{\partial y} + u_z \frac{\partial u_x}{\partial z} = -\frac{1}{\rho}\frac{\partial p}{\partial x} + g_x$$

$$\frac{\partial u_y}{\partial t} + u_x \frac{\partial u_y}{\partial x} + u_y \frac{\partial u_y}{\partial y} + u_z \frac{\partial u_y}{\partial z} = -\frac{1}{\rho}\frac{\partial p}{\partial y} + g_y$$

$$\frac{\partial u_z}{\partial t} + u_x \frac{\partial u_z}{\partial x} + u_y \frac{\partial u_z}{\partial y} + u_z \frac{\partial u_z}{\partial z} = -\frac{1}{\rho}\frac{\partial p}{\partial z} + g_z$$

$$\frac{\partial u_x}{\partial x} + \frac{\partial u_y}{\partial y} + \frac{\partial u_z}{\partial z} = 0.$$

To fully define a model of a physical problem, appropriate initial and boundary conditions need to be specified. For example, for flow inside a container one must have no flow across the boundary, hence one applies $\vec{u} \cdot \vec{n} = 0$ on $\partial\Omega$.

The Euler equations can be hugely simplified by assuming that the *vorticity*, $\vec{w} := \nabla \times \vec{u}$ is zero. This is called *irrotational flow*, and is a reasonable assumption in a number of important applications, for example, when considering the motion of small amplitude water waves. For irrotational flow, taking any fixed origin O the line integral

$$\phi(P) := \int_O^P \vec{u} \cdot \mathrm{d}\vec{x}$$

uniquely defines the value of a *fluid potential* ϕ at the point P. This is a consequence of Stokes' Theorem: if C is a closed curve in three-dimensional space and S is any surface that forms a "cap" on this curve (think of a soap bubble blown from any particular shape of plastic loop) then

$$\int_C \vec{u} \cdot \mathrm{d}\vec{x} = \int_S \nabla \times \vec{u} \cdot \vec{n}\,\mathrm{d}S.$$

In other words, the line integral defining $\phi(P)$ is independent of the actual path taken from O to P. Differentiation of this line integral in each coordinate direction then gives

$$\vec{u} = -\nabla\phi.$$

Notice that choosing a different origin O simply changes ϕ by a constant value in space. The value of this constant clearly does not alter $\vec{u}$ even though there may

be a different constant for every time t. (The minus sign is our convention—
other authors take a positive sign.) Combining the definition of the potential
with the incompressibility condition $\nabla \cdot \vec{u} = 0$ yields *Laplace's equation*

$$-\nabla^2 \phi = 0 \quad \text{in } \Omega.$$

This, in turn, is a particular case of the *Poisson equation*

$$-\nabla^2 \phi = f \quad \text{in } \Omega$$

that is the subject of Chapters 1 and 2.

   For such *potential flow* problems, if the body forces are conservative so that
$\vec{f} = -\nabla \Xi$ for some scalar potential Ξ, then the pressure can be recovered using
a classical construction as follows. Substituting the vector identity

$$(\vec{u} \cdot \nabla)\vec{u} = \tfrac{1}{2}\nabla(\vec{u} \cdot \vec{u}) - \vec{u} \times (\nabla \times \vec{u})$$

into the Euler equations and enforcing the condition that $\nabla \times \vec{u} = 0$ gives

$$\frac{\partial \vec{u}}{\partial t} + \frac{1}{\rho}\nabla p_T = 0 \quad \text{in } \Omega,$$

where $p_T = \tfrac{1}{2}\rho\vec{u} \cdot \vec{u} + p + \rho\Xi$ is called the *total pressure*. If we rewrite this as

$$\nabla\left(-\frac{\partial \phi}{\partial t} + \frac{p_T}{\rho}\right) = 0 \quad \text{in } \Omega$$

and integrate with respect to the spatial variables, we see that

$$-\frac{\partial \phi}{\partial t} + \frac{p_T}{\rho} = h(t),$$

where h is an arbitrary function of time only. Since $h(t)$ is constant in space,
one can consider the value of ϕ to be changed by this constant and so without
any loss of generality can take $h(t) = 0$. Thus, to summarize, having computed
the potential ϕ by solving Laplace's equation together with appropriate boundary conditions, the velocity $\vec{u}$ and pressure p can be explicitly computed via
$\vec{u} = -\nabla\phi$, and

$$p = p_T - \frac{1}{2}\rho\vec{u} \cdot \vec{u} - \rho\Xi = \rho\left(\frac{\partial \phi}{\partial t} - \frac{1}{2}\vec{u} \cdot \vec{u} - \Xi\right),$$

respectively.

   For a "real" viscous fluid, each small volume of fluid is not only acted on by
pressure forces (*normal stresses*), but also by *tangential stresses* (also called *shear*

stresses). Thus, as in the inviscid case the normal stresses are due to pressure giving rise to a force on the volume D of fluid,

$$\int_{\partial D} p(-\vec{n})\,\mathrm{d}S,$$

which can be written using the *unit diagonal tensor*

$$\vec{\mathbf{I}} = \begin{bmatrix} 1 & 0 & 0 \\ 0 & 1 & 0 \\ 0 & 0 & 1 \end{bmatrix}$$

as

$$\int_{\partial D} -p\vec{\mathbf{I}}\vec{n}\,\mathrm{d}S.$$

(The tensor vector product is like a matrix vector product.) The shear stresses on the other hand can act in any direction at the different points on ∂D so that a full tensor

$$\vec{\mathbf{T}} = \begin{bmatrix} T_{xx} & T_{xy} & T_{xz} \\ T_{yx} & T_{yy} & T_{yz} \\ T_{zx} & T_{zy} & T_{zz} \end{bmatrix}$$

is needed, and the force due to the shear stresses is given by

$$\int_{\partial D} \vec{\mathbf{T}}\vec{n}\,\mathrm{d}S.$$

Applying the Divergence theorem to each row of the shear stress tensor $\vec{\mathbf{T}}$, we arrive at the shear forces

$$\int_D \nabla\cdot\vec{\mathbf{T}}\,\mathrm{d}\Omega,$$

where $\nabla\cdot\vec{\mathbf{T}}$ is the vector

$$\begin{bmatrix} \dfrac{\partial T_{xx}}{\partial x} + \dfrac{\partial T_{xy}}{\partial y} + \dfrac{\partial T_{xz}}{\partial z} \\[2ex] \dfrac{\partial T_{yx}}{\partial x} + \dfrac{\partial T_{yy}}{\partial y} + \dfrac{\partial T_{yz}}{\partial z} \\[2ex] \dfrac{\partial T_{zx}}{\partial x} + \dfrac{\partial T_{zy}}{\partial y} + \dfrac{\partial T_{zz}}{\partial z} \end{bmatrix}.$$

A *Newtonian fluid* is characterized by the fact that the shear stress tensor is a linear function of the *rate of strain tensor*,

$$\vec{\epsilon} := \frac{1}{2}\left[\nabla\vec{u} + (\nabla\vec{u})^T\right] = \frac{1}{2}\begin{bmatrix} 2\dfrac{\partial u_x}{\partial x} & \dfrac{\partial u_x}{\partial y} + \dfrac{\partial u_y}{\partial x} & \dfrac{\partial u_x}{\partial z} + \dfrac{\partial u_z}{\partial x} \\[2mm] \dfrac{\partial u_y}{\partial x} + \dfrac{\partial u_x}{\partial y} & 2\dfrac{\partial u_y}{\partial y} & \dfrac{\partial u_y}{\partial z} + \dfrac{\partial u_z}{\partial y} \\[2mm] \dfrac{\partial u_z}{\partial x} + \dfrac{\partial u_x}{\partial z} & \dfrac{\partial u_z}{\partial y} + \dfrac{\partial u_y}{\partial z} & 2\dfrac{\partial u_z}{\partial z} \end{bmatrix}.$$

In this book, only such Newtonian fluids are considered. These include most common fluids such as air and water (and hence blood, etc.). Any fluid satisfying a nonlinear stress-strain relationship is called a non-Newtonian fluid. For a list of possibilities, see the book by Joseph [115]. For a Newtonian fluid, we have that

$$\mathbf{\vec{T}} = \mu\vec{\epsilon} + \lambda\text{Trace}(\vec{\epsilon})\mathbf{\vec{I}},$$

where μ and λ are parameters describing the "stickiness" of the fluid. For an incompressible fluid, the parameter λ is unimportant because $\text{Trace}(\vec{\epsilon}) = \nabla \cdot \vec{u} = 0$. However the *molecular viscosity*, μ, which is a fluid property measuring the resistance of the fluid to shearing, gives rise to the viscous shear forces

$$\mu\nabla \cdot \vec{\epsilon} = \mu\begin{bmatrix} \nabla^2 u_x + \left(\dfrac{\partial}{\partial x}\right)(\nabla \cdot \vec{u}) \\[2mm] \nabla^2 u_y + \left(\dfrac{\partial}{\partial y}\right)(\nabla \cdot \vec{u}) \\[2mm] \nabla^2 u_z + \left(\dfrac{\partial}{\partial z}\right)(\nabla \cdot \vec{u}) \end{bmatrix} = \mu\begin{bmatrix} \nabla^2 u_x \\[1mm] \nabla^2 u_y \\[1mm] \nabla^2 u_z \end{bmatrix} = \mu\nabla^2\vec{u}.$$

Application of Newton's Second Law of Motion in the case of a Newtonian fluid then gives

$$\int_D \rho\left(\frac{\partial\vec{u}}{\partial t} + (\vec{u} \cdot \nabla)\vec{u}\right) = \int_{\partial D} -p\mathbf{\vec{I}}\vec{n}\,dS + \int_D \mu\nabla^2\vec{u}\,d\Omega + \int_D \rho\vec{f}\,d\Omega.$$

We can convert the pressure term into an integral over D exactly as is done in the case of an ideal fluid above. Then, using the fact that D is an arbitrary region in the flow leads to the following generalization of the Euler equations,

$$\frac{\partial\vec{u}}{\partial t} + (\vec{u} \cdot \nabla)\vec{u} = -\frac{1}{\rho}\nabla p + \nu\nabla^2\vec{u} + \vec{f} \quad \text{in } \Omega$$

$$\nabla \cdot \vec{u} = 0 \quad \text{in } \Omega,$$

where $\nu := \mu/\rho$ is called the *kinematic viscosity*.

The focus of this book is *steady* flow problems. In this case, the time-derivative term on the left-hand side is zero. Then, absorbing the constant density into the pressure $(p \leftarrow p/\rho)$, leads to the "steady-state" *Navier–Stokes equations*,

$$-\nu\nabla^2\vec{u} + \vec{u}\cdot\nabla\vec{u} + \nabla p = \vec{f} \quad \text{in } \Omega$$
$$\nabla\cdot\vec{u} = 0 \quad \text{in } \Omega,$$

which when combined with boundary conditions forms one of the most general models of incompressible viscous fluid flow. This system is the focus of Chapters 7 and 8.

As observed above, the Navier–Stokes equations are nonlinear. However, simplification can be made in situations where the velocity is small or the flow is tightly confined. In such situations, a good approximation is achieved by dropping the quadratic (nonlinear) term from the Navier–Stokes equations and absorbing the constant ν into the velocity $(\vec{u} \leftarrow \nu\vec{u})$, giving the *Stokes equations*

$$-\nabla^2\vec{u} + \nabla p = \vec{f} \quad \text{in } \Omega$$
$$\nabla\cdot\vec{u} = 0 \quad \text{in } \Omega.$$

There is little loss of generality if $\vec{f}$ is set to zero. A conservative body force (e.g. that due to gravity when the fluid is supported below) is the gradient of a scalar field, that is, $\vec{f} = -\nabla\Xi$, and thus it can be incorporated into the system by redefining the pressure $(p \leftarrow p + \Xi)$. The numerical solution of Stokes equations in this form is described in Chapters 5 and 6.

Another linearization of the Navier–Stokes equations replaces the term $\vec{u}\cdot\nabla\vec{u}$ with $\vec{w}\cdot\nabla\vec{u}$ where $\vec{w}$ is a known vector field (often called the *wind* in this context). If the pressure term is dropped from the momentum equation, the resulting *Convection–Diffusion equation*

$$-\nu\nabla^2\vec{u} + \vec{w}\cdot\nabla\vec{u} = \vec{f},$$

is an important equation for various reasons. As well as describing many significant physical processes like the transport and diffusion of pollutants, it turns out to be very important to have efficient methods for this equation in order to get good numerical solution strategies for the Navier–Stokes equations. As given here, this is just three uncoupled scalar equations for the separate velocity components and so is no more or less general than the equation for convection and diffusion of a scalar quantity u with a scalar forcing term f,

$$-\nu\nabla^2 u + \vec{w}\cdot\nabla u = f \quad \text{in } \Omega.$$

Numerical methods for the solution of this equation together with suitable boundary conditions are described in Chapters 3 and 4.

1

THE POISSON EQUATION

The Poisson equation

$$-\nabla^2 u = f \tag{1.1}$$

is the simplest and the most famous elliptic partial differential equation. The source (or load) function f is given on some two- or three-dimensional domain denoted by $\Omega \subset \mathbb{R}^2$ or $\mathbb{R}^3$. A solution u satisfying (1.1) will also satisfy boundary conditions on the boundary $\partial\Omega$ of Ω; for example

$$\alpha u + \beta \frac{\partial u}{\partial n} = g \quad \text{on } \partial\Omega, \tag{1.2}$$

where $\partial u/\partial n$ denotes the directional derivative in the direction normal to the boundary $\partial\Omega$ (conventionally pointing outwards) and α and β are constant, although variable coefficients are also possible. In a practical setting, u could represent the temperature field in Ω subject to the external heat source f. Other important physical models include gravitation, electromagnetism, elasticity and inviscid fluid mechanics, see Ockendon et al. [138, chap. 5] for motivation and discussion.

The combination of (1.1) and (1.2) together is referred to as a *boundary value problem*. If the constant β in (1.2) is zero, then the boundary condition is of Dirichlet type, and the boundary value problem is referred to as the Dirichlet problem for the Poisson equation. Alternatively, if the constant α is zero, then we correspondingly have a Neumann boundary condition, and a Neumann problem. A third possibility is that Dirichlet conditions hold on part of the boundary $\partial\Omega_D$, and Neumann conditions (or indeed *mixed* conditions where α and β are both nonzero) hold on the remainder $\partial\Omega \backslash \partial\Omega_D$.

The case $\alpha = 0$, $\beta = 1$ in (1.2) demands special attention. First, since $u = $ constant satisfies the *homogeneous* problem with $f = 0$, $g = 0$, it is clear that a solution to a Neumann problem can only be unique up to an additive constant. Second, integrating (1.1) over Ω using Gauss's theorem gives

$$-\int_{\partial\Omega} \frac{\partial u}{\partial n} = -\int_{\Omega} \nabla^2 u = \int_{\Omega} f; \tag{1.3}$$

thus a necessary condition for the existence of a solution to a Neumann problem is that the source and boundary data satisfy the *compatibility* condition

$$\int_{\partial\Omega} g + \int_{\Omega} f = 0. \tag{1.4}$$

1.1   Reference problems

The following examples of two-dimensional Poisson problems will be used to illustrate the power of the finite element approximation techniques that are developed in the remainder of the chapter. Since these problems are all of Dirichlet type (i.e. the boundary condition associated with (1.1) is of the form $u = g$ on $\partial\Omega$), the problem specification involves the shape of the domain Ω, the source data f and the boundary data g. The examples are posed on one of two domains: a square $\Omega_\square = (-1, 1) \times (-1, 1)$, or an L-shaped domain $\Omega_\vdash$ consisting of the complement in $\Omega_\square$ of the quadrant $(-1, 0] \times (-1, 0]$.

1.1.1 Example: Square domain $\Omega_\square$, constant source function $f(x) \equiv 1$, zero boundary condition.

This problem represents a simple diffusion model for the temperature distribution $u(x, y)$ in a square plate. The specific source term in this example models uniform heating of the plate, and the boundary condition models the edge of the plate being kept at an ice-cold temperature. The simple shape of the domain enables the solution to be explicitly represented. Specifically, using separation of variables it can be shown that

$$u(x, y) = \frac{(1 - x^2)}{2} - \frac{16}{\pi^3} \sum_{\substack{k=1 \\ k \text{ odd}}}^{\infty} \left\{ \frac{\sin(k\pi(1 + x)/2)}{k^3 \sinh(k\pi)} \right.$$

$$\left. \times (\sinh(k\pi(1 + y)/2) + \sinh(k\pi(1 - y)/2)) \right\}. \tag{1.5}$$

Series solutions of this type can only be found in the case of geometrically simple domains. Moreover, although such solutions are aesthetically pleasing to mathematicians, they are rather less useful in terms of computation. These are the raisons d'etre for approximation strategies such as the finite element method considered in this monograph.

A finite element solution (computed using our IFISS software) approximating the exact solution u is illustrated in Figure 1.1. The accuracy of the computed solution is explored in Computational Exercise 1.1.

1.1.2 Example: L-shaped domain $\Omega_\vdash$, constant source function $f(x) \equiv 1$, zero boundary condition.

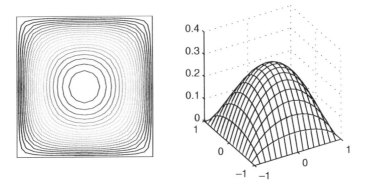

FIG. 1.1. Contour plot (left) and three-dimensional surface plot (right) of a finite element solution of Example 1.1.1.

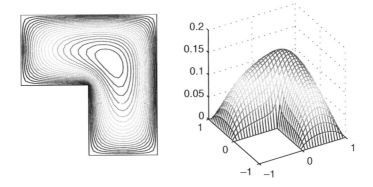

FIG. 1.2. Contour plot (left) and three-dimensional surface plot (right) of a finite element solution of Example 1.1.2.

A typical finite element solution is illustrated in Figure 1.2 (and is again easily computed using our IFISS software, see Computational Exercise 1.2). Notice that the contours are very close together around the corner at the origin, suggesting that the temperature is rapidly varying in this vicinity. A more careful investigation shows that the underlying Poisson problem has a *singularity* — the solution u is closely approximated at the origin by the function

$$u_{\Box}^*(r, \theta) = r^{2/3} \sin((2\theta + \pi)/3), \qquad (1.6)$$

where r represents the radial distance from the origin, and θ the angle with the vertical axis. This singular behavior is identified more precisely in Example 1.1.4. Here we simply note that radial derivatives of $u_{\Box}^*$ (and by implication those of u) are unbounded at the origin. See Strang & Fix [187, chap. 8] for further discussion of this type of function.

In order to assess the accuracy of approximations to the solution of boundary value problems in this and subsequent chapters, it will be convenient to refer to *analytic* test problems — these have an exact solution that can be explicitly computed at all points in the domain. Examples 1.1.3 and 1.1.4 are in this category.

1.1.3 Example: Square domain $\Omega_\square$, analytic solution.

This analytic test problem is associated with the following solution of Laplace's equation (i.e. (1.1) with $f = 0$),

$$u^*(x, y) = \frac{2(1 + y)}{(3 + x)^2 + (1 + y)^2}. \qquad (1.7)$$

Note that this function is perfectly smooth since the domain $\Omega_\square$ excludes the point $(-3, -1)$. A finite element approximation to u^* is given in Figure 1.3. For future reference we note that the boundary data g is given by the finite element interpolant of u^* on $\partial\Omega_\square$. We will return to this example when we consider finite element approximation errors in Section 1.5.1.

1.1.4 Example: L-shaped domain Ω_Γ, analytic solution.

This analytic test problem is associated with the singular solution u^*_Γ introduced in Example 1.1.2. A typical finite element approximation to u^*_Γ is given in Figure 1.4. Note that although u^*_Γ satisfies (1.1) with $f = 0$, see Problem 1.1, u^*_Γ is not smooth enough to meet the strict definition of a classical solution given in the next section. We will return to this example when discussing a posteriori error estimation in Section 1.5.2.

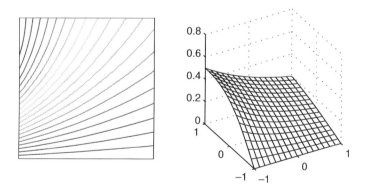

FIG. 1.3. Contour plot (left) and three-dimensional surface plot (right) of a finite element solution of Example 1.1.3.

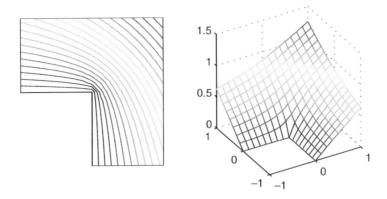

FIG. 1.4. Contour plot (left) and three-dimensional surface plot (right) of a
    finite element solution of Example 1.1.4.

1.2 Weak formulation

A sufficiently smooth function u satisfying both (1.1) and (1.2) is known as a
classical solution to the boundary value problem, see Renardy & Rogers [156].
For a Dirichlet problem, u is a classical solution only if it has continuous second
derivatives in Ω (i.e. u is in $C^2(\Omega)$) and is continuous up to the boundary (u is
in $C^0(\overline{\Omega})$); see Braess [19, p. 34] for further details. In cases of non-smooth
domains or discontinuous source functions, the function u satisfying (1.1)–(1.2)
may not be smooth (or *regular*) enough to be regarded as a classical solution. As
we have observed, on the non-convex domain $\Omega_{\vdash}$ of Example 1.1.4, the solution
$u_{\vdash}^*$ does not even have a square integrable second derivative (see Problem 1.20
and the discussion in Section 1.5.1). Alternatively, suppose that the source
function is discontinuous, say $f = 1$ on $\{(x, y) \,|\, 0 < x < 1\} \subset \Omega_{\square}$ and $f = 0$ on
$\{(x, y) \,|\, -1 < x < 0\}$, which corresponds to a weight placed on part of an elastic
membrane. Since f is discontinuous in the x direction, the second partial derivative of the solution u with respect to x is discontinuous, and hence u cannot
be in $C^2(\Omega)$, and there is no classical solution. For such problems, which arise
from perfectly reasonable mathematical models, an alternative description of
the boundary value problem is required. Since this alternative description is less
restrictive in terms of the admissible data it is called a *weak formulation*.

   To derive a weak formulation of a Poisson problem, we require that for an
appropriate set of *test functions* v,

$$\int_{\Omega} (\nabla^2 u + f) v = 0. \tag{1.8}$$

This formulation exists provided that the integrals are well defined. If u is
a classical solution then it must also satisfy (1.8). If v is sufficiently smooth

however, then the smoothness required of u can be reduced by using the derivative of a product rule and the divergence theorem

$$-\int_{\Omega} v\nabla^2 u = \int_{\Omega} \nabla u \cdot \nabla v - \int_{\Omega} \nabla \cdot (v\nabla u)$$

$$= \int_{\Omega} \nabla u \cdot \nabla v - \int_{\partial\Omega} v\frac{\partial u}{\partial n},$$

so that

$$\int_{\Omega} \nabla u \cdot \nabla v = \int_{\Omega} vf + \int_{\partial\Omega} v\frac{\partial u}{\partial n}. \tag{1.9}$$

The point here is that the Problem (1.9) may have a solution u, called a *weak solution*, that is not smooth enough to be classical solution. If a classical solution does exist then (1.9) is equivalent to (1.1)–(1.2) and the weak solution is classical.

The case of a Neumann problem ($\alpha = 0$, $\beta = 1$ in (1.2)) is particularly straightforward. Substituting (1.2) into (1.9) gives the following formulation: find u defined on Ω such that

$$\int_{\Omega} \nabla u \cdot \nabla v = \int_{\Omega} vf + \int_{\partial\Omega} vg \tag{1.10}$$

for all suitable test functions v.

We need to address an important question at this point, namely, in what sense are the weak solution u and the test functions v in (1.10) meaningful? This is essentially a question of *where* to look to find the solution u, and what is meant by "all suitable v". To provide an answer we use the space of functions that are square-integrable in the sense of Lebesgue

$$L_2(\Omega) := \left\{ u : \Omega \to \mathbb{R} \,\middle|\, \int_{\Omega} u^2 < \infty \right\}, \tag{1.11}$$

and make use of the L_2 measure

$$\|u\| := \left(\int_{\Omega} u^2 \right)^{1/2}. \tag{1.12}$$

The integral on the left-hand side of (1.10) will be well defined if all first derivatives are in $L_2(\Omega)$; for example, if Ω is a two-dimensional domain and $\partial u/\partial x, \partial u/\partial y \in L_2(\Omega)$ with $\partial v/\partial x, \partial v/\partial y \in L_2(\Omega)$, then using the

Cauchy–Schwarz inequality,

$$\int_\Omega \nabla u \cdot \nabla v = \int_\Omega \left(\frac{\partial u}{\partial x}\right)\left(\frac{\partial v}{\partial x}\right) + \int_\Omega \left(\frac{\partial u}{\partial y}\right)\left(\frac{\partial v}{\partial y}\right)$$

$$\leq \left\|\frac{\partial u}{\partial x}\right\|\left\|\frac{\partial v}{\partial x}\right\| + \left\|\frac{\partial u}{\partial y}\right\|\left\|\frac{\partial v}{\partial y}\right\| < \infty.$$

Similarly, the integrals on the right-hand side of (1.10) will certainly be well-defined if $f \in L_2(\Omega)$ and $g \in L_2(\partial\Omega)$.[1]

To summarize, if $\Omega \subset \mathbb{R}^2$ then the *Sobolev space* $\mathcal{H}^1(\Omega)$ given by

$$\mathcal{H}^1(\Omega) := \left\{ u : \Omega \to \mathbb{R} \,\middle|\, u, \frac{\partial u}{\partial x}, \frac{\partial u}{\partial y} \in L_2(\Omega) \right\}$$

is the space where a weak solution of (1.10) naturally exists, and this space is also the natural home for the test functions v. For clarity of exposition, further discussion of such technical issues is postponed until Section 1.5.

We now return to (1.9) and consider other types of boundary conditions. In general, we only need to look for weak solutions among those functions that satisfy the Dirichlet boundary conditions. (Engineers call Dirichlet boundary conditions "essential conditions" whereas Neumann conditions are "natural conditions" for the Laplacian.) To fix ideas, in the remainder of the chapter we restrict our attention to the following generic boundary value problem:

Find u such that

$$-\nabla^2 u = f \quad \text{in } \Omega \tag{1.13}$$

$$u = g_D \text{ on } \partial\Omega_D \quad \text{and} \quad \frac{\partial u}{\partial n} = g_N \text{ on } \partial\Omega_N, \tag{1.14}$$

where $\partial\Omega_D \cup \partial\Omega_N = \partial\Omega$ and $\partial\Omega_D$ and $\partial\Omega_N$ are distinct.

We assume that $\int_{\partial\Omega_D} ds \neq 0$, so that (1.14) does not represent a Neumann condition. Then we define *solution* and *test* spaces by

$$\mathcal{H}_E^1 := \{ u \in \mathcal{H}^1(\Omega) \,|\, u = g_D \text{ on } \partial\Omega_D \}, \tag{1.15}$$

$$\mathcal{H}_{E_0}^1 := \{ v \in \mathcal{H}^1(\Omega) \,|\, v = 0 \text{ on } \partial\Omega_D \}, \tag{1.16}$$

respectively. We should emphasize the difference between the two spaces: the Dirichlet condition from (1.14) is built into the definition of the solution space $\mathcal{H}_E^1$, whereas functions in the test space $\mathcal{H}_{E_0}^1$ are zero on the Dirichlet portion of the boundary. This is in contrast to the Neumann case where the solution and

---

[1]The boundary term can be shown to be well-defined using the *trace inequality* given in Lemma 1.5 in Section 1.5.1.

the test functions are not restricted on the boundary. Notice that the solution space is not closed under addition so strictly speaking it is not a vector space.

From (1.9) it is clear that any function u that satisfies (1.13) and (1.14) is also a solution of the following weak formulation:

Find $u \in \mathcal{H}_E^1$ such that

$$\int_\Omega \nabla u \cdot \nabla v = \int_\Omega vf + \int_{\partial \Omega_N} vg_N \quad \text{for all } v \in \mathcal{H}_{E_0}^1. \qquad (1.17)$$

We reiterate a key point here; a classical solution of a Poisson problem has to be twice differentiable in Ω — this is a much more stringent requirement than square integrability of first derivatives. Using (1.17) instead as the starting point enables us to look for approximate solutions that only need satisfy the smoothness requirement and the essential boundary condition embodied in (1.15). The case of a Poisson problem with a mixed boundary condition (1.2) is explored in Problem 1.2.

1.3 The Galerkin finite element method

We now develop the idea of approximating u by taking a finite-dimensional subspace of the solution space $\mathcal{H}_E^1$. The starting point is the weak formulation (1.15)–(1.17) of the generic problem (1.13)–(1.14). To construct an approximation method, we assume that $S_0^h \subset \mathcal{H}_{E_0}^1$ is a finite n-dimensional vector space of test functions for which $\{\phi_1, \phi_2, \ldots, \phi_n\}$ is a convenient basis. Then, in order to ensure that the Dirichlet boundary condition in (1.15) is satisfied, we extend this basis set by defining additional functions $\phi_{n+1}, \ldots, \phi_{n+n_\partial}$ and select fixed coefficients $\mathbf{u}_j, j = n+1, \ldots, n+n_\partial$, so that the function $\sum_{j=n+1}^{n+n_\partial} \mathbf{u}_j \phi_j$ interpolates the boundary data g_D on $\partial \Omega_D$. The finite element approximation $u_h \in S_E^h$ is then uniquely associated with the vector $\mathbf{u} = (\mathbf{u}_1, \mathbf{u}_2, \ldots, \mathbf{u}_n)^T$ of real coefficients in the expansion

$$u_h = \sum_{j=1}^{n} \mathbf{u}_j \phi_j + \sum_{j=n+1}^{n+n_\partial} \mathbf{u}_j \phi_j. \qquad (1.18)$$

The functions $\phi_i, i = 1, \ldots, n$ in the first sum in (1.18) define a set of *trial functions*. (In a finite element context they are often called *shape* functions.)

The construction (1.18) cleverly simplifies the characterization of discrete solutions when faced with difficult-to-satisfy essential boundary data,[2] for example, when solving test problems like that in Example 1.1.3.

---

[2]But it complicates the error analysis, see Section 1.5; if the data g_D is approximated then $S_E^h \not\subset \mathcal{H}_E^1$.

The construction of the space S_E^h is achieved above by ensuring that the specific choice of trial functions in (1.18) coincides with the choice of test functions that form the basis for S_0^h, and is generally referred to as the *Galerkin* (or more precisely *Bubnov–Galerkin*) approximation method. A more general approach is to construct approximation spaces for (1.15) and (1.16) using different trial and test functions. This alternative is called a *Petrov–Galerkin* approximation method, and a specific example will be discussed in Chapter 3.

The result of the Galerkin approximation is a finite-dimensional version of the weak formulation: find $u_h \in S_E^h$ such that

$$\int_\Omega \nabla u_h \cdot \nabla v_h = \int_\Omega v_h f + \int_{\partial \Omega_N} v_h g_N \quad \text{for all } v_h \in S_0^h. \tag{1.19}$$

For computations, it is convenient to enforce (1.19) for each basis function; then it follows from (1.18) that (1.19) is equivalent to finding $\mathbf{u}_j, j = 1, \dots, n$ such that

$$\sum_{j=1}^n \mathbf{u}_j \int_\Omega \nabla \phi_j \cdot \nabla \phi_i = \int_\Omega \phi_i f + \int_{\partial \Omega_N} \phi_i g_N - \sum_{j=n+1}^{n+n_\partial} \mathbf{u}_j \int_\Omega \nabla \phi_j \cdot \nabla \phi_i \tag{1.20}$$

for $i = 1, \dots, n$. This can be written in matrix form as the linear system of equations

$$A\mathbf{u} = \mathbf{f} \tag{1.21}$$

with

$$A = [a_{ij}], \quad a_{ij} = \int_\Omega \nabla \phi_j \cdot \nabla \phi_i, \tag{1.22}$$

and

$$\mathbf{f} = [\boldsymbol{f}_i], \quad \boldsymbol{f}_i = \int_\Omega \phi_i f + \int_{\partial \Omega_N} \phi_i g_N - \sum_{j=n+1}^{n+n_\partial} \mathbf{u}_j \int_\Omega \nabla \phi_j \cdot \nabla \phi_i. \tag{1.23}$$

The system of linear equations (1.21) is called the *Galerkin system*, and the function u_h computed by substituting the solution of (1.21) into (1.18) is the *Galerkin solution*. The matrix A is also referred to as the *stiffness matrix*.

The Galerkin coefficient matrix (1.22) is clearly symmetric (in contrast, using different test and trial functions necessarily leads to a nonsymmetric system matrix), and it is also positive-definite. To see this, consider a general coefficient vector $\mathbf{v}$ corresponding to a specific function $v_h = \sum_{j=1}^{n} \mathbf{v}_j \phi_j \in S_0^h$, so that

$$
\begin{aligned}
\mathbf{v}^T A \mathbf{v} &= \sum_{j=1}^{n} \sum_{i=1}^{n} \mathbf{v}_j a_{ji} \mathbf{v}_i \\
&= \sum_{j=1}^{n} \sum_{i=1}^{n} \mathbf{v}_j \left(\int_{\Omega} \nabla \phi_j \cdot \nabla \phi_i \right) \mathbf{v}_i \\
&= \int_{\Omega} \left(\sum_{j=1}^{n} \mathbf{v}_j \nabla \phi_j \right) \cdot \left(\sum_{i=1}^{n} \mathbf{v}_i \nabla \phi_i \right) \\
&= \int_{\Omega} \nabla v_h \cdot \nabla v_h \\
&\geq 0.
\end{aligned}
$$

Thus we see that A is at least semi-definite. Definiteness follows from the fact that $\mathbf{v}^T A \mathbf{v} = 0$ if and only if $\nabla v_h = 0$, that is, if and only if v_h is constant in Ω. Since $v_h \in S_0^h$, it is continuous up to the boundary and is zero on $\partial \Omega_D$, thus $\nabla v_h = 0$ implies $v_h = 0$. Finally, since the test functions are a basis for S_0^h we have that $v_h = 0$ implies $\mathbf{v} = \mathbf{0}$.

Once again the Neumann problem (1.10) requires special consideration. The Galerkin matrix is only semi-definite in this case and has a null space of vectors $\mathbf{v}$ corresponding to functions $\nabla v_h = 0$. In this situation it is essential to constrain the subspace $S_h \subset \mathcal{H}^1$ by choosing a set of trial functions $\{\phi_j\}, j = 1, \ldots, n$ that define a *partition of unity*, that is, every vector in S_h must be associated with a coefficient vector $v_h = \sum_{j=1}^{n} \mathbf{v}_j \phi_j$ satisfying

$$
\sum_{j=1}^{n} \phi_j = 1. \tag{1.24}
$$

The construction (1.24) ensures that if v_h is a constant function, say $v_h \equiv \alpha$, then v_h is associated with a discrete vector that satisfies $\mathbf{v}_j = \alpha$ for all the coefficients. This means that the null space of the Galerkin matrix associated with (1.10) is one-dimensional, consisting of constant coefficient vectors. Notice that the solvability of the discrete Neumann system (the analogue of (1.21)) requires that the null space of the Galerkin matrix A be orthogonal to the right-hand side vector $\mathbf{f}$, that is, we require that $(1, \ldots, 1)^T \mathbf{f} = 0$ with

$$
\mathbf{f} = [\boldsymbol{f}_i], \qquad \boldsymbol{f}_i = \int_{\Omega} \phi_i f + \int_{\partial \Omega} \phi_i g. \tag{1.25}
$$

Using the property (1.24) shows that the discrete Neumann problem is solvable if and only if the underlying boundary value problem is well posed in the sense that (1.4) holds.

Returning to the general case (1.19), it is clear that the choices of S_E^h and S_0^h are central in that they determine whether or not u_h has any relation to the weak solution u. The inclusions $S_E^h \subset \mathcal{H}_E^1$ and $S_0^h \subset \mathcal{H}_{E_0}^1$ lead to *conforming* approximations; more general *nonconforming* approximation spaces containing specific discontinuous functions are also possible, see for example [19, pp. 104–106], but these are not considered here. The general desire is to choose S_E^h and S_0^h so that approximation to any required accuracy can be achieved if the dimension n is large enough. That is, it is required that the error $\|u - u_h\|$ reduces rapidly as n is increased, and moreover that the computational effort associated with solving (1.21) is acceptable — the choice of basis is critical in this respect. These issues are addressed in Section 1.5 and in Chapter 2.

The mathematical motivation for finite element approximation is the observation that a smooth function can often be approximated to arbitrary accuracy using *piecewise polynomials*. Starting from the Galerkin system (1.19), the idea is to choose basis functions $\{\phi_j\}$ in (1.18) that are locally nonzero on a mesh of triangles ($\mathbb{R}^2$) or tetrahedra ($\mathbb{R}^3$) or a grid of rectangles or bricks. We discuss two-dimensional elements first.

1.3.1   *Triangular finite elements* ($\mathbb{R}^2$)

For simplicity, we assume that $\Omega \subset \mathbb{R}^2$ is polygonal (as is often the case in practice), so that we are able to *tile* (or *tessellate*) the domain with a set of triangles $\triangle_k, k = 1, \ldots, K$, defining a *triangulation* $\mathcal{T}_h$. This means that vertices of neighboring triangles coincide and that

- $\bigcup_k \overline{\triangle}_k = \overline{\Omega}$,
- $\triangle_\ell \cap \triangle_m = \emptyset$ for $\ell \neq m$.

The points where triangle vertices meet are called *nodes*. Surrounding any node is a *patch* of triangles that each have that node as a vertex (see Figure 1.5). If we label the nodes $j = 1, \ldots, n$, then for each j, we define a basis function ϕ_j that is nonzero only on that patch. The simplest choice here (leading to a conforming approximation) is the $\boldsymbol{P_1}$ or piecewise linear basis function: ϕ_j is a linear function on each triangle, which takes the value one at the node point j and zero at all other node points on the mesh. Notice that ϕ_j is clearly continuous on Ω (see Figure 1.6). Moreover, although ϕ_j has discontinuities in slope across element boundaries, it is smooth enough that $\phi_j \in \mathcal{H}^1(\Omega)$, and so it leads to a conforming approximation space $S_0^h = \mathrm{span}(\phi_1, \phi_2, \ldots, \phi_n)$ for use with (1.19).

In terms of approximation, the precise choice of basis for the space is not important; for practical application however, the availability of a locally defined basis such as this one is crucial. Having only three basis functions that are not identically zero on a given triangle means that the construction of the Galerkin

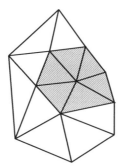

FIG. 1.5. A triangular mesh with a patch shaded.

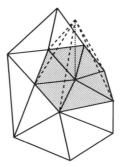

FIG. 1.6. A $\boldsymbol{P}_1$ basis function.

matrix A in (1.21) is easily automated. Another important point is that the Galerkin matrix has a well-defined *sparse* structure: $a_{ij} \neq 0$ only if the node points labeled i and j lie on the same edge of a triangular element. This is important for the development of efficient methods for solving the linear system (1.21), see Chapter 2.

Summarizing, $\boldsymbol{P}_1$ approximation can be characterized by saying that the overall approximation is continuous, and that on any element with vertices i, j and k there are only the three basis functions ϕ_i, ϕ_j and ϕ_k that are not identically zero. Within an element, ϕ_i is a linear function that takes the value one at node i and zero at nodes j and k . This local characterization is convenient for implementation of the finite element method (see Section 1.4) and it is also useful for the description of piecewise polynomial approximation spaces of higher degree.

For piecewise quadratic (or $\boldsymbol{P}_2$) approximation it is convenient to introduce additional nodes at the midpoint of each edge. Thus on each triangle there are six nodes, giving six basis functions that are not identically zero (recall that quadratic functions are of the form $ax^2 + bxy + cy^2 + dx + ey + f$ and thus have six coefficients). As in the linear case, we choose basis functions that have the

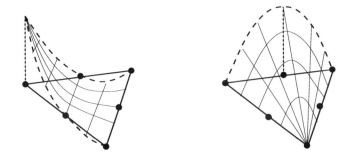

FIG. 1.7. P_2 basis functions of vertex type (left) and edge type (right).

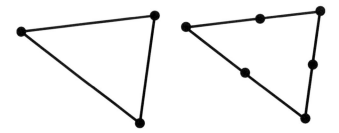

FIG. 1.8. Representation of P_1 (left) and P_2 (right) elements.

value one at a single node and zero at the other nodes as illustrated in Figure 1.7, see also Problem 1.3. These define a global approximation space of piecewise quadratic functions on the triangulation $\mathcal{T}_h$. Note that there are now "edge" as well as "vertex" functions, and that continuity across edges is guaranteed since there is a unique univariate quadratic (parabola) that takes given values at three points.

The illustration in Figure 1.8 is a convenient way to represent the P_1 and P_2 triangular elements. In Section 1.5 we will show that there can be advantages in the use of higher order approximations, in terms of the accuracy of approximation. The construction of higher order approximations (P_m with $m \geq 3$) is a straightforward generalization, see [19, pp. 65ff].

1.3.2 Quadrilateral elements ($\mathbb{R}^2$)

Although they are less flexible than triangle elements, it is often convenient to consider grids made up of rectangular (or more general quadrilateral) elements. For the simplest domains such as $\Omega_\square$ or $\Omega_\vdash$ in Section 1.1, it is clearly trivial to tile using square or rectangular elements. For more general domains, it is possible to use rectangles in the interior and then use triangles to match up to the boundary.

The simplest conforming quadrilateral element for a Poisson problem is the bilinear Q_1 element defined as follows. On a rectangle, each function is of the form $(ax+b)(cy+d)$ (hence bilinear). Again, for each of the four basis functions ϕ_j that are not identically zero on an element, the four coefficients are defined by the conditions that ϕ_j has the value one at vertex j and zero at all other vertices. For example, on an element $x \in [0, h]$, $y \in [0, h]$ the *element basis functions* are

$$(1 - x/h)(1 - y/h), \quad x/h(1 - y/h), \quad xy/h^2, \quad (1 - x/h)y/h,$$

starting with the function that is one at the origin and then moving anticlockwise. The global basis function on a patch of four elements is shown in Figure 1.9. Note that the Q_1 element has the additional "twist" term xy, which is not present in the P_1 triangle, and this generally gives the approximate solution some nonzero curvature on each element. Notice however, that when restricted to an edge the Q_1 element behaves like the P_1 triangle since it varies linearly. In both cases the approximation is continuous but has a discontinuous normal derivative. The upshot is that the Q_1 rectangle is in $\mathcal{H}^1(\Omega)$ and is hence conforming for (1.19).

In the case of arbitrary quadrilaterals, straightforward bilinear approximation as described above does not lead to a conforming approximation. A bilinear function is generally quadratic along an edge that is not aligned with a coordinate axis, and so it is not uniquely defined by its value at the two end points. This difficulty can be overcome by defining the approximation through an *isoparametric transformation*. The idea is to define the element basis functions

$$
\begin{aligned}
\chi_1(\xi, \eta) &= (\xi - 1)(\eta - 1)/4 \\
\chi_2(\xi, \eta) &= -(\xi + 1)(\eta - 1)/4 \\
\chi_3(\xi, \eta) &= (\xi + 1)(\eta + 1)/4 \\
\chi_4(\xi, \eta) &= -(\xi - 1)(\eta + 1)/4
\end{aligned}
\tag{1.26}
$$

on a *reference* element $\xi \in [-1, 1]$, $\eta \in [-1, 1]$, and then to map to any general quadrilateral with vertex coordinates (x_ν, y_ν), $\nu = 1, 2, 3, 4$ by the change

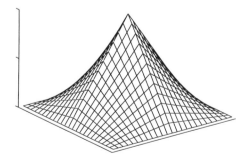

FIG. 1.9. A typical Q_1 basis function.

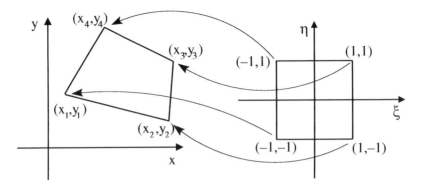

FIG. 1.10. Isoparametric mapping of Q_1 element.

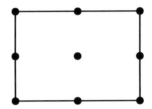

FIG. 1.11. Representation of Q_2 element.

of variables

$$x(\xi, \eta) - \sum_{\nu=1}^{4} x_\nu \chi_\nu(\xi, \eta), \quad y(\xi, \eta) = \sum_{\nu=1}^{4} y_\nu \chi_\nu(\xi, \eta), \tag{1.27}$$

see Figure 1.10. The outcome is that the mapped element basis function defined on the general element through (1.27) is linear along each element edge, and so it will connect continuously to the adjacent quadrilateral element whose basis will be defined isoparametrically based on its own vertex positions, see Problem 1.4. Element mappings are more fully discussed in Section 1.4. Notice that when using triangles one could employ a similar isoparametric transformation to a reference triangle based on the P_1 basis, see Section 1.4.1 for details.

Higher order approximations are defined analogously. For example, we can define a biquadratic finite element approximation on rectangles by introducing four additional mid-side node points, together with a ninth node at the centroid, as illustrated in the pictorial representation of Figure 1.11. In this case there are four vertex functions, four edge functions and one internal (or *bubble*) function in the element basis. The resulting approximation — which on each rectangle is of the form $(ax^2 + bx + c)(dy^2 + ey + f)$ — is a linear combination of the nine terms $1, x, y, x^2, xy, y^2, x^2y, xy^2, x^2y^2$ and is called Q_2. Note that just as Q_1 approximation is a complete linear polynomial together with the xy term of a bivariate quadratic, Q_2 has all six terms of a complete quadratic plus the two cubic terms $x^2 y, xy^2$ and the single quartic term $x^2 y^2$.

Clearly Q_2 approximation on rectangles is continuous (for exactly the same reason as P_2) and so is conforming for (1.19). Q_2 approximation may also be employed on arbitrary quadrilaterals through use of the bilinear mapping (1.27) (in which case the mapping is *subparametric*). Another point worth noting is that triangles and quadrilaterals may be used together in a conforming approximation space. For example, P_2 and Q_2 both have quadratic variation along edges and so can be used together.

Higher-degree piecewise polynomials may also be defined on rectangles (and thus on quadrilaterals) but other possibilities also present themselves; for example, by excluding the centroid node one is left with eight degrees of freedom, which allows the construction of a basis including all Q_2 terms except for the $x^2 y^2$ term, see for example [19, pp. 66ff]. Such an element is a member of the "serendipity" family.

1.3.3 Tetrahedral elements ($\mathbb{R}^3$)

The natural counterpart to triangular elements in three dimensions are tetrahedral (or simplex) elements. Any polyhedral region $\Omega \subset \mathbb{R}^3$ can be completely filled with tetrahedra $\triangle_k$ where each triangular face is common to only two tetrahedra, or else is part of the boundary $\partial \Omega$. Thus in a manner analogous to how triangles are treated, we define nodes at the vertices of the faces of the tetrahedra, and we define a P_1 basis function ϕ_j that is only nonzero on the set of tetrahedra for which node j is a vertex of one of its faces.

For each node j in a tessellation of Ω, we define $\phi_j(x, y, z)$ to be a linear function (i.e. of the form $a + bx + cy + dz$) on each tetrahedral element satisfying the interpolation condition

$$\phi_j(\text{node } i) = \begin{cases} 1 & \text{when } i = j, \\ 0 & \text{when } i \neq j. \end{cases} \tag{1.28}$$

Each basis function ϕ_j is continuous, thus ensuring a conforming approximation space: $S_0^h = \text{span}(\phi_1, \phi_2, \ldots, \phi_n) \subset \mathcal{H}_{E_0}^1$, where n is the number of nodes, as before. Note that there are precisely four basis functions that are nonzero on any particular tetrahedral element, corresponding to the four coefficients needed to define the linear approximation in the element. This also leads to a convenient implementation, exactly as in the triangular case.

Higher order tetrahedral elements are defined by introducing additional nodes. For example, the P_2 element has additional mid-edge nodes as depicted in Figure 1.12. This gives ten nodes in each element, matching the ten coefficients needed to define a trivariate quadratic polynomial (of the form $a + bx + cy + dz + ex^2 + fy^2 + gz^2 + hxy + kxz + lyz$). On any triangular face (which defines a plane, $\hat{a}x + \hat{b}y + \hat{c}z = \hat{d}$), one of the variables, z say, can be eliminated in terms of a linear combination of $1, x$ and y, to give a bivariate quadratic (in x, y) that is uniquely determined by its value at the six nodes of the $\boldsymbol{P}_2$ triangular element. As a result, continuity across inter-element faces, and hence a conforming approximation for (1.19), is assured.

More generally, $\boldsymbol{P}_m$ elements corresponding to continuous piecewise mth degree trivariate polynomial approximation are defined by an obvious generalization of the bivariate triangular analogue. All such tetrahedral elements have a gradient whose normal component is discontinuous across inter-element faces.

1.3.4 Brick elements ($\mathbb{R}^3$)

Three-dimensional approximation on cubes (or more generally *bricks*, which have six rectangular faces) is realized by taking the *tensor product* of lower dimensional elements. Thus the simplest conforming element for (1.19) is the trilinear $\boldsymbol{Q}_1$ element that takes the form $(ax + b)(cy + d)(ez + f)$ on each brick. Written as a linear combination, there are eight terms $1, x, y, z, xy, xz, yz, xyz$ and the eight coefficients are determined using the eight corner nodes, as illustrated in Figure 1.13. As a result, adopting the standard definition of a trilinear basis satisfying (1.28) there are precisely eight basis functions that are not identically zero within each element.

The recipe for higher order approximation on bricks is obvious. The $\boldsymbol{Q}_2$ triquadratic represented in Figure 1.13 has twenty-seven nodes; there are eight

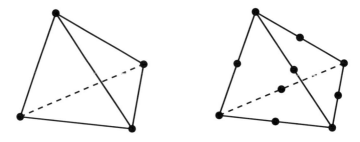

FIG. 1.12. Representation of $\boldsymbol{P}_1$ and $\boldsymbol{P}_2$ tetrahedral elements.

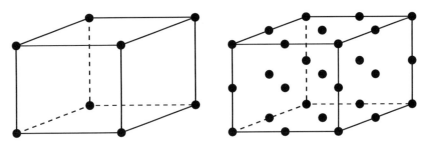

FIG. 1.13. Representation of $\boldsymbol{Q}_1$ and $\boldsymbol{Q}_2$ brick elements.

corner basis functions, twelve mid-edge basis functions, six mid-face basis functions and a single bubble function associated with the node at the centroid. Finally, we note that elements with six quadrilateral faces can be defined analogously to the two-dimensional case via a trilinear parametric mapping to the unit cube.

1.4   Implementation aspects

The computation of a finite element approximation consists of the following tasks. These are all built into the IFISS software.

(a) Input of the data on $\Omega, \partial\Omega_N, \partial\Omega_D$ defining the problem to be solved.

(b) Generation of a grid or mesh of elements.

(c) Construction of the Galerkin system.

(d) Solution of the discrete system, using a linear solver that exploits the sparsity of the finite element coefficient matrix.

(e) A posteriori error estimation.

In this section we focus on the core aspect (c) of setting up the discrete Galerkin system (1.21). Other key aspects, namely, the solution of this system and a posteriori error analysis, are treated in Chapter 2 and Section 1.5 respectively. Postprocessing of the solution is also required in general. This typically involves visualization and the calculation of derived quantities (e.g. boundary derivatives). A posteriori error analysis is particularly important. If the estimated errors are larger than desired, then the approximation space may be increased in dimension, either through local mesh subdivision (*h-refinement*), or by increasing the order of the local polynomial basis (*p-refinement*). An acceptable solution may then be calculated by cycling through steps (b)–(e) in an efficient way that builds on the existing structure, until the required error tolerance is satisfied.

The key idea in the implementation of finite element methodology is to consider everything "elementwise", that is, locally one element at a time. In effect the discrete problem is broken up; for example, (1.20) is rewritten as

$$\sum_{j=1}^{n} \mathbf{u}_j \int_{\Omega} \nabla\phi_j \cdot \nabla\phi_i = \sum_{j=1}^{n} \mathbf{u}_j \left\{ \sum_{\triangle_k \in \mathcal{T}_h} \int_{\triangle_k} \nabla\phi_j \cdot \nabla\phi_i \right\}. \qquad (1.29)$$

Notice that when forming the sum over the elements in (1.29), we need only take account of those elements where the basis functions ϕ_i and ϕ_j are both nonzero. This means that entries a_{ij} and f_i in the Galerkin system (1.21) can be computed by calculating contributions from each of the elements, and then gathering (or *assembling*) them together.

If the kth element has n_k local degrees of freedom, then there are n_k basis functions that are not identically zero on the element. For example, in the case of a mesh made up entirely of P_1 triangles, we have $n_k = 3$ for all elements, so that in each $\triangle_k$ there are three *element* basis functions associated with the restriction of three different *global* basis functions ϕ_j. In the case of a mesh containing a mixture of Q_2 rectangles and P_2 triangles, we have $n_k = 9$ if element k is a rectangle and $n_k = 6$ otherwise. In all cases the local functions form an (element) basis set

$$\Xi_k := \{\psi_{k,1}, \psi_{k,2}, \ldots, \psi_{k,n_k}\}, \tag{1.30}$$

so that the solution within the element takes the form

$$u_h|_k = \sum_{i=1}^{n_k} \mathbf{u}_i^{(k)} \psi_{k,i}. \tag{1.31}$$

Using triangular elements, for example, and localizing (1.22) and (1.23), we need to compute a set of $n_k \times n_k$ element matrices A_k and a set of n_k-vectors $\mathbf{f}_k$ such that

$$A_k = [a_{ij}^{(k)}], \quad a_{ij}^{(k)} = \int_{\triangle_k} \nabla \psi_{k,i} \cdot \nabla \psi_{k,j}, \tag{1.32}$$

$$\mathbf{f}_k = [\mathbf{f}_i^{(k)}], \quad \mathbf{f}_i^{(k)} = \int_{\triangle_k} f \, \psi_{k,i} + \int_{\partial \Omega_N \cap \partial \triangle_k} g_N \, \psi_{k,i}. \tag{1.33}$$

The matrix A_k is referred to as the *element stiffness matrix* (*local stiffness matrix*) associated with element $\triangle_k$. Its construction for the cases of triangular and quadrilateral elements is addressed in Sections 1.4.1 and 1.4.2. (A completely analogous construction is required for $\mathbb{R}^3$, see Hughes [108, chap. 3].) Notice that for computational convenience the essential boundary condition has not been enforced in (1.33). This is the standard implementation; essential conditions are usually imposed after the assembly of the element contributions into the Galerkin matrix has been completed. We will return to this point in the discussion of the assembly process in Section 1.4.3.

1.4.1   Triangular element matrices

The first stage in the computation of the element stiffness matrix A_k is to map from a reference element $\triangle_*$ onto the given element $\triangle_k$, as illustrated in Figure 1.14. For straight sided triangles the local–global mapping is defined for all points $(x, y) \in \triangle_k$ and is given by

$$x(\xi, \eta) = x_1 \chi_1(\xi, \eta) + x_2 \chi_2(\xi, \eta) + x_3 \chi_3(\xi, \eta) \tag{1.34}$$

$$y(\xi, \eta) = y_1 \chi_1(\xi, \eta) + y_2 \chi_2(\xi, \eta) + y_3 \chi_3(\xi, \eta), \tag{1.35}$$

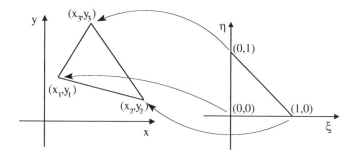

FIG. 1.14. Isoparametric mapping of P_1 element.

where

$$\chi_1(\xi, \eta) = 1 - \xi - \eta$$
$$\chi_2(\xi, \eta) = \xi \qquad (1.36)$$
$$\chi_3(\xi, \eta) = \eta$$

are the P_1 basis functions defined on the reference element. We note in passing that elements with curved sides can be generated using the analogous mapping defined by the P_2 reference element basis functions illustrated in Figure 1.7.

Clearly, the map from the reference element onto $\triangle_k$ is (and has to be) differentiable. Thus, given a differentiable function $\varphi(\xi, \eta)$, we can transform derivatives via

$$\begin{bmatrix} \dfrac{\partial \varphi}{\partial \xi} \\[2mm] \dfrac{\partial \varphi}{\partial \eta} \end{bmatrix} = \begin{bmatrix} \dfrac{\partial x}{\partial \xi} & \dfrac{\partial y}{\partial \xi} \\[2mm] \dfrac{\partial x}{\partial \eta} & \dfrac{\partial y}{\partial \eta} \end{bmatrix} \begin{bmatrix} \dfrac{\partial \varphi}{\partial x} \\[2mm] \dfrac{\partial \varphi}{\partial y} \end{bmatrix}. \qquad (1.37)$$

The *Jacobian matrix* in (1.37) may be simply calculated by substituting (1.36) into (1.34)–(1.35) and differentiating to give

$$J_k = \frac{\partial(x, y)}{\partial(\xi, \eta)} = \begin{bmatrix} x_2 - x_1 & y_2 - y_1 \\ x_3 - x_1 & y_3 - y_1 \end{bmatrix}. \qquad (1.38)$$

Thus in this simple case, we see that J_k is a constant matrix over the reference element, and that the determinant

$$|J_k| = \begin{vmatrix} x_2 - x_1 & y_2 - y_1 \\ x_3 - x_1 & y_3 - y_1 \end{vmatrix} = \begin{vmatrix} 1 & x_1 & y_1 \\ 1 & x_2 & y_2 \\ 1 & x_3 & y_3 \end{vmatrix} = 2|\triangle_k| \qquad (1.39)$$

is simply the ratio of the area of the mapped element $\triangle_k$ to that of the reference element $\triangle_*$. The fact that $|J_k(\xi, \eta)| \neq 0$ for all points $(\xi, \eta) \in \triangle_*$ is very important; it ensures that the inverse mapping from $\triangle_k$ onto the reference element is uniquely defined and is differentiable. This means that the derivative transformation (1.37) can be inverted to give

$$
\begin{bmatrix} \dfrac{\partial \varphi}{\partial x} \\[2mm] \dfrac{\partial \varphi}{\partial y} \end{bmatrix} = \begin{bmatrix} \dfrac{\partial \xi}{\partial x} & \dfrac{\partial \eta}{\partial x} \\[2mm] \dfrac{\partial \xi}{\partial y} & \dfrac{\partial \eta}{\partial y} \end{bmatrix} \begin{bmatrix} \dfrac{\partial \varphi}{\partial \xi} \\[2mm] \dfrac{\partial \varphi}{\partial \eta} \end{bmatrix}. \tag{1.40}
$$

Thus we see that derivatives of functions defined on $\triangle_k$ satisfy

$$
\begin{aligned}
\frac{\partial \xi}{\partial x} &= \frac{1}{|J_k|}\frac{\partial y}{\partial \eta}, & \frac{\partial \eta}{\partial x} &= -\frac{1}{|J_k|}\frac{\partial y}{\partial \xi}, \\[2mm]
\frac{\partial \xi}{\partial y} &= -\frac{1}{|J_k|}\frac{\partial x}{\partial \eta}, & \frac{\partial \eta}{\partial y} &= \frac{1}{|J_k|}\frac{\partial x}{\partial \xi}.
\end{aligned} \tag{1.41}
$$

Given the basis functions on the master element $\psi_{*,i};\ i = 1,\dots,n_k$, (see, e.g. Problem 1.3), the $\boldsymbol{P}_m$ element stiffness matrix A_k in (1.32) is easily computed:

$$
\begin{aligned}
a_{ij}^{(k)} &= \int_{\triangle_k} \frac{\partial \psi_{k,i}}{\partial x}\frac{\partial \psi_{k,j}}{\partial x} + \frac{\partial \psi_{k,i}}{\partial y}\frac{\partial \psi_{k,j}}{\partial y}\, dx\, dy \quad i,j = 1,\dots,n_k \\[2mm]
&= \int_{\triangle_*} \left\{ \frac{\partial \psi_{*,i}}{\partial x}\frac{\partial \psi_{*,j}}{\partial x} + \frac{\partial \psi_{*,i}}{\partial y}\frac{\partial \psi_{*,j}}{\partial y} \right\} |J_k|\, d\xi\, d\eta.
\end{aligned} \tag{1.42}
$$

In the specific case of the linear mapping given by (1.36), it is convenient to define the following coefficients:

$$
\begin{aligned}
b_1 &= y_2 - y_3; & b_2 &= y_3 - y_1; & b_3 &= y_1 - y_2; \\
c_1 &= x_3 - x_2; & c_2 &= x_1 - x_3; & c_3 &= x_2 - x_1;
\end{aligned} \tag{1.43}
$$

in which case (1.38)–(1.41) implies that

$$
\begin{bmatrix} \dfrac{\partial \varphi}{\partial x} \\[2mm] \dfrac{\partial \varphi}{\partial y} \end{bmatrix} = \frac{1}{2|\triangle_k|} \begin{bmatrix} b_2 & b_3 \\ c_2 & c_3 \end{bmatrix} \begin{bmatrix} \dfrac{\partial \varphi}{\partial \xi} \\[2mm] \dfrac{\partial \varphi}{\partial \eta} \end{bmatrix}. \tag{1.44}
$$

Combining (1.44) with (1.42) gives the general form of the stiffness matrix expressed in terms of the local derivatives of the element basis functions:

$$
\begin{aligned}
a_{ij}^{(k)} &= \int_{\triangle_*} \left(b_2 \frac{\partial \psi_{*,i}}{\partial \xi} + b_3 \frac{\partial \psi_{*,i}}{\partial \eta} \right)\left(b_2 \frac{\partial \psi_{*,j}}{\partial \xi} + b_3 \frac{\partial \psi_{*,j}}{\partial \eta} \right) \frac{1}{|J_k|}\, d\xi\, d\eta \\[2mm]
&\quad + \int_{\triangle_*} \left(c_2 \frac{\partial \psi_{*,i}}{\partial \xi} + c_3 \frac{\partial \psi_{*,i}}{\partial \eta} \right)\left(c_2 \frac{\partial \psi_{*,j}}{\partial \xi} + c_3 \frac{\partial \psi_{*,j}}{\partial \eta} \right) \frac{1}{|J_k|}\, d\xi\, d\eta.
\end{aligned} \tag{1.45}
$$

With the simplest linear approximation, that is, $\psi_{*,i} = \chi_i$ (see (1.36)), the local derivatives $\partial \psi_{*,i}/\partial \xi$, $\partial \psi_{*,i}/\partial \eta$ are constant, so the local stiffness matrix is trivial to compute (see Problem 1.5).

From a practical perspective, the simplest way of effecting the local–global transformation given by (1.36) is to define local element functions using *triangular* or *barycentric* coordinates (see Problem 1.6).

1.4.2 *Quadrilateral element matrices*

In the case of quadrilateral elements (and rectangular elements in particular), the stiffness matrix A_k is typically computed by mapping as in Figure 1.10 from a reference element $\Box_*$ onto the given element $\Box_k$, and then using quadrature. For quadrilaterals the local–global mapping is defined for all points $(x, y) \in \Box_k$ and is given by

$$x(\xi, \eta) = x_1 \chi_1(\xi, \eta) + x_2 \chi_2(\xi, \eta) + x_3 \chi_3(\xi, \eta) + x_4 \chi_4(\xi, \eta) \tag{1.46}$$
$$y(\xi, \eta) = y_1 \chi_1(\xi, \eta) + y_2 \chi_2(\xi, \eta) + y_3 \chi_3(\xi, \eta) + y_4 \chi_4(\xi, \eta), \tag{1.47}$$

where

$$\chi_1(\xi, \eta) = (\xi - 1)(\eta - 1)/4$$
$$\chi_2(\xi, \eta) = -(\xi + 1)(\eta - 1)/4$$
$$\chi_3(\xi, \eta) = (\xi + 1)(\eta + 1)/4$$
$$\chi_4(\xi, \eta) = -(\xi - 1)(\eta + 1)/4$$

are the $\boldsymbol{Q}_1$ basis functions defined on the reference element (see Figure 1.10).

The map from the reference element onto $\Box_k$ is differentiable, and derivatives are defined via (1.37), as in the triangular case. The big difference here is that the entries in the Jacobian matrix are *linear* functions of the coordinates (ξ, η) (cf. (1.38))

$$J_k = \frac{\partial(x, y)}{\partial(\xi, \eta)} = \begin{bmatrix} \sum_{j=1}^{4} x_j \dfrac{\partial \chi_j}{\partial \xi} & \sum_{j=1}^{4} y_j \dfrac{\partial \chi_j}{\partial \xi} \\[2mm] \sum_{j=1}^{4} x_j \dfrac{\partial \chi_j}{\partial \eta} & \sum_{j=1}^{4} y_j \dfrac{\partial \chi_j}{\partial \eta} \end{bmatrix}. \tag{1.48}$$

Note that the determinant $|J_k|$ is always a linear function of the coordinates, see Problem 1.7. In simple terms, the mapped element must have straight edges. If the mapped element $\Box_k$ is a parallelogram then the Jacobian turns out to be a constant matrix.

A sufficient condition for a well-defined inverse mapping ($|J_k(\xi, \eta)| > 0$ for all points $(\xi, \eta) \in \Box_*$) is that the mapped element be convex. In this case, derivatives on $\square_k$ can be computed[3] using (1.40), with

$$\frac{\partial\xi}{\partial x} = \frac{1}{|J_k|}\sum_{j=1}^{4} y_j \frac{\partial\chi_j}{\partial\eta}, \qquad \frac{\partial\eta}{\partial x} = -\frac{1}{|J_k|}\sum_{j=1}^{4} y_j \frac{\partial\chi_j}{\partial\xi},$$

$$\frac{\partial\xi}{\partial y} = -\frac{1}{|J_k|}\sum_{j=1}^{4} x_j \frac{\partial\chi_j}{\partial\eta}, \qquad \frac{\partial\eta}{\partial y} = \frac{1}{|J_k|}\sum_{j=1}^{4} x_j \frac{\partial\chi_j}{\partial\xi}, \tag{1.49}$$

and the Q_m element stiffness matrix is computed via

$$a_{ij}^{(k)} = \int_{\square_*} \left\{ \frac{\partial\psi_{*,i}}{\partial x}\frac{\partial\psi_{*,j}}{\partial x} + \frac{\partial\psi_{*,i}}{\partial y}\frac{\partial\psi_{*,j}}{\partial y} \right\} |J_k|\, d\xi\, d\eta. \tag{1.50}$$

Note that if general quadrilateral elements are used then the integrals in (1.50) involve rational functions of polynomials.

Gauss quadrature is almost always used to evaluate the definite integrals that arise in the calculation of the element matrices A_k and the vectors $\mathbf{f}_k$. Quadrilateral elements are particularly amenable to quadrature because integration rules can be constructed by taking tensor products of the standard one-dimensional Gauss rules. This is the approach adopted in the IFISS software. The definite integral (1.50) is approximated by the summation

$$\bar{a}_{ij}^{(k)} = \sum_{s=1}^{m}\sum_{t=1}^{m} w_{st}|J_k(\xi_s,\eta_t)| \left\{ \frac{\partial\psi_{*,i}}{\partial x}\frac{\partial\psi_{*,j}}{\partial x} + \frac{\partial\psi_{*,i}}{\partial y}\frac{\partial\psi_{*,j}}{\partial y} \right\}\Bigg|_{(\xi_s,\eta_t)},$$

where the quadrature points (ξ_s,η_t) are those associated with one of the Gauss tensor-product hierarchy illustrated in Figure 1.15. The quadrature weights w_{st} are computed by taking the tensor product of the weights associated with the classical one-dimensional rule, see [108, pp. 141–145] for further details.

In one dimension, all polynomials of degree $2m-1$ can be integrated exactly using the classical m point Gauss rule. This is *optimal* in the sense that any

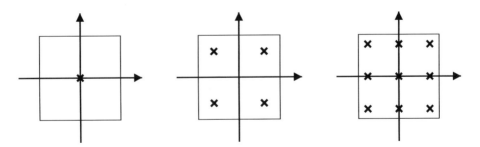

FIG. 1.15. Sampling points for 1×1, 2×2 and 3×3 Gauss quadrature rules.

---

[3] Using `deriv.m` and `qderiv.m` in IFISS.

rule with m points has precisely $2m$ free parameters (namely, the weights and positions of the quadrature points). Although the tensor-product rules are not optimal in this sense, the $m \times m$ rule does have the nice property that it exactly integrates all $\boldsymbol{Q}_{2m-1}$ functions. This means that in the case of grids of rectangular (or more generally parallelogram) elements, the bilinear element matrix A_k can be exactly computed using the 2×2 rule, see Problem 1.9. Similarly, the biquadratic element matrix A_k can be exactly integrated (for a rectangular element) if the 3×3 rule is used.

The element source vector (1.33) is also typically computed using quadrature. For example, the interior contribution to the source vector (1.33)

$$\boldsymbol{f}_i^{(k)} = \int_{\square*} f \psi_{*,i} \, |J_k| \mathrm{d}\xi \, \mathrm{d}\eta, \tag{1.51}$$

can be approximated via[4]

$$\bar{\boldsymbol{f}}_i^{(k)} = \sum_{s=1}^{m} \sum_{t=1}^{m} w_{st} f(\xi_s, \eta_t) \, \psi_{*,i}(\xi_s, \eta_t) \, |J_k(\xi_s, \eta_t)|. \tag{1.52}$$

The 2×2 rule would generally be used in the case of bilinear approximation, and the 3×3 rule if the approximation is biquadratic. Gauss integration rules designed for triangular elements are tabulated in [108, pp. 173–174].

1.4.3 Assembly of the Galerkin system

The assembly of the element contributions A_k and $\mathbf{f}_k$ into the Galerkin system is a reversal of the localization process illustrated in Figure 1.16.

The main computational issue is the need for careful bookkeeping to ensure that the element contributions are added into the correct locations in the coefficient matrix A and the vector $\mathbf{f}$. The simplest way of implementing the process is to represent the mapping between local and global entities using a connectivity matrix. For example, in the case of the mesh of $\boldsymbol{P}_1$ triangles illustrated in

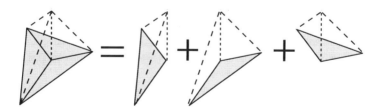

FIG. 1.16. Assembly of $\boldsymbol{P}_1$ global basis function from component element functions.

---

[4] Using gauss_source.m in IFISS.

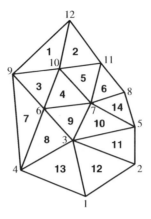

FIG. 1.17. Nodal and element numbering for the mesh in Figure 1.5.

Figure 1.17 we introduce the connectivity matrix defined by

$$\mathbf{P}^T = \begin{array}{c} \begin{array}{cccccccccccccc} 1 & 2 & 3 & 4 & 5 & 6 & 7 & 8 & 9 & 10 & 11 & 12 & 13 & 14 \end{array} \\ \begin{bmatrix} 9 & 12 & 9 & 6 & 10 & 11 & 4 & 4 & 6 & 5 & 5 & 2 & 1 & 8 \\ 10 & 10 & 6 & 7 & 7 & 7 & 6 & 3 & 3 & 7 & 3 & 3 & 3 & 7 \\ 12 & 11 & 10 & 10 & 11 & 8 & 9 & 6 & 7 & 3 & 2 & 1 & 4 & 5 \end{bmatrix} \end{array},$$

so that the index $j = P(k,i)$ specifies the global node number of local node i in element k, and thus identifies the coefficient $\mathbf{u}_i^{(k)}$ in (1.31) with the global coefficient $\mathbf{u}_j$ in the expansion (1.18) of u_h. Given P, the matrices A_k and vectors $\mathbf{f}_k$ for the mesh in Figure 1.17 can be assembled into the Galerkin system matrix and vector using a set of nested loops.

```
k = 1:14
  j = 1:3
    i = 1:3
      Agal(P(k,i),P(k,j)) = Agal(P(k,i),P(k,j)) + A(k,i,j)
    endloop i
    fgal(P(k,j)) = fgal(P(k,j)) + f(k,j)
  endloop j
endloop k
```

A few observations are appropriate here. First, in a practical implementation, the Galerkin matrix Agal will be stored in an appropriate sparse format. Second, it should be apparent that as the elements are assembled in order above, then for any node s say, a stage will be reached when subsequent assemblies do not affect node s (i.e. the sth row and column of the Galerkin matrix). When this stage is reached the variable is said to be fully summed; for example, variable 6 is fully summed after assembly of element **9**. This observation motivates the development of specialized direct solvers (known as *frontal solvers*) whereby the assembly process is intertwined with Gaussian elimination. In essence, as soon as a variable becomes fully summed, row operations can be performed to make entries below the diagonal zero and the modified row can then be saved for subsequent back-substitution, see for example Johnson [111, pp. 117–120].

It should also be emphasized that the intuitive element-by-element assembly embodied in the loop structure above is likely to be very inefficient; the inner loop involves indirect addressing and is too short to allow effective vectorization. The best way of generating efficient finite element code[5] is to work with blocks of elements and to reorder the loops so that the element loop k is the innermost. For real efficiency the number of elements in a block should be set so that all required data can fit into cache memory.

We now turn our attention to the imposition of essential boundary conditions on the assembled Galerkin system (1.21). We assume here that the basis functions are of Lagrangian type, that is, each basis function ϕ_j has a node $x_j \in \overline{\Omega}$ associated with it such that

$$\phi_j(x_j) = 1, \quad \phi_j(x_i) = 0 \qquad \text{for all nodes } x_i \neq x_j.$$

This property is depicted for the P_2 basis functions in Figure 1.7. It follows from this assumption that for $x_j \in \partial\Omega_D$, $u_h(x_j) = \mathbf{u}_j$, where the required value of $\mathbf{u}_j$ is interpolated from the Dirichlet boundary data. See Ciarlet [44, Section 2.2] for treatment of more general basis functions.

Now consider how to impose this condition at node 5 of the mesh in Figure 1.17. Suppose that a preliminary version of the Galerkin matrix A is constructed via (1.22) for $1 \leq i, j \leq n + n_\partial$, and that in addition, all the contributions $\int_\Omega \phi_i f$ have been assembled into the right-hand side vector $\mathbf{f}$. There are then two things needed to specify the system (1.21) via (1.20) and (1.23): the given value of $\mathbf{u}_5$ must be included in the definition of the vector $\mathbf{f}$ of (1.23), and the fifth row and column of the preliminary Galerkin matrix must be deleted (since ϕ_5 is being removed from the space of test functions). The first step can be achieved by multiplying the fifth column of A by the specified boundary value $\mathbf{u}_5$ and then subtracting the result from $\mathbf{f}$. An alternative technique[6] is to retain the imposed degree of freedom in the Galerkin system by modifying the row and column (5, here) of the Galerkin matrix corresponding to the boundary node so that the diagonal value is unity and the off-diagonal entries are set to zero, and then setting the corresponding value of $\mathbf{f}$ to the boundary value $\mathbf{u}_5$. Notice that the modified Galerkin matrix thus has a multiple eigenvalue of unity, with multiplicity equal to the number of nodes on the Dirichlet part of the boundary.

---

[5] Embodied in the IFISS routines `femq1_diff.m` and `femq2_diff.m`.
[6] Embodied in the IFISS routine `nonzerobc.m`.

Finally we remark that it is easiest to treat any nonzero Neumann boundary conditions in the system (1.21) after the assembly process and the imposition of essential boundary conditions has been completed. At this stage, the boundary contribution in (1.23) can be assembled by running through the boundary edges on $\partial\Omega_N$ and evaluating the component edge contributions using standard (one-dimensional) Gauss quadrature.

1.5   Theory of errors

Our starting point is the generic problem (1.13)–(1.14). The associated weak formulation is the following: find $u \in \mathcal{H}_E^1$ such that

$$\int_\Omega \nabla u \cdot \nabla v = \int_\Omega vf + \int_{\partial\Omega} v\, g_N \quad \text{for all } v \in \mathcal{H}_{E_0}^1, \tag{1.53}$$

with spaces $\mathcal{H}_E^1$ and $\mathcal{H}_{E_0}^1$ given by (1.15) and (1.16), respectively.

To simplify the notation we follow the established convention of not distinguishing between scalar-valued functions (e.g. $u : \Omega \to \mathbb{R}$) and vector-valued functions (e.g. $\vec{u} : \Omega \to \mathbb{R}^d$) as long as there is no ambiguity. In general, a bold typeface is used to represent a space of vector-valued functions, and norms and inner products are to be interpreted componentwise.

Definition 1.1 ($\mathbf{L}_2(\Omega)$ inner product and norm). Let $L_2(\Omega)$ denote the space of square-integrable scalar-valued functions defined on Ω, see (1.11), with associated inner product $(\cdot, \cdot)$. The space $\mathbf{L}_2(\Omega)$ of square-integrable vector-valued functions defined on Ω consists of functions with each component in $L_2(\Omega)$, and has inner product

$$(\vec{u}, \vec{v}) := \int_\Omega \vec{u} \cdot \vec{v},$$

and norm

$$\|\vec{u}\| := (\vec{u}, \vec{u})^{1/2}.$$

For example, for two-dimensional vectors $\vec{u} = (u_x, u_y)$ and $\vec{v} = (v_x, v_y)$, $(\vec{u}, \vec{v}) = (u_x, v_x) + (u_y, v_y)$ and $\|\vec{u}\|^2 = \|u_x\|^2 + \|u_y\|^2$.

Our first task is to establish that a weak solution is uniquely defined. To this end, we assume two weak solutions satisfying (1.53); $u_1 \in \mathcal{H}_E^1$ and $u_2 \in \mathcal{H}_E^1$ say, and then try to establish that $u_1 = u_2$ everywhere. Subtracting the two variational equations shows that $u_1 - u_2 \in \mathcal{H}_{E_0}^1$ satisfies the equation

$$\int_\Omega \nabla(u_1 - u_2) \cdot \nabla v = 0 \quad \text{for all } v \in \mathcal{H}_{E_0}^1. \tag{1.54}$$

Substituting $v = u_1 - u_2$ then shows that $\|\nabla(u_1 - u_2)\| = 0$, and this implies that $u_1 - u_2$ is a constant function. To make progress the case of a pure Neumann problem needs to be excluded. We can then use the additional fact that $u_1 = u_2$ on the Dirichlet part of the boundary. The following lemma holds the key to this.

Lemma 1.2 (Poincaré–Friedrichs inequality). *Assume that $\Omega \subset \mathbb{R}^2$ is contained in a square with side length L (and, in the case $\int_{\partial \Omega_N} ds \neq 0$, that it has a sufficiently smooth boundary). Given that $\int_{\partial \Omega_D} ds \neq 0$, it follows that*

$$\|v\| \leq L\,\|\nabla v\| \quad \text{for all } v \in \mathcal{H}^1_{E_0}.$$

L is called the Poincaré constant.

This inequality is discussed in many texts on finite element error analysis; for example, [19, pp. 30–31], [28, pp. 128–130]. Establishing the inequality in the simplest case of a square domain Ω is a worthy exercise, see Problem 1.12. Making the choice $v = u_1 - u_2$ in Lemma 1.2 implies that the solution is unique.

Returning to (1.53), we identify the left-hand side with the bilinear form $a : \mathcal{H}^1(\Omega) \times \mathcal{H}^1(\Omega) \to \mathbb{R}$, and the right-hand side with the linear functional $\ell : \mathcal{H}^1(\Omega) \to \mathbb{R}$, so that

$$a(u, v) := (\nabla u, \nabla v); \qquad \ell(v) := (f, v) + (g_N, v)_{\partial \Omega_N}; \qquad (1.55)$$

and restate the problem as:

Find $u \in \mathcal{H}^1_E$ such that

$$a(u, v) = \ell(v) \quad \text{for all } v \in \mathcal{H}^1_{E_0}. \qquad (1.56)$$

The corresponding discrete problem (1.19) is then given by:

Find $u_h \in S^h_E$ such that

$$a(u_h, v_h) = \ell(v_h) \quad \text{for all } v_h \in S^h_0. \qquad (1.57)$$

Assuming that the approximation is conforming, $S^h_E \subset \mathcal{H}^1_E$ and $S^h_0 \subset \mathcal{H}^1_{E_0}$, our task here is to estimate the quality of the approximation $u_h \approx u$.

We will outline the conventional (a priori) analysis of the approximation error arising using finite element approximation spaces in (1.57) in Section 1.5.1. Such error bounds are asymptotic in nature, and since they involve the true solution u they are not readily computable. We go on to discuss computable error bounds (usually referred to as a posteriori estimates) in Section 1.5.2.

1.5.1   A priori error bounds

To get a handle on the error, we can simply pick a generic $v \in \mathcal{H}_{E_0}^1$ and subtract $a(u_h, v)$ from (1.56) to give

$$a(u, v) - a(u_h, v) = \ell(v) - a(u_h, v).$$

This is the basic equation for the error: our assumption that $S_E^h \subset \mathcal{H}_E^1$ implies[7] that $e = u - u_h \in \mathcal{H}_{E_0}^1$ and satisfies

$$a(e, v) = \ell(v) - a(u_h, v) \quad \text{for all } v \in \mathcal{H}_{E_0}^1. \tag{1.58}$$

Note that $e \in \mathcal{H}_{E_0}^1$ since $S_E^h \subset \mathcal{H}_E^1$.

We now make explicit use of the fact that the underlying bilinear form $a(\cdot, \cdot)$ defines an inner product over the space $\mathcal{H}_{E_0}^1 \times \mathcal{H}_{E_0}^1$, with an associated (*energy*) norm $\|\nabla u\| = a(u, u)^{1/2}$, see Problem 1.10. The starting point is the *Galerkin orthogonality property*: taking $v_h \in S_0^h$ in (1.58) and using (1.57) we have that

$$a(u - u_h, v_h) = 0 \quad \text{for all } v_h \in S_0^h. \tag{1.59}$$

In simple terms, the error $e \in \mathcal{H}_{E_0}^1$ is orthogonal to the subspace S_0^h, with respect to the energy inner product. An immediate consequence of (1.59) is the *best approximation property* established below.

Theorem 1.3. $\|\nabla u - \nabla u_h\| = \min\{\|\nabla u - \nabla v_h\|: v_h \in S_E^h\}.$

Proof   Let $v_h \in S_E^h$, and note that $u - u_h \in \mathcal{H}_{E_0}^1$ so by definition

$$
\begin{aligned}
\|\nabla(u - u_h)\|^2 &= a(u - u_h, u - u_h) \\
&= a(u - u_h, u - v_h + v_h - u_h) \\
&= a(u - u_h, u - v_h) + a(u - u_h, v_h - u_h) \\
&= a(u - u_h, u - v_h) \quad \text{(using Galerkin orthogonality)} \\
&\leq \|\nabla(u - u_h)\| \, \|\nabla(u - v_h)\| \quad \text{(using Cauchy–Schwarz).}
\end{aligned}
$$

Hence, for either $\|\nabla(u - u_h)\| = 0$ or $\|\nabla(u - u_h)\| \neq 0$, we have that

$$\|\nabla(u - u_h)\| \leq \|\nabla(u - v_h)\| \quad \text{for all } v_h \in S_E^h. \tag{1.60}$$

Notice that the minimum is achieved since $u_h \in S_E^h$.   □

The energy error bound (1.60) is appealingly simple, and moreover in the case $g_D = 0$ it leads to the useful characterization that

$$\|\nabla(u - u_h)\|^2 = \|\nabla u\|^2 - \|\nabla u_h\|^2, \tag{1.61}$$

---

[7]Recall that in practice, the essential boundary condition is interpolated, see (1.18), so that $u_h \neq g_D$ on $\partial \Omega_D$ whenever the boundary data g_D is not a polynomial. In such cases the error $u - u_h$ must be estimated using a more refined *nonconforming* analysis, see Brenner & Scott [28, pp. 195ff] for details.

see Problem 1.11. (If u is known analytically, then (1.61) can be used to calculate the error in the energy norm without using elementwise integration.) The synergy between Galerkin orthogonality (1.59) and best approximation (1.60) is a reflection of the fact that u_h is the *projection* of u into the space S_E^h. This property will be exploited in Chapter 2, where fast solution algorithms are developed for the discrete problem (1.57).

The remaining challenge is to derive bounds on the error $u - u_h$ with respect to other norms, in particular, that associated with the *Hilbert space*[8] $\mathcal{H}^1(\Omega)$ introduced in Section 1.2. A formal definition is the following.

Definition 1.4 ($\mathcal{H}^1(\Omega)$ norm). Let $\mathcal{H}^1(\Omega)$ denote the set of functions u in $L_2(\Omega)$ possessing generalized[9] first derivatives. An inner product on $\mathcal{H}^1(\Omega)$ is given by

$$(u, v)_{1,\Omega} := (u, v) + (\nabla u, \nabla v), \qquad (1.62)$$

and this induces the associated norm

$$\|u\|_{1,\Omega} := (\|u\|^2 + \|D^1 u\|^2)^{1/2} \qquad (1.63)$$

where $D^1 u$ denotes the sum of squares of the first derivatives; for a two-dimensional domain Ω,

$$\|D^1 u\|^2 := \int_\Omega \left(\left(\frac{\partial u}{\partial x}\right)^2 + \left(\frac{\partial u}{\partial y}\right)^2 \right).$$

An important property of functions v in $\mathcal{H}^1(\Omega)$ is that they have a well-defined restriction to the boundary $\partial\Omega$. (This is an issue because functions in $\mathcal{H}^1(\Omega)$ need not be continuous.) The theoretical basis for this assertion is the following lemma.

Lemma 1.5 (Trace inequality). *Given a bounded domain Ω with a sufficiently smooth (e.g. polygonal) boundary $\partial\Omega$, a constant $C_{\partial\Omega}$ exists such that*

$$\|v\|_{\partial\Omega} \le C_{\partial\Omega} \|v\|_{1,\Omega} \quad \text{for all } v \in \mathcal{H}^1(\Omega).$$

Notice that, in contrast, there is no constant C such that $\|v\|_{\partial\Omega} \le C \|v\|$ for every v in $L_2(\Omega)$, hence associating boundary values with $L_2(\Omega)$ functions is not meaningful. The proof of Lemma 1.5 is omitted, for details see Braess [19, pp. 48ff]. Applications of the trace inequality will be found in later sections.

Extending the energy error estimate of Theorem 1.3 to a general error bound in $\mathcal{H}^1(\Omega)$ is a simple consequence of the Poincaré–Friedrichs inequality.

---

[8]This means a vector space with an inner-product, which contains the limits of every Cauchy sequence that is defined with respect to the norm $\|\cdot\|_{1,\Omega}$.

[9]This includes functions like $|x|$ that are differentiable except at a finite number of points. To keep the exposition simple, we omit a formal definition; for details see [19, p. 28] or [28, pp. 24–27].

Proposition 1.6. *Let Ω satisfy the assumptions in Lemma 1.2. Then there is a constant C_Ω independent of v, such that*

$$\|\nabla v\| \le \|v\|_{1,\Omega} \le C_\Omega \|\nabla v\| \quad \text{for all } v \in \mathcal{H}^1_{E_0}. \tag{1.64}$$

Proof See Problem 1.13.   □

We are now ready to state a *quasi-optimal* error bound that reflects the fact that $\|u - u_h\|_{1,\Omega}$ is proportional to the best possible approximation from the space S^h_E.

Theorem 1.7. *Let Ω satisfy the assumptions in Lemma 1.2. Then*

$$\|u - u_h\|_{1,\Omega} \le C_\Omega \min_{v_h \in S^h_E} \|u - v_h\|_{1,\Omega}. \tag{1.65}$$

Proof Note that $u - v_h \in \mathcal{H}^1_{E_0}$ if $v_h \in S^h_E$. Combining (1.60) with (1.64) then gives (1.65).   □

The best approximation error bound (1.60) is quite general in the sense that it is valid for any problem where the bilinear form $a(\cdot, \cdot)$ in (1.56) defines an inner product over the test space $\mathcal{H}^1_{E_0}$. The line of analysis above is not valid however, if the bilinear form $a(\cdot, \cdot)$ in the variational formulation is not symmetric, as, for example, for the convection–diffusion equation; see Chapter 3. In such cases, a priori error bounds in the underlying function space must be established using a different theoretical argument — typically using the coercivity and continuity of the underlying bilinear form over the space $\mathcal{H}^1_{E_0}$, see Problem 1.14. Further details are given in Chapter 3.

We now develop the general error bound (1.65) in the case of the finite element approximation spaces that were introduced in Sections 1.3.1 and 1.3.2. We first consider the simplest case of triangular elements using $\boldsymbol{P}_1$ (piecewise linear) approximation. That is, given a partitioning of the domain $\mathcal{T}_h$ consisting of triangular elements $\triangle_k$ we make the specific choice $S^h_0 = X^1_h$, where

$$X^1_h := \{v \in C^0(\Omega), v = 0 \text{ on } \partial\Omega_D; v|_\triangle \in \boldsymbol{P}_1, \forall\triangle \in \mathcal{T}_h\}. \tag{1.66}$$

We will state the error bound in the form of a theorem. We also need a couple of preliminary definitions.

Definition 1.8 ($\mathcal{H}^2(\Omega)$ norm). The set of functions $u \in \mathcal{H}^1(\Omega)$ that also possess generalized second derivatives can be identified with the Sobolev space $\mathcal{H}^2(\Omega)$. More precisely, $\mathcal{H}^2(\Omega) \subset \mathcal{H}^1(\Omega)$ is a Hilbert space that is complete with respect to the norm

$$\|u\|_{2,\Omega} := \left(\|u\|^2_{1,\Omega} + \|D^2 u\|^2 \right)^{1/2},$$

where D^2u denotes the sum of squares of second derivatives. More specifically, in the case of a two-dimensional domain Ω

$$\left\| D^2u \right\|^2 := \int_\Omega \left(\left(\frac{\partial^2 u}{\partial x^2} \right)^2 + \left(\frac{\partial^2 u}{\partial x \partial y} \right)^2 + \left(\frac{\partial^2 u}{\partial y^2} \right)^2 \right).$$

Definition 1.9 ($\mathcal{H}^2$ regularity). The variational problem (1.56) is said to be $\mathcal{H}^2$–*regular* if there exists a constant C_Ω such that for every $f \in L_2(\Omega)$, there is a solution $u \in \mathcal{H}_E^1$ that is also in $\mathcal{H}^2(\Omega)$ such that

$$\left\| u \right\|_{2,\Omega} \leq C_\Omega \left\| f \right\|.$$

Theorem 1.10. *If the variational problem* (1.56) *is solved using a mesh of linear triangular elements, so that* $S_0^h = X_h^1$ *in* (1.57), *and if a minimal angle condition is satisfied (see* Definition 1.15), *then there exists a constant* C_1 *such that*

$$\left\| \nabla(u - u_h) \right\| \leq C_1\, h\, \left\| D^2u \right\|, \tag{1.67}$$

where $\left\| D^2u \right\|$ *measures the* $\mathcal{H}^2$–*regularity of the target solution, and* h *is the length of the longest triangle edge in the mesh.*

Notice that if (1.56) is $\mathcal{H}^2$-regular, then (1.67) implies that the finite element solution u_h converges to the exact solution u in the limit $h \to 0$. The fact that the right-hand side of (1.67) is proportional to h is referred to as *first order* (or *linear*) convergence. Furthermore, Proposition 1.6 implies that if Lemma 1.2 is valid, then the order of convergence in $\mathcal{H}^1$ is the same as the order of convergence in the energy norm.

The issue of $\mathcal{H}^2$-regularity is central to the proof of Theorem 1.10, the first step of which is to break the bound (1.60) into pieces by introducing an appropriate interpolant $\pi_h u$ from the approximation space S_E^h. Making the specific choice $v_h = \pi_h u$ in (1.60) then gives

$$\left\| \nabla(u - u_h) \right\| \leq \left\| \nabla(u - \pi_h u) \right\|. \tag{1.68}$$

What is important here[10] is that $u \in \mathcal{H}^2(\Omega) \subset C^0(\Omega)$ so the simple piecewise linear interpolant $\pi_h u$, satisfying $\pi_h u(\mathbf{x}_i) = u(\mathbf{x}_i)$ at every vertex $\mathbf{x}_i$ of the triangulation, is a well-defined function in S_E^h (since $\pi_h u \in X_h^1$ in the case of zero boundary data). The localization of the error is now immediate since (1.68) can be broken up into elementwise error bounds

$$\left\| \nabla(u - \pi_h u) \right\|^2 = \sum_{\triangle_k \in \mathcal{T}_h} \left\| \nabla(u - \pi_h u) \right\|_{\triangle_k}^2. \tag{1.69}$$

---

[10]The relationship between continuous functions and Sobolev spaces is dependent on the domain Ω; if Ω is one-dimensional then $\mathcal{H}^1(\Omega) \subset C^0(\Omega)$, but for two-dimensional domains functions exist that are not bounded (hence not continuous) yet still have square integrable first derivatives, see [19, pp. 31–32].

The problem of estimating the overall error is now reduced to one of approximation theory — we need good estimates for the interpolation error on a typical element.

It is at this point that the local–global mapping in Section 1.4 plays an important role. Rather than estimating the error for every individual element, the idea is to map the element interpolation error from (1.69) onto the reference element, since the error can easily be bounded there in terms of derivatives of the interpolated function. This type of construction is referred to as a *scaling argument*. Each of the three stages in the process is summarized below in the form of a lemma. Note that h_k denotes the length of the longest edge of $\triangle_k$, and $\bar{u}$ denotes the mapped function defined on the reference element $\triangle*$.

Lemma 1.11. $\|\nabla(u - \pi_h u)\|^2_{\triangle_k} \leq 2\dfrac{h_k^2}{|\triangle_k|} \|\nabla(\bar{u} - \pi_h\bar{u})\|^2_{\triangle*}$.

Proof Define $e_k = (u - \pi_h u)|_{\triangle_k}$ and let $\bar{e}_k$ denote the mapped function defined on $\triangle*$. By definition

$$\|\nabla e_k\|^2_{\triangle_k} = \int_{\triangle_k} \left(\frac{\partial e_k}{\partial x}\right)^2 + \left(\frac{\partial e_k}{\partial y}\right)^2 \, dx\, dy$$

$$= \int_{\triangle*} \left(\left(\frac{\partial \bar{e}_k}{\partial x}\right)^2 + \left(\frac{\partial \bar{e}_k}{\partial y}\right)^2\right) 2|\triangle_k| \, d\xi d\eta, \qquad (1.70)$$

where the derivatives satisfy (1.44); in particular the first term is of the form

$$\left(\frac{\partial \bar{e}_k}{\partial x}\right)^2 = \frac{1}{4|\triangle_k|^2} \left(b_2 \frac{\partial \bar{e}_k}{\partial \xi} + b_3 \frac{\partial \bar{e}_k}{\partial \eta}\right)^2,$$

with b_2 and b_3 defined by (1.43). Using the facts that $(a + b)^2 \leq 2(a^2 + b^2)$ and $|b_i| \leq h_k$, we get the bound

$$2|\triangle_k| \left(\frac{\partial \bar{e}_k}{\partial x}\right)^2 \leq \frac{h_k^2}{|\triangle_k|} \left(\left(\frac{\partial \bar{e}_k}{\partial \xi}\right)^2 + \left(\frac{\partial \bar{e}_k}{\partial \eta}\right)^2\right).$$

The second term in (1.70) can be bounded in exactly the same way $(|c_i| \leq h_k)$. Summing the terms gives the stated result.   □

The following bound is a special case of a general estimate for interpolation error in Sobolev spaces known as the *Bramble–Hilbert* lemma. In simple terms, the error due to linear interpolation in an unit triangle measured in the energy norm is bounded by the L^2 norm of the second derivative of the interpolation error.

Lemma 1.12.

$$\|\nabla(\bar{u} - \pi_h\bar{u})\|_{\triangle*} \leq C \left\|D^2(\bar{u} - \pi_h\bar{u})\right\|_{\triangle*} \equiv C \left\|D^2\bar{u}\right\|_{\triangle*}. \qquad (1.71)$$

Whilst proving the analogous result in $\mathbb{R}^1$ is a straightforward exercise, see Problem 1.15, the proof of (1.71) is technical and so is omitted. An accessible discussion can be found in [19, pp. 75–76], and a complete and rigorous treatment is given in [28, chap. 4].

Lemma 1.13. $\left\| D^2\bar{u} \right\|^2_{\triangle_*} \leq 18h_k^2 \dfrac{h_k^2}{|\triangle_k|} \left\| D^2 u \right\|^2_{\triangle_k}.$

Proof By definition,

$$\left\| D^2\bar{u} \right\|^2_{\triangle_*} = \int_{\triangle_*} \left(\frac{\partial^2 \bar{u}}{\partial \xi^2} \right)^2 + \left(\frac{\partial^2 \bar{u}}{\partial \xi \partial \eta} \right)^2 + \left(\frac{\partial^2 \bar{u}}{\partial \eta^2} \right)^2 d\xi d\eta$$

$$= \int_{\triangle_k} \left(\left(\frac{\partial^2 u}{\partial \xi^2} \right)^2 + \left(\frac{\partial^2 u}{\partial \xi \partial \eta} \right)^2 + \left(\frac{\partial^2 u}{\partial \eta^2} \right)^2 \right) \frac{1}{2|\triangle_k|} \, dx \, dy, \qquad (1.72)$$

where the derivatives are mapped using (1.37); in particular the first term is of the form

$$\left(\frac{\partial}{\partial \xi} \left(\frac{\partial u}{\partial \xi} \right) \right)^2 = \left(c_3 \frac{\partial}{\partial x} \left(\frac{\partial u}{\partial \xi} \right) - b_3 \frac{\partial}{\partial y} \left(\frac{\partial u}{\partial \xi} \right) \right)^2$$

$$= \left(c_3^2 \frac{\partial^2 u}{\partial x^2} - 2c_3 b_3 \frac{\partial^2 u}{\partial x \partial y} + b_3^2 \frac{\partial^2 u}{\partial y^2} \right)^2$$

$$\leq 3 \left(c_3^4 \left(\frac{\partial^2 u}{\partial x^2} \right)^2 + 4 c_3^2 b_3^2 \left(\frac{\partial^2 u}{\partial x \partial y} \right)^2 + b_3^4 \left(\frac{\partial^2 u}{\partial y^2} \right)^2 \right)$$

$$\leq 12 h_k^4 \left(\left(\frac{\partial^2 u}{\partial x^2} \right)^2 + \left(\frac{\partial^2 u}{\partial x \partial y} \right)^2 + \left(\frac{\partial^2 u}{\partial y^2} \right)^2 \right). \qquad (1.73)$$

The second and third terms in (1.72) can be bounded in exactly the same way. Summing the three terms gives the stated result. $\qquad \square$

The bound in Lemma 1.11 (and that in Lemma 1.13) involves the triangle aspect ratio $h_k^2/|\triangle_k|$. Keeping the aspect ratio small is equivalent to a minimum angle condition, as is shown in the following, see Figure 1.18.

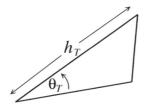

FIG. 1.18. Minimum angle condition.

Proposition 1.14. *Given any triangle, we have the equivalence relation*

$$\frac{h_T^2}{4}\sin\theta_T \leq |\triangle_T| \leq \frac{h_T^2}{2}\sin\theta_T, \tag{1.74}$$

where $0 < \theta_T \leq \pi/3$ *is the smallest of the interior angles.*

Proof See Problem 1.16.   □

The result (1.74) shows that bounding the aspect ratio is equivalent to ensuring that the minimum interior angle is bounded away from zero. Combining (1.74) with the bounds in Lemmas 1.11–1.13, we see that the interpolation error bound (1.69) satisfies

$$\|\nabla(u - \pi_h u)\|^2 \leq C \sum_{\triangle_k \in T_h} \frac{1}{\sin^2\theta_k} h_k^2 \left\| D^2 u \right\|_{\triangle_k}^2. \tag{1.75}$$

The bound (1.75) can be further simplified by making the assumption that the mesh refinement is shape regular as follows.

Definition 1.15 (Minimum angle condition). A sequence of triangular grids $\{T_h\}$ is said to be *shape regular* if there exists a minimum angle $\theta_* \neq 0$ such that every element in T_h satisfies $\theta_T \geq \theta_*$.

In particular, shape regularity ensures that $1/\sin\theta_k \leq 1/\sin\theta_*$ for all triangles in T_h, so that (1.75) simplifies to

$$\|\nabla(u - \pi_h u)\|^2 \leq C(\theta_*) \sum_{\triangle_k \in T_h} h_k^2 \left\| D^2 u \right\|_{\triangle_k}^2. \tag{1.76}$$

Noting that $h_k \leq h$ for all triangles $\triangle_k$ gives the desired uniform bound (i.e. independent of the triangulation)

$$\|\nabla(u - \pi_h u)\|^2 \leq Ch^2 \sum_{\triangle_k \in T_h} \left\| D^2 u \right\|_{\triangle_k}^2 = Ch^2 \left\| D^2 u \right\|^2.$$

Combining with (1.68) then gives the error bound (1.67) in Theorem 1.10. We also note that the less stringent *maximum angle condition*, which requires all angles to be uniformly bounded away from π, can also be used to obtain these results; see Krizek [125].

A similar argument can be used to establish a bound for the L_2 interpolation error associated with the function u itself. In particular, for a mesh of linear elements the following result can be readily established.

Proposition 1.16. $\|u - \pi_h u\|^2 \leq C \sum_{\triangle_k \in T_h} h_k^4 \left\| D^2 u \right\|_{\triangle_k}^2.$

Proof See Problem 1.17.   □

Shape regularity is not required here, since derivatives are not mapped from $\triangle_k$ to the reference element.

We now consider the analogue of Theorem 1.10 in the case of grids of rectangular elements using Q_1 approximation. (Recall from Problem 1.8 that the Jacobian reduces to a constant diagonal matrix in this case.) The analogues of Lemmas 1.11 and 1.13 are given below.

Proposition 1.17. *Given a rectangular element* $\square_k$, *with horizontal and vertical edges of lengths* hx, hy, *respectively, let* π_h^1 *be the standard bilinear interpolant, which agrees with the underlying function at the four vertices. Then*

$$\left\|\nabla(u - \pi_h^1 u)\right\|_{\square_k}^2 \leq \max\left\{\frac{hx}{hy}, \frac{hy}{hx}\right\} \left\|\nabla(\bar{u} - \pi_h^1 \bar{u})\right\|_{\square_*}^2, \tag{1.77}$$

$$\left\|D^2 \bar{u}\right\|_{\square_*}^2 \leq h_k^2 \max\left\{\frac{hx}{hy}, \frac{hy}{hx}\right\} \left\|D^2 u\right\|_{\square_k}^2, \tag{1.78}$$

where $h_k = \max\{hx, hy\}$.

Proof See Problem 1.18. $\qquad\square$

Notice that the rectangle aspect ratio $\beta_T = \max\{hx/hy, hy/hx\}$ plays the same role as the triangle aspect ratio in Lemmas 1.11 and 1.13.

Definition 1.18 (Aspect ratio condition). A sequence of rectangular grids $\{\mathcal{T}_h\}$ is said to be *shape regular* if there exists a maximum rectangle edge ratio β_* such that every element in $\mathcal{T}_h$ satisfies $1 \leq \beta_T \leq \beta_*$.

A second key point is that the analogue of Lemma 1.12 also holds in this case,

$$\left\|\nabla(\bar{u} - \pi_h^1 \bar{u})\right\|_{\square_*} \leq C \left\|D^2(\bar{u} - \pi_h^1 \bar{u})\right\|_{\square_*} \equiv C \left\|D^2 \bar{u}\right\|_{\square_*}. \tag{1.79}$$

Combining the bounds (1.77), (1.79) and (1.78) gives the anticipated error estimate.

Theorem 1.19. *If the variational problem (1.57) is solved using a mesh of bilinear rectangular elements, and if the aspect ratio condition is satisfied (see Definition 1.18), then there exists a constant* C_1 *such that*

$$\left\|\nabla(u - u_h)\right\| \leq C_1 h \left\|D^2 u\right\|, \tag{1.80}$$

where h *is the length of the longest edge in* $\mathcal{T}_h$.

Remark 1.20. If the degree of element distortion is small, a similar bound to (1.80) also holds in the case of Q_1 approximation on grids of isoparametrically mapped quadrilateral elements. For grids of parallelograms, given an appropriate definition of shape regularity (involving a minimum angle and an aspect ratio condition) the convergence bound is identical to (1.80), see [19, Theorem 7.5].

The construction of the error estimate (1.80) via the intermediate results (1.77), (1.79) and (1.78) provides the basis for establishing error bounds when higher-order (P_m, Q_m, with $m \geq 2$) approximation spaces are used.

Theorem 1.21. *Using a higher-order finite element approximation space P_m or Q_m with $m \geq 2$ leads to the higher-order convergence bound*

$$\|\nabla(u - u_h)\| \leq C_m h^m \|D^{m+1}u\|. \tag{1.81}$$

In other words, we get mth order convergence as long as the regularity of the target solution is good enough. Note that $\|D^{m+1}u\| < \infty$ if and only if the $(m+1)$st generalized derivatives of u are in $L_2(\Omega)$.

For example, using biquadratic approximation on a square element grid, we have the following analogue of Proposition 1.17.

Proposition 1.22. *For a grid of square elements $\square_k$ with edges of length h, let π_h^2 be the standard biquadratic interpolant, which agrees with the underlying function at nine points, see Figure 1.11. Then*

$$\left\|\nabla(u - \pi_h^2 u)\right\|_{\square_k}^2 \leq \left\|\nabla(\bar{u} - \pi_h^2 \bar{u})\right\|_{\square_*}^2, \tag{1.82}$$

$$\left\|D^3 \bar{u}\right\|_{\square_*}^2 \leq h^4 \left\|D^3 u\right\|_{\square_k}^2. \tag{1.83}$$

Proof See Problem 1.19.   □

Combining (1.82) and (1.83) with the reference element bound given by the Bramble–Hilbert lemma (in this case bounding in terms of the third derivatives; cf. Lemma 1.12)

$$\left\|\nabla(\bar{u} - \pi_h^2 \bar{u})\right\|_{\square_*} \leq C \left\|D^3(\bar{u} - \pi_h^2 \bar{u})\right\|_{\square_*} \equiv C \left\|D^3 \bar{u}\right\|_{\square_*} \tag{1.84}$$

leads to (1.81) with $m = 2$.

To conclude this section, we will use the problems in Examples 1.1.3 and 1.1.4 to illustrate that the orders of convergence suggested by the error bounds (1.80) and (1.81) are typical of the behavior of the error as the grid is successively refined. An assessment of the orders of convergence that is obtained for the problems in Examples 1.1.1 and 1.1.2 are given as Computational Exercises 1.1 and 1.2 respectively. Results for the problem in Example 1.1.3 are given in Table 1.1. The error measure E_h used here is the difference between the exact and the discrete energy, that is

$$E_h = |\, \|\nabla u\|^2 - \|\nabla u_h\|^2 \,|^{1/2}. \tag{1.85}$$

If zero essential boundary conditions are imposed, then $\|\nabla u\| \geq \|\nabla u_h\|$ and E_h is identical to the energy error $\|\nabla(u - u_h)\|$, see Problem 1.11. Notice how the Q_1 errors in Table 1.1 decrease by a factor of two for every successive refinement,[11] whereas the Q_2 errors ultimately decrease by a factor of four. The outcome is that biquadratic elements are more accurate than bilinear elements — in fact they

---

[11]ℓ is the *grid parameter* specification in the IFISS software that is associated with the tabulated entry.

Table 1.1 *Energy error E_h for Example 1.1.3: ℓ is the grid refinement level; h is $2^{1-\ell}$ for Q_1 approximation, and $2^{2-\ell}$ for Q_2 approximation.*

ℓ	Q_1	Q_2	n
2	5.102×10^{-2}	6.537×10^{-3}	9
3	2.569×10^{-2}	2.368×10^{-3}	49
4	1.287×10^{-2}	5.859×10^{-4}	225
5	6.437×10^{-3}	1.460×10^{-4}	961
6	3.219×10^{-3}	3.646×10^{-5}	3969

Table 1.2 *Energy error E_h for Example 1.1.4.*

ℓ	Q_1	Q_2	n
2	1.478×10^{-1}	9.860×10^{-2}	33
3	9.162×10^{-2}	6.207×10^{-2}	161
4	5.714×10^{-2}	3.909×10^{-2}	705
5	3.577×10^{-2}	2.462×10^{-2}	2945

are generally more cost-effective wherever the underlying solution is sufficiently smooth if reasonable accuracy is required. For example, the Q_2 solution on the coarsest grid has approximately $1/4^3$ of the degrees of freedom of the Q_1 solution on the second finest grid, yet both are of comparable accuracy.

If the weak solution is not smooth, then the superiority of the Q_2 approximation method over the simpler Q_1 method is not so clear. To illustrate this, the energy differences E_h computed in the case of the singular problem in Example 1.1.4 are tabulated in Table 1.2. Notice that — in contrast to the behavior in Table 1.1 — the Q_1 and Q_2 errors both decrease by a factor of approximately $2^{2/3} \approx 1.5874$ with every successive refinement of the grid. The explanation for this is that the solution regularity is between $\mathcal{H}^1$ and $\mathcal{H}^2$ in this case,[12] see Problem 1.20. The upshot is that in place of (1.80) and (1.81), the following convergence bound is the best that can be achieved (for all $\varepsilon > 0$):

$$\|\nabla(u - u_h)\| \leq C_{\mathbf{m}}(\epsilon)\, h^{2/3-\varepsilon} \tag{1.86}$$

using approximation of arbitrary order $m \geq 1$!

When solving problems like those in Examples 1.1.3 and 1.1.4, it is natural to try to design rectangular or triangular meshes that concentrate the degrees

---

[12]Introducing Sobolev spaces with fractional indices as in Johnson [111, pp. 92–94], it may be shown that $u \in \mathcal{H}^{5/3-\varepsilon}$.

Table 1.3 *Energy error E_h for stretched grid solutions of Example 1.1.4: $\ell = 4$.*

α	Q_1	Q_2
1	9.162×10^{-2}	6.207×10^{-2}
5/4	6.614×10^{-2}	3.723×10^{-2}
3/2	7.046×10^{-2}	2.460×10^{-2}
2	1.032×10^{-1}	2.819×10^{-2}

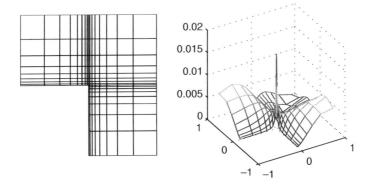

FIG. 1.19. Stretched level 4 grid with $\alpha = 3/2$ (left) for Example 1.1.4 and surface plot (right) of the estimated error using Q_1 approximation (see Section 1.5.2).

of freedom in the neighborhood of the singularity. The motivation for doing this is the intermediate bound (1.76), which suggests that it is important to try to balance the size of h_k with that of $\left\|D^2u\right\|_{\triangle_k}$. Roughly speaking, h_k should be small in those elements where the derivatives of u are large. To illustrate the idea, Table 1.3 lists the errors E_h obtained when solving the problem in Example 1.1.4 using tensor-product grids that are geometrically stretched towards the singularity, with successive element edges a factor α times longer than the adjacent edge, see Figure 1.19. Notice that comparing the results in Table 1.3 with those in Table 1.2, we see that an appropriately stretched grid of Q_2 elements with 161 degrees of freedom, gives better accuracy than that obtained using a uniform grid with 2945 degrees of freedom — the challenge here is to determine the optimal stretching a priori!

1.5.2 *A posteriori error bounds*

The fact that physically interesting problems typically have singularities is what motivates the concept of a posteriori error estimation. Specifically, given a finite element subdivision $\mathcal{T}_h$ and a solution u_h, we want to compute a local (element) error estimator η_T such that $\|\nabla\eta_T\|$ approximates the local energy error $\|\nabla(u - u_h)\|_T$ for every element T in $\mathcal{T}_h$. An important factor is the requirement that η_T should be cheap to compute — as a rule of thumb, the computational work should scale linearly as the number of elements is increased — yet there should be guaranteed accuracy in the sense that the estimated global error should give an upper bound on the exact error, so that

$$\|\nabla(u - u_h)\|^2 \equiv \sum_{T\in\mathcal{T}_h} \|\nabla(u - u_h)\|_T^2 \le C(\theta_*) \sum_{T\in\mathcal{T}_h} \eta_T^2 \qquad (1.87)$$

with a constant C that depends only on shape regularity. If, in addition to satisfying (1.87), η_T provides a lower bound for the exact local error

$$\eta_T \le C(\theta_{\omega_T}) \|\nabla(u - u_h)\|_{\omega_T}, \qquad (1.88)$$

where ω_T typically represents a local patch of elements adjoining T, then the estimator η_T is likely to be effective if it is used to drive an adaptive refinement process. For the problem in Example 1.1.4, such a process will give rise to successive meshes that are selectively refined in the vicinity of the singularity so as to equidistribute the error among all elements and enhance overall cost effectiveness.

The two key aspects of error estimation are *localization* and *approximation*. The particular strategy that is built into the IFISS software is now described. The starting point is the characterization (1.58) of the error $e = u - u_h \in \mathcal{H}_{E_0}^1$:

$$a(e, v) = \ell(v) - a(u_h, v) \quad \text{for all } v \in \mathcal{H}_{E_0}^1. \qquad (1.89)$$

For simplicity, it is assumed here that Neumann data is homogeneous, so that $\ell(v) = (f, v)$. Using the shorthand notation $(u, v)_T := \int_T uv$ and $a(u, v)_T := \int_T \nabla u \cdot \nabla v$ to represent the localized L_2 and energy inner products respectively, the error equation (1.89) may be broken up into element contributions

$$\sum_{T\in\mathcal{T}_h} a(e, v)_T = \sum_{T\in\mathcal{T}_h} (f, v)_T - \sum_{T\in\mathcal{T}_h} a(u_h, v)_T. \qquad (1.90)$$

Integrating by parts elementwise then gives

$$-a(u_h, v)_T = (\nabla^2 u_h, v)_T - \sum_{E\in\mathcal{E}(T)} \langle \nabla u_h \cdot \vec{n}_{E,T}, v \rangle_E, \qquad (1.91)$$

where $\mathcal{E}(T)$ denotes the set of edges (faces in $\mathbb{R}^3$) of element T, $\vec{n}_{E,T}$ is the outward normal with respect to E, $\langle\cdot,\cdot\rangle_E$ is the L_2 inner product on E, and $\nabla u_h \cdot \vec{n}_{E,T}$ is the discrete (outward-pointing) normal flux. The finite element approximation typically has a discontinuous normal derivative across inter-element boundaries. Consequently it is convenient to define the *flux jump* across edge or face E adjoining elements T and S as

$$\left[\!\!\left[\frac{\partial v}{\partial n} \right]\!\!\right] := (\nabla v|_T - \nabla v|_S) \cdot \vec{n}_{E,T} = (\nabla v|_S - \nabla v|_T) \cdot \vec{n}_{E,S}, \qquad (1.92)$$

and then to *equidistribute* the flux jump contribution in (1.90) to the adjoining
elements in equal proportion (with an appropriate modification for elements that
have one or more edges/faces adjoining $\partial\Omega$):

$$\sum_{T\in\mathcal{T}_h} a(e,v)_T = \sum_{T\in\mathcal{T}_h}\left[(f+\nabla^2 u_h,v)_T - \frac{1}{2}\sum_{E\in\mathcal{E}(T)}\left\langle\left[\!\left[\frac{\partial u_h}{\partial n}\right]\!\right],v\right\rangle_E\right]. \quad (1.93)$$

It is evident from the structure of the right-hand side of equation (1.93) that
e has two distinct components; these are the (element) *interior residual* $R_T :=$
$\{f+\nabla^2 u_h\}|_T$, and the (inter-element) *flux jump* $R_E := [\![\partial u_h/\partial n]\!]$. Notice also
that if u_h agrees with the classical solution everywhere then both R_T and R_E
are identically zero. The residual terms R_T and R_E enter either implicitly or
explicitly into the definition of many finite element error estimators.

In the remainder of this section we concentrate on the specific case of S_0^h
being defined by the $\boldsymbol{P}_1$ or $\boldsymbol{Q}_1$ approximation over a triangular or rectangular
element subdivision. The appeal of these lowest order methods is their simplicity;
the flux jump is piecewise constant in the $\boldsymbol{P}_1$ case, and in both cases the interior
residual $R_T = f|_T$ is independent of u_h and thus can be computed a priori. As a
further simplification, R_T can be approximated by a constant R_T^0 by projecting
f onto the space of piecewise constant functions.

To define a consistent flux jump operator with respect to elements adjoining
$\partial\Omega$, some additional notation is needed. We let $\mathcal{E}_h = \cup_{T\in\mathcal{T}_h}\mathcal{E}(T)$ denote the set
of all edges split into interior and boundary edges via

$$\mathcal{E}_h := \mathcal{E}_{h,\Omega}\cup\mathcal{E}_{h,D}\cup\mathcal{E}_{h,N};$$

where $\mathcal{E}_{h,\Omega} := \{E\in\mathcal{E}_h: E\subset\Omega\}$, $\mathcal{E}_{h,D} := \{E\in\mathcal{E}_h: E\subset\partial\Omega_D\}$ and $\mathcal{E}_{h,N} := \{E\in\mathcal{E}_h: E\subset\partial\Omega_N\}$. We then define the operator

$$R_E^* = \begin{cases} \frac{1}{2}[\![\partial u_h/\partial n]\!] & E\in\mathcal{E}_{h,\Omega} \\ -\nabla u_h\cdot\vec{n}_{E,T} & E\in\mathcal{E}_{h,N} \\ 0 & E\in\mathcal{E}_{h,D}. \end{cases}$$

The fact that the exact error e is characterized by the enforcement of (1.93)
over the space $\mathcal{H}_{E_0}^1$ provides us with a handle for estimating the local error
in each element T. Specifically, if a suitable (finite-dimensional) approximation
space, $\mathcal{Q}_T$ say, is constructed, then an approximation to $e|_T$ can be obtained
by enforcing (1.93) elementwise. Specifically, a function $e_T\in\mathcal{Q}_T$ is computed
such that

$$(\nabla e_T,\nabla v)_T = (R_T^0,v)_T - \sum_{E\in\mathcal{E}(T)}\langle R_E^*,v\rangle_E, \quad (1.94)$$

for all $v\in\mathcal{Q}_T$, and the local error estimator is the energy norm of e_T

$$\eta_T = \|\nabla e_T\|_T. \quad (1.95)$$

Making an appropriate choice of approximation space $\mathcal{Q}_T$ in (1.94) is clearly crucial. A clever choice (due to Bank & Weiser [9]) is the "correction" space

$$\mathcal{Q}_T = Q_T \oplus B_T \tag{1.96}$$

consisting of edge and interior bubble functions, respectively;

$$Q_T = \text{span } \{\psi_E \colon E \in \mathcal{E}(T) \cap (\mathcal{E}_{h,\Omega} \cup \mathcal{E}_{h,N})\} \tag{1.97}$$

where $\psi_E : T \to \mathbb{R}$ is the quadratic (or biquadratic) edge-bubble that is zero on the other two (or three) edges of T. B_T is the space spanned by interior cubic (or biquadratic) bubbles ϕ_T such that $0 \le \phi_T \le 1$, $\phi_T = 0$ on ∂T and $\phi_T = 1$ only at the centroid. The upshot is that for each triangular (or rectangular) element a 4×4 (or 5×5) system of equations must be solved to compute e_T.[13]

A feature of the choice of space (1.97) is that $(\nabla v, \nabla v)_T > 0$ for all functions v in $\mathcal{Q}_T$ (intuitively, a constant function in T cannot be represented as a linear combination of bubble functions), so the local problem (1.94) is well posed. This means that the element matrix systems are all non-singular, see Problem 1.21. This is important for a typical element T that has no boundary edges, since the local problem (1.94) represents a weak formulation of the Neumann problem:

$$-\nabla^2 e_T = f \quad \text{in } T \tag{1.98}$$

$$\frac{\partial e_T}{\partial n} = -\frac{1}{2} \left[\!\left[\frac{\partial u_h}{\partial n} \right]\!\right] \quad \text{on } E \in \mathcal{E}(T), \tag{1.99}$$

suggesting that a compatibility condition cf. (1.4)

$$\int_T f - \frac{1}{2} \sum_{E \in \mathcal{E}(T)} \int_E \left[\!\left[\frac{\partial u_h}{\partial n} \right]\!\right] = 0, \tag{1.100}$$

needs to be satisfied in order to ensure the existence of e_T. The difficulty associated with the need to enforce (1.100) is conveniently circumvented by the choice (1.97).

To illustrate the effectiveness of this very simple error estimation procedure, the analytic test problem in Example 1.1.3 is discretized using uniform grids of Q_1 elements, and a comparison between the exact energy error $\|\nabla e\|$ and the estimated global error $\eta = \left(\sum_{T \in \mathcal{T}_h} \eta_T^2\right)^{1/2}$ is given in Table 1.4. The close agreement between the estimated and exact errors is quite amazing.[14] Another virtue of the estimator illustrated by Table 1.4 is the fact that the *global effectivity index* $X_\eta := \eta / \|\nabla e\|$ converges to unity as $h \to 0$. This property is usually referred to as *asymptotic exactness*.

The results in Table 1.4 suggest that the estimator η_T satisfies the required error bound (1.87) (with a proportionality constant $C(\theta_*)$) that is close to unity if

---

[13]This is embodied in the IFISS routine diffpost_p.m.

[14]Although performance deteriorates using stretched meshes, the agreement between exact and estimated errors is quite acceptable, see Computational Exercise 1.3.

Table 1.4 *Comparison of estimated and exact errors for Example 1.1.3.*

ℓ	$\|\nabla(u - u_h)\|$	η	X_η
2	5.032×10^{-2}	4.954×10^{-2}	0.9845
3	2.516×10^{-2}	2.511×10^{-2}	0.9980
4	1.258×10^{-2}	1.257×10^{-2}	0.9992
5	6.291×10^{-3}	6.288×10^{-3}	0.9995

the elements are not too distorted). A precise result is stated below. This should also be compared with the a priori error bound given in Theorem 1.19.

Theorem 1.23. *If the variational problem (1.57) is solved using a mesh of bilinear rectangular elements, and if the rectangle aspect ratio condition is satisfied with β_* given in Definition 1.18, then the estimator $\eta_T \equiv \|\nabla e_T\|_T$ computed via (1.94) using the approximation space (1.97) gives the bound*

$$\|\nabla(u - u_h)\| \leq C(\beta_*) \left(\sum_{T \in \mathcal{T}_h} \eta_T^2 + h^2 \sum_{T \in \mathcal{T}_h} \|R_T - R_T^0\|_T^2 \right)^{1/2}, \qquad (1.101)$$

where h is the length of the longest edge in $\mathcal{T}_h$.

Remark 1.24. If f is a piecewise constant function then the consistency error term $\|R_T - R_T^0\|_T$ is identically zero. Otherwise, if f is smooth, this term represents a high-order perturbation. In any case the estimator η_T is reliable; for further details see Verfürth [204].

A proof of Theorem 1.23 is outlined below. An important difference between the a priori bound (1.80) and the a posteriori bound (1.101) is that $\mathcal{H}^2$-regularity is not assumed in the latter case. This adds generality (since the bound (1.101) applies even if the problem is singular) but raises the technical issue within the proof of Theorem 1.23 of having to approximate a possibly discontinuous $\mathcal{H}^1$ function. Since point values of $\mathcal{H}^1(\Omega)$ functions are not defined for $\Omega \subset \mathbb{R}^2$, an alternative to interpolation using local averaging over neighborhoods of the vertices of the subdivision is required. This leads to the quasi-interpolation estimates (due to Clément [45]) given in the following lemma. For a detailed discussion, see Brenner & Scott [28, pp. 118–120].

Lemma 1.25. *Given $e \in \mathcal{H}^1_{E_0}$ there exists a quasi-interpolant $e_h^* \in S_0^h$ such that,*

$$\|e - e_h^*\|_T \leq C_1(\beta_{\tilde{\omega}_T}) \, h_T \, \|\nabla e\|_{\tilde{\omega}_T} \quad \text{for all } T \in \mathcal{T}_h, \qquad (1.102)$$

$$\|e - e_h^*\|_E \leq C_2(\beta_{\tilde{\omega}_T}) \, h_E^{1/2} \, \|\nabla e\|_{\tilde{\omega}_T} \quad \text{for all } E \in \mathcal{E}_h, \qquad (1.103)$$

where $\tilde{\omega}_T$ is the patch of all the neighboring elements that have at least one vertex connected to a vertex of element T.

Notice that the constants in (1.102) and (1.103) depend only on the maximum aspect ratio over all elements in the patch. The proof of Theorem 1.23 will also require so-called *local inverse estimates*. A typical example is given in the lemma below. A proof for low-order basis functions is provided; see [28, Section 4.5], [44, Section 3.2] for more general analysis.

Lemma 1.26. *Given a polynomial function u_k defined in a triangular or rectangular element T, a constant C exists, depending only on the element aspect ratio, such that*

$$\|\nabla u_k\|_T \leq C h_T^{-1} \|u_k\|_T, \tag{1.104}$$

where h_T is the length of the longest edge of T.

Proof This is a standard scaling argument of the type used in the proof of Lemma 1.11. In the case of triangular elements with P_1 (linear) basis functions, the argument of that proof gives

$$\|\nabla u_k\|_{\triangle_k}^2 \leq 2 \frac{h_k^2}{|\triangle_k|} \|\nabla \bar{u}_k\|_{\triangle_*}^2. \tag{1.105}$$

Note that

$$\bar{u}_k \mapsto \|\nabla \bar{u}_k\|_{\triangle_*}, \quad \bar{u}_k \mapsto \|\bar{u}_k\|_{\triangle_*}$$

constitute a seminorm and norm, respectively, on finite-dimensional spaces. It follows that, as in the equivalence of norms on finite-dimensional spaces,

$$\|\nabla \bar{u}_k\|_{\triangle_*} \leq C \|\bar{u}_k\|_{\triangle_*}. \tag{1.106}$$

Mapping back to the original element, we have

$$\|\bar{u}_k\|_{\triangle_*}^2 = \frac{1}{2|\triangle_k|} \|u_k\|_{\triangle_k}^2, \tag{1.107}$$

and combining (1.105), (1.106), (1.107) with (1.74) gives the stated result. The proof for a rectangular element is left as an exercise, see Problem 1.22. □

Returning to the proof of Theorem 1.23, the first step is to use Galerkin orthogonality (1.59), the error equation (1.89) and the definition of R_E^*:

$$\|\nabla e\|^2 = a(e, e)$$
$$= a(e, e - e_h^*) \quad \text{(setting } v_h = e_h^* \text{ in (1.59))}$$
$$= \ell(e - e_h^*) - a(u_h, e - e_h^*) \quad \text{(setting } v = e - e_h^* \text{ in (1.89))}$$
$$= \sum_{T \in \mathcal{T}_h} \left\{ (R_T, e - e_h^*)_T - \sum_{E \in \mathcal{E}(T)} \langle R_E^*, e - e_h^* \rangle_E \right\} \quad \text{(using (1.91))}$$
$$\leq C(\beta_*) \sum_{T \in \mathcal{T}_h} \left\{ h_T \|R_T\|_T \|\nabla e\|_{\tilde{\omega}_T} + \sum_{E \in \mathcal{E}(T)} h_E^{1/2} \|R_E^*\|_E \|\nabla e\|_{\tilde{\omega}_T} \right\}$$
$$\leq C(\beta_*) \left(\sum_{T \in \mathcal{T}_h} \|\nabla e\|_{\tilde{\omega}_T}^2 \right)^{1/2} \left(\sum_{T \in \mathcal{T}_h} \left\{ h_T \|R_T\|_T + \sum_E h_E^{1/2} \|R_E^*\|_E \right\}^2 \right)^{1/2}.$$

For a rectangular subdivision, the union of the patches $\tilde{\omega}_T$ covers Ω at most nine times, thus $\sum_{T \in \mathcal{T}_h} \|\nabla e\|_{\tilde{\omega}_T}^2 \leq 9 \|\nabla e\|^2$. Noting that $(a + b)^2 \leq 2a^2 + 2b^2$ then leads to the following *residual estimator* error bound

$$\|\nabla(u - u_h)\| \leq C(\beta_*) \left(\sum_{T \in \mathcal{T}_h} \left\{ h_T^2 \|R_T\|_T^2 + \sum_{E \in \mathcal{E}(T)} h_E \|R_E^*\|_E^2 \right\} \right)^{1/2}. \quad (1.108)$$

Remark 1.27. The combination of the interior residual and flux jump terms on the right-hand side of (1.108) can be used to define a simple *explicit* estimator $\bar{\eta}_T$, see [204]. In practice, the far superior accuracy of the local problem estimator (1.95) outweighs the computational cost incurred in solving the local problems (1.94) so the use of the cheaper estimator $\bar{\eta}_T$ in place of η_T is not recommended.

To show that the residual bound (1.108) implies the bound (1.101), we take the trivial bound $\|R_T\|_T \leq \|R_T^0\|_T + \|R_T - R_T^0\|_T$, and exploit the fact that R_T^0 and R_E^* are piecewise constant to show that the terms $h_T^2 \|R_T^0\|_T^2$ and $h_E \|R_E^*\|_E^2$ on the right-hand side of (1.108) are individually bounded by η_T^2. The interior residual term is dealt with first. For $T \in \mathcal{T}_h$, we note that $R_T^0|_T \in P_0$ and define $w_T = R_T^0 \phi_T \in B_T \subset Q_T$. It follows that

$$\|R_T^0\|_T^2 = C(R_T^0, w_T)_T = C(\nabla e_T, \nabla w_T)_T \quad \text{(setting } v = w_T \text{ in (1.94))}$$
$$\leq C \|\nabla e_T\|_T \|\nabla w_T\|_T$$
$$\leq C \|\nabla e_T\|_T h_T^{-1} \|w_T\|_T \quad \text{(applying (1.104))}$$
$$\leq C h_T^{-1} \|\nabla e_T\|_T \|R_T^0\|_T,$$

where in the last step we use the fact that $0 \leq \phi_T \leq 1$. This gives

$$h_T \left\| R_T^0 \right\|_T \leq C \left\| \nabla e_T \right\|_T. \tag{1.109}$$

The jump term is handled in the same way. For an interior edge, we define ω_E to be the union of the two elements adjoining edge $E \in \mathcal{E}(T)$, and define $w_E = R_E \psi_E \in Q_T \subset \mathcal{Q}_T$. From (1.94) we then have that

$$\left\| R_E^* \right\|_E^2 \leq C < R_E^*, w_E >_E = C \sum_{T' \subset \omega_E} \left[-(\nabla e_{T'}, \nabla w_E)_{T'} + (R_{T'}^0, w_E)_{T'} \right],$$

which, when combined with the scaling results $\left\| w_E \right\|_{T'} \leq h_E^{1/2} \left\| w_E \right\|_E$ and $\left\| \nabla w_E \right\|_{T'} \leq h_E^{-1/2} \left\| w_E \right\|_E$, leads to the desired bound

$$h_E^{1/2} \left\| R_E^* \right\|_E \leq C \sum_{T' \subset \omega_E} \left\| \nabla e_{T'} \right\|_{T'}. \tag{1.110}$$

Combining (1.109) and (1.110) with (1.108) then gives the upper bound (1.101) in Theorem 1.23.

The remaining issue is whether or not the estimated error η_T gives a lower bound on the local error. A precise statement is given below.

Proposition 1.28. *If the variational problem* (1.57) *is solved using a grid of bilinear rectangular elements, and if the rectangle aspect ratio condition is satisfied, then the estimator* $\eta_T \equiv \left\| \nabla e_T \right\|_T$ *computed via* (1.94) *using the approximation space* (1.97) *gives the bound*

$$\eta_T \leq C(\beta_{\omega_T}) \left\| \nabla (u - u_h) \right\|_{\omega_T}, \tag{1.111}$$

where ω_T *represents the patch of five elements that have at least one boundary edge* E *from the set* $\mathcal{E}(T)$.

Proof See Problem 1.23.   □

To illustrate the usefulness of a posteriori error estimation, plots of the estimated error e_T associated with computed solutions u_h to the problems in Examples 1.1.3 and 1.1.4, are presented in Figures 1.20 and 1.21, respectively. The structure of the error can be seen to be very different in these two cases. Whereas the error distribution is a smooth function when solving Example 1.1.3, the effect of the singularity on the error distribution is very obvious in Figure 1.21. Moreover, comparing this error distribution, which comes from a uniform grid, with that in Figure 1.19 (derived from a stretched grid with the same number of degrees of freedom) clearly suggests that the most effective way of increasing accuracy at minimal cost is to perform local refinement in the neighborhood of the corner. These issues are explored further in Computational Exercise 1.4.

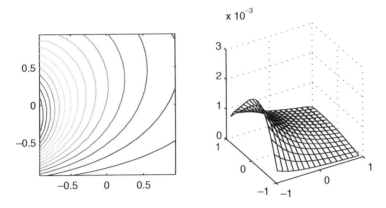

FIG. 1.20. Contour plot (left) and three-dimensional surface plot (right) of the estimated error associated with the finite element solution to Example 1.1.3 given in Figure 1.3.

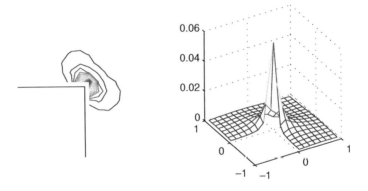

FIG. 1.21. Contour plot (left) and three-dimensional surface plot (right) of the estimated error associated with the finite element solution to Example 1.1.4 given in Figure 1.4.

1.6   Matrix properties

In this section, we describe some properties of the matrices arising from finite element discretization of the Poisson equation. These results will be used in the next chapter to analyze the behavior of iterative solution algorithms applied to the discrete systems of equations.

Let $\langle \mathbf{v}, \mathbf{w} \rangle = \mathbf{v}^T \mathbf{w}$ denote the Euclidean inner product on $\mathbb{R}^n$, with associated norm $\|\mathbf{v}\| = \langle \mathbf{v}, \mathbf{v} \rangle^{1/2}$. We begin with the observation that for *any* symmetric positive-definite matrix A of order n, the bilinear form given by

$$\langle \mathbf{v}, \mathbf{w} \rangle_A := \langle A\mathbf{v}, \mathbf{w} \rangle \tag{1.112}$$

defines an inner product on $\mathbb{R}^n$ with associated norm $\|\mathbf{v}\|_A = \langle \mathbf{v}, \mathbf{v} \rangle_A^{1/2}$. Given that A, the discrete Laplacian operator introduced in Section 1.3, is indeed symmetric and positive-definite, the inner product (1.112) and norm are well-defined in this case. Any vector $\mathbf{v} \in \mathbb{R}^n$ uniquely corresponds to a finite element function $v_h \in S_0^h$, and in particular there is a unique correspondence between the finite element solution u_h and the solution $\mathbf{u} = (\mathbf{u}_1, \mathbf{u}_2, \ldots, \mathbf{u}_n)^T$ to the matrix equation (1.21). If v_h and w_h are two functions in S_0^h, with coefficient vectors $\mathbf{v}$ and $\mathbf{w}$ respectively in $\mathbb{R}^n$, then the bilinear form derived from the Poisson equation, that is, $a(\cdot, \cdot)$ of (1.55), satisfies

$$a(v_h, w_h) = \int_\Omega \nabla v_h \cdot \nabla w_h = \langle \mathbf{v}, \mathbf{w} \rangle_A. \tag{1.113}$$

In simple terms, there is a one-to-one correspondence between the bilinear form $a(\cdot, \cdot)$ defined on the function space S_0^h, and the discrete inner product (1.112).

Recall from (1.22) that the discrete Laplacian can be viewed as the *Grammian matrix* of the basis $\{\phi_j\}$ associated with the inner product $a(\cdot, \cdot)$. It will also turn out to be useful to identify the Grammian with respect to the L_2-inner product,

$$Q = [q_{ij}], \qquad q_{ij} = \int_\Omega \phi_j \phi_i. \tag{1.114}$$

With this definition, it follows that for $v_h, w_h \in S_0^h$,

$$(v_h, w_h) = \langle Q\mathbf{v}, \mathbf{w} \rangle.$$

An immediate consequence is that Q is symmetric positive-definite, and the inner product $\langle \cdot, \cdot \rangle_Q$ defined by it constitutes a representation in $\mathbb{R}^n$ of the L_2-inner product in S_0^h. The matrix Q in (1.114) is referred to as the *mass matrix*.

The key property of the mass matrix is the following.

Proposition 1.29. *For P_1 or Q_1 approximation on a subdivision in $\mathbb{R}^2$ for which a shape regularity condition holds (as given in Definitions 1.15 and 1.18), the mass matrix Q approximates the scaled identity matrix in the sense that*

$$\underline{c}\underline{h}^2 \leq \frac{\langle Q\mathbf{v}, \mathbf{v} \rangle}{\langle \mathbf{v}, \mathbf{v} \rangle} \leq C h^2 \tag{1.115}$$

for all $\mathbf{v} \in \mathbb{R}^n$. Here $\underline{h} = \min_{\triangle_k \in \mathcal{T}_h} h_k$ and $h = \max_{\triangle_k \in \mathcal{T}_h} h_k$. The constants c and C are independent of both $\underline{h}$ and h.

Proof See Problem 1.24.   □

The bound (1.115) can be further refined by making the assumption that the subdivision is *quasi-uniform*.

Definition 1.30 (Quasi-uniform subdivision). A sequence of triangular grids $\{\mathcal{T}_h\}$ is said to be *quasi-uniform* if there exists a constant $\rho > 0$ such that $\underline{h} \geq \rho h$ for every grid in the sequence.

For a quasi-uniform subdivision in $\mathbb{R}^2$ of shape regular elements, the bound (1.115) simplifies:

$$ch^2 \le \frac{\langle Q\mathbf{v}, \mathbf{v}\rangle}{\langle \mathbf{v}, \mathbf{v}\rangle} \le Ch^2 \quad \text{for all } \mathbf{v} \in \mathbb{R}^n. \tag{1.116}$$

Remark 1.31. If the subdivision is quasi-uniform, then the bound (1.116) holds for any degree of approximation, $\boldsymbol{P}_m$, $\boldsymbol{Q}_m$ with $m \ge 2$ (see Problem 1.25). However, the constants c and C depend on m.

The bound (1.116) depends on the spatial dimension. For tetrahedral or brick elements on a quasi-uniform discretization of a domain in $\mathbb{R}^3$, the corresponding bound is

$$ch^3 \le \frac{\langle Q\mathbf{v}, \mathbf{v}\rangle}{\langle \mathbf{v}, \mathbf{v}\rangle} \le Ch^3 \quad \text{for all } \mathbf{v} \in \mathbb{R}^n. \tag{1.117}$$

The mass matrix is a fundamental component of finite element analysis, arising naturally, for example, in the study of time-dependent problems. Here, however, it is only making a "cameo appearance" for the purposes of developing bounds on the eigenvalues of the discrete Laplacian A. The mass matrix will resurface later in Chapters 6 and 8.

One other property of the mass matrix will be useful in the next chapter. Given a Poisson problem (1.13), (1.14) and an approximation space S_E^h, the finite element solution u_h in S_E^h satisfying (1.19) is identical to that where the source function f is replaced by its projection $f_h \in S_0^h$ with respect to the L_2-norm. This is simply because f_h so defined satisfies $(f - f_h, w_h) = 0$ for every $w_h \in S_0^h$, so that

$$\int_\Omega \phi_i\, f_h = \int_\Omega \phi_i\, f \tag{1.118}$$

for each i, and thus there is no change in (1.23) when f_h is used instead of f. Note that if f_h in (1.118) is expressed in terms of the basis set $\{\phi_i\}_{i=1}^n$ then the coefficients are determined by solving the linear system $Q\mathbf{x} = \mathbf{f}$, where Q is the mass matrix, see Problem 1.26.

Another fundamental concept used for the analysis of matrix computations is the *condition number* of a matrix,

$$\kappa = \kappa(A) := \|A\|\,\|A^{-1}\|,$$

where the matrix norm is

$$\|A\| := \max_{\mathbf{v} \ne 0} \frac{\|A\mathbf{v}\|}{\|\mathbf{v}\|}.$$

When A is a symmetric positive-definite matrix, $\|A\| = \lambda_{\max}(A)$, the largest eigenvalue of A, and $\|A^{-1}\| = 1/\lambda_{\min}(A)$. Consequently, the condition number is

$$\kappa(A) = \lambda_{\max}(A)/\lambda_{\min}(A).$$

For direct solution methods, the size of the condition number is usually related to the number of accurate decimal places in a computed solution (see Higham [105]). In the next chapter, the convergence behavior of iterative solution methods will be precisely characterized in terms of $\kappa(A)$. In anticipation of this development, bounds on the condition number of the discrete Laplacian A are derived here. Two alternative approaches that can be used to establish such bounds will be described.

The first approach uses tools developed in Section 1.5 and is applicable in the case of arbitrarily shaped domains and non-uniform grids.

Theorem 1.32. *For P_1 or Q_1 approximation on a shape regular, quasi-uniform subdivision of $\mathbb{R}^2$, the Galerkin matrix A in (1.21) satisfies*

$$ch^2 \leq \frac{\langle A\mathbf{v}, \mathbf{v}\rangle}{\langle \mathbf{v}, \mathbf{v}\rangle} \leq C \quad \text{for all } \mathbf{v} \in \mathbb{R}^n. \tag{1.119}$$

Here h is the length of the longest edge in the mesh or grid, and c and C are positive constants that are independent of h. In terms of the condition number, $\kappa(A) \leq C_* h^{-2}$ where $C_* = C/c$.

Proof Suppose that λ is an eigenvalue of A, that is, $A\mathbf{v} = \lambda\mathbf{v}$ for some eigenvector $\mathbf{v}$. Then $\lambda = \langle A\mathbf{v}, \mathbf{v}\rangle/\langle\mathbf{v}, \mathbf{v}\rangle$, and it follows that

$$\min_{\mathbf{v} \in \mathbb{R}^n} \frac{\langle A\mathbf{v}, \mathbf{v}\rangle}{\langle\mathbf{v}, \mathbf{v}\rangle} \leq \lambda \leq \max_{\mathbf{v} \in \mathbb{R}^n} \frac{\langle A\mathbf{v}, \mathbf{v}\rangle}{\langle\mathbf{v}, \mathbf{v}\rangle}. \tag{1.120}$$

For any $\mathbf{v} \in \mathbb{R}^n$, let v_h denote the corresponding function in S_0^h. The Poincaré-Friedrichs inequality (Lemma 1.2) implies that there is a constant c_Ω that is independent of the mesh parameter h such that

$$c_\Omega \|v_h\|^2 \leq \|\nabla v_h\|^2 = a(v_h, v_h)$$

for all $v_h \in S_0^h$. Rewriting in terms of matrices gives

$$c_\Omega \langle Q\mathbf{v}, \mathbf{v}\rangle \leq \langle A\mathbf{v}, \mathbf{v}\rangle \quad \text{for all } \mathbf{v} \in \mathbb{R}^n.$$

Combining the left-hand inequality of (1.116) and the characterization (1.120) shows that the smallest eigenvalue of A is bounded below by a quantity of order h^2.

For a bound on the largest eigenvalue of A, we turn to the local inverse estimate derived in Lemma 1.26, which states that for the restriction of v_h to an element T,

$$\|\nabla v_h\|_T^2 \leq Ch_T^{-2} \|v_h\|_T^2.$$

Summing over all the elements and using the quasi-uniformity bound $h_T^{-1} \leq Ch^{-1}$ together with the right-hand inequality of (1.116) gives

$$\langle A\mathbf{v}, \mathbf{v}\rangle = a(v_h, v_h) \leq Ch^{-2}\|v_h\|^2 \leq C \langle\mathbf{v}, \mathbf{v}\rangle.$$

Thus, the bound on the largest eigenvalue is independent of h. $\qquad\square$

Remark 1.33. The Galerkin matrix bound (1.119) holds for any degree of approximation, $\boldsymbol{P}_m$, $\boldsymbol{Q}_m$ with $m \geq 2$. The constants c and C depend on m.

Remark 1.34. With tetrahedral or brick elements on a quasi-uniform discretization of a domain in $\mathbb{R}^3$, the corresponding bound is

$$ch^3 \leq \frac{\langle A\mathbf{v}, \mathbf{v}\rangle}{\langle \mathbf{v}, \mathbf{v}\rangle} \leq Ch \quad \text{for all } \mathbf{v} \in \mathbb{R}^n. \tag{1.121}$$

This leads to an identical bound, $\kappa(A) \leq C_* h^{-2}$ on the condition number of the discrete Laplacian in arbitrary dimensions.

The second way of obtaining eigenvalue bounds is with Fourier analysis. This avoids the use of functional analytic tools, but the assumptions on the mesh are more restrictive than those expressed in Theorem 1.32.

A typical result is worked out in Problem 1.27. We use a double index notation to refer to the nodes ordered in a so-called "lexicographic" order as illustrated in Figure 1.22. Consider the concrete case of Example 1.1.1 discretized using $\boldsymbol{Q}_1$ approximation; Ω is a square of size $L = 2$, and a uniform $k \times k$ grid is used, so that the matrix dimension is $n = (k-1)^2$ with $k = L/h$. The analysis leads to the explicit identification of *all* of the eigenvalues:

$$\lambda^{(r,s)} = \frac{8}{3} - \frac{2}{3}\left(\cos\frac{r\pi}{k} + \cos\frac{s\pi}{k}\right) - \frac{4}{3}\cos\frac{r\pi}{k}\cos\frac{s\pi}{k}, \quad r,s = 1,\ldots,k-1 \tag{1.122}$$

together with the associated eigenvectors $\mathbf{U}^{(r,s)}$:

$$\mathbf{U}_{i,j}^{(r,s)} = \sin\frac{ri\pi}{k}\sin\frac{sj\pi}{k}, \tag{1.123}$$

where the index $i, j = 1, \ldots, k - 1$ refers to the grid location.

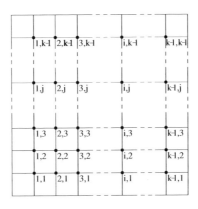

FIG. 1.22. Lexicographic ordering of node points with double index.

From (1.122), we see that the extreme eigenvalues of the Q_1 discrete Laplacian are thus

$$\lambda_{\min} = \lambda^{(1,1)} = \frac{8}{3} - \frac{4}{3}\cos\frac{\pi}{k} - \frac{4}{3}\cos^2\frac{\pi}{k} = \frac{2\pi^2}{L^2}h^2 + \mathcal{O}(h^4),$$

$$\lambda_{\max} = \lambda^{(1,k-1)} = \lambda^{(k-1,1)} = \frac{8}{3} + \frac{4}{3}\cos^2\frac{\pi}{k} = 4 - \frac{4\pi^2}{3L^2}h^2 + \mathcal{O}(h^4),$$

and the condition number is

$$\kappa(A) = \frac{2L^2}{\pi^2}h^{-2} - \frac{1}{6} + \mathcal{O}(h^2). \tag{1.124}$$

Notice that the bound of Theorem 1.32 is tight in this case. (See also Computational Exercise 1.6.) Analogous estimates can also be established in the three-dimensional case, see Problem 1.28. Fourier analysis will be used in later chapters to give insight in other contexts, for example, to explore the convergence properties of multigrid methods (Section 2.5), and to investigate discrete approximations that exhibit high frequency oscillations in cases where the continuous solution is non-oscillatory (Section 3.5).

Problems

1.1. Show that the function $u(r, \theta) = r^{2/3}\sin((2\theta + \pi)/3)$ satisfies Laplace's equation expressed in polar coordinates;

$$\frac{\partial^2 u}{\partial r^2} + \frac{1}{r}\frac{\partial u}{\partial r} + \frac{1}{r^2}\frac{\partial^2 u}{\partial \theta^2} = 0.$$

1.2. Show that a solution u satisfying the Poisson equation and a mixed condition $\alpha u + \frac{\partial u}{\partial n} = 0$ on the boundary $\partial\Omega$, where $\alpha > 0$ is a constant, also satisfies the following weak formulation: find $u \in \mathcal{H}^1(\Omega)$ such that

$$\int_\Omega \nabla u \cdot \nabla v + \alpha \int_{\partial\Omega} uv = \int_\Omega vf \quad \text{for all } v \in \mathcal{H}^1(\Omega).$$

Show that $c(u, v) := \int_\Omega \nabla u \cdot \nabla v + \alpha \int_{\partial\Omega} uv$ defines an inner product over $\mathcal{H}^1(\Omega)$, and hence establish that a solution of the weak formulation is uniquely defined.

1.3. Construct the P_2 basis functions for the element with vertices $(0,0)$, $(1,0)$ and $(0,1)$ illustrated in Figure 1.7.

1.4. For the pair of elements illustrated, show that the bilinear function that takes on the value one at vertex P and zero at the other vertices gives different values at the midpoint M on the common edge (and is hence discontinuous).

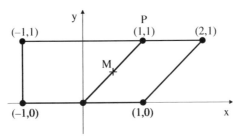

Show that the corresponding isoparametrically mapped bilinear function (defined via (1.26) and (1.27)) is continuous along the common edge.

1.5. By substituting (1.36) into (1.45), show that the P_1 stiffness matrix is given by

$$A_k(i,j) = \frac{1}{4|\triangle_k|}(b_i b_j + c_i c_j),$$

where b_i and c_i, $i = 1,2,3$ are defined in (1.43).

1.6. A generic point P in a triangle is parameterized by three triangular (or *barycentric*) coordinates L_1, L_2 and L_3, which are simple ratios of the triangle areas illustrated $(L_1, L_2, L_3) \equiv (|\triangle_{3P2}|/|\triangle|, |\triangle_{1P3}|/|\triangle|, |\triangle_{2P1}|/|\triangle|)$.

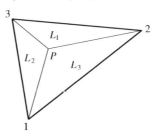

By construction, show that

$$L_i = \frac{1}{2|\triangle|}(a_i + b_i x + c_i y),$$

where a_i satisfies $\sum_{i=1}^{3} a_i = 2|\triangle|$, and b_i and c_i are given by (1.43). Check that the functions L_i are linear nodal basis functions (so that $L_i \equiv \psi_{k,i}$, see Section 1.4.1), and hence verify the formula for the P_1 stiffness matrix given in Problem 1.5.

1.7. Show that the determinant of the Q_1 Jacobian matrix (1.48) is a linear function of the coordinates (ξ, η). Verify that the Q_1 Jacobian (1.48) is a constant matrix if the mapped element is a parallelogram.

1.8. Show that the Q_1 Jacobian (1.48) is a diagonal matrix if the mapped element is a rectangle aligned with the coordinate axes. Compute the Q_1 stiffness matrix in this case (assume that the horizontal and vertical sides are of length hx and hy, respectively).

1.9. Given the Gauss points $\xi_s = \pm 1/\sqrt{3}$ and $\eta_t = \pm 1/\sqrt{3}$ as illustrated in Figure 1.15, show that if f is bilinear, that is, $f(\xi, \eta) = (a + b\xi)(c + d\eta)$ where a, b, c and d are constants, then

$$\int_{-1}^{1} \int_{-1}^{1} f\,d\xi\,d\eta = \sum_s \sum_t f(\xi_s, \eta_t).$$

1.10. Show that if $\int_{\partial \Omega_D} ds \neq 0$ then the bilinear form $a(\cdot, \cdot)$ in (1.55) defines an inner product over the space $\mathcal{H}^1_{E_0} \times \mathcal{H}^1_{E_0}$.

1.11. Show that, if u and u_h satisfy (1.56) and (1.57) respectively in the case of zero Dirichlet data (so that $\mathcal{H}^1_E = \mathcal{H}^1_{E_0}$), then the error in energy satisfies

$$\|\nabla(u - u_h)\|^2 = \|\nabla u\|^2 - \|\nabla u_h\|^2.$$

1.12. Given a square domain $\Omega = [0, L] \times [0, L]$, show that

$$\int_\Omega u^2 \leq \frac{L^2}{2} \int_\Omega \left(\left| \frac{\partial u}{\partial x} \right|^2 + \left| \frac{\partial u}{\partial y} \right|^2 \right)$$

for any function $u \in \mathcal{H}^1(\Omega)$ that is zero everywhere on the boundary. (Hint: $u(x, y) = u(0, y) + \int_0^x \frac{\partial u}{\partial x}(\xi, y)\,d\xi$.)

1.13. Prove Proposition 1.6. (Hint: use the definition (1.63).)

1.14. Let V_h be a finite element subspace of $V := \mathcal{H}^1(\Omega)$. Define a bilinear form $a(\cdot, \cdot)$ on $V \times V$, and let $u \in V$ and $u_h \in V_h$ satisfy

$$a(u, v) = (f, v) \quad \text{for all } v \in V$$
$$a(u_h, v_h) = (f, v_h) \quad \text{for all } v_h \in V_h$$

respectively. If there exist positive constants γ and Γ such that

$$a(v, v) \geq \gamma \|v\|^2_{1,\Omega} \qquad \text{for all } v \in V$$
$$|a(u, v)| \leq \Gamma \|u\|_{1,\Omega} \|v\|_{1,\Omega} \quad \text{for all } u, v \in V,$$

show that there exists a positive constant $C(\gamma, \Gamma)$ such that

$$\|u - u_h\|_{1,\Omega} \leq C \inf_{v_h \in V_h} \|u - v_h\|_{1,\Omega}.$$

1.15. For any function $w(x)$ defined on $[0, 1]$, let Πw be the linear interpolant satisfying $\Pi w(0) = w(0)$ and $\Pi w(1) = w(1)$. Use Rolle's theorem to show that

$e = w - \Pi w$ satisfies

$$\int_0^1 (e')^2 \mathrm{d}x \le \frac{1}{2} \int_0^1 (e'')^2 \mathrm{d}x.$$

1.16. Prove Proposition 1.14. (Hint: use simple trigonometric identities.)

1.17. Prove Proposition 1.16.

1.18. Prove Proposition 1.17. (Hint: follow the proofs of Lemma 1.11 and Lemma 1.13.)

1.19. Prove Proposition 1.22. (Hint: do Problem 1.18 first.)

1.20. Given that $u \in \mathcal{H}^s(\Omega) \Leftrightarrow \|D^s u\| < \infty$, show that the function $u(r, \theta) = r^{2/3} \sin((2\theta + \pi)/3)$ defined on the pie-shaped domain Ω where $0 \le r \le 1$ and $-\pi/2 \le \theta \le \pi$ is in $\mathcal{H}^1(\Omega)$, but is not in $\mathcal{H}^2(\Omega)$.

1.21. Show that if $\mathcal{Q}_T$ is the space (1.97) of edge and bubble functions then there exists a unique solution to the local problem (1.94).

1.22. Show that the inverse estimate (1.104) holds in the case of a rectangular element T.

1.23. Prove Proposition 1.28. (Hint: take $v = e_T$ in the local problem (1.94), and then choose w_T and w_E as in the proof of Theorem 1.23 with a view to separately bounding the interior and jump residual terms by $\|\nabla(u - u_h)\|_T$.)

1.24. Prove Proposition 1.29. (Hint: given that Q is the mass matrix (1.114) and that $u_h = \sum_{j=1}^n \mathbf{u}_j \phi_j$ is a finite element function on $\Omega \subset \mathbb{R}^2$, show that $\|u_h\|_\Omega^2 = \langle Q\mathbf{u}, \mathbf{u} \rangle$. Write $\langle Q\mathbf{u}, \mathbf{u} \rangle = \sum_{k \in \mathcal{T}_h} \langle Q^{(k)} \mathbf{v}_k, \mathbf{v}_k \rangle$ where $Q^{(k)}$ is the element mass matrix for element k, that is, $Q^{(k)} = [q_{ij}]$ with $q_{ij} = \int_{\Delta_k} \phi_j \phi_i$. Then prove that for a shape regular element, $\underline{c}h_k^2 \langle \mathbf{v}_k, \mathbf{v}_k \rangle \le \langle Q^{(k)} \mathbf{v}_k, \mathbf{v}_k \rangle \le \bar{c}h_k^2 \langle \mathbf{v}_k, \mathbf{v}_k \rangle$ for all functions $\mathbf{v}_k$.)

1.25. Prove that the mass matrix bound (1.116) holds for $\mathbf{Q}_2$ approximation on a uniform grid of square elements.

1.26. For any $f \in L_2(\Omega)$, let f_h be the L_2 projection into S_0^h. Writing $f_h = \sum_{j=1}^n \bar{\mathbf{f}}_j \phi_j$, show that the coefficient vector $\bar{\mathbf{f}} = (\bar{\mathbf{f}}_1, \bar{\mathbf{f}}_2, \ldots, \bar{\mathbf{f}}_n)^T$ is the solution of

$$Q\bar{\mathbf{f}} = \mathbf{f},$$

where Q is the mass matrix (1.114) and $\mathbf{f} = [\mathbf{f}_i]$ with $\mathbf{f}_i = \int_\Omega \phi_i f$.

1.27. In the double index notation indicated by Figure 1.22 and with $\mathbf{U}_{i,j}$ denoting the value of u_h at the lattice point i, j, the Galerkin system of equations derived from $\mathbf{Q}_1$ approximation on a uniform square grid can be written as

$$\frac{8}{3}\mathbf{U}_{i,j} - \frac{1}{3}\mathbf{U}_{i+1,j+1} - \frac{1}{3}\mathbf{U}_{i+1,j} - \frac{1}{3}\mathbf{U}_{i+1,j-1} - \frac{1}{3}\mathbf{U}_{i,j+1} - \frac{1}{3}\mathbf{U}_{i,j-1}$$

$$- \frac{1}{3}\mathbf{U}_{i-1,j+1} - \frac{1}{3}\mathbf{U}_{i-1,j} - \frac{1}{3}\mathbf{U}_{i-1,j-1} = h^2\mathbf{f}_{i,j}$$

with $\mathbf{U}_{k,j}, \mathbf{U}_{0,j}, \mathbf{U}_{i,0}, \mathbf{U}_{i,k}$ given by the Dirichlet boundary condition. The eigenvalues $\lambda^{r,s}$ of the Galerkin matrix therefore satisfy

$$\frac{8}{3}\mathbf{U}_{i,j}^{r,s} - \frac{1}{3}\mathbf{U}_{i+1,j+1}^{r,s} - \frac{1}{3}\mathbf{U}_{i+1,j}^{r,s} - \frac{1}{3}\mathbf{U}_{i+1,j-1}^{r,s} - \frac{1}{3}\mathbf{U}_{i,j+1}^{r,s} - \frac{1}{3}\mathbf{U}_{i,j-1}^{r,s}$$

$$- \frac{1}{3}\mathbf{U}_{i-1,j+1}^{r,s} - \frac{1}{3}\mathbf{U}_{i-1,j}^{r,s} - \frac{1}{3}\mathbf{U}_{i-1,j-1}^{r,s} = \lambda^{r,s}\,\mathbf{U}_{i,j}^{r,s}$$

for $r, s = 1, \ldots, k - 1$. Verify that the vector $\mathbf{U}^{r,s}$ with entries

$$\mathbf{U}_{i,j}^{r,s} = \sin\frac{ri\pi}{k}\sin\frac{sj\pi}{k}, \quad i, j = 1, \ldots, k - 1$$

is an eigenvector for arbitrary $r, s = 1, \ldots, k-1$, and hence that the corresponding eigenvalue is

$$\lambda^{r,s} = \frac{8}{3} - \frac{2}{3}\left(\cos\frac{r\pi}{k} + \cos\frac{s\pi}{k}\right) - \frac{4}{3}\cos\frac{r\pi}{k}\cos\frac{s\pi}{k}, \quad r, s = 1, \ldots, k - 1.$$

1.28. In triple index notation with $\mathbf{U}_{i,j,k}$ denoting the value of u_h at the lattice point i, j, k, $i = 1, \ldots, l - 1$, $j = 1, \ldots, l - 1$, $k = 1, \ldots, l - 1$ show that the Galerkin system derived from trilinear approximation of the Poisson equation with Dirichlet boundary conditions on a uniform grid of cube elements with side length h can be written as

$$\frac{8h}{3}\mathbf{U}_{i,j,k} - \frac{h}{6}\left(\mathbf{U}_{i,j+1,k-1} + \mathbf{U}_{i,j-1,k-1} + \mathbf{U}_{i+1,j,k-1} + \mathbf{U}_{i-1,j,k-1}\right)$$

$$- \frac{h}{12}\left(\mathbf{U}_{i+1,j+1,k-1} + \mathbf{U}_{i+1,j-1,k-1} + \mathbf{U}_{i-1,j+1,k-1} + \mathbf{U}_{i-1,j-1,k-1}\right)$$

$$- \frac{h}{6}\left(\mathbf{U}_{i+1,j+1,k} + \mathbf{U}_{i+1,j-1,k} + \mathbf{U}_{i-1,j+1,k} + \mathbf{U}_{i-1,j-1,k}\right)$$

$$- \frac{h}{12}\left(\mathbf{U}_{i+1,j+1,k+1} + \mathbf{U}_{i+1,j-1,k+1} + \mathbf{U}_{i-1,j+1,k+1} + \mathbf{U}_{i-1,j-1,k+1}\right)$$

$$- \frac{h}{6}\left(\mathbf{U}_{i,j+1,k+1} + \mathbf{U}_{i,j-1,k+1} + \mathbf{U}_{i+1,j,k+1} + \mathbf{U}_{i-1,j,k+1}\right) = h^2\mathbf{f}_{i,j,k}$$

for $i, j, k = 1, \ldots, l - 1$ with $\mathbf{U}_{i,j,k}$ given by the Dirichlet boundary conditions when any of i, j or k is 0 or l.

1.29. This builds on Problems 1.27 and 1.28. Show that

$$\mathbf{U}_{i,j,k}^{r,s,t} = \sin\frac{ri\pi}{l}\sin\frac{sj\pi}{l}\sin\frac{tk\pi}{l}, \quad i, j, k = 1, \ldots, l - 1$$

is an eigenvector of the Galerkin matrix in Problem 1.28 for $r, s, t = 1, \ldots, l-1$, and that the eigenvalues $\lambda^{r,s,t}$ satisfying

$$\frac{8h}{3} \mathbf{U}_{i,j,k}^{r,s,t} - \frac{h}{6} \left(\mathbf{U}_{i,j+1,k-1}^{r,s,t} + \mathbf{U}_{i,j-1,k-1}^{r,s,t} + \mathbf{U}_{i+1,j,k-1}^{r,s,t} + \mathbf{U}_{i-1,j,k-1}^{r,s,t} \right)$$

$$- \frac{h}{12} \left(\mathbf{U}_{i+1,j+1,k-1}^{r,s,t} + \mathbf{U}_{i+1,j-1,k-1}^{r,s,t} + \mathbf{U}_{i-1,j+1,k-1}^{r,s,t} + \mathbf{U}_{i-1,j-1,k-1}^{r,s,t} \right)$$

$$- \frac{h}{6} \left(\mathbf{U}_{i+1,j+1,k}^{r,s,t} + \mathbf{U}_{i+1,j-1,k}^{r,s,t} + \mathbf{U}_{i-1,j+1,k}^{r,s,t} + \mathbf{U}_{i-1,j-1,k}^{r,s,t} \right)$$

$$- \frac{h}{12} \left(\mathbf{U}_{i+1,j+1,k+1}^{r,s,t} + \mathbf{U}_{i+1,j-1,k+1}^{r,s,t} + \mathbf{U}_{i-1,j+1,k+1}^{r,s,t} + \mathbf{U}_{i-1,j-1,k+1}^{r,s,t} \right)$$

$$- \frac{h}{6} \left(\mathbf{U}_{i,j+1,k+1}^{r,s,t} + \mathbf{U}_{i,j-1,k+1}^{r,s,t} + \mathbf{U}_{i+1,j,k+1}^{r,s,t} + \mathbf{U}_{i-1,j,k+1}^{r,s,t} \right) = \lambda^{r,s,t} \, \mathbf{U}_{i,j,k}^{r,s,t}$$

are therefore

$$\lambda^{r,s,t} = \frac{8h}{3} - \frac{2h}{3} \left(\cos \frac{r\pi}{l} \cos \frac{s\pi}{l} + \cos \frac{r\pi}{l} \cos \frac{t\pi}{l} + \cos \frac{s\pi}{l} \cos \frac{t\pi}{l} \right)$$

$$- \frac{2h}{3} \cos \frac{r\pi}{l} \cos \frac{s\pi}{l} \cos \frac{t\pi}{l}, \quad r, s, t = 1, \ldots, l-1.$$

Computational exercises

Two specific domains are built into IFISS by default, $\Omega_\square \equiv (-1,1) \times (-1,1)$ and $\Omega_{\hookleftarrow} \equiv \Omega_\square \backslash \{(-1,0) \times (-1,0)\}$. Numerical solutions to a Dirichlet problem defined on $\Omega_\square$ or $\Omega_{\hookleftarrow}$ can be computed by running `square_diff` or `ell_diff` as appropriate, with source data f and boundary data g specified in function m-files `../diffusion/specific_rhs` and `../diffusion/specific_bc`, respectively. Running the driver `diff_testproblem` sets up the data files `specific_rhs` and `specific_bc` associated with the reference problems in Examples 1.1.1–1.1.4.

1.1. Consider Example 1.1.1 with a typical solution illustrated in Figure 1.1. Evaluating the series solution (1.5) the maximum value of u is given by $u(0,0) = 0.294685413126$. Tabulate a set of computed approximations $u_h(0,0)$ to $u(0,0)$ using uniform 8×8, 16×16 and 32×32 grids with bilinear and biquadratic approximation. Then, by computing $|u(0,0) - u_h(0,0)|$, estimate the order of convergence that is achieved in each case.

1.2. Consider Example 1.1.2 with a typical solution illustrated in Figure 1.2. Tabulate a set of computed approximations u_h^* to the (unknown) maximum value of u^* using a sequence of uniform grids. By comparing successive approximations $|u_h^* - u_{h/2}^*|$, estimate the order of convergence that is achieved using bilinear and biquadratic approximation.

1.3. Write a function that postprocesses a Q_1 solution, u_h, and computes the global error $\|\nabla(u - u_h)\|$ for the analytic test problem in Example 1.1.3. Hence, verify the results given in Table 1.4. Then use your function to generate a table of estimated and exact errors for the set of stretched element grids that is automatically generated by IFISS.

1.4. Consider Example 1.1.2 with a typical solution illustrated in Figure 1.2. Tabulate the estimated error η for a sequence of uniform square grids, and hence estimate the order of convergence. Then change the source function from $f = 1$ to $f = xy$ and repeat the experiment. Can you explain the difference in the order of convergence?

1.5. Quadrilateral elements are also built into IFISS. Specifically, the function `quad_diff` can be used to solve problems defined on general quadrilateral domains with a Neumann condition on the right-hand boundary. By setting the source function to unity and the Dirichlet boundary data to zero, the effect of geometry on models of the deflection of an elastic membrane stretched over the bow-tie shaped domain illustrated below can be explored. (The Neumann condition acts as a symmetry condition, so only half of the bow-tie needs to be considered.)

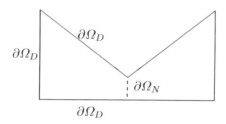

1.6. Using the matlab `eig` function, compute the eigenvalues of the coefficient matrix for Example 1.1.1 with $k = 8$ using Q_1 approximation on a uniform grid, and verify that there are $(k - 1)^2$ eigenvalues given by the analytic expression (1.122) together with $4k$ eigenvalues of unity (corresponding to the Dirichlet boundary nodes). Then, use the matlab `eigs` function to compute the maximum and minimum eigenvalues of the Q_1 stiffness matrix on a sequence of uniformly refined grids, and verify that the condition number grows like $8/(\pi^2 h^2)$ in the limit $h \to 0$.

2

SOLUTION OF DISCRETE POISSON PROBLEMS

The coefficient matrix of the linear system arising from finite element discretization of the Poisson equation is symmetric positive-definite. For the discretizations discussed in Chapter 1, it is also sparse. That is, only a very small proportion of its entries is nonzero. In this chapter, we discuss effective iterative methods for solving such systems, concentrating on two general approaches, *Krylov subspace* methods and *multigrid* methods.

Before proceeding, we briefly discuss the alternative to iterative algorithms, namely, *direct methods* based on Gaussian elimination (see, e.g. Golub & Van Loan [87]). There are very effective versions of this strategy, so-called *sparse elimination* methods, which use sophisticated techniques to exploit sparsity in the coefficient matrix to reduce computational requirements. These include *frontal methods* as mentioned in Section 1.4.3 (which were developed by early practitioners of the finite element method, in particular Irons [109]), and *reordering strategies*, which reorder the rows and columns of the coefficient matrix so that the matrix factors arising from the elimination process are made sparse to the extent possible. Further advantage is obtained for symmetric positive-definite systems using Cholesky factorization. A precise statement of the applicability and effectiveness of such methods is highly dependent on the matrix structure. Generally, direct methods work well for systems with thousands of degrees of freedom, but they require infeasible computational resources for linear systems of much larger dimension. That is, the number of nonzeros in the factors obtained from elimination, as well as the computational time required to perform the elimination, are too large. The practical limit for feasibility is often that direct sparse methods are competitive for two-dimensional partial differential equation problems but iterative methods are required in three dimensions. For an overview of these ideas, see Duff et al. [52] or George & Liu [83].

A feature of iterative methods is that they can take full advantage of sparsity of the coefficient matrix. In particular, their storage requirements typically depend only on the number of nonzeros in the matrix. The aim then becomes to make convergence as fast as possible. In this chapter, we consider symmetric positive-definite systems, our model being the discrete Poisson equation (1.21). Methods for nonsymmetric systems are considered in Chapter 4, and methods for symmetric indefinite systems are discussed in Chapter 6.

2.1 The conjugate gradient method

The conjugate gradient method (CG) developed by Hestenes & Stiefel [104] is the most well known of the general family of *Krylov subspace* methods. The utility of this class of methods lies in the observation that the sparsity of the coefficient matrix A in (1.21) enables the product with any vector, $\mathbf{x}$ say, to be computed very cheaply: if $A \in \mathbb{R}^{n \times n}$ has at most ℓ nonzero entries in any row then $A\mathbf{x}$ can be computed in only $n\ell$ flops.[1] For a logically rectangular grid of Q_1 elements for example, $\ell = 9$. Once the vector $A\mathbf{x}$ is computed, then for a further $n\ell$ operations $A(A\mathbf{x})$ can be found; similarly $A^3\mathbf{x}, A^4\mathbf{x} \dots$. Thus, it is easy to compute members of the *Krylov subspace*

$$\mathcal{K}_k(A, \mathbf{x}) := \text{span}\{\mathbf{x}, A\mathbf{x}, A^2\mathbf{x}, \dots, A^{k-1}\mathbf{x}\}$$

by taking appropriate linear combinations of these vectors. This is a poorly conditioned basis, and we will discuss the Lanczos procedure for computing an orthogonal basis in Section 2.4. Note that $\mathcal{K}_k(A, \mathbf{x})$ is the linear span of k vectors and so will be a k-dimensional subspace of $\mathbb{R}^n$ when these vectors are linearly independent.

One is then led to consider how well the solution $\mathbf{u}$ of the linear system (1.21) can be approximated from $\mathcal{K}_k(A, \mathbf{x})$ for any particular vector $\mathbf{x}$ and the different (increasing) values of $k = 1, 2, \dots$. Noting that any member, $\mathbf{y}$ say, of $\mathcal{K}_k(A, \mathbf{x})$ is of the form

$$\mathbf{y} = \sum_{j=0}^{k-1} \alpha_j A^j \mathbf{x},$$

for some coefficients $\alpha_j, j = 0, 1, \dots, k-1$, we can alternatively write

$$\mathbf{y} = q_{k-1}(A)\mathbf{x}$$

where $q_{k-1}(t)$ is the real polynomial $q_{k-1}(t) = \sum_{j=0}^{k-1} \alpha_j t^j$ of degree $k-1$. That is, $\mathbf{y}$ is specified by the coefficients of the polynomial q_{k-1}. Since we are at liberty to choose any polynomial (i.e. any member of $\mathcal{K}_k(A, \mathbf{x})$), the key question is now rephrased as: how well can $\mathbf{u}(= A^{-1}\mathbf{f})$ be approximated by a vector of the form $q_{k-1}(A)\mathbf{x}$?

Let us first consider what might be a convenient choice for $\mathbf{x}$. If we are prepared to take k as large as n, then the answer can be discerned from the *Cayley–Hamilton theorem*: any matrix satisfies its own characteristic equation $c(A) = 0$, where $c(z)$ is the characteristic polynomial $\prod_{j=1}^n (\lambda_j - z)$ with $\lambda_j, j = 1, 2, \dots, n$ the eigenvalues of A. Thus,

$$c(A) = (-1)^n A^n + \dots - \left(\sum_{i=1}^n \prod_{j=1, j \neq i}^n \lambda_j \right) A + \left(\prod_{j=1}^n \lambda_j \right) I = 0,$$

---

[1] A *flop* is a single multiply and addition, the basic floating point operation on a computer.

and so using the fact that for a non-singular matrix, $\det(A) = \prod_{j=1}^{n} \lambda_j \neq 0$, premultiplication by A^{-1} and rearrangement gives

$$\left((-1)^{n-1} A^{n-1} + \cdots + \left(\sum_{i=1}^{n} \prod_{j=1, j \neq i}^{n} \lambda_j \right) I \right) \Big/ (\det(A)) = A^{-1}. \qquad (2.1)$$

That is,

$$\mathbf{u} = A^{-1}\mathbf{f} = q_{n-1}(A)\mathbf{f},$$

where q_{n-1} is the polynomial on the left-hand side of (2.1), and we see that $\mathbf{x} = \mathbf{f}$ is a natural choice. More generally, if $\mathbf{u}^{(0)}$ is a starting vector, and $\mathbf{r}^{(0)} = \mathbf{f} - A\mathbf{u}^{(0)}$ is the associated *residual*, then

$$\mathbf{u} - \mathbf{u}^{(0)} = A^{-1}\mathbf{r}^{(0)} = q_{n-1}(A)\mathbf{r}^{(0)}. \qquad (2.2)$$

That is, given a nonzero starting vector $\mathbf{u}^{(0)}$, the natural way to generate a solution is to compute a correction to $\mathbf{u} - \mathbf{u}^{(0)}$ using $\mathbf{x} = \mathbf{r}^{(0)}$. In practice, $\mathbf{u}^{(0)}$ is often chosen to be the zero vector.

It remains to decide how to choose the sequence of polynomials $q_k, k = 0, 1, 2, \ldots$ in order that the *iterates*

$$\mathbf{u}^{(k)} = \mathbf{u}^{(0)} + q_{k-1}(A)\mathbf{r}^{(0)} \in \mathbf{u}^{(0)} + \mathcal{K}_k(A, \mathbf{r}^{(0)}) \qquad (2.3)$$

are successively closer to $A^{-1}\mathbf{f}$. A particularly simple example of such an iteration, which involves just one multiplication of a vector by A at each step, is the *Richardson iteration*

$$\mathbf{u}^{(k)} = \mathbf{u}^{(k-1)} + \alpha_{k-1}\mathbf{r}^{(k-1)}, \qquad (2.4)$$

where α_{k-1} is a parameter to be determined. It is easy to see that $\mathbf{u}^{(k)}$ has the form shown in (2.3). If $\alpha_k \equiv \alpha$ is constant for all k, then the error vector $\mathbf{u} - \mathbf{u}^{(k)}$ satisfies

$$\mathbf{u} - \mathbf{u}^{(k)} = (I - \alpha A)^k (\mathbf{u} - \mathbf{u}^{(0)}).$$

In this case, the iteration is convergent provided $|1 - \alpha\lambda| < 1$ for all eigenvalues λ of A, and the optimal choice for α (which makes the algebraically largest and smallest eigenvalues of $I - \alpha A$ equal in magnitude) is

$$\alpha = \frac{2}{\lambda_{\min}(A) + \lambda_{\max}(A)}. \qquad (2.5)$$

One way to choose the parameter α adaptively, giving the so-called *steepest descent method*, is to require that the A-norm of the error $\mathbf{u} - \mathbf{u}^{(k)}$ is locally minimized:

$$\alpha_{k-1} = \langle \mathbf{r}^{(k-1)}, \mathbf{r}^{(k-1)} \rangle / \langle A\mathbf{r}^{(k-1)}, \mathbf{r}^{(k-1)} \rangle. \tag{2.6}$$

For an analysis of the choices (2.5) and (2.6), see Problem 2.1.

To motivate a more effective approach, we recall the a priori error analysis of finite element discretization presented in Section 1.5. In particular, Theorem 1.3 shows that the Galerkin solution u_h is the best approximation to the PDE solution $u \in \mathcal{H}_E^1$ with respect to the energy norm, that is,

$$\|\nabla(u - u_h)\|^2 = a(u - u_h, u - u_h) = \min_{v_h \in S_E^h} \|\nabla(u - v_h)\|^2.$$

Using an expansion of the form (1.18), let the iterate $\mathbf{u}^{(k)} \in \mathbb{R}^n$ define a function $u_h^{(k)} \in S_E^h$, so that $u_h - u_h^{(k)} \in S_0^h$. Our goal is to estimate the error between $u_h^{(k)}$ and u in the energy norm. Using Galerkin orthogonality ($a(u - u_h, v_h) = 0$ for all $v_h \in S_0^h$) gives

$$\begin{aligned}
a(u - u_h^{(k)}, u - u_h^{(k)}) &= a(u - u_h + u_h - u_h^{(k)}, u - u_h + u_h - u_h^{(k)}) \\
&= a(u - u_h, u - u_h) + a(u_h - u_h^{(k)}, u_h - u_h^{(k)}) \\
&= \|\nabla(u - u_h)\|^2 + \|\nabla(u_h - u_h^{(k)})\|^2.
\end{aligned}$$

Moreover, the characterization (1.113)

$$\|\nabla(u_h - u_h^{(k)})\|^2 = \langle \mathbf{u} - \mathbf{u}^{(k)}, \mathbf{u} - \mathbf{u}^{(k)} \rangle_A = \|\mathbf{u} - \mathbf{u}^{(k)}\|_A^2$$

shows that the overall error is exactly the sum of the finite element discretization error and the A-norm of the iteration error:

$$\|\nabla(u - u_h^{(k)})\|^2 = \|\nabla(u - u_h)\|^2 + \|\mathbf{u} - \mathbf{u}^{(k)}\|_A^2. \tag{2.7}$$

This result has significant consequences. We know that u_h is the best approximation from S_E^h to $u \in \mathcal{H}_E^1$. A Krylov subspace method computes, via an iterate $\mathbf{u}^{(k)}$, a function $u_h^{(k)}$ in the subspace of S_E^h associated with $\mathcal{K}_k(A, \mathbf{r}^{(0)})$. We now have the additional fact that $u_h^{(k)}$ is the best approximation to u from this subspace if and only if $\|\mathbf{u} - \mathbf{u}^{(k)}\|_A$ is minimized over the Krylov space, that is, q_{k-1} in (2.3) is such that the A-norm of the error is minimal.

The steepest descent method achieves a local version of this minimization using a one-dimensional subspace of $\mathcal{K}_k(A, \mathbf{r}^{(0)})$. The optimal method, minimizing $\|\mathbf{u} - \mathbf{u}^{(k)}\|_A$ over the k-dimensional Krylov space, is the *conjugate gradient method.*

> **Algorithm 2.1:** THE CONJUGATE GRADIENT METHOD
> Choose $\mathbf{u}^{(0)}$, compute $\mathbf{r}^{(0)} = \mathbf{f} - A\mathbf{u}^{(0)}$, set $\mathbf{p}^{(0)} = \mathbf{r}^{(0)}$
> for $k = 0$ until convergence do
> $\quad \alpha_k = \langle \mathbf{r}^{(k)}, \mathbf{r}^{(k)} \rangle / \langle A\mathbf{p}^{(k)}, \mathbf{p}^{(k)} \rangle$
> $\quad \mathbf{u}^{(k+1)} = \mathbf{u}^{(k)} + \alpha_k \mathbf{p}^{(k)}$
> $\quad \mathbf{r}^{(k+1)} = \mathbf{r}^{(k)} - \alpha_k A\mathbf{p}^{(k)}$
> $\quad$ <Test for convergence>
> $\quad \beta_k = \langle \mathbf{r}^{(k+1)}, \mathbf{r}^{(k+1)} \rangle / \langle \mathbf{r}^{(k)}, \mathbf{r}^{(k)} \rangle$
> $\quad \mathbf{p}^{(k+1)} = \mathbf{r}^{(k+1)} + \beta_k \mathbf{p}^{(k)}$
> enddo

The computational work of one iteration is two inner products, three vector updates and one matrix-vector product.

In Algorithm 2.1, the choice of the scalar α_k is such that the new iterate minimizes the A-norm of the error among all choices along the "direction vector" $\mathbf{p}^{(k)}$, that is

$$\|\mathbf{u} - \mathbf{u}^{(k+1)}\|_A = \min_{\alpha_k} \|\mathbf{u} - (\mathbf{u}^{(k)} + \alpha_k \mathbf{p}^{(k)})\|_A = \min_{\mathbf{v} = \mathbf{u}^{(k)} + \alpha \mathbf{p}^{(k)}} \|\mathbf{u} - \mathbf{v}\|_A.$$

For a proof of this assertion, see Problem 2.2. More importantly, the one-dimensional minimization is also a k-dimensional one: $\mathbf{u}^{(k)}$ is the unique vector in the translated Krylov space $\mathbf{u}^{(0)} + \mathcal{K}_k(A, \mathbf{r}^{(0)})$ for which the A-norm of the error is minimized—exactly the desired property. The proof makes use of the following lemma.

Lemma 2.1. *For any k such that $\mathbf{u}^{(k)} \neq \mathbf{u}$, the vectors generated by the conjugate gradient method satisfy*

(i) $\langle \mathbf{r}^{(k)}, \mathbf{p}^{(j)} \rangle = \langle \mathbf{r}^{(k)}, \mathbf{r}^{(j)} \rangle = 0, \ j < k$,

(ii) $\langle A\mathbf{p}^{(k)}, \mathbf{p}^{(j)} \rangle = 0, \quad j < k$,

(iii) $\mathrm{span}\{\mathbf{r}^{(0)}, \mathbf{r}^{(1)}, \ldots, \mathbf{r}^{(k-1)}\} = \mathrm{span}\{\mathbf{p}^{(0)}, \mathbf{p}^{(1)}, \ldots, \mathbf{p}^{(k-1)}\} = \mathcal{K}_k(A, \mathbf{r}^{(0)})$.

Proof See Problem 2.3. □

Such a minimization in an inner product space is implied by, and is in fact equivalent to, a corresponding (Galerkin!) orthogonality condition (see Problem 2.4). We establish both characterizations in the following theorem.

Theorem 2.2. *The iterate $\mathbf{u}^{(k)}$ generated by the conjugate gradient method is the unique member of $\mathbf{u}^{(0)} + \mathcal{K}_k(A, \mathbf{r}^{(0)})$ for which the A-norm of the error is minimized, or alternatively, the residual $\mathbf{f} - A\mathbf{u}^{(k)}$ is orthogonal to $\mathcal{K}_k(A, \mathbf{r}^{(0)})$.*

Proof Recall that solving the system $A\mathbf{u} = \mathbf{f}$ is equivalent to computing the correction $\mathbf{c} = \mathbf{u} - \mathbf{u}^{(0)}$ by solving the system $A\mathbf{c} = \mathbf{r}^{(0)}$, where $\mathbf{r}^{(0)} = \mathbf{f} - A\mathbf{u}^{(0)}$ is the initial residual. Let $\mathbf{c}^{(k)} = \mathbf{u}^{(k)} - \mathbf{u}^{(0)} \in \mathcal{K}_k(A, \mathbf{r}^{(0)})$ denote the iterated correction where $\mathbf{u}^{(k)}$ is the kth conjugate gradient iterate. The error at the kth step satisfies

$$\mathbf{u} - \mathbf{u}^{(k)} = \mathbf{c} - \mathbf{c}^{(k)},$$

and the residual satisfies

$$\mathbf{r}^{(k)} = A(\mathbf{u} - \mathbf{u}^{(k)}) = A(\mathbf{c} - \mathbf{c}^{(k)}).$$

It therefore follows from Lemma 2.1, (i) and (iii) that

$$\langle \mathbf{c} - \mathbf{c}^{(k)}, \mathbf{v} \rangle_A = \langle A(\mathbf{c} - \mathbf{c}^{(k)}), \mathbf{v} \rangle = \langle \mathbf{r}^{(k)}, \mathbf{v} \rangle = 0 \quad \text{for all } \mathbf{v} \in \mathcal{K}_k(A, \mathbf{r}^{(0)}). \quad (2.8)$$

In any inner product space, Galerkin orthogonality implies optimality: consider $\tilde{\mathbf{c}} = \mathbf{c}^{(k)} + \epsilon \mathbf{v} \in \mathcal{K}_k(A, \mathbf{r}^{(0)})$, then

$$\|\mathbf{c} - \tilde{\mathbf{c}}\|_A^2 = \langle \mathbf{c} - \mathbf{c}^{(k)}, \mathbf{c} - \mathbf{c}^{(k)} \rangle_A - 2\epsilon \langle \mathbf{c} - \mathbf{c}^{(k)}, \mathbf{v} \rangle_A + \epsilon^2 \langle \mathbf{v}, \mathbf{v} \rangle_A$$

so that (2.8) implies that $\|\mathbf{c} - \tilde{\mathbf{c}}\|_A$ is minimized when $\epsilon = 0$. That is, $\|\mathbf{c} - \tilde{\mathbf{c}}\|_A$ is minimized for $\tilde{\mathbf{c}} \in \mathcal{K}_k(A, \mathbf{r}^{(0)})$ by the choice $\tilde{\mathbf{c}} = \mathbf{c}^{(k)}$, or, what is the same thing, $\|\mathbf{u} - \tilde{\mathbf{u}}\|_A$ is minimized for $\tilde{\mathbf{u}} \in \mathbf{u}^{(0)} + \mathcal{K}_k(A, \mathbf{r}^{(0)})$ by $\tilde{\mathbf{u}} = \mathbf{u}^{(k)}$.

To establish uniqueness, suppose there exists $\tilde{\mathbf{c}} \in \mathcal{K}_k(A, \mathbf{r}^{(0)})$ such that

$$\langle \mathbf{c} - \tilde{\mathbf{c}}, \mathbf{v} \rangle_A = 0 \quad \text{for all } \mathbf{v} \in \mathcal{K}_k(A, \mathbf{r}^{(0)}).$$

Combining this with (2.8) yields

$$\langle \mathbf{c}^{(k)} - \tilde{\mathbf{c}}, \mathbf{v} \rangle_A = 0 \quad \text{for all } \mathbf{v} \in \mathcal{K}_k(A, \mathbf{r}^{(0)}).$$

Making the particular choice $\mathbf{v} = \mathbf{c}^{(k)} - \tilde{\mathbf{c}}$ in this relation establishes that $\tilde{\mathbf{c}} = \mathbf{c}^{(k)}$. $\qquad \square$

Remark 2.3. The symbiosis between Galerkin approximation for the Poisson equation via finite elements and Galerkin approximation for the algebraic system via CG is deeply satisfying. It comes from the serendipitous fact that CG is optimal with respect to just the right norm, as dictated by (2.7). The choice of a relevant norm (or "energy") and its effect on both approximation and solution algorithms can be just as important (but is not as elegantly resolved) for the problems discussed in later Chapters 3–4 and 5–6.

2.1.1 *Convergence analysis*

Bounds on the rate of convergence of the sequence of iterates $\{\mathbf{u}^{(k)}: k = 0, 1, 2, \ldots\}$ to the solution $\mathbf{u}$ are obtained for the CG method through use of the polynomial representation (2.3). Writing $\mathbf{e}^{(k)}$ for the error in the kth iterate $\mathbf{u} - \mathbf{u}^{(k)}$, and substituting for $\mathbf{u}^{(k)}$ from (2.3), gives

$$\mathbf{e}^{(k)} = \mathbf{e}^{(0)} - q_{k-1}(A)\mathbf{r}^{(0)} = (I - Aq_{k-1}(A))\mathbf{e}^{(0)},$$

since $A\mathbf{e}^{(0)} = \mathbf{r}^{(0)}$. We can write this as

$$\mathbf{e}^{(k)} = p_k(A)\mathbf{e}^{(0)}, \quad (2.9)$$

where $p_k(z)(= 1 - zq_{k-1}(z))$ is a polynomial of degree k with constant term equal to 1 (so that $p_k(0) = 1$). Now Theorem 2.2 implies that $\|\mathbf{e}^{(k)}\|_A$ is minimized over the Krylov subspace, so we have a characterization of the polynomial p_k that corresponds to the kth CG iterate:

$$\|\mathbf{e}^{(k)}\|_A = \min_{p_k \in \Pi_k, p_k(0)=1} \|p_k(A)\mathbf{e}^{(0)}\|_A. \tag{2.10}$$

Here Π_k is the set of real polynomials of degree k. Simplification is achieved by expanding $\mathbf{e}^{(0)}$ in terms of the eigenvector basis of A:

$$\mathbf{e}^{(0)} = \sum_{j=1}^{n} \gamma_j \mathbf{v}^{(j)},$$

where $\mathbf{v}^{(j)}$ satisfies $A\mathbf{v}^{(j)} = \lambda_j \mathbf{v}^{(j)}$ for the eigenvalue λ_j. Thus, (2.10) becomes

$$\|\mathbf{e}^{(k)}\|_A = \min_{p_k \in \Pi_k, p_k(0)=1} \left\| \sum_{j=1}^{n} \gamma_j p_k(\lambda_j) \mathbf{v}^{(j)} \right\|_A$$

$$\leq \min_{p_k \in \Pi_k, p_k(0)=1} \left\| \max_j |p_k(\lambda_j)| \sum_{j=1}^{n} \gamma_j \mathbf{v}^{(j)} \right\|_A$$

$$= \min_{p_k \in \Pi_k, p_k(0)=1} \max_j |p_k(\lambda_j)| \; \|\mathbf{e}^{(0)}\|_A. \tag{2.11}$$

The bound (2.11) is not useful in this form since knowledge of all of the eigenvalues of A generally requires much more work than the mere solution of a single linear system! However, (2.11) is more helpful in the case where one can find a convenient inclusion set for all of the eigenvalues. In particular if $\lambda_j \in [a, b]$ for all j with $a > 0$ (the best values are $a = \lambda_{\min}(A)$ and $b = \lambda_{\max}(A)$) then

$$\frac{\|\mathbf{e}^{(k)}\|_A}{\|\mathbf{e}^{(0)}\|_A} \leq \min_{p_k \in \Pi_k, p_k(0)=1} \max_{z \in [a,b]} |p_k(z)|. \tag{2.12}$$

The polynomial that achieves the minimization in (2.12) is known (see Rivlin [158]): it is the shifted and scaled Chebyshev polynomial

$$\chi_k(t) = \left[\tau_k \left(\frac{b+a}{b-a} - \frac{2t}{b-a} \right) \right] \Big/ \left[\tau_k \left(\frac{b+a}{b-a} \right) \right], \tag{2.13}$$

where the standard Chebyshev polynomials for $k = 0, 1, 2, \ldots$ are given by

$$\tau_k(t) \equiv \begin{cases} \cos(k \cos^{-1} t) & \text{for } t \in [-1, 1] \\ \cosh(k \cosh^{-1} t) & \text{for } t > 1 \\ (-1)^k \tau_k(-t) & \text{for } t < -1. \end{cases} \tag{2.14}$$

Using (2.14) and the recurrence

$$\tau_{k+1}(t) = 2t\tau_k(t) - \tau_{k-1}(t) \tag{2.15}$$

it can be shown that (see Problem 2.5)

$$\tau_k(t) = \frac{1}{2}\left[(t + \sqrt{t^2 - 1})^k + (t - \sqrt{t^2 - 1})^k\right]. \tag{2.16}$$

Thus if

$$\kappa = \kappa(A) = \lambda_{\max}(A)/\lambda_{\min}(A)$$

then for $a = \lambda_{\min}(A)$ and $b = \lambda_{\max}(A)$,

$$\tau_k\left(\frac{b+a}{b-a}\right) = \tau_k\left(\frac{\kappa+1}{\kappa-1}\right) = \frac{1}{2}\left[\left(\frac{\sqrt{\kappa}-1}{\sqrt{\kappa}+1}\right)^k + \left(\frac{\sqrt{\kappa}+1}{\sqrt{\kappa}-1}\right)^k\right] \geq \frac{1}{2}\left(\frac{\sqrt{\kappa}+1}{\sqrt{\kappa}-1}\right)^k.$$

Noting that as the argument in the numerator of (2.13) lies in $[-1, 1]$ so also does its value, the following (classical) convergence theorem is established.

Theorem 2.4. *After k steps of the conjugate gradient method, the iteration error $\mathbf{e}^{(k)} = \mathbf{u} - \mathbf{u}^{(k)}$ satisfies the bound*

$$\|\mathbf{e}^{(k)}\|_A \leq 2\left(\frac{\sqrt{\kappa}-1}{\sqrt{\kappa}+1}\right)^k \|\mathbf{e}^{(0)}\|_A. \tag{2.17}$$

The bound (2.17) intuitively leads to the notion that if a matrix has small condition number then convergence of CG will be rapid. The converse is not true however! A system matrix of order n with two multiple eigenvalues at a and b can have arbitrarily large condition number, yet since a polynomial of degree two, say $p_2(z)$, can be constructed that satisfies $p_2(a) = 0$ and $p_2(b) = 0$, then the tighter bound (2.11) implies that CG will converge to the solution of such a system in at most two iterations! This issue is explored further in Computational Exercise 2.1. The convergence bound (2.17) is better than the analogous bound for the steepest descent method, see Problem 2.1.

2.1.2 Stopping criteria

Now suppose that the objective is to run the CG process until the relative error satisfies a pre-selected tolerance, ϵ, that is, $\|\mathbf{e}^{(k)}\|_A/\|\mathbf{e}^{(0)}\|_A \leq \epsilon$. In light of the bound in Theorem 2.4, it suffices for the inequality

$$\epsilon \geq 2\left(\frac{\sqrt{\kappa}-1}{\sqrt{\kappa}+1}\right)^k = 2\left(1 - \frac{2/\sqrt{\kappa}}{1 + 1/\sqrt{\kappa}}\right)^k$$

to hold. Taking the logarithm of both sides of the inequality and using the fact that $\log\left(1 - \frac{2/\sqrt{\kappa}}{1+1/\sqrt{\kappa}}\right) \approx -2/\sqrt{\kappa}$ for large κ, gives the characterization

$$k \approx \tfrac{1}{2}|\log(\epsilon/2)|\sqrt{\kappa}. \qquad (2.18)$$

The asymptotic estimate (2.18) provides a bound on the number of iterations required to meet a given stopping criterion. In cases where (2.17) is a tight bound, associated iteration counts will be proportional to the square root of the condition number of the coefficient matrix.

For a discrete Laplacian on a quasi-uniform mesh in $\mathbb{R}^2$ or $\mathbb{R}^3$, $\kappa(A) = \mathcal{O}(h^{-2})$ (see Theorem 1.32, and Remarks 1.33 and 1.34), and the estimate (2.18) suggests that for uniformly refined grids, the number of CG iterations required to meet a fixed tolerance will approximately double with each grid refinement. To demonstrate that this is indeed what happens in practice, CG iteration counts are tabulated in Table 2.1 associated with the solution of systems arising from Examples 1.1.1 and 1.1.4, with a starting vector $\mathbf{u}^{(0)} = 0$. It is clear that CG exhibits the anticipated behavior in both cases.

The "stopping tolerance", ϵ, has been viewed up to this point as being an arbitrary small quantity. However, if we make use of the specific finite element matrix properties discussed in Section 1.6, then it is possible to develop good heuristics for choosing ϵ so that the solution obtained by applying CG to the system (1.21) is as accurate as needed. Moreover, although this discussion is in the context of CG iteration, we should emphasize that the stopping strategy that is developed also applies to the preconditioned versions of CG that are discussed in Section 2.2—it also has relevance for the multigrid iteration discussed in Section 2.5.

Our starting point is the characterization (2.7) of the overall error in terms of the sum of the squares of the discretization error and the algebraic error associated with the iteration. For smooth (i.e. H^2–regular) problems, a priori

Table 2.1 *Number of* CG *iterations needed to satisfy* $\|\mathbf{r}^{(k)}\|/\|\mathbf{f}\| \leq 10^{-4}$ *for* Q_1 *discretization of Examples* 1.1.1 *and* 1.1.4: ℓ *is the grid refinement level;* h *is* $2^{1-\ell}$.

ℓ	Example	
	1.1.1	1.1.4
3	7	9
4	15	16
5	30	31
6	60	59
7	121	113

error analysis (see Theorem 1.19) provides a bound on the Q_1 discretization error of the form $C_1 h \|D^2 u\|$. This implies that

$$\frac{\|\nabla(u - u_h)\|}{\|\nabla u\|} \le C_1 h \frac{\|D^2 u\|}{\|\nabla u\|}.$$

Since the overall error in (2.7) is limited by the discretization error, a reasonable strategy is to stop the CG iteration when the relative algebraic error satisfies

$$\frac{\|\mathbf{e}^{(k)}\|_A}{\|\mathbf{u}\|_A} \le \tau h, \tag{2.19}$$

for $\tau < C_1 C_u$, with $C_u = \|D^2 u\|/\|\nabla u\|$ an $O(1)$ constant. Note that for problems with singularities, such as Example 1.1.4, where the discretization error on uniform grids decreases like h^α with $\alpha < 1$, the strategy (2.19) is a conservative choice.[2] Of course, we don't know the constants C_u and C_1 in general, but this is an uncertainty that we might be prepared to live with. In the case of Examples 1.1.3 and 1.1.4, the results in Tables 1.1 and 1.2 suggest that C_1 is in the range $O(10^{-1})$–$O(10^{-2})$. A safe strategy is then to choose τ on the order of 10^{-4}.

One immediate difficulty with implementing the strategy (2.19) in Algorithm 2.1 is that the energy error $\|\mathbf{e}^{(k)}\|_A$ is not available.[3] To address this difficulty, we can exploit the following relationship between energy and residual errors

$$\frac{\|\mathbf{e}^{(k)}\|_A}{\|\mathbf{e}^{(0)}\|_A} \le \sqrt{\kappa(A)} \frac{\|\mathbf{r}^{(k)}\|}{\|\mathbf{r}^{(0)}\|}, \tag{2.20}$$

see Problem 2.6. In particular, setting $\mathbf{u}^{(0)} = 0$, we have that $\mathbf{e}^{(0)} = \mathbf{u}$ and $\mathbf{r}^{(0)} = \mathbf{f}$. For Q_1 approximation on a sequence of uniform grids, we know from Theorem 1.32 that $\kappa(A) \le C_* h^{-2}$, and thus the criterion (2.19) is guaranteed to be satisfied if the CG iteration is stopped once the relative residual error $\|\mathbf{r}^{(k)}\|/\|\mathbf{f}\|$ is smaller than $\tau h^2 / \sqrt{C_*}$. For the problem in Example 1.1.1, the explicit value of $\kappa(A)$ given by (1.124) suggests that $C_* \approx 2\ell^2/\pi^2 \approx 1$, thus a practical stopping criterion (ensuring (2.19)) is

$$\frac{\|\mathbf{r}^{(k)}\|}{\|\mathbf{r}^{(0)}\|} \le 10^{-4} h^2. \tag{2.21}$$

We also remark that the inequality of (2.20) is likely to be pessimistic, providing a built-in safety factor.

With the CG stopping parameter defined as in (2.21), the influence of the mesh parameter on the number of CG iterations is changed. However, a glance

---

[2]With a non-quasi-uniform sequence of grids, the stopping criterion could be related to the smallest element diameter, $\underline{h}$, rather than h.

[3]Indeed, we circumvented this point in generating Table 2.1, by using the standard residual norm stopping test that is built into the matlab function pcg.m.

Table 2.2 *Number of* CG *iterations needed to satisfy* $\|\mathbf{r}^{(k)}\|/\|\mathbf{f}\| \le 10^{-4}h^2$ *for* $\mathbf{Q}_1$ *discretization of Examples 1.1.1 and 1.1.4.*

| | Example | | |
ℓ	1.1.1	1.1.4	$10^{-4}h^2$
3	8	10	6.3×10^{-6}
4	18	19	1.6×10^{-6}
5	37	40	3.9×10^{-7}
6	78	83	9.8×10^{-8}
7	161	171	2.4×10^{-8}

at (2.18) indicates that impact of this change is small: the bound on the number of iterations is now proportional to $h^{-1}|\log h|$ instead of h^{-1}. This point is also explored in Table 2.2, which shows the iteration counts associated with the stopping test (2.21) for the systems arising from Examples 1.1.1 and 1.1.4. Notice that the iteration counts are bigger than in Table 2.1, but since the logarithm is a slowly growing function, the growth trends are essentially similar. Alternatively, the additional logarithmic growth factor can be eliminated altogether by using coarse grids to provide a more accurate starting vector (see Braess [19, p. 234]). The extension of this strategy to the case of using a higher order discretization (for which the discretization error reduces more rapidly as the grid is refined, see Theorem 1.21), is addressed in Computational Exercise 2.2.

2.2　Preconditioning

The results in Tables 2.1 and 2.2 show the typical behavior for CG solution of the Galerkin system. That is, for a finer discretization there is not only more work in carrying out a single iteration, but also more iterations are required for convergence. It would be ideal if the number of iterations required to satisfy the stopping criterion did not grow under mesh refinement, so that the computational work would grow linearly with the dimension of the discrete system. *Preconditioning* is usually employed in order to achieve this ideal or to get closer to it.

　　The basic idea is to construct a matrix (or a linear process), M say, that approximates the coefficient matrix A but for which it requires little work to apply the action of the inverse of M, that is, to compute $M^{-1}\mathbf{v}$ for given $\mathbf{v}$. One may then think of solving

$$M^{-1}A\mathbf{u} = M^{-1}\mathbf{f} \tag{2.22}$$

instead of $A\mathbf{u} = \mathbf{f}$; they clearly have the same solution. If M is a good approximation of A, then it might be expected that CG iteration will be more rapidly convergent for the preconditioned system (2.22) than for the original system, and the overall computational work may be significantly reduced. For very large problems, preconditioning may be necessary to make computation feasible.

To use the CG method, we must ensure that the preconditioned coefficient matrix is symmetric. This means we need to consider a variant of (2.22), which will lead to the *preconditioned (or generalized) conjugate gradient method*. Given a symmetric positive-definite matrix M, let $M = HH^T$, and consider the system

$$H^{-1}AH^{-T}\mathbf{v} = H^{-1}\mathbf{f}, \quad \mathbf{v} = H^T\mathbf{u}. \tag{2.23}$$

The coefficient matrix $H^{-1}AH^{-T}$ is symmetric positive-definite, so that CG is applicable. We emphasize that this is a formalism; in practice, as is evident in Algorithm 2.2, the only thing needed is to be able to compute the action of the inverse of M. In particular, H is not required, nor indeed is an explicit representation of M. For (2.23), H can be taken to be the square root of M or the lower triangular Cholesky factor of M — in either case positive-definiteness is important.[4]

A very important property of CG is that regardless of the preconditioning, the A-norm of the error decreases monotonically. To see this, consider the system (2.23) for which the preconditioned CG iteration error is given by

$$\begin{aligned}
\|\mathbf{v} - \mathbf{v}^{(k)}\|^2_{H^{-1}AH^{-T}} &= (\mathbf{v} - \mathbf{v}^{(k)})^T H^{-1}AH^{-T}(\mathbf{v} - \mathbf{v}^{(k)}) \\
&= (H^T(\mathbf{u} - \mathbf{u}^{(k)}))^T H^{-1}AH^{-T}(H^T(\mathbf{u} - \mathbf{u}^{(k)})) \\
&= (\mathbf{u} - \mathbf{u}^{(k)})^T A(\mathbf{u} - \mathbf{u}^{(k)}) = \|\mathbf{u} - \mathbf{u}^{(k)}\|^2_A.
\end{aligned}$$

Thus, indeed, $\|\mathbf{u} - \mathbf{u}^{(k)}\|_A$ is minimized with respect to a norm that is independent of the particular choice of M. (The Krylov space in this case is that defined by the preconditioned matrix $H^{-1}AH^{-T}$.) Moreover, the error characterization (2.7) as the sum of the discretization error and the algebraic iteration error, is independent of the preconditioning used.

The theory from Section 2.1.1 shows that convergence of the preconditioned CG iteration depends on the eigenvalues of $H^{-1}AH^{-T}$, which are identical to the eigenvalues of $M^{-1}A$ because of the similarity transformation $H^{-T}(H^{-1}AH^{-T})H^T = M^{-1}A$. Thus, introducing the loose notation

$$\kappa(M^{-1}A) := \lambda_{\max}(H^{-1}AH^{-T})/\lambda_{\min}(H^{-1}AH^{-T}), \tag{2.24}$$

we see that (2.18) suggests that $\frac{1}{2}\sqrt{\kappa(M^{-1}A)}|\log \epsilon/2|$ preconditioned CG iterations will be needed in order to reduce the A-norm of the error by ϵ. In particular, if a preconditioner M can be found such that $\kappa(M^{-1}A)$ is bounded independently of h then, for a fixed convergence tolerance ϵ, the required number of iterations will not increase when more accurate solutions are sought using more refined grids.

---

[4]If $M = Q\Lambda Q^T$ is the diagonalization of M, then $M^{1/2} = Q\Lambda^{1/2}Q^T$ where $\Lambda^{1/2}$ is a diagonal matrix comprising the (positive) square roots of the eigenvalues.

The improved performance achieved through preconditioning is best expressed in terms of the "clustering of eigenvalues". Contrary to popular wisdom, it is not actually necessary to reduce the condition number by preconditioning to achieve fast convergence. If a polynomial can be constructed that is "small at the eigenvalues", then the bound (2.11) implies that fast convergence will be achieved. This will be true, for example, if the eigenvalues of the preconditioned matrix cluster about two points, see Computational Exercise 2.1.

The simplest way to derive the preconditioned CG algorithm is to start with the standard CG algorithm for the preconditioned system (2.23), introduce the preconditioned residual $\mathbf{z}^{(k)} = M^{-1}\mathbf{r}^{(k)}$ for each k, and then convert back to unpreconditioned direction vectors and the like (see Problem 2.7). The outcome is the following algorithm.

Algorithm 2.2: THE PRECONDITIONED CONJUGATE GRADIENT METHOD
Choose $\mathbf{u}^{(0)}$, compute $\mathbf{r}^{(0)} = \mathbf{f} - A\mathbf{u}^{(0)}$, solve $M\mathbf{z}^{(0)} = \mathbf{r}^{(0)}$, set $\mathbf{p}^{(0)} = \mathbf{z}^{(0)}$
for $k = 0$ until convergence do
$\qquad \alpha_k = \langle \mathbf{z}^{(k)}, \mathbf{r}^{(k)} \rangle / \langle A\mathbf{p}^{(k)}, \mathbf{p}^{(k)} \rangle$
$\qquad \mathbf{u}^{(k+1)} = \mathbf{u}^{(k)} + \alpha_k \mathbf{p}^{(k)}$
$\qquad \mathbf{r}^{(k+1)} = \mathbf{r}^{(k)} - \alpha_k A\mathbf{p}^{(k)}$
$\qquad$ <Test for convergence>
$\qquad$ Solve $M\mathbf{z}^{(k+1)} = \mathbf{r}^{(k+1)}$
$\qquad \beta_k = \langle \mathbf{z}^{(k+1)}, \mathbf{r}^{(k+1)} \rangle / \langle \mathbf{z}^{(k)}, \mathbf{r}^{(k)} \rangle$
$\qquad \mathbf{p}^{(k+1)} = \mathbf{z}^{(k+1)} + \beta_k \mathbf{p}^{(k)}$
enddo

Note the requirement at each iteration to solve $M\mathbf{z}^{(k+1)} = \mathbf{r}^{(k+1)}$. This is the only additional work compared to Algorithm 2.1. There are two extreme choices of M: if $M = I$ then $\mathbf{z}^{(k)} = \mathbf{r}^{(k)}$ for each k, but trivially $\kappa(M^{-1}A) = \kappa(A)$, so that one has the (unpreconditioned) CG method; otherwise, if $M = A$ then $\mathbf{z}^{(0)} = A^{-1}(\mathbf{f} - A\mathbf{u}^{(0)}) = \mathbf{u} - \mathbf{u}^{(0)}$ and so $\alpha_1 = 1$ and $\mathbf{u}^{(1)} = \mathbf{u}$. In the latter case, the system is solved by computing the correction as in (2.2).

The central idea in preconditioning is to employ an operator M satisfying two basic properties:

I. M should be a good approximation to A in the sense that the action of its inverse on A should have the effect of clustering the eigenvalues. In this case, Algorithm 2.2 will converge to a prescribed tolerance in fewer iterations than Algorithm 2.1.

II. The solution of $M\mathbf{z}^{(k+1)} = \mathbf{r}^{(k+1)}$ should be inexpensive in terms of computer resources. Ideally, the work and memory requirements of one iteration of Algorithm 2.2 should be comparable with that of one iteration of Algorithm 2.1.

A number of preconditioning approaches have been developed that, to varying degrees, meet these competing criteria. It should be obvious that no "best" preconditioner exists, but what is less clear is that really effective preconditioning is often at the heart of efficient solution algorithms, in particular when solving three-dimensional PDE problems. Since this is a rapidly developing research area, we simply outline four of the many possible approaches.

(i) *Diagonal scaling.* The simple choice $M = \mathtt{diag}(A)$ leads to diagonal scaling, and is also called Jacobi preconditioning. Condition II is clearly satisfied, since computing the action of M^{-1} involves only n divisions, but such simple scaling, though useful for stretched grids, generally reduces κ by only a small constant (independent of h). See Computational Exercise 2.3.

(ii) *Incomplete factorization.* The main reason frontal elimination methods become infeasibly expensive for three-dimensional Poisson problems is the large amount of "fill", that is, there is a very significant storage requirement associated with the nonzero entries of the triangular factors produced in the elimination phase. On the other hand, for any matrix that is available in triangular factored form, a corresponding linear system can be efficiently solved by forward and backward substitution. To balance these issues, a generally applicable class of algebraic preconditioners has been developed based on the idea of *incomplete triangular (LU) factorization*. Much of the development of these widely used methods stems from the work of Meijerink & van der Vorst [130]. For symmetric matrices, incomplete Cholesky factorization is used, for which $U = L^{T}$.

The basic principle is to either preselect some sparsity pattern outside which any nonzero entries that would arise in the factors L or U are dropped, or else to specify some threshold magnitude and drop all entries in L and U that are smaller than this. The key issue here is to choose the threshold so that the preconditioner is not too dense, but not too sparse! A common choice is to select the allowed fill to exactly match the sparsity pattern of the original matrix A, so that only a replica of the data structure employed for storing A is needed. For symmetric A this variant is called Incomplete Cholesky(0) or IC(0) and is depicted in Algorithm 2.3.

Algorithm 2.3: IC(0) FACTORIZATION
```
for i = 1 until n do
    m = min{k | a_ik ≠ 0}
    for j = m until i - 1 do
        if a_ij ≠ 0 then
```
$$l_{ij} \leftarrow a_{ij} - \sum_{k=m}^{j-1} l_{ik} l_{jk} / l_{jj}$$
```
        endif
    enddo
```
$$l_{ii} = \left(a_{ii} - \sum_{k=m}^{i-1} l_{ik} l_{ik} \right)^{1/2}$$
```
enddo
```

Because of the triangular factored form, criterion II is satisfied. Typically, incomplete factorizations cluster many eigenvalues and are reasonably effective preconditioners. Another attractive feature is that they are defined purely algebraically and require no geometric information. The performance of IC(0) preconditioning is explored in Computational Exercise 2.3.

Although there is a wealth of empirical evidence that IC preconditioning dramatically improves the convergence rate of CG applied to discrete Poisson problems, rigorous analysis is quite limited. One known result is that the asymptotic behavior of the condition number using IC(0) preconditioning is unchanged: $\kappa(M^{-1}A) = \mathcal{O}(h^{-2})$. In contrast, a variant of Algorithm 2.3 known as the *Modified Incomplete Cholesky* factorization, wherein a constraint on the row sums of M is imposed, can be shown to have a better condition number bound: $\kappa(M^{-1}A) = \mathcal{O}(h^{-1})$, see Gustafsson [96] and Computational Exercise 2.4. An accessible and comprehensive discussion of this class of preconditioners is given by Meurant [131].

(iii) *Domain decomposition.* For parallel computation, the idea of solving a number of subproblems (or at least smaller dimensional problems) presents a possible preconditioning method. The principal idea in this context is called *domain decomposition*: the domain Ω is partitioned into subdomains $\Omega_k, k = 1, \ldots, n_k$ and, with appropriate conditions specified on the subdomain boundaries, subproblems on these separate subdomains are solved in parallel (either using a sparse direct method or some other iteration). Although there are many wrinkles that need to be dealt with, effective preconditioners exist that are based on this idea, and there is a strong underpinning in the sense of rigorous analysis. See Chan & Mathew [39], Quarteroni & Valli [153], and Smith et al. [179] for details.

(iii) *Multigrid.* Multigrid methods (which are described in detail in later sections) can be very effective linear system solvers for discrete Poisson problems. Correspondingly, a single multigrid cycle (defined in Section 2.5.2) can be an extremely effective preconditioner even for problems where it is not clear how to construct convergent multigrid iterations. We return to this theme in Chapter 6.

Remark 2.5. A useful concept when discussing preconditioners for both the discrete Poisson problem and for more complicated partial differential equation problems is *spectral equivalence*. If the symmetric and positive-definite matrix A arises from approximation of a partial differential operator on a computational mesh of mesh size h, we say that a symmetric and positive-definite matrix M of the same dimension is *spectrally equivalent* to A if there are positive constants c, C independent of h such that

$$c \leq \frac{\langle A\mathbf{x}, \mathbf{x}\rangle}{\langle M\mathbf{x}, \mathbf{x}\rangle} \leq C \quad \text{for all } \mathbf{x} \neq \mathbf{0}. \tag{2.25}$$

In later sections it will be established that appropriate multigrid cycles are spectrally equivalent to the Galerkin matrix for the Poisson problem.

For future reference, it can often be helpful to consider preconditioners in the context of matrix *splittings*. The splitting $A = M - R$ leads naturally to the *simple* or *stationary* iteration

$$M\mathbf{u}^{(k)} = R\mathbf{u}^{(k-1)} + \mathbf{f}, \quad \mathbf{u}^{(0)} \text{arbitrary} \tag{2.26}$$

for the solution of $A\mathbf{u} = \mathbf{f}$. If such an iteration converges, then $\mathbf{u}$ is the fixed point. Subtracting (2.26) from $M\mathbf{u} = R\mathbf{u} + \mathbf{f}$ leads to

$$\mathbf{e}^{(k)} = M^{-1}R\mathbf{e}^{(k-1)} = (I - M^{-1}A)^k \mathbf{e}^{(0)}, \tag{2.27}$$

where $\mathbf{e}^{(k)} = \mathbf{u} - \mathbf{u}^{(k)}$ as above. The iteration (2.27) is convergent if and only if all the eigenvalues of $I - M^{-1}A$ are less than one in modulus (see Problem 2.8), or equivalently, if the eigenvalues λ of $M^{-1}A$ satisfy $0 < \lambda < 2$. Convergence will generally be slow if there are eigenvalues $\lambda \approx 0, 2$. The connection here is that preconditioned CG can also be viewed as an *acceleration* of an underlying stationary iteration based on the splitting $A = M - R$, see Problem 2.9. That is, CG generates an optimal polynomial as in (2.9) rather than the specific polynomial $(1 - z)^k$ determined by (2.27). Note that the CG iteration will converge provided that both A and M are symmetric positive-definite, even if the corresponding simple iteration (2.26) is divergent. On the other hand, the choice of an effective preconditioner M leading to fast CG convergence generally (but not always) corresponds to having a convergent stationary iteration.

An alternative representation of the simple iteration (2.26) is

$$\mathbf{u}^{(k)} = \mathbf{u}^{(k-1)} + M^{-1}(\mathbf{f} - A\mathbf{u}^{(k-1)}). \tag{2.28}$$

This can be rewritten as

$$\mathbf{e}^{(k)} = \mathbf{e}^{(k-1)} - M^{-1}\mathbf{r}^{(k-1)},$$

where $\mathbf{r}^{(k-1)} = \mathbf{f} - A\mathbf{u}^{(k-1)}$. Comparison with (2.4) shows that (2.28) is a simple Richardson iteration for the system $M^{-1}A\mathbf{u} = M^{-1}\mathbf{f}$. We will explore the connection between splittings and preconditioning in more depth in Chapter 4.

2.3 Singular systems are not a problem

In the case where only Neumann (derivative) boundary conditions are prescribed (so that $\partial\Omega_N = \partial\Omega$), the Galerkin matrix is singular as described in Section 1.3. The null space (kernel) consists of constant vectors, that is span$\{\mathbf{1}\}$ where $\mathbf{1} = (1, \dots, 1)^T$. Nevertheless, for a well-posed problem the linear system is always consistent (see (1.4) and (1.25)). There are infinitely many solutions but any two solutions differ only by a scalar multiple of $\mathbf{1}$. It is common to look for the solution with zero mean value, that is the solution orthogonal to $\mathbf{1}$. The question then arises: are iterative methods able to compute solutions of such a consistent singular system?

In the case of a simple iteration (2.26), provided M is non-singular, the answer is certainly yes. To see this, we note that $A\mathbf{1} = \mathbf{0}$ is the same as $M\mathbf{1} = N\mathbf{1}$ so that $M^{-1}N\mathbf{1} = \mathbf{1}$. Thus it follows from (2.27) that any component of $\mathbf{e}^{(0)}$ in the direction of $\mathbf{1}$ will remain unchanged under iteration. Thus, if all other eigenvalues of $M^{-1}N$ are inside the unit disc then the iteration will certainly convergence to a particular solution of the consistent linear system. If the solution with zero mean is required, then the average value can be subtracted to produce this solution.

The analysis of CG iteration for singular consistent systems is not quite so straightforward, but the outcome is essentially the same: for a symmetric and positive semi-definite matrix, convergence will occur to a consistent solution provided a non-singular preconditioner is employed. It is helpful to use a zero starting vector. In particular, the speed of convergence will depend only on the nonzero eigenvalues (see van der Vorst [199, section 10]). Thus for a Neumann problem the *effective condition number* is λ_n/λ_2 if the eigenvalues are numbered in ascending order with $\lambda_1 = 0$.

2.4   The Lanczos and minimum residual methods

The CG iteration method is closely related to the Lanczos method for estimating eigenvalues of symmetric matrices. This connection has been of significant use in the development of iterative solution algorithms for systems of equations with nonsymmetric and indefinite coefficient matrices such as are required for the problems described in the following chapters of this book. In particular, the Lanczos method (for symmetric matrices) and the related Arnoldi method (for nonsymmetric matrices) have the important property that they explicitly calculate orthogonal bases for Krylov subspaces and so, as we describe below, they lead to solution algorithms for linear systems. Details of the connection between conjugate gradient and Lanczos methods are given in this section. For more on the history of these ideas, see Golub & O'Leary [88].

The Lanczos algorithm is defined as follows. Let $\mathbf{v}^{(1)}$ be a vector such that $\|\mathbf{v}^{(1)}\| = 1$, and let $\mathbf{v}^{(0)} = \mathbf{0}$. An orthogonal basis for $\mathcal{K}_k(A, \mathbf{v}^{(1)})$ can be constructed by the recurrence

$$\gamma_{j+1}\mathbf{v}^{(j+1)} = A\mathbf{v}^{(j)} - \delta_j\mathbf{v}^{(j)} - \gamma_j\mathbf{v}^{(j-1)}, \quad 1 \leq j \leq k, \qquad (2.29)$$

where $\delta_j = \langle A\mathbf{v}^{(j)}, \mathbf{v}^{(j)}\rangle$, and γ_{j+1} is chosen so that $\|\mathbf{v}^{(j+1)}\| = 1$ (see Problem 2.10). Note that $\delta_j > 0$ for positive-definite A, whereas the sign of γ_{j+1} is not prescribed.

Let $V_k = [\mathbf{v}^{(1)}, \mathbf{v}^{(2)}, \ldots, \mathbf{v}^{(k)}]$ denote the matrix containing $\mathbf{v}^{(j)}$ in its jth column, $j = 1, \ldots, k$, and let T_k denote the symmetric tridiagonal matrix

$$\texttt{tridiag}[\gamma_j, \delta_j, \gamma_{j+1}], \quad 1 \leq j \leq k.$$

Then (2.29) is equivalent to the relation

$$AV_k = V_kT_k + \gamma_{k+1}[\mathbf{0}, \ldots, \mathbf{0}, \mathbf{v}^{(k+1)}], \qquad (2.30)$$

from which follows

$$V_k^T A V_k = T_k. \qquad (2.31)$$

The key point here is that the Lanczos algorithm computes the orthonormal set $\{\mathbf{v}^{(j)}, j = 1, \ldots, k\}$, which is a basis for the Krylov subspace $\mathcal{K}_k(A, \mathbf{v}^{(1)})$ (see Problem 2.10). For the eigenvalue problem (which is not our concern here), the eigenvalues of T_k are estimates for the eigenvalues of A.

In this construction, the scalar δ_j is uniquely defined so that the new vector $\gamma_{j+1}\mathbf{v}^{(j+1)}$ is orthogonal to $\mathbf{v}^{(j)}$, and the computed vector $\gamma_{j+1}\mathbf{v}^{(j+1)}$ is then uniquely defined by quantities constructed at steps j and $j-1$. Thus, the vectors and scalars are uniquely defined up to the signs of the off-diagonal entries of T_k. In the following, it will be convenient to assume that γ_{j+1} is chosen to be negative.

Now consider the recurrences satisfied by the residuals $\{\mathbf{r}^{(j)}\}$ and direction vectors $\{\mathbf{p}^{(j)}\}$ in the CG Algorithm 2.1. We have

$$\mathbf{r}^{(j+1)} = \mathbf{r}^{(j)} - \alpha_j A(\mathbf{r}^{(j)} + \beta_{j-1}\mathbf{p}^{(j-1)})$$

$$= -\alpha_j A\mathbf{r}^{(j)} + \left(1 + \frac{\alpha_j \beta_{j-1}}{\alpha_{j-1}}\right)\mathbf{r}^{(j)} - \frac{\alpha_j \beta_{j-1}}{\alpha_{j-1}}\mathbf{r}^{(j-1)},$$

or in matrix form

$$AR_k = R_k S_k - 1/\alpha_{k-1}[\mathbf{0}, \ldots, \mathbf{0}, \mathbf{r}^{(k)}]. \tag{2.32}$$

Here,

$$R_k = [\mathbf{r}^{(0)}, \ldots, \mathbf{r}^{(k-1)}]$$

and

$$S_k = \texttt{tridiag}\left[-\frac{1}{\alpha_{j-1}}, \frac{1}{\alpha_j} + \frac{\beta_{j-1}}{\alpha_{j-1}}, -\frac{\beta_j}{\alpha_j}\right].$$

Note that S_k is similar to a symmetric matrix $\widetilde{T}_k$ via a diagonal similarity transformation $\widetilde{T}_k = \Delta_k S_k \Delta_k^{-1}$, where

$$\Delta_k = \texttt{diag}(\|\mathbf{r}^{(0)}\|, \|\mathbf{r}_1^{(1)}\|, \ldots, \|\mathbf{r}^{(k-1)}\|).$$

Postmultiplying (2.32) by Δ_k^{-1} and letting

$$\widetilde{V}_k = R_k \Delta_k^{-1} = [\tilde{\mathbf{v}}^{(1)}, \tilde{\mathbf{v}}^{(2)}, \ldots, \tilde{\mathbf{v}}^{(k)}]$$

denote the matrix of normalized residuals leads to the equivalent relation

$$A\widetilde{V}_k = \widetilde{V}_k \widetilde{T}_k - \frac{\sqrt{\beta_{k-1}}}{\alpha_{k-1}}[\mathbf{0}, \ldots, \mathbf{0}, \tilde{\mathbf{v}}^{(k+1)}], \tag{2.33}$$

where $\tilde{\mathbf{v}}^{(k+1)} = \mathbf{r}^{(k)}/\|\mathbf{r}^{(k)}\|$. This means that the normalized residuals generated by CG comprise an orthonormal set of vectors that satisfy a three-term recurrence identical in form to the Lanczos algorithm (2.29), (2.30). Moreover, if $\mathbf{v}^{(1)} = \mathbf{r}^{(0)}/\|\mathbf{r}^{(0)}\|$, then the uniqueness of this construction implies that the normalized residuals are precisely the Lanczos vectors.

The CG iterate $\mathbf{u}^{(k)}$ can also be recovered directly from the Lanczos vectors. By Theorem 2.2, $\mathbf{u}^{(k)}$ is the unique vector in $\mathbf{u}^{(0)} + \mathcal{K}_k(A, \mathbf{r}^{(0)})$ with residual orthogonal to $\mathcal{K}_k(A, \mathbf{r}^{(0)})$. Alternatively, this is the requirement that

$$\mathbf{u}^{(k)} = \mathbf{u}^{(0)} + V_k \mathbf{y}^{(k)}, \tag{2.34}$$

where $\mathbf{y}^{(k)} = (y_1^{(k)}, \ldots, y_k^{(k)})^T$ is a vector of dimension k such that $V_k^T \mathbf{r}^{(k)} = \mathbf{0}$. By virtue of (2.30), the residual of $\mathbf{u}^{(k)}$ satisfies

$$\mathbf{r}^{(k)} = \mathbf{r}^{(0)} - A V_k \mathbf{y}^{(k)} = V_k(\|\mathbf{r}^{(0)}\|\mathbf{e}_1 - T_k \mathbf{y}^{(k)}) - \gamma_{k+1} y_k^{(k)} \mathbf{v}^{(k+1)}, \tag{2.35}$$

where $\mathbf{e}_1 = (1, 0, \ldots, 0)^T$ is the unit vector of dimension k. Orthogonality is imposed by choosing $\mathbf{y}^{(k)}$ to satisfy

$$\mathbf{0} = V_k^T \mathbf{r}^{(k)} = \|\mathbf{r}^{(0)}\|\mathbf{e}_1 - T_k \mathbf{y}^{(k)}, \tag{2.36}$$

that is, by solving a tridiagonal system of equations. Note that by (2.35) and (2.36),

$$\mathbf{r}^{(k)} = -\gamma_{k+1} y_k^{(k)} \mathbf{v}^{(k+1)}, \quad \|\mathbf{r}^{(k)}\| = |\gamma_{k+1} y_k^{(k)}|; \tag{2.37}$$

that is, the norm of the residual is automatically available as a by-product of the construction of the CG iterate.

Another important Krylov subspace method derived from the Lanczos algorithm is the minimum residual method (MINRES) derived by Paige & Saunders [142]. This method is applicable to any symmetric system; that is, MINRES is a robust algorithm for *indefinite* coefficient matrices as well as symmetric positive-definite matrices. The point here is that the representation (2.31) guarantees positive-definiteness of T_k when A is positive-definite, but since T_k may be singular for indefinite A, it may not be possible to solve the system of (2.36) in this case. The MINRES strategy circumvents this difficulty by minimizing the Euclidean norm of the residual $\|\mathbf{r}^{(k)}\|$ using the least squares solution to (2.35). This gives iterates

$$\mathbf{u}^{(k)} \in \mathbf{u}^{(0)} + \mathcal{K}_k(A, \mathbf{r}^{(0)})$$

and corresponding residuals

$$\mathbf{r}^{(k)} \in \mathbf{r}^{(0)} + \mathrm{span}\{A\mathbf{r}^{(0)}, \ldots, A^k \mathbf{r}^{(0)}\}$$

characterized by the following property:

$$\|\mathbf{r}^{(k)}\| \leq \|\mathbf{s}\| \quad \text{for all} \quad \mathbf{s} \in \mathbf{r}^{(0)} + \mathrm{span}\{A\mathbf{r}^{(0)}, \ldots, A^k \mathbf{r}^{(0)}\}. \tag{2.38}$$

To derive an implementation of the strategy, we rewrite (2.35) as

$$\|\mathbf{r}^{(k)}\| = \left\| V_{k+1}\left(\|\mathbf{r}^{(0)}\|\mathbf{e}_1 - \hat{T}_k \mathbf{y}^{(k)}\right) \right\|, \tag{2.39}$$

where $\hat{T}_k \in \mathbb{R}^{k+1 \times k}$ is the tridiagonal matrix T_k with an additional final row $[0, \ldots, 0, \gamma_{k+1}]$, and with $\mathbf{e}_1$ here representing the unit vector in $k+1$ dimensions.

Then, since V_{k+1} has orthonormal columns, the least squares solution $\mathbf{y}^{(k)}$ can readily be found by performing a QR factorization of $\hat{T}_k$ employing k Givens rotations. In fact a single new rotation at each iteration can update the QR factorization from the previous iteration (see Fischer [74, p. 179] for details). The following implementation of MINRES is the result.

Algorithm 2.4: THE MINRES METHOD
$\mathbf{v}^{(0)} = \mathbf{0}, \mathbf{w}^{(0)} = \mathbf{0}, \mathbf{w}^{(1)} = \mathbf{0}$
Choose $\mathbf{u}^{(0)}$, compute $\mathbf{v}^{(1)} = \mathbf{f} - A\mathbf{u}^{(0)}$, set $\gamma_1 = \|\mathbf{v}^{(1)}\|$
Set $\eta = \gamma_1, s_0 = s_1 = 0, c_0 = c_1 = 1$
for $j = 1$ until convergence do
$\qquad \mathbf{v}^{(j)} = \mathbf{v}^{(j)}/\gamma_j$
$\qquad \delta_j = \langle A\mathbf{v}^{(j)}, \mathbf{v}^{(j)} \rangle$
$\qquad \mathbf{v}^{(j+1)} = A\mathbf{v}^{(j)} - \delta_j \mathbf{v}^{(j)} - \gamma_j \mathbf{v}^{(j-1)}$ (Lanczos process)
$\qquad \gamma_{j+1} = \|\mathbf{v}^{(j+1)}\|$
$\qquad \alpha_0 = c_j\delta_j - c_{j-1}s_j\gamma_j$ (update QR factorization)
$\qquad \alpha_1 = \sqrt{\alpha_0^2 + \gamma_{j+1}^2}$
$\qquad \alpha_2 = s_j\delta_j + c_{j-1}c_j\gamma_j$
$\qquad \alpha_3 = s_{j-1}\gamma_j$
$\qquad c_{j+1} = \alpha_0/\alpha_1; \ s_{j+1} = \gamma_{j+1}/\alpha_1$ (Givens rotation)
$\qquad \mathbf{w}^{(j+1)} = (\mathbf{v}^{(j)} - \alpha_3\mathbf{w}^{(j-1)} - \alpha_2\mathbf{w}^{(j)})/\alpha_1$
$\qquad \mathbf{u}^{(j)} = \mathbf{u}^{(j-1)} + c_{j+1}\eta\mathbf{w}^{(j+1)}$
$\qquad \eta = -s_{j+1}\eta$
$\qquad$ <Test for convergence>
enddo

Comparing Algorithm 2.4 with Algorithm 2.1, it is evident that MINRES is only slightly more work than CG, requiring two inner products, five vector updates and a single matrix-vector product at each iteration.

The parameter α_0 essentially determines the residual reduction profile of MINRES. If $\alpha_0 = 0$ then $c_{j+1} = 0$ and the iteration makes no progress at the jth step: $\mathbf{u}^{(j)} = \mathbf{u}^{(j-1)}$. Such a stagnation can be traced back to a symmetry of the eigenvalues of A about the origin. Fischer in [74] shows that the CG iterate $\mathbf{u}_{CG}^{(j)}$ can be recovered from the MINRES iterate $\mathbf{u}^{(j)}$ just before the convergence test via

$$\mathbf{u}_{CG}^{(j)} = \mathbf{u}^{(j)} - s_{j+1}\eta\mathbf{w}^{(j+1)}/c_{j+1}.$$

The immediate consequence is that stagnation of MINRES corresponds to the breakdown of the CG algorithm applied to the same system. This is very important in the context of solving discrete Stokes problems, since a staircase convergence pattern (no progress at every odd iteration) is often observed when using multigrid preconditioning. We defer further discussion until Chapter 6.

For now, using the characterization (2.38) and following the convergence analysis in Section 2.1.1 gives a convergence bound in terms of the eigenvalue interval $\lambda_j \in [a, b]$:

$$\frac{\|\mathbf{r}^{(k)}\|}{\|\mathbf{r}^{(0)}\|} \leq \min_{p_k \in \Pi_k, p_k(0)=1} \max_{z \in [a,b]} |p_k(z)|. \qquad (2.40)$$

Note that this generalizes the CG theory since the MINRES bound (2.40) remains valid in the case $0 \in [a, b]$. Note that in the positive-definite case the bounds (2.12) and (2.40) differ only in the definition of the error norm. Indeed, when solving discrete Poisson problems the convergence of MINRES is almost identical to that of CG, see Computational Exercise 2.5.

Readers interested in more detailed algebraic descriptions of Krylov subspace methods are referred to van der Vorst [199], Saad [164], Meurant [131], Hackbusch [98], Greenbaum [89] or Axelsson [6].

2.5   Multigrid

Multigrid methods are at the heart of the most effective iterative solvers for discrete Poisson problems. If multigrid parameters are appropriately chosen, then it is possible to generate a discrete solution with a residual of prescribed accuracy with computational work proportional to the dimension of the discrete system. That is, the convergence rate of a well-tuned multigrid iteration does not depend on the mesh size, h. In this sense, multigrid is an *optimal* solver for discrete Poisson problems. The multigrid procedure described in this section requires some geometric information. If access to such geometric data is difficult, purely algebraic multigrid processes can be employed instead, see Briggs et al. [37, chap. 8] or Trottenberg et al. [196, App. A] for further details.

Multigrid is a recursive version of a procedure most easily understood in the context of a "two-grid" algorithm. The essence of multigrid is a decomposition of a grid function into its components in two subspaces. The two ingredients of multigrid are an iteration process that rapidly reduces the error components in one subspace, and an approximate inverse operator for the second subspace. The two-grid method achieves this decomposition by having a *coarse grid*, which might be a grid of square (or brick in $\mathbb{R}^3$) elements with width $2h$, and a corresponding *coarse grid space*, S^{2h}, the set of bilinear (or trilinear) functions defined on the coarse grid. If the solution is required on a grid of width h, then the second subspace, the *fine grid correction space* B^h, is the span of the bilinear (or trilinear) basis functions associated with those nodes that are not nodes of the coarse grid. This gives the representation $S^h = S^{2h} + B^h$. Figure 2.1 illustrates the basis functions in a one-dimensional slice along a particular grid line.

Let S^h be of dimension n_h and S^{2h} of dimension n_{2h}. In two dimensions, $n_h \approx 4n_{2h}$, and in three dimensions $n_h \approx 8n_{2h}$. The decomposition of a function in S^h into coarse grid and fine grid components depends on two *grid transfer* operators, a *prolongation* operator I_{2h}^h that maps S^{2h} to S^h, and a *restriction* operator mapping S^h to S^{2h}.

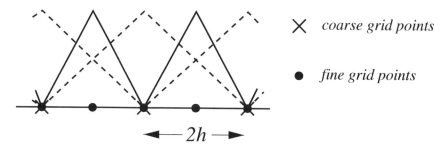

$\times$ *coarse grid points*

$\bullet$ *fine grid points*

FIG. 2.1. Coarse grid space basis (dashed line) and fine grid correction space basis (full line).

Prolongation is defined through the natural inclusion of S^{2h} into S^h,

$$I^h_{2h} v_{2h} = v_{2h} \tag{2.41}$$

for all functions $v_{2h} \in S^{2h}$. A representation of I^h_{2h} is determined by the finite element bases. Suppose that ϕ^{2h}_j is a nodal basis function for S^{2h}. Then since $S^{2h} \subset S^h$, we must have

$$\phi^{2h}_j = \sum_{i=1}^{n_h} p_{ij} \phi^h_i \tag{2.42}$$

for some $\{p_{ij}\}$, where $\{\phi^h_i, i = 1, \ldots, n_h\}$ is the nodal basis for S^h. Let $v_{2h} \in S^{2h}$, with associated coefficient vector $\mathbf{v}^{2h}$. It follows that

$$v_{2h} = \sum_{j=1}^{n_{2h}} \mathbf{v}^{2h}_j \phi^{2h}_j = \sum_{j=1}^{n_{2h}} \mathbf{v}^{2h}_j \sum_{i=1}^{n_h} p_{ij} \phi^h_i$$

$$= \sum_{i=1}^{n_h} \left(\sum_{j=1}^{n_{2h}} p_{ij} \mathbf{v}^{2h}_j \right) \phi^h_i = \sum_{i=1}^{n_h} \left[P \mathbf{v}^{2h} \right]_i \phi^h_i,$$

where

$$P = [p_{ij}], \quad i = 1, \ldots, n_h, \ j = 1, \ldots, n_{2h}$$

is called the *prolongation matrix*. That is, if $\mathbf{v}^{2h} \in \mathbb{R}^{n_{2h}}$ is the coefficient vector that represents v_{2h} in S^{2h}, then $\mathbf{v}^h = P \mathbf{v}^{2h}$ is the vector that represents v_{2h} in S^h.

For one-dimensional problems with basis functions numbered from left to right along interior nodes, a coarse grid basis function ϕ^{2h}_j is the weighted average $\frac{1}{2}\phi^h_{2j-1} + \phi^h_{2j} + \frac{1}{2}\phi^h_{2j+1}$ (see Figure 2.2). This means that the jth column of P, corresponding to coarse grid node j, has the form

$$P_{:,j} = [0, \ldots, 0, \ 1/2, \ 1, \ 1/2, \ 0, \ldots, 0]^T.$$

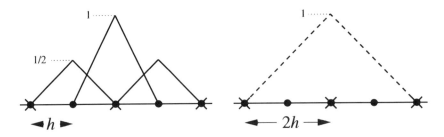

FIG. 2.2. Prolongation: any coarse grid basis function is a member of the fine grid space and is a weighted sum of three fine grid basis functions.

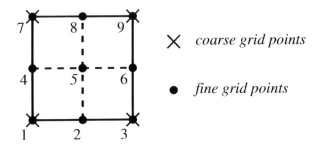

$\times$ coarse grid points

$\bullet$ fine grid points

FIG. 2.3. Prolongation for Q_1 approximation: values at coarse grid nodes $1, 3, 7, 9$ are prolonged to fine grid nodes $1, \ldots, 9$.

Observe also that the coefficients of the fine grid basis functions are the interpolants of the coarse grid functions at the nodes. Because of this, the inclusion (2.41) is also referred to as the *interpolation property*.

The definition of the prolongation matrix P given in (2.42) is independent of the spatial dimension. For Q_1 approximation on square grids, the general structure of P is illustrated for the single-element coarse grid in Figure 2.3. The matrix representing the prolongation in this case is

$$P = \begin{pmatrix} 1 & 0 & 0 & 0 \\ 1/2 & 1/2 & 0 & 0 \\ 0 & 1 & 0 & 0 \\ 1/2 & 0 & 1/2 & 0 \\ 1/4 & 1/4 & 1/4 & 1/4 \\ 0 & 1/2 & 0 & 1/2 \\ 0 & 0 & 1 & 0 \\ 0 & 0 & 1/2 & 1/2 \\ 0 & 0 & 0 & 1 \end{pmatrix},$$

and the interpolation property corresponds to bilinear interpolation.

Given the $n_h \times n_{2h}$ matrix P associated with the prolongation operator I^h_{2h}, the $n_{2h} \times n_h$ matrix $R = P^T$ analogously defines a restriction operator I^{2h}_h from S^h to S^{2h}. That is, given any $v_h \in S^h$ with associated coordinate vector $\mathbf{v}^h$, $I^{2h}_h v_h = v_{2h}$ where $\mathbf{v}^{2h} = R\mathbf{v}^h$. For the trivial grid illustrated in Figure 2.3,

$$R = \begin{pmatrix} 1 & 1/2 & 0 & 1/2 & 1/4 & 0 & 0 & 0 & 0 \\ 0 & 1/2 & 1 & 0 & 1/4 & 1/2 & 0 & 0 & 0 \\ 0 & 0 & 0 & 1/2 & 1/4 & 0 & 1 & 1/2 & 0 \\ 0 & 0 & 0 & 0 & 1/4 & 1/2 & 0 & 1/2 & 1 \end{pmatrix}.$$

We highlight a useful property of these choices of restriction operator and restriction matrix. If $f \in L_2(\Omega)$ is some source function in (1.1), then the definition of the right-hand side vector

$$\mathbf{f}^h = [\boldsymbol{f}^h_i, i = 1, \ldots, n_h], \quad \boldsymbol{f}^h_i = \int_\Omega \phi^h_i f + \int_{\partial\Omega_N} \phi^h_i g_N,$$

associated with a zero Dirichlet boundary condition, leads to a corresponding coarse grid function

$$\mathbf{f}^{2h} = [\boldsymbol{f}^{2h}_j, j = 1, \ldots, n_{2h}], \quad \boldsymbol{f}^{2h}_j = \int_\Omega \phi^{2h}_j f + \int_{\partial\Omega_N} \phi^{2h}_j g_N$$

$$= \sum_i p_{ij} \left(\int_\Omega \phi^h_i f + \int_{\partial\Omega_N} \phi^h_i g_N \right)$$

$$= \sum_j r_{ij} \left(\int_\Omega \phi^h_j f + \int_{\partial\Omega_N} \phi^h_j g_N \right).$$

Thus we have the natural characterization

$$\mathbf{f}^{2h} = R\mathbf{f}^h. \tag{2.43}$$

Now that the grid transfer operators and their matrix representations have been defined, the central idea of multigrid can be described. Consider the $n_h \times n_h$ linear system $A\mathbf{u} = \mathbf{f}$, representing the Galerkin finite element system on the fine grid. For a starting vector $\mathbf{u}^{(0)}$, the algebraic error $\mathbf{e}^{(0)} := \mathbf{u} - \mathbf{u}^{(0)}$ is written as $\mathbf{e}^{(0)} = P\mathbf{e}^{(0)}_{S^{2h}} + \mathbf{e}^{(0)}_{B^h}$ where $\mathbf{e}_{S^{2h}}$ is a vector of coefficients with respect to the basis $\{\phi^{2h}_i\}$ of a function in S^{2h}, and $\mathbf{e}_{B^h}$ is a vector of coefficients with respect to the basis $\{\phi^h_j\}$ of a function in the fine grid correction space (see Figure 2.1).

The first ingredient of multigrid is an iterative method — known as the *smoother* — that rapidly reduces the fine grid component of the error. Many simple iterations of the form (2.26) are effective in this regard, that is, after k steps (typically two or three) the smoothed error function $e^{(k)}_h \in S^h$ corresponding to the coefficient vector

$$\mathbf{e}^{(k)} = (I - M^{-1}A)^k \mathbf{e}^{(0)} = (P\mathbf{e}^{(k)}_{S^{2h}} + \mathbf{e}^{(k)}_{B^h}) \tag{2.44}$$

will have almost no component in the fine grid correction space B^h. Writing (2.44) in terms of grid functions, we have that

$$e_h^{(k)} = I_{2h}^h e_{2h}^{(k)} + b_h^{(k)}, \tag{2.45}$$

and so if $b_h^{(k)}$ is the zero function, then (2.41) implies

$$e_h^{(k)} = I_{2h}^h e_{2h}^{(k)} = e_{2h}^{(k)}. \tag{2.46}$$

This means that, if the smoother is ideal, then the error $e_h^{(k)}$ can be perfectly well represented in the coarse grid space. In terms of (2.44), if $\mathbf{e}_{B^h}^{(k)} \approx \mathbf{0}$, then the vector $\mathbf{e}^{(k)} = (I - M^{-1}A)^k \mathbf{e}^{(0)}$ is essentially a prolongation of a vector associated with a coarse grid function in S^{2h}. This puts us in a promising position: the fine grid error vector $\mathbf{e}^{(k)}$ can be restricted to the coarse grid without losing any information.

Although it is convenient to describe the smoothing iteration in terms of its effect on the algebraic error, in practice the smoother is applied to the residual. That is, instead of restricting the smoothed error vector $e_h^{(k)}$, we need to work with the *smoothed residual* vector associated with $r_h^{(k)} \in S^h$. This is computable from (2.44) via

$$\mathbf{r}^{(k)} = A\mathbf{e}^{(k)} = A(I - M^{-1}A)^k \mathbf{e}^{(0)} = (I - AM^{-1})^k(\mathbf{f} - A\mathbf{u}^{(0)}).$$

Restricting the smoothed residual to the coarse grid then gives the (short) vector

$$\bar{\mathbf{r}} = R\mathbf{r}^{(k)} = P^T \mathbf{r}^{(k)} = P^T(I - AM^{-1})^k(\mathbf{f} - A\mathbf{u}^{(0)}).$$

Note that $P^T A \mathbf{e} = P^T \mathbf{r}$, and so setting $\mathbf{e} = P\bar{\mathbf{e}}$ gives the coarse grid problem

$$P^T AP\bar{\mathbf{e}} = \bar{\mathbf{r}}, \tag{2.47}$$

where $\bar{\mathbf{e}}$ is usually referred to as the *coarse grid correction*. The matrix $\bar{A} = P^T AP$ is a coarse grid representation of the underlying Laplacian operator and is called the *coarse grid operator*.

Specifically, the natural interpolation and projection associated with (2.42) gives

$$\int_\Omega \nabla \phi_i^{2h} \cdot \nabla \phi_k^{2h} = \int_\Omega \left(\sum_j p_{ij} \nabla \phi_j^h \right) \cdot \left(\sum_m p_{km} \nabla \phi_m^h \right)$$

$$= \sum_j p_{ij} \sum_m p_{km} \int_\Omega \nabla \phi_j^h \cdot \nabla \phi_m^h,$$

which is a componentwise statement of the matrix equivalence $A^{2h} = P^T A^h P$. In other words, the coarse grid operator, $\bar{A}$, is here the standard Galerkin coefficient matrix A^{2h} defined on the coarse grid. For the PDE problems in later chapters this is not necessarily the case: employing (2.47) defines the *Galerkin*

coarse grid operator which may differ from the corresponding discretization on
the coarse grid.

The solution of (2.47)—the *coarse grid correction*—is the second ingredient
of multigrid. The prolonged solution of (2.47) represents a good correction to
the coarse grid component of the error, and thus we update

$$\mathbf{u}^{(k)} \leftarrow \mathbf{u}^{(k)} + P\bar{\mathbf{e}}.$$

Putting these components together gives the basic two-grid algorithm:

> **Algorithm 2.5:** TWO-GRID ITERATION
> Choose $\mathbf{u}^{(0)}$
> for $i = 0$ until convergence do
>   for k steps $\mathbf{u}^{(i)} \leftarrow (I - M^{-1}A)\,\mathbf{u}^{(i)} + M^{-1}\mathbf{f}$ (smoothing)
> $\bar{\mathbf{r}} = P^T(\mathbf{f} - A\mathbf{u}^{(i)})$ (restrict residual)
> $\bar{A}\bar{\mathbf{e}} = \bar{\mathbf{r}}$ (solve for coarse grid correction)
> $\mathbf{u}^{(i)} \leftarrow \mathbf{u}^{(i)} + P\bar{\mathbf{e}}$ (prolong and update)
> $\mathbf{u}^{(i+1)} \leftarrow \mathbf{u}^{(i)}$ (update for next iteration)
> enddo

Writing this in terms of the error $\mathbf{e}^{(i)} = \mathbf{u} - \mathbf{u}^{(i)}$ we have

$$\mathbf{e}^{(i+1)} = (A^{-1} - P\bar{A}^{-1}P^T)A(I - M^{-1}A)^k\mathbf{e}^{(i)}, \qquad (2.48)$$

where the solitary A arises because the smoothed residual is A times the
smoothed error (see Problems 2.11 and 2.12).

In the next section we establish a rigorous result showing that the iteration
(2.48) is a contraction and, more crucially, that the contraction factor does not
depend on h. This means that two-grid convergence is achieved in a number
of iterations that is independent of the dimension of the discrete problem. To
motivate this analysis we first consider an intuitive explanation for the success of
the two-grid algorithm. Writing $\mathbf{e}^{(i)} = P\mathbf{e}_{S^{2h}} + \mathbf{e}_{B^h}$, we assume that the smoother
is ideal so that all of the component $\mathbf{e}_{B^h}$ is removed. That is,

$$(I - M^{-1}A)^k\mathbf{e}^{(i)} = P\mathbf{v}^{2h},$$

where $\mathbf{v}^{2h}$ is the n_{2h}-vector of coefficients of a function in S^{2h}. Now consider the
matrix

$$(A^{-1} - P\bar{A}^{-1}P^T)A = I - P\bar{A}^{-1}P^T A$$

with $\bar{A} = P^T AP$. A simple calculation shows that

$$(I - P\bar{A}^{-1}P^T A)^2 = I - P\bar{A}^{-1}P^T A.$$

This means that $I - P\bar{A}^{-1}P^T A$ is a *projection matrix* with eigenvalues 0 and 1.
As a consequence, $\mathbb{R}^{n_h}$ can be divided into two subspaces, one of which is anni-
hilated by $I - P\bar{A}^{-1}P^T A$ (it maps every vector in this space to zero), and
an orthogonal complement, which is left unchanged. Now, if the first of these

subspaces happened to correspond exactly to smooth vectors of the form $P\mathbf{v}^{2h}$, and the latter to the space $\mathbf{e}_{B^h}$, then the error after a single smoothing and coarse grid correction step would be identically zero. In reality of course, this event is unlikely; first, the smoother does not remove all fine grid correction space components; and second, the null space of $I - P\bar{A}^{-1}P^T A$ is not exactly the coarse grid space S^{2h}. Nevertheless, there is grounds for optimism. Regarding the second point, for a Neumann problem in one dimension, the prolongation matrix is given by

$$
P = \begin{pmatrix}
1 & 0 & \cdots & & & & \\
1/2 & 1/2 & 0 & \cdots & & & \\
0 & 1 & 0 & \cdots & & & \\
0 & 1/2 & 1/2 & 0 & \cdots & & \\
0 & 0 & 1 & 0 & \cdots & & \\
0 & 0 & 1/2 & 1/2 & 0 & \cdots & \\
0 & 0 & 0 & 1 & 0 & \cdots & \\
& & & & \ddots & & \\
& & \cdots & 0 & 1/2 & 1/2 \\
& & \cdots & & 0 & 1
\end{pmatrix}, \tag{2.49}
$$

and it can be readily verified that the null space of the associated restriction matrix is given by

$$
\mathtt{null}(P^T) = \mathrm{span} \left\{ \begin{pmatrix} 1/2 \\ -1 \\ 1/2 \\ 0 \\ 0 \\ \vdots \\ 0 \\ 0 \end{pmatrix}, \begin{pmatrix} 0 \\ 0 \\ 1/2 \\ -1 \\ 1/2 \\ 0 \\ \vdots \\ 0 \end{pmatrix}, \ldots, \begin{pmatrix} 0 \\ 0 \\ \vdots \\ 0 \\ 0 \\ 1/2 \\ -1 \\ 1/2 \end{pmatrix} \right\}.
$$

A typical function in the null space is illustrated in Figure 2.4. It seems entirely likely that such rapidly oscillating functions also lie in the nullspace of $P^T A$, as required.

One characteristic of the basic two-grid method is that the iteration matrix in (2.48) is *not symmetric* with respect to the underlying A inner product. To construct a symmetric two-grid iteration matrix *post-smoothing* must be introduced. That is, after the correction step, further iterations of the transposed smoother $(I - (M^T)^{-1}A)$ are applied with the updated iterate as the starting vector. (This is of relevance when point Gauss–Seidel smoothing is used, since transposing the smoother corresponds to reversing the ordering.) This leads to

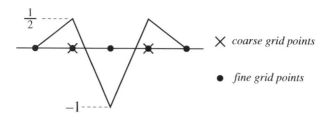

FIG. 2.4. Oscillatory function from the null space of the restriction operator.

the following algorithm:

Algorithm 2.6: GENERALIZED TWO-GRID ITERATION
Choose $\mathbf{u}^{(0)}$
for $i = 0$ until convergence do
    for k steps $\mathbf{u}^{(i)} \leftarrow (I - M^{-1}A)\,\mathbf{u}^{(i)} + M^{-1}\mathbf{f}$ (pre-smoothing)
    $\bar{\mathbf{r}} = P^T(\mathbf{f} - A\mathbf{u}^{(i)})$ (restrict residual)
    $\bar{A}\bar{\mathbf{e}} = \bar{\mathbf{r}}$ (solve for coarse grid correction)
    $\mathbf{u}^{(i)} \leftarrow \mathbf{u}^{(i)} + P\bar{\mathbf{e}}$ (prolong and update)
    for m steps $\mathbf{u}^{(i)} \leftarrow (I - (M^T)^{-1}A)\,\mathbf{u}^{(i)} + (M^T)^{-1}\mathbf{f}$ (post-smoothing)
    $\mathbf{u}^{(i+1)} \leftarrow \mathbf{u}^{(i)}$ (update for next iteration)
enddo

The corresponding two-grid iteration matrix is defined by

$$\mathbf{e}^{(i+1)} = (I - (M^T)^{-1}A)^m (A^{-1} - P\bar{A}^{-1}P^T) A (I - M^{-1}A)^k \mathbf{e}^{(i)}. \qquad (2.50)$$

(See Problem 2.13.) Moreover, symmetry with respect to the A inner product is assured if the number of pre-smoothing steps and the number of post-smoothing steps are the same (see Problem 2.14). The importance of this is the following observation.

Remark 2.6. If the two-grid iteration matrix is *symmetric* then performing a fixed number of two-grid iterations generates a viable preconditioner for CG or MINRES.

2.5.1   Two-grid convergence theory

To establish reduction of the iteration error in (2.48) by a factor independent of h, we describe an analytic framework that was originally introduced by Hackbusch [97]. The basic idea is to consider the two ingredients of multigrid independently, and to verify

- the *smoothing property*:

$$\|A(I - M^{-1}A)^k \mathbf{y}\| \le \eta(k)\|\mathbf{y}\|_A \quad \text{with } \eta(k) \to 0 \text{ as } k \to \infty, \qquad (2.51)$$

  for all vectors $\mathbf{y} \in \mathbb{R}^n$, and

- the *approximation property*:

$$\|(A^{-1} - P\bar{A}^{-1}P^T)\mathbf{y}\|_A \le C\|\mathbf{y}\|, \tag{2.52}$$

for all vectors $\mathbf{y} \in \mathbb{R}^n$. What makes this tricky is that both η and C must be independent of the grid size h. Note that here $\|\cdot\|$ is the regular Euclidean (or ℓ_2-) norm, but other choices of compatible norm could be used instead. Our use of the underlying A-norm to measure the iteration error $\|\mathbf{e}^{(i)}\|_A$ is consistent with the characterization in Section 2.1, and in particular the statement (2.7). It is also worth noting that the definitions (2.51) and (2.52) are only valid in the context of two-dimensional Poisson problems — we will generalize these definitions to cover three-dimensional problems at the end of this section.

Before establishing (2.51) and (2.52), we will demonstrate how they can be combined to give a convergence bound for the basic two-grid method applied to a two-dimensional Poisson problem.

Theorem 2.7. *If the smoothing property and the approximation properties* (2.51) *and* (2.52) *are satisfied, then the basic two-grid iteration in Algorithm 2.5 converges and the contraction rate is independent of* h.

Proof From (2.48)

$$\begin{aligned}
\|\mathbf{e}^{(i+1)}\|_A &= \|(A^{-1} - P\bar{A}^{-1}P^T)A(I - M^{-1}A)^k \mathbf{e}^{(i)}\|_A \\
&\le C\|A(I - M^{-1}A)^k \mathbf{e}^{(i)}\| \quad \text{(approximation property)} \\
&\le C\eta(k)\|\mathbf{e}^{(i)}\|_A \quad \text{(smoothing property)}.
\end{aligned}$$

Hence, since $\eta(k) \to 0$ as $k \to \infty$ there exists a minimal number of smoothing steps k^*, independent of h, such that $\|\mathbf{e}^{(i+1)}\|_A \le \gamma\|\mathbf{e}^{(i)}\|_A$ with $\gamma < 1$. $\qquad\square$

The approximation property is discussed first. We make use of the discretization error analysis presented in Section 1.5.1, and the properties of the mass matrix discussed in Section 1.6. Before doing so, we need a bound on the right-hand side of the discrete problem $A\mathbf{x} = \mathbf{f}$ in terms of the right-hand side of the Poisson problem $-\nabla^2 u = f$. Note that given the L_2 projection property (1.118), there is no loss of generality in assuming that $f = f_h \in S_0^h$.

Lemma 2.8. *Given a quasi-uniform subdivision in* $\mathbb{R}^2$ *of shape regular elements, discrete data* $\mathbf{f} \in \mathbb{R}^n$ *corresponding to* $f \in S_0^h$ *satisfies*

$$h\|f\| \le \frac{1}{\sqrt{c}}\|\mathbf{f}\|,$$

where c *is the lower bound in* (1.116).

Proof Since $f \in S_0^h$ we define $f = \sum_{j=1}^n \hat{f}_j \phi_j$, where $\hat{\mathbf{f}} = (\hat{f}_1, \hat{f}_2, \ldots, \hat{f}_n)^T$ is the vector of nodal coefficients. This means that

$$\|f\|^2 = \sum_{j=1}^n \sum_{i=1}^n \hat{f}_j \hat{f}_i \int_\Omega \phi_j \phi_i = \langle Q\hat{\mathbf{f}}, \hat{\mathbf{f}} \rangle,$$

where Q is the mass matrix. Similarly, defining $\mathbf{f} = (f_1, f_2, \ldots, f_n)^T$ so that

$$f_i = \int_\Omega f_h \phi_i = \sum_{j=1}^n \hat{f}_j \int_\Omega \phi_j \phi_i = (Q\hat{\mathbf{f}})_i$$

we see that $\|\mathbf{f}\|^2 = \langle Q\hat{\mathbf{f}}, Q\hat{\mathbf{f}} \rangle$. Setting $\hat{\mathbf{g}} = Q^{1/2}\hat{\mathbf{f}}$ and using the mass matrix eigenvalue bound (1.116) then gives

$$ch^2 \leq \frac{\langle Q\hat{\mathbf{g}}, \hat{\mathbf{g}} \rangle}{\langle \hat{\mathbf{g}}, \hat{\mathbf{g}} \rangle} = \frac{\langle Q\hat{\mathbf{f}}, Q\hat{\mathbf{f}} \rangle}{\langle Q\hat{\mathbf{f}}, \hat{\mathbf{f}} \rangle}. \qquad \square$$

Theorem 2.9. *Under the assumptions of Lemma 2.8, and assuming an $\mathcal{H}^2$-regular Poisson problem (see Definition 1.9), the approximation property (2.52) is satisfied.*

Proof Given the Poisson problem $-\nabla^2 u = f$, with $f \in S_0^h$, we let $\mathbf{u}$ be the vector of coefficients in the expansion of the fine grid solution $u_h \in S^h$ satisfying $\mathbf{u} = A^{-1}\mathbf{f}$. Furthermore, the matrix equivalence $\bar{A} = P^T A P$, together with the right-hand side projection property (2.43) means that $\bar{\mathbf{u}}$ satisfying $\bar{\mathbf{u}} = \bar{A}^{-1} P^T \mathbf{f}$ determines the coefficients of the coarse grid solution $u_{2h} \in S^{2h}$ of the same problem. Thus, setting $\mathbf{y} = \mathbf{f}$ and using (1.113) gives

$$\|(A^{-1} - P\bar{A}^{-1}P^T)\mathbf{y}\|_A = \|\mathbf{u} - P\bar{\mathbf{u}}\|_A$$
$$= \langle \mathbf{u} - P\bar{\mathbf{u}}, \mathbf{u} - P\bar{\mathbf{u}} \rangle_A^{1/2}$$
$$= a(u_h - I_{2h}^h u_{2h}, u_h - I_{2h}^h u_{2h})^{1/2}.$$

Applying the interpolation property (2.41), and introducing u, the weak solution of the underlying Poisson problem, gives

$$\|(A^{-1} - P\bar{A}^{-1}P^T)\mathbf{y}\|_A = a(u_h - u_{2h}, u_h - u_{2h})^{1/2}$$
$$= \|\nabla u_h - \nabla u_{2h}\|$$
$$\leq \|\nabla(u_h - u)\| + \|\nabla(u - u_{2h})\|.$$

Finally, using either the Q_1 or P_1 approximation error bound (Theorem 1.19 or Theorem 1.10) as appropriate, together with $\mathcal{H}^2$-regularity gives

$$\|(A^{-1} - P\bar{A}^{-1}P^T)\mathbf{y}\|_A \leq C_1\,h\,\|D^2u\| + C_1\,2h\,\|D^2u\|$$
$$\leq 3\,C_1\,C_\Omega\,h\,\|f\|$$
$$\leq 3\,C_1C_\Omega/\sqrt{c}\,\|\mathbf{y}\|,$$

where the final inequality follows from Lemma 2.8. This establishes (2.52) with $C = 3C_1C_\Omega/\sqrt{c}$. $\qquad\square$

We will establish the smoothing property for the simplest smoother, corresponding to the choice $M = \theta I$ for some $\theta \in \mathbb{R}$. This covers the important case of damped Jacobi smoothing when the diagonal of A is constant (e.g. when solving a Dirichlet problem using a uniform grid of bilinear elements). For a discussion of other standard smoothers see Trottenberg et al. [196].

Theorem 2.10. *If $M = \theta I$ and the eigenvalues of $I - M^{-1}A$ are contained in the interval $[-\sigma, 1]$ with σ such that $0 \leq \sigma < 1$ independent of h, then the smoothing property* (2.51) *holds.*

Proof We expand the vector $\mathbf{y}$ in terms of the (orthogonal) eigenvector basis of the symmetric matrix $I - \frac{1}{\theta}A$. That is, we write $\mathbf{y} = \sum_i c_i\mathbf{z}_i$, where $(I - \frac{1}{\theta}A)\mathbf{z}_i = \lambda_i\mathbf{z}_i$, with $\langle \mathbf{z}_i, \mathbf{z}_j \rangle = 0$ for $i \neq j$. Rearranging gives $A\mathbf{z}_i = \theta(1 - \lambda_i)\mathbf{z}_i$, and thus $A(I - M^{-1}A)^k\mathbf{y} = \sum_i c_i\,\theta\,\lambda_i^k\,(1 - \lambda_i)\,\mathbf{z}_i$ and

$$\|A(I - M^{-1}A)^k\mathbf{y}\|^2 = \sum_i c_i^2\,\theta^2\,\lambda_i^{2k}\,(1 - \lambda_i)^2\,\langle \mathbf{z}_i, \mathbf{z}_i \rangle.$$

For $\lambda_i \in [-\sigma, 1]$, the quantity $\lambda_i^{2k}\,(1 - \lambda_i)$ is maximized either at the stationary point $\lambda_i = 2k/(2k + 1)$, or when $\lambda_i = -\sigma$. Moreover, since

$$\left(\frac{2k}{2k + 1}\right)^{2k}\frac{1}{(2k + 1)} = \frac{1}{2k\left(1 + \frac{1}{2k}\right)^{2k+1}} \leq \frac{1}{2ke}$$

and $\left(1 + \frac{1}{2k}\right)^{2k+1}$ monotonically reduces to $e\,(= 2.718...)$ as $k \to \infty$, we have

$$\max_{\lambda_i \in [-\sigma, 1]} \lambda_i^{2k}\,(1 - \lambda_i) \leq \max\left\{\frac{1}{2ke}, \sigma^{2k}(1 + \sigma)\right\},$$

and

$$\|A(I - M^{-1}A)^k\mathbf{y}\|^2 \leq \max\left\{\frac{\theta}{2ke}, \theta\sigma^{2k}(1 + \sigma)\right\}\sum_i c_i^2\,\theta\,(1 - \lambda_i)\,\langle \mathbf{z}_i, \mathbf{z}_i \rangle$$
$$= \max\left\{\frac{\theta}{2ke}, \theta\sigma^{2k}(1 + \sigma)\right\}\|\mathbf{y}\|_A^2.$$

Hence, (2.51) is satisfied with $\eta(k) = \max\{\sqrt{\theta/(2ke)}, \sigma^k \sqrt{\theta(1+\sigma)}\} \to 0$ as $k \to \infty$. ☐

A similar argument can be used to establish that post-smoothing using $M^T = \theta I$ is at worst benign.

Corollary 2.11. *Under the conditions of Theorem* 2.10,

$$\|(I - (M^T)^{-1}A)^m \mathbf{y}\|_A = \|(I - M^{-1}A)^m \mathbf{y}\|_A \leq \|\mathbf{y}\|_A \quad \text{for all vectors } \mathbf{y} \in \mathbb{R}^n.$$

Proof

$$\|(I - M^{-1}A)^m \mathbf{y}\|_A^2 = \left\|\left(I - \frac{1}{\theta}A\right)^m \mathbf{y}\right\|_A^2$$

$$= \sum_i c_i^2 \lambda_i^{2m} \theta (1 - \lambda_i)\langle \mathbf{z}_i, \mathbf{z}_i \rangle.$$

Since, by assumption $\lambda_i \in [-\sigma, 1]$ for all i, we have $\lambda_i^{2m} \leq 1$, and the result follows. ☐

An immediate consequence of Corollary 2.11 is that the generalized two-grid iteration (2.50) is convergent with a contraction rate independent of h.

Note that from (2.27), a simple iteration is convergent if and only if the eigenvalues of $I - M^{-1}A$ satisfy $|\lambda| < 1$, but this condition is not sufficient to satisfy the conditions of Theorem 2.10. The stronger condition, that no eigenvalue approaches -1 as $h \to 0$, is required here. In practice, this means that the damping parameter θ needs to be chosen carefully.

To identify good values of θ, it is convenient to use discrete Fourier analysis. As a concrete example, consider Example 1.1.1 discretized using Q_1 approximation on a uniform $n \times n$ square grid, so that A has exact eigenvalues given in (1.122) with n in place of k. The eigenvalues of the shifted matrix $I - A/\theta$ are then given by

$$\lambda^{(r,s)} = 1 - \frac{8}{3\theta} + \frac{2}{3\theta}\left(\cos\frac{r\pi}{n} + \cos\frac{s\pi}{n}\right) + \frac{4}{3\theta}\cos\frac{r\pi}{n}\cos\frac{s\pi}{n}; \qquad (2.53)$$

for $r, s = 1, \ldots, n-1$. The specific choice $\theta = \frac{8}{3}$ in (2.53) corresponds to standard Jacobi smoothing, $M = \mathrm{diag}(A)$, and the resulting eigenvalues are

$$\lambda^{(r,s)} = \frac{1}{4}\left(\cos\frac{r\pi}{n} + \cos\frac{s\pi}{n}\right) + \frac{1}{2}\cos\frac{r\pi}{n}\cos\frac{s\pi}{n}, \quad r, s = 1, \ldots, n-1. \quad (2.54)$$

Writing $\lambda_i = \lambda^{(r,s)}$ we have

$$(I - M^{-1}A)^k \mathbf{y} = \sum_i c_i \lambda_i^k \mathbf{z}_i,$$

and we see that eigenvector components corresponding to eigenvalues $\lambda^{(r,s)}$ near zero are rapidly reduced by the smoothing iteration. In the case (2.54) low frequencies (smooth eigenvectors) corresponding to small values of both r and s

correspond to eigenvalues near unity, and so these are damped out very slowly. This is the intrinsically slow convergence of Jacobi iteration that the coarse grid correction step of multigrid is able to circumvent.

Looking at (2.54) from the point of view of the smoothing property of Jacobi iteration, we find that the critical eigenvalues are given by $r = n - 1$, $s = 1$ (and $r = 1$, $s = n - 1$) associated with eigenvectors with the highest frequency representable on the grid in the x-direction (and y-direction). Noting that $\cos(n - 1)\frac{\pi}{n} = -1 + \mathcal{O}(h^2)$ and $\cos\frac{\pi}{n} = 1 - \mathcal{O}(h^2)$, we have

$$\lambda^{(n-1,1)} = \frac{1}{4}\left(\cos(n - 1)\frac{\pi}{n} + \cos\frac{\pi}{n}\right) + \frac{1}{2}\cos(n - 1)\frac{\pi}{n}\cos\frac{\pi}{n} \approx -\frac{1}{2} + \mathcal{O}(h^2)$$

so that powers of $\lambda^{(n-1,1)}$ and $\lambda^{(1,n-1)}$ oscillate, but also decrease exponentially by a factor of two. This implies that the parameter σ in Theorem 2.10 is essentially $1/2$, which in turn suggests that (undamped) Jacobi iteration is an effective smoother in this case. (This is somewhat surprising—undamped Jacobi does not work as a smoother in the frequently studied case of a standard five point finite difference approximation of the Laplacian.)

Returning to the damped Jacobi iteration eigenvalues (2.53), the "problem" $r = n - 1$, $s = 1$ eigenvector can be damped even faster if θ is increased. Indeed, making the specific choice $\theta = 4$ reduces the corresponding eigenvalue to $\mathcal{O}(h^2)$. Choosing a large value of θ is not an optimal strategy though. As θ is increased, other eigenvalues associated with rapidly oscillating eigenvectors approach unity, and so smoothing is less effective overall. This is explored in Computational Exercise 2.6.

The standard paradigm used to identify the *optimal* smoothing parameter θ^* is to ensure that all eigenvectors with $r > n/2$ or $s > n/2$ correspond to eigenvalues in an interval $[-\beta, \beta]$ where β is as small as possible.

Proposition 2.12. *Using Q_1 approximation on a uniform square grid, the optimal[5] damping parameter for Jacobi smoothing is $\theta^* = 3 = \frac{9}{8}\mathrm{diag}(A)$.*

Proof See Problem 2.16 $\qquad\qquad\qquad\qquad\qquad\qquad\qquad\qquad\qquad$ □

We end this section with a discussion of three-dimensional Poisson problems. Basically, the proof of two-grid convergence is very similar to that in $\mathbb{R}^2$; the main difference is that the smoothing and approximation properties must be scaled to reflect the fact that mass matrix eigenvalues in $\mathbb{R}^3$ are scaled differently from those in $\mathbb{R}^2$. To this end, the smoothing and approximation properties, (2.51) and (2.52), need to be modified in $\mathbb{R}^3$ to read

$$\|A(I - M^{-1}A)^k\mathbf{y}\| \le \eta(k)h^{1/2}\|\mathbf{y}\|_A \quad \text{with } \eta(k) \to 0 \quad \text{as } k \to \infty, \qquad (2.55)$$

---

[5]This is the default value used in the IFISS software.

and

$$\|(A^{-1} - P\bar{A}^{-1}P^T)\mathbf{y}\|_A \leq Ch^{-1/2}\|\mathbf{y}\|, \qquad (2.56)$$

respectively. With these modified definitions, establishing an h independent contraction rate immediately follows from the proof of Theorem 2.7.

To see where (2.56) comes from, we note the following generalization of Lemma 2.8.

Lemma 2.13. *In* $\mathbb{R}^3$*, the analogue of Lemma 2.8 is*

$$h^{3/2}\|f_h\| \leq \frac{1}{\sqrt{c}}\|\mathbf{f}\|,$$

where c is the lower bound in (1.117).

Proof The proof is identical to that of Lemma 2.8 except for the application of (1.117) in place of (1.116). □

Establishing the smoothing property (2.55) in the case of damped Jacobi iteration can be done with exactly the same argument used in Theorem 2.10. The only difference is that whereas θ is $\mathcal{O}(1)$ in $\mathbb{R}^2$, the right-hand bound in (1.121) shows that $\theta = \mathcal{O}(h)$ in $\mathbb{R}^3$. Satisfying (2.55) is thus achieved by defining η independent of h via

$$\max\{\sqrt{\theta/2ke}, \sigma^k\sqrt{\theta(1+\sigma)}\} = h^{1/2}\eta(k)$$

so that $\eta(k) \to 0$ as $k \to \infty$.

Performing a discrete Fourier analysis of Q_1 approximation for a uniform subdivision of cubic elements of edge length h, we find that the parameter $\theta^* = 8h/3$ (corresponding to unrelaxed Jacobi iteration) turns out to be the optimal choice. The proof is left as an exercise, see Problems 1.29 and 2.17.

2.5.2 *Extending two-grid to multigrid*

Once the motivational principle of two-grid iteration is established, the idea of multigrid is obvious: use a two-grid iteration (with a further level of coarse grid) to solve the coarse grid correction equation $\bar{A}\bar{e} = \bar{r}$. This gives a three-grid method! In general, recursively solving for successive coarse grid corrections is what characterizes true multigrid. The recursion is terminated using a direct elimination method to solve the coarse grid problem once the dimension is sufficiently small.

In principle, a variety of different grid cycling strategies are possible. The two simplest strategies are the V- and W-cycle. The naming convention is obvious from the diagrammatic representation of the two alternatives shown in Figure 2.5. The square symbols represent coarse level exact solves and the circles represent pre-smoothing and restriction or prolongation and post-smoothing. A single

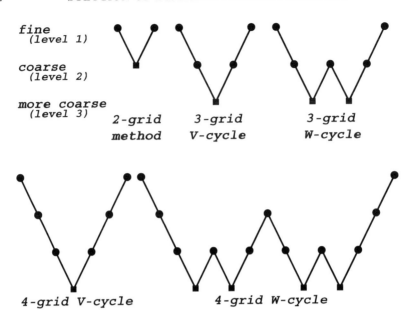

FIG. 2.5. Multigrid V-cycle and W-cycle strategies.

multigrid V-cycle is given by the following recursive algorithm:

Algorithm 2.7: defines $\mathbf{u} \leftarrow$ V-cycle$(A, \mathbf{f}, \mathbf{u}, \texttt{level})$
    for k steps $\mathbf{u} \leftarrow (I - M^{-1}A)\mathbf{u} + M^{-1}\mathbf{f}$ (pre-smoothing)
    if coarsest level
        solve $A\mathbf{u} = \mathbf{f}$
    else
        $\bar{\mathbf{r}} = P^T(\mathbf{f} - A\mathbf{u})$ (restrict residual)
        $\bar{\mathbf{e}} \leftarrow \mathbf{0}$
        $\bar{\mathbf{e}} \leftarrow$ V-cycle$(\bar{A}, \bar{\mathbf{r}}, \bar{\mathbf{e}}, \texttt{level}+1)$ (recursive coarse grid correction)
        $\mathbf{u} \leftarrow \mathbf{u} + P\bar{\mathbf{e}}$ (prolong and update)
    endif
    for m steps $\mathbf{u} \leftarrow (I - (M^T)^{-1}A)\mathbf{u} + (M^T)^{-1}\mathbf{f}.$ (post-smoothing)

The W-cycle algorithm is very similar to the V-cycle, the only difference being that the recursive coarse grid correction step in Algorithm 2.7 is executed twice in succession rather than just once. As is evident from Figure 2.5, more time is spent smoothing on coarser grids using W-cycles, suggesting that low frequency (smooth) components of the error may be handled more effectively. The main attraction of the W-cycle is the theoretical underpinning—compared to the V-cycle, it is relatively straightforward to show that the number of W-cycles required for convergence to a fixed error tolerance is independent of h.

Theorem 2.14. *Given an $\mathcal{H}^2$-regular problem, and a quasi-uniform subdivision in $\mathbb{R}^2$ or $\mathbb{R}^3$ of size h, and solving $A\mathbf{u} = \mathbf{f}$ using an ℓ-level multigrid W-cycle, there exists a minimal number of pre-smoothing steps k^* independent of h, and a contraction factor η_ℓ bounded away from unity independent of h, such that*

$$\|\mathbf{u} - \mathbf{u}^{(i+1)}\|_A \le \eta_\ell \|\mathbf{u} - \mathbf{u}^{(i)}\|_A. \tag{2.57}$$

Here, $\mathbf{u}^{(i)}$ is the ℓ-level multigrid iterate computed after i cycles.

Proof The basic idea of the proof is to estimate the ℓ-level contraction factor η_ℓ in terms of the two-level contraction η_2 using an induction argument. We thus assume that (2.57) is valid using multigrid at levels $2, 3, \ldots, \ell - 1$, and then try to establish (2.57) using multigrid at level ℓ.

We start with notation. Let $\mathbf{u}_k^{(i)}$ be the vector resulting from k pre-smoothing steps at the finest level with starting vector $\mathbf{u}^{(i)}$. We also let $\bar{\mathbf{u}}^{(i+1)}$ represent the iterate that would be computed using an exact coarse grid correction $\bar{\mathbf{e}}$ at the finest level, that is

$$\bar{\mathbf{u}}^{(i+1)} = \mathbf{u}_k^{(i)} + P\bar{\mathbf{e}}. \tag{2.58}$$

The ℓ-level multigrid iterate that is actually computed is given by

$$\mathbf{u}^{(i+1)} = \mathbf{u}_k^{(i)} + P\hat{\mathbf{e}}, \tag{2.59}$$

where $\hat{\mathbf{e}}$ is the result of the two $(\ell - 1)$-level coarse grid correction steps associated with the W-cycle algorithm. Denoting the multigrid iteration error by $\mathbf{e}^{(i)} = \mathbf{u} - \mathbf{u}^{(i)}$, we use (2.59) and eliminate $\mathbf{u}_k^{(i)}$ using (2.58) to give

$$\mathbf{e}^{(i+1)} = \mathbf{u} - \mathbf{u}^{(i+1)} = \mathbf{u} - \bar{\mathbf{u}}^{(i+1)} + P(\bar{\mathbf{e}} - \hat{\mathbf{e}}), \tag{2.60}$$

so that

$$\|\mathbf{e}^{(i+1)}\|_A \le \|\mathbf{u} - \bar{\mathbf{u}}^{(i+1)}\|_A + \|P(\bar{\mathbf{e}} - \hat{\mathbf{e}})\|_A. \tag{2.61}$$

The first term on the right side of (2.61) is the iteration error associated with the two-grid method, which can be bounded by $\eta_2\|\mathbf{e}^{(i)}\|_A$ using Theorem 2.7. Writing $A = A^h$ for clarity, and noting that $\bar{\mathbf{e}}, \hat{\mathbf{e}}$ are vectors of coefficients for functions $\bar{e}^{2h}, \hat{e}^{2h} \in S^{2h}$, respectively, the second term on the right-hand side of (2.61) can be simplified using (1.113) and (2.41):

$$\begin{aligned}
\|P(\bar{\mathbf{e}} - \hat{\mathbf{e}})\|_{A^h} &= \langle P(\bar{\mathbf{e}} - \hat{\mathbf{e}}), P(\bar{\mathbf{e}} - \hat{\mathbf{e}})\rangle_{A^h}^{1/2} \\
&= a(I_{2h}^h(\bar{e}^{2h} - \hat{e}^{2h}), I_{2h}^h(\bar{e}^{2h} - \hat{e}^{2h}))^{1/2} \\
&= a(\bar{e}^{2h} - \hat{e}^{2h}, \bar{e}^{2h} - \hat{e}^{2h})^{1/2} \\
&= \langle \bar{\mathbf{e}} - \hat{\mathbf{e}}, \bar{\mathbf{e}} - \hat{\mathbf{e}}\rangle_{A^{2h}}^{1/2} = \|\bar{\mathbf{e}} - \hat{\mathbf{e}}\|_{A^{2h}}. \tag{2.62}
\end{aligned}$$

We now note that, by hypothesis, solving the coarse grid correction problem $\bar{A}\bar{\mathbf{e}} = \bar{\mathbf{r}}$, using a W-cycle algorithm with initial vector $\mathbf{0}$ gives the approximate solution $\hat{\mathbf{e}}$. Moreover, noting that the W-cycle is characterized by two recursive coarse grid correction steps, and applying the inductive hypothesis gives the estimate

$$\|\bar{\mathbf{e}} - \hat{\mathbf{e}}\|_{A^{2h}} \leq \eta_{\ell-1}^2 \|\bar{\mathbf{e}} - \mathbf{0}\|_{A^{2h}}. \tag{2.63}$$

To bound the right-hand side of (2.63), we repeat the argument (2.62), then use (2.58), and bound the pre-smoothed error using Corollary 2.11,

$$\begin{aligned}
\|\bar{\mathbf{e}}\|_{A^{2h}} &= \|P\bar{\mathbf{e}}\|_{A^h} \\
&= \|\bar{\mathbf{u}}^{(i+1)} - \mathbf{u}_k^{(i)}\|_{A^h} \\
&= \|\mathbf{u} - \mathbf{u}_k^{(i)} - (\mathbf{u} - \bar{\mathbf{u}}^{(i+1)})\|_{A^h} \\
&\leq \|\mathbf{u} - \mathbf{u}_k^{(i)}\|_{A^h} + \|\mathbf{u} - \bar{\mathbf{u}}^{(i+1)}\|_{A^h} \\
&\leq \|\mathbf{u} - \mathbf{u}^{(i)}\|_{A^h} + \eta_2 \|\mathbf{u} - \mathbf{u}^{(i)}\|_{A^h}. \tag{2.64}
\end{aligned}$$

Combining (2.61), (2.62), (2.63) and (2.64) gives

$$\|\mathbf{u} - \mathbf{u}^{(i+1)}\|_A \leq \left(\eta_2 + \eta_{\ell-1}^2 (1 + \eta_2)\right) \|\mathbf{u} - \mathbf{u}^{(i)}\|_A,$$

which establishes (2.57) with a contraction factor

$$\eta_\ell \leq \eta_2 + \eta_{\ell-1}^2 (1 + \eta_2). \tag{2.65}$$

To complete the proof it must be verified that the recurrence formula (2.65) leads to an estimate of η_ℓ that is independent of ℓ. This is certainly the case if the two-grid contraction η_2 is small enough. (See Problem 2.18.) $\square$

Establishing that the multigrid V-cycle algorithm contraction is uniformly bounded away from unity requires a more technical argument. A precise result is stated below.

Theorem 2.15. *Given an $\mathcal{H}^2$-regular problem, and a quasi-uniform subdivision of size h in $\mathbb{R}^2$ or $\mathbb{R}^3$, and solving $A\mathbf{u} = \mathbf{f}$ using a symmetric ℓ-level multigrid V-cycle with a damped Jacobi smoother, there exists a contraction factor $\rho_\ell < 1$ such that*

$$\|\mathbf{u} - \mathbf{u}^{(i+1)}\|_A \leq \rho_\ell \|\mathbf{u} - \mathbf{u}^{(i)}\|_A, \quad \rho_\ell \leq \rho_\infty := C/(C + k), \tag{2.66}$$

where C is independent of ℓ, and $k = m$ is the number of smoothing steps.

The original proof of V-cycle convergence is due to Braess & Hackbusch [20]. An accessible proof of Theorem 2.15 is given in Brenner & Scott [28, chap. 6].

Theorem 2.15 establishes uniform convergence of Algorithm 2.7 in the simple case $m = k = 1$, that is, with just one pre-smoothing and one post-smoothing step at each level. Numerical results for the problem in Example 1.1.1 given in Table 2.3 show just how effective the symmetric V-cycle multigrid method is. In all cases, the smoother is damped Jacobi with the optimal factor, the

Table 2.3 *Number of V-cycles needed to satisfy* $\|\mathbf{r}^{(k)}\|/\|\mathbf{f}\| \leq 10^{-4}$ *for uniform grid, Q_1 discretization of Example 1.1.1, with estimated residual reduction per cycle in parentheses. The number of damped Jacobi pre-smoothing and post-smoothing steps is given by the pair k–m. ℓ is the grid refinement level; h is $2^{1-\ell}$.*

ℓ	1–0	1–1	2–0	n
3	7 (0.25)	4 (0.10)	4 (0.09)	49
4	13 (0.43)	4 (0.11)	5 (0.11)	225
5	20 (0.57)	4 (0.12)	6 (0.14)	961
6	31 (0.66)	4 (0.14)	7 (0.15)	3969
7	45 (0.73)	5 (0.16)	8 (0.15)	16129

Table 2.4 *Number of V-cycles needed to satisfy* $\|\mathbf{r}^{(k)}\|/\|\mathbf{f}\| \leq 10^{-4}h^2$ *for uniform grid, Q_1 discretization of Example 1.1.1.*

ℓ	1–0	1–1	2–0	$10^{-4}h^2$
3	9	5	5	6.5×10^{-6}
4	17	6	7	1.6×10^{-6}
5	30	7	9	3.9×10^{-7}
6	48	8	11	9.8×10^{-8}
7	71	9	13	2.4×10^{-8}

starting vector is zero, and the coarsest grid corresponds to $\ell = 2$. An estimate of the achieved residual reduction per cycle is given in parentheses. Fast grid-independent convergence is clearly observed using two pre-smoothing steps. The 1–1 symmetric V-cycle is even faster. On the other hand, a single pre-smoothing step is not sufficient to achieve h-independent convergence. In contrast, with Gauss–Seidel smoothing in place of damped Jacobi, fast convergence is obtained using a 1–0 smoothing strategy, see Computational Exercise 2.7.

Table 2.4 gives corresponding iteration counts for a more appropriate h-dependent stopping tolerance (as in Table 2.2). In this case there is growth in the iteration count as h is decreased. This increase in work as h is reduced can be avoided by using a nested iteration strategy; that is, by projecting the computed solution obtained on successive coarse grids and using it as a starting guess for the next finer grid. This strategy is illustrated in Figure 2.6 and is referred to as *full multigrid*. Such an approach realizes the ideal of solving the underlying

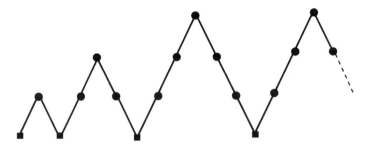

FIG. 2.6. *Nested iteration strategy.*

Table 2.5 *Number of V-cycles needed to satisfy* $\|\mathbf{r}^{(k)}\|/\|\mathbf{f}\| \leq 10^{-4}$ *for uniform grid,* $\mathbf{Q}_1$ *discretization of Example 1.1.2 with estimated residual reduction per cycle in parentheses.* ℓ *is the grid refinement level;* h *is* $2^{1-\ell}$.

ℓ	1–1	2–0	n
3	4 (0.11)	5 (0.12)	33
4	4 (0.11)	5 (0.12)	161
5	4 (0.12)	6 (0.14)	705
6	4 (0.13)	7 (0.16)	2945
7	5 (0.15)	8 (0.16)	12033

problem with a computational complexity that is proportional to the dimension of the system. For a more complete discussion, see Braess [19, pp. 234ff.] or Brenner & Scott [28, pp. 164–165].

Our technical assumption that the underlying Poisson problem is $\mathcal{H}^2$ regular is not a restriction in practice. Indeed, numerical results in Table 2.5 corresponding to Example 1.1.2, whose solution is singular and not in $\mathcal{H}^2$, are essentially identical to those in Table 2.3. The analysis of V-cycle multigrid convergence without a requirement of $\mathcal{H}^2$ regularity was pioneered by Bramble & Pasciak, see [22] and [23], and this paved the way for the development of purely algebraic arguments for establishing the approximation property. Such arguments have been used to extend multigrid theory to discrete convection–diffusion problems. Further details are given in Chapter 4.

Remark 2.16. We have restricted our attention in this section to non-singular systems, although essentially all the results carry over to consistent singular systems. In later chapters we will be concerned with multigrid solution of the Poisson equation with Neumann boundary conditions, which has a single zero eigenvalue corresponding to the constant eigenvector $\mathbf{1}$. For a consistent system, the residual $\mathbf{r}^{(k)}$ is orthogonal to this eigenvector, and multigrid will work provided the restricted residual $R\mathbf{r}^{(k)}$ is orthogonal to the coarse grid version of this vector, $\mathbf{1}_c$. This is equivalent to the requirement that $P\mathbf{1}_c$ be parallel to $\mathbf{1}$, which is true when prolongation is done by interpolation; this can be seen from (2.49) for the one-dimensional case. For more on singular systems, see Hackbusch [97, Chapter 12] or Wesseling [214, Section 6.4].

Remark 2.17. Throughout this section we have assumed lowest order finite element approximation. For higher order elements, multigrid procedures based on the principles of smoothing and coarse grid correction could be devised, but it is computationally more convenient to use $\mathbf{Q}_1$ multigrid processes. For example for $\mathbf{Q}_2$ approximation running the $\mathbf{Q}_1$ multigrid procedure on the mesh which is defined by subdividing each element into four subelements with nodes at the vertices leads to a divergent iteration. However one V-cycle of $\mathbf{Q}_1$ multigrid is a very effective preconditioner[6] for use with CG or MINRES (because the $\mathbf{Q}_1$ multigrid operator is spectrally equivalent to the $\mathbf{Q}_2$ Laplacian). We return to this issue in Section 8.3.2. See also Computational Exercise 2.10.

Problems

2.1. Suppose that $\mathbf{u}$ satisfies $A\mathbf{u} = \mathbf{f}$ where A is a symmetric positive-definite matrix. Let $\{\mathbf{u}^{(k)}\}$ be the sequence of iterates generated by the Richardson iteration with $\alpha_{k-1} \equiv \alpha$ defined by (2.5). Show that the error $\mathbf{e}^{(k)} = \mathbf{u} - \mathbf{u}^{(k)}$ satisfies the bound

$$\|\mathbf{e}^{(k)}\| \leq \left(\frac{\kappa - 1}{\kappa + 1}\right)^k \|\mathbf{e}^{(0)}\|.$$

Now let $\{\mathbf{u}^{(k)}\}$ be generated by the steepest descent method, that is, with α_{k-1} given by (2.6), and show that the error in this case satisfies

$$\|\mathbf{e}^{(k)}\|_A \leq \left(\frac{\kappa - 1}{\kappa + 1}\right)^k \|\mathbf{e}^{(0)}\|_A.$$

Hint: Use the Kantorovich inequality

$$\frac{\langle \mathbf{v}, \mathbf{v} \rangle^2}{\langle A\mathbf{v}, \mathbf{v} \rangle \langle A^{-1}\mathbf{v}, \mathbf{v} \rangle} \geq \frac{4\lambda_{\min}(A)\lambda_{\max}(A)}{(\lambda_{\min}(A) + \lambda_{\max}(A))^2}.$$

2.2. By expanding $\|\mathbf{u} - (\mathbf{u}^{(k)} + \alpha\mathbf{p}^{(k)})\|_A^2$ and using simple calculus, show that the value $\alpha = \mathbf{p}^{(k)^T}\mathbf{r}^{(k)}/\mathbf{p}^{(k)^T}A\mathbf{p}^{(k)}$ is minimizing. Use the result of Lemma 2.1 to further show that $\mathbf{p}^{(k)^T}\mathbf{r}^{(k)} = \mathbf{r}^{(k)^T}\mathbf{r}^{(k)}$.

---

[6]This is the approach used in the IFISS software.

2.3. For a chosen $\mathbf{u}^{(0)}$, if $\mathbf{r}^{(0)} = \mathbf{f} - A\mathbf{u}^{(0)}$ and $\mathbf{p}^{(0)} = \mathbf{r}^{(0)}$, and for $k = 0, 1, \ldots$

$$\alpha_k = \langle \mathbf{p}^{(k)}, \mathbf{r}^{(k)} \rangle / \langle A\mathbf{p}^{(k)}, \mathbf{p}^{(k)} \rangle$$

(1) $\mathbf{u}^{(k+1)} = \mathbf{u}^{(k)} + \alpha_k \mathbf{p}^{(k)}$

(2) $\mathbf{r}^{(k+1)} = \mathbf{f} - A\mathbf{u}^{(k+1)}$

$$\beta_k = -\langle A\mathbf{r}^{(k+1)}, \mathbf{p}^{(k)} \rangle / \langle A\mathbf{p}^{(k)}, \mathbf{p}^{(k)} \rangle$$

(3) $\mathbf{p}^{(k+1)} = \mathbf{r}^{(k+1)} + \beta_k \mathbf{p}^{(k)}$

show that (2) and (1) imply

$$\mathbf{r}^{(k+1)} = \mathbf{r}^{(k)} - \alpha_k A\mathbf{p}^{(k)}.$$

Prove that the definition of α_k implies $\langle \mathbf{r}^{(k+1)}, \mathbf{p}^{(j)} \rangle = 0$ for $j = k$ and that the definition of β_k implies $\langle A\mathbf{p}^{(k+1)}, \mathbf{p}^{(j)} \rangle = 0$ for $j = k$. Prove also that $\langle \mathbf{r}^{(k+1)}, \mathbf{r}^{(j)} \rangle = 0$ for $j = k$. Now by employing induction in k for $k = 1, 2, \ldots$, prove these three assertions for $j = 1, 2, \ldots, k-1$. (The inductive assumption will be that

$$\langle \mathbf{r}^{(k)}, \mathbf{p}^{(j)} \rangle = 0, \quad \langle \mathbf{r}^{(k)}, \mathbf{r}^{(j)} \rangle = 0, \quad \langle A\mathbf{p}^{(k)}, \mathbf{p}^{(j)} \rangle = 0, \qquad j = 0, 1, \ldots, k-1$$

and you may wish to tackle the assertions in this order.)

2.4. If V is a linear space endowed with an inner product $\langle \cdot, \cdot \rangle$, and $U \subset V$ is a finite-dimensional subspace, show that for any given $f \in V$, $u \in U$ minimizes $\|f - w\| = \langle f - w, f - w \rangle^{1/2}$ over all $w \in U$ if and only if $\langle f - u, w \rangle = 0$ for all $w \in U$.

2.5. By considering the trigonometric and hyperbolic identities

$$\cos(k \pm 1)\theta = \cos k\theta \cos\theta \mp \sin k\theta \sin\theta$$

$$\cosh(k \pm 1)\theta = \cosh k\theta \cosh\theta \pm \sinh k\theta \sinh\theta,$$

prove that the Chebyshev polynomials $\tau_k(t)$ satisfy the three-term recurrence

$$\tau_{k+1}(t) = 2t\tau_k(t) - \tau_{k-1}(t) \tag{2.67}$$

(for the cases $|t| \leq 1$ and $|t| > 1$, respectively). Then by induction or otherwise prove that

$$\tau_k(t) = \frac{1}{2}\left[\left(t + \sqrt{t^2 - 1}\right)^k + \left(t - \sqrt{t^2 - 1}\right)^k\right]. \tag{2.68}$$

2.6. Derive inequality (2.20).

2.7. Derive the preconditioned CG Algorithm 2.2 by considering the unpreconditioned CG Algorithm 2.1 applied to

$$H^{-1}AH^{-T}\mathbf{v} = H^{-1}\mathbf{f}, \quad \mathbf{v} = H^T\mathbf{u}.$$

2.8. Prove that the iteration

$$\mathbf{x}^{(k)} = T\mathbf{x}^{(k-1)} = T^k\mathbf{x}^{(0)}$$

converges to the zero vector as $k \to \infty$, if and only if all the eigenvalues of T are less than one in modulus.

2.9. Consider a symmetric coefficient matrix A. Show that if the splitting matrix M is also symmetric, then the iteration matrix $S = I - M^{-1}A$ is symmetric with respect to the A inner product; that is

$$\langle S\mathbf{x}, \mathbf{y}\rangle_A = \langle \mathbf{x}, S\mathbf{y}\rangle_A.$$

This means that S (and indeed S^k, where k is the number of iteration steps) may be used as a preconditioner for CG.

2.10. Consider the recurrence

$$\gamma_{j+1}\mathbf{v}^{(j+1)} = A\mathbf{v}^{(j)} - \delta_j\mathbf{v}^{(j)} - \gamma_j\mathbf{v}^{(j-1)}, \quad 1 \le j \le k,$$

where $\mathbf{v}^{(1)}$ is arbitrary with $\|\mathbf{v}^{(1)}\| = 1$, $\mathbf{v}^{(0)} = 0$, $\delta_j = (A\mathbf{v}^{(j)}, \mathbf{v}^{(j)})$, and γ_j is chosen so that $\|\mathbf{v}^{(j)}\| = 1$. Prove that this procedure generates an orthonormal basis for the Krylov subspace $\mathcal{K}_k(A, \mathbf{v}^{(1)})$. (Hint: use induction and note that for a symmetric matrix A

$$\langle A\mathbf{v}^{(j)}, \mathbf{v}^{(j-1)}\rangle = \langle \mathbf{v}^{(j)}, A\mathbf{v}^{(j-1)}\rangle,$$

and also that $A\mathbf{v}^{(j-1)} = \gamma_j\mathbf{v}^{(j)} + \mathbf{w}$, with $\mathbf{w} \in \text{span}\{\mathbf{v}^{(1)}, \dots \mathbf{v}^{(j-1)}\}$.)

2.11. Consider the basic two-grid iteration for $A\mathbf{u} = \mathbf{f}$. Denote by $\mathbf{u}_s^{(i)}$ the result of k smoothing steps on the current iterate $\mathbf{u}^{(i)}$. Show that the next two-grid iterate is

$$\mathbf{u}^{(i+1)} = \mathbf{u}_s^{(i)} + P\bar{A}^{-1}P^T(\mathbf{f} - A\mathbf{u}_s^{(i)}).$$

By further showing that

$$\mathbf{f} - A\mathbf{u}_s^{(i)} = A(I - M^{-1}A)^k(\mathbf{u} - \mathbf{u}^{(i)}),$$

deduce that

$$\mathbf{u} - \mathbf{u}^{(i+1)} = (A^{-1} - P\bar{A}^{-1}P^T)A(I - M^{-1}A)^k(\mathbf{u} - \mathbf{u}^{(i)}).$$

2.12. Show that the two-grid iteration residual vector $\mathbf{r}^{(i)} = A\mathbf{e}^{(i)}$ satisfies

$$\mathbf{r}^{(i+1)} = A(A^{-1} - P\bar{A}^{-1}P^T)(I - AM^{-1})^k\mathbf{r}^{(i)}.$$

(Hint: start from the result in Problem 2.11.)

2.13. Building further on Problem 2.11, show that if m post-smoothing iterations with splitting matrix M^T are performed then

$$\mathbf{u} - \mathbf{u}^{(i+1)} = (I - (M^T)^{-1}A)^m(A^{-1} - P\bar{A}^{-1}P^T)A(I - M^{-1}A)^k(\mathbf{u} - \mathbf{u}^{(i)}).$$

2.14. By writing the generalized two-grid iteration matrix

$$(I - (M^T)^{-1}A)^m (A^{-1} - P\bar{A}^{-1}P^T)A(I - M^{-1}A)^k$$

in the form $I - M_{MG}^{-1}A$, show that M_{MG} is a symmetric matrix with respect to the A inner product if $k = m$. (Hint: do Problem 2.9 first.)

2.15. Writing

$$G^{\text{pre}} = (A^{-1} - P\bar{A}^{-1}P^T)A(I - M^{-1}A)^k$$

for the two-grid iteration matrix when only pre-smoothing is used and

$$G^{\text{post}} = (I - (M^T)^{-1}A)^m (A^{-1} - P\bar{A}^{-1}P^T)A$$

for the two-grid iteration matrix when only post-smoothing is used, show that

$$G^{\text{post}}G^{\text{pre}} = (I - (M^T)^{-1}A)^m (A^{-1} - P\bar{A}^{-1}P^T)A(I - M^{-1}A)^k$$

is the standard two-grid iteration matrix with m post-smoothing and k pre-smoothing iterations. It follows that if $\|G^{\text{pre}}\| < 1$ and $\|G^{\text{post}}\| < 1$ then the standard two-grid iteration is convergent in the same norm.

2.16. Show that the optimal relaxation factor, θ^*, for Jacobi smoothing for a Poisson problem for a grid of square bilinear elements is $\theta^* = 3 = \frac{9}{8}\text{diag}(A)$ and that for this value the high frequency eigenvalues $r > n/2$ or $s > n/2$ all lie in $[-\frac{1}{3}, \frac{1}{3}]$. This corresponds to a relaxed Jacobi iteration matrix $I - \frac{8}{9}\text{diag}(A)^{-1}A$.

2.17. Building on Problem 1.29, show that $\theta^* = 8h/3 = \text{diag}(A)$ is the optimal value for smoothing for this three-dimensional problem, and that for this value the high frequency eigenvalues $r > n/2$ or $s > n/2$ or $t > n/2$ all lie in $[-\frac{1}{2}, \frac{1}{2}]$. Note that this corresponds to the undamped Jacobi iteration matrix $I - \text{diag}(A)^{-1}A$.

2.18. Suppose that the two-grid contraction factor satisfies $\eta_2 \leq 1/5$. Show by induction that the W-cycle recurrence formula $\eta_\ell \leq \eta_2 + \eta_{\ell-1}^2(1 + \eta_2)$ implies that

$$\eta_\ell \leq \tfrac{5}{3}\eta_2 \leq \tfrac{1}{3}, \quad \text{for } \ell = 3, 4, \ldots.$$

Computational exercises

Two iterative solver drivers are built into IFISS; a preconditioned Krylov subspace solver it_solve, and a multigrid solver mg_solve. Either of these may be called after having run the driver diff_testproblem. The function it_solve offers the possibility of using either CG or MINRES as a solver. In either case, the function calls the built-in matlab routines, pcg or minres. Incomplete factorization preconditioning is achieved by calling the built-in matlab function cholinc.

2.1. The function `cg_test` solves a discrete system $W\mathbf{x} = \mathbf{r}$ by CG iteration for four different coefficient matrices $W \in \mathbb{R}^{100 \times 100}$. Verify the assertion that clustering of eigenvalues enhances CG convergence by computing the condition number and eigenvalues of each of these matrices, and comparing the iteration counts.

2.2. Devise an appropriate stopping tolerance for terminating the `matlab` CG iteration applied to Q_2 discretization of Example 1.1.3. (Hint: use the results in Table 1.1.) Compare the resulting iteration counts with those in Table 2.2.

2.3. Using the function `it_solve`, solve Examples 1.1.1 and 1.1.4 by CG iteration using diagonal and IC(0) preconditioning. Construct a table of iteration counts for a sequence of uniform grids and compare your results with those given for the unpreconditioned CG algorithm in Tables 2.1 and 2.2. Repeat the exercise using a sequence of stretched grids for each problem. What differences (if any) do you notice in the performance of the preconditioners?

2.4. Replace the call to `cholinc` in the function `it_solve` with a call to the IFISS routine `milu0` so as to solve Example 1.1.1 by CG iteration using modified incomplete Cholesky preconditioning. Construct a table of iteration counts for a sequence of uniform grids, and compare the performance with those obtained using standard IC(0) in Computational Exercise 2.3. Then, using the `matlab` `eigs` function, compute the extremal eigenvalues of the preconditioned system for both modified and unmodified incomplete Cholesky preconditioning for the same sequence of grids.

2.5. Using the function `it_solve`, solve Example 1.1.1 for a sequence of uniform grids using MINRES iteration, and compare the convergence curves with those obtained using CG iteration. Then, use IC(0) preconditioning and compare the iteration counts with those obtained using CG in Exercise 2.3. Is MINRES better than CG?

2.6. Using the function `mg_solve`, solve Example 1.1.1 by multigrid V-cycle iteration using Q_1 elements on uniform grids, using the default damped Jacobi smoother with one pre-smoothing and one post-smoothing sweep. Then, try changing the damping factor ω (defined in terms of θ by $\theta = (1/\omega)\text{diag}(A)$) from the theoretically optimal value 8/9 to 1 and 7/9, respectively, in the IFISS routine `solvers/mg_smooth.m`. What is the effect on V-cycle convergence?

2.7. Solve the problem in Exercise 2.6 using point Gauss–Seidel smoothing in place of damped Jacobi. Compare iteration counts with those for Jacobi smoothing given in Table 2.3.

2.8. By executing the coarse grid correction twice in `solvers/mg_iter.m` compute a table of convergence results for the multigrid W-cycle to compare with those of Table 2.3. Compare timings (or `flop` counts) in both cases so as to determine which cycling strategy is better overall.

2.9. Use it_solve to compute a table of results for the problem in Exercise 2.6 using CG preconditioned with one multigrid V-cycle. Compare timings (or flop counts) with those obtained using multigrid as a solver. Does the relative performance of the two strategies change if the singular problem in Example 1.1.4 is solved instead?

2.10. Using the function mg_solve, solve Example problem 1.1.1 by multigrid V-cycle iteration using Q_2 elements on a uniform grid using the default damped Jacobi smoother with one pre-smoothing and one post-smoothing sweep. Then use it_solve to compute a table of results analogous to Table 2.3 using CG preconditioned with one multigrid V-cycle.

3

THE CONVECTION–DIFFUSION EQUATION

We next consider the convection–diffusion equation

$$-\epsilon \nabla^2 u + \vec{w} \cdot \nabla u = f, \tag{3.1}$$

where $\epsilon > 0$. This equation arises in numerous models of flows and other physical phenomena. The unknown function u may represent the concentration of a pollutant being transported (or "convected") along a stream moving at velocity $\vec{w}$ and also subject to diffusive effects. Alternatively, it may represent the temperature of a fluid moving along a heated wall, or the concentration of electrons in models of semiconductor devices. This equation is also a fundamental subproblem for models of incompressible flow, considered in Chapter 7, where $\vec{u}$ is a vector-valued function representing flow velocity and ϵ is a viscosity parameter. Typically, diffusion is a less significant physical effect than convection: on a windy day the smoke from a chimney moves in the direction of the wind and any spreading due to molecular diffusion is small. This implies that, for most practical problems, $\epsilon \ll |\vec{w}|$. This chapter is concerned with the properties of finite element discretization of the convection–diffusion equation, and Chapter 4 with effective algorithms for solving the discrete linear equation systems that arise from the discretization process.

The boundary value problem that is considered is equation (3.1) posed on a two-dimensional or three-dimensional domain Ω, together with boundary conditions on $\partial \Omega = \partial \Omega_D \cup \partial \Omega_N$ given by (1.14), that is,

$$u = g_D \text{ on } \partial \Omega_D, \quad \frac{\partial u}{\partial n} = g_N \text{ on } \partial \Omega_N. \tag{3.2}$$

We will assume, as is commonly the case, that the flow characterized by $\vec{w}$ is incompressible, that is, $\text{div } \vec{w} = 0$. The domain boundary $\partial \Omega$ will be subdivided according to its relation with the velocity field $\vec{w}$: if $\vec{n}$ denotes the outward-pointing normal to the boundary, then

$$\partial \Omega_+ = \{x \in \partial \Omega \,|\, \vec{w} \cdot \vec{n} > 0\}, \quad \text{the } \textit{outflow boundary},$$

$$\partial \Omega_0 = \{x \in \partial \Omega \,|\, \vec{w} \cdot \vec{n} = 0\}, \quad \text{the } \textit{characteristic boundary},$$

$$\partial \Omega_- = \{x \in \partial \Omega \,|\, \vec{w} \cdot \vec{n} < 0\}, \quad \text{the } \textit{inflow boundary}.$$

The presence of the first-order convection term $\vec{w} \cdot \nabla u$ gives the convection–diffusion equation a decidedly different character from that of the Poisson equation. Under the assumption that $\epsilon/(|\vec{w}|L)$ is small, where L is a

113

characteristic length scale associated with (3.1), the solution to (3.1) in most of the domain tends to be close to the solution, $\hat{u}$, of the hyperbolic equation

$$\vec{w} \cdot \nabla \hat{u} = f. \tag{3.3}$$

Let us first briefly consider this alternative equation, which will be referred to as the *reduced problem*. The differential operator of (3.3) is of lower order than that of (3.1), and $\hat{u}$ generally cannot satisfy all the boundary conditions imposed on u. To see this, consider the *streamlines* or *characteristic curves* associated with $\vec{w}$ as illustrated in Figure 3.1. These are defined to be the parameterized curves $\vec{c}(s)$ that have tangent vector $\vec{w}(\vec{c}(s))$ at every point on $\vec{c}$. The characterization $d\vec{c}/ds = \vec{w}$ implies that, for a fixed streamline, the solution to the reduced problem satisfies the ordinary differential equation

$$\frac{d}{ds} [\hat{u}(\vec{c}(s))] = f(\vec{c}(s)). \tag{3.4}$$

Equivalently, if $f = 0$, the reduced solution is constant along streamlines. Suppose further that the parameterization is such that $\vec{c}(s_0)$ lies on the inflow boundary $\partial\Omega_-$ for some s_0. If $u(\vec{c}(s_0))$ is used to specify an initial condition for the differential equation (3.4), and if, in addition, for some $s_1 > s_0$, $\vec{c}(s_1)$ intersects another point on $\partial\Omega$ (say on the outflow boundary $\partial\Omega_+$), then the boundary value $\hat{u}(\vec{c}(s_1))$ is determined by solving (3.4). This value need not have any relation to the corresponding value taken on by u, which is determined by (3.2). Notice also that in the case of a discontinuous boundary condition on $\partial\Omega_-$, the characterization (3.4) implies that $\hat{u}$ will have a discontinuity propagating into Ω along the streamline that originates at the point of discontinuity on the inflow boundary.

Because of these phenomena, it often happens that the solution u to the convection–diffusion equation has a steep gradient in a portion of the domain. For example, u may be close to $\hat{u}$ in most of Ω, but along a streamline going to an outflow boundary where u and $\hat{u}$ differ, u will exhibit a steep gradient in order to satisfy the boundary condition. (The following one-dimensional problem offers a lot of insight here: given ϵ and $w > 0$, find $u(x)$ such that $-\epsilon u'' + wu' = 1$ for $x \in (0, L)$, with $u(0) = 0, u(L) = 0$; see Problem 3.1.) In such a situation, the problem defined by (3.1) and boundary conditions (3.2) is said to be *singularly perturbed*, and the solution u has an exponential *boundary layer*. The diffusion in

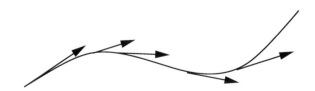

FIG. 3.1. Streamline $\vec{c}(s)$ associated with vector field $\vec{w}$.

equation (3.1) may also lead to steep gradients transverse to streamlines where u is smoother than $\hat{u}$. For example, for a discontinuous boundary condition on $\partial\Omega_-$ as discussed above, the diffusion term in (3.1) leads to a smoothing of the discontinuity inside Ω. In this instance the solution u is continuous but rapidly varying across an *internal layer* that follows the streamline emanating from the discontinuity on the inflow boundary. The presence of layers of both types makes it difficult to construct accurate discrete approximations in cases when convection is dominant.

Equally significant, when $\vec{w} \neq \vec{0}$, the boundary value problem (3.1)–(3.2) is not self-adjoint. (This means that $\int_\Omega (\mathcal{L}u)v \neq \int_\Omega u(\mathcal{L}v)$, where the differential operator of (3.1) is denoted by $\mathcal{L} = -\epsilon\nabla^2 + \vec{w}\cdot\nabla$.) As a result, the coefficient matrix derived from discretization is invariably nonsymmetric — in contrast to the Poisson problem where the coefficient matrix is always symmetric positive-semidefinite. Non-symmetry in turn affects the choice and performance of iterative solution algorithms for solving the discrete problems, and different techniques from those discussed in Chapter 2 must be used to achieve effective performance.

In discussing convection–diffusion equations, it is useful to have a quantitative measure of the relative contributions of convection and diffusion. This can be done by normalizing equation (3.1) with respect to the size of the domain and the magnitude of the velocity. Thus, as above, let L denote a characterizing length scale for the domain Ω; for example, L can be the Poincaré constant of Lemma 1.2. In addition, let the velocity $\vec{w}$ be specified as $\vec{w} = W\vec{w}_*$ where W is a positive constant and $|\vec{w}_*|$ is normalized to have value unity in some measure $|\cdot|$. If points in Ω are denoted by $\vec{x}$, then $\vec{\xi} = \vec{x}/\mathrm{L}$ denotes elements of a normalized domain. With $u_*(\vec{\xi}) = u(\mathrm{L}\vec{\xi})$ on this domain, (3.1) can be rewritten as

$$-\nabla^2 u_* + \left(\frac{\mathrm{WL}}{\epsilon}\right) \vec{w}_* \cdot \nabla u_* = \frac{\mathrm{L}^2}{\epsilon} f, \tag{3.5}$$

and the relative contributions of convection and diffusion can be encapsulated in the *Peclet number*:

$$\mathcal{P} := \frac{\mathrm{WL}}{\epsilon}. \tag{3.6}$$

If $\mathcal{P} \leq 1$, equation (3.5) is diffusion-dominated and relatively benign. In contrast, the construction of accurate approximations and the design of effective solvers in the convection-dominated case, (3.5) with $\mathcal{P} \gg 1$, will be shown to be fraught with difficulty.

3.1   Reference problems

Here and subsequently, we will refer to the velocity vector $\vec{w}$ as *the wind*. Several examples of two-dimensional convection–diffusion problems will be used to illustrate the effect of the wind direction and strength on properties of solutions, and on the quality of finite element discretizations. The problems are all posed on the square domain $\Omega \equiv \Omega_{\Box} = (-1,1) \times (-1,1)$, with wind of order unity $\|\vec{w}\|_\infty = O(1)$ and zero source term $f = 0$. Since the Peclet number is inversely proportional to ϵ the problems are convection-dominated if $\epsilon \ll 1$. The quality and accuracy of discretizations of these problems will be discussed in Sections 3.3 and 3.4, and special issues relating to solution algorithms will be considered in Chapter 4.

3.1.1 Example: Analytic solution, zero source term, constant vertical wind, exponential boundary layer.

The function

$$u(x,y) = x\left(\frac{1 - e^{(y-1)/\epsilon}}{1 - e^{-2/\epsilon}}\right) \tag{3.7}$$

satisfies equation (3.1) with $\vec{w} = (0,1)$ and $f = 0$. Dirichlet conditions on the boundary $\partial\Omega$ are determined by (3.7) and satisfy

$$u(x,-1) = x, \quad u(x,1) = 0,$$
$$u(-1,y) \approx -1, \quad u(1,y) \approx 1,$$

where the latter two approximations hold except near $y = 1$. The streamlines are given by the vertical lines $c(s) = (\alpha, s)$ where $\alpha \in (-1,1)$ is constant, giving a flow in the vertical direction. On the characteristic boundaries $x = \pm 1$, the boundary values vary dramatically near $y = 1$, changing from (essentially) -1 to 0 on the left and from $+1$ to 0 on the right. For small ϵ, the solution u is very close to that of the reduced problem $\hat{u} \equiv x$ except near the outflow boundary $y = 1$, where it is zero.

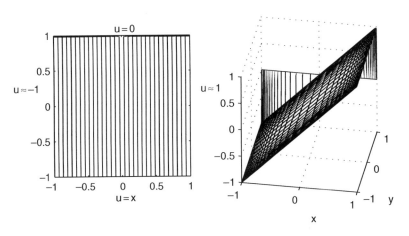

FIG. 3.2. Contour plot (left) and three-dimensional surface plot (right) of an accurate finite element solution of Example 3.1.1, for $\epsilon = 1/200$.

The dramatic change in the value of u near $y = 1$ constitutes a boundary layer. In this example, the layer is determined by the function $e^{(1-y)/\epsilon}$ and has width proportional to ϵ. Figure 3.2 shows a contour plot and three-dimensional rendering of the solution for $\epsilon = 1/200$. Using asymptotic expansions (see Eckhaus [53]), it can be shown in general that boundary layers arising from "hard" Dirichlet conditions on the outflow boundary can be represented using exponential functions in local coordinates. Following Roos et al. [159, Section III.1.3], we refer to such layers as *exponential boundary layers*. For $\|\vec{w}\|_\infty \neq 1$, they have width inversely proportional to the Peclet number.

3.1.2 Example: Zero source term, variable vertical wind, characteristic boundary layers

In this example the wind is vertical $\vec{w} = (0, 1 + (x + 1)^2/4)$ but increases in strength from left to right. Dirichlet boundary values apply on the inflow and characteristic boundary segments; u is set to unity on the inflow boundary, and decreases to zero quadratically on the right wall, and cubically on the left wall, see Figure 3.3. A zero Neumann condition on the top boundary ensures that there is no exponential boundary layer in this case. The fact that the reduced solution $\hat{u} \equiv 1$ is incompatible with the specified values on the characteristic boundary, generates *characteristic layers* on each side. These layers are typical of so-called *shear layers* that commonly arise in fluid flow models. The width of shear layers is proportional to $\sqrt{\epsilon}$ rather than ϵ, so they are less intimidating than exponential layers—this can be seen by comparing the solution in Figure 3.3 with that in Figure 3.2. The reason that the layer on the right is sharper than the layer on the left is that the wind is twice as strong along the associated boundary.

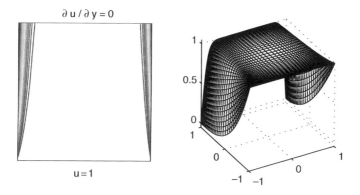

FIG. 3.3. Contour plot (left) and three-dimensional surface plot (right) of an accurate finite element solution of Example 3.1.2, for $\epsilon = 1/200$.

3.1.3 Example: Zero source term, constant wind at a 30° angle to the left of vertical, downstream boundary layer and interior layer.

In this example, the wind is a constant vector $\vec{w} = \left(-\sin\frac{\pi}{6}, \cos\frac{\pi}{6}\right)$. Dirichlet boundary conditions are imposed everywhere on $\partial\Omega$, with values either zero or unity with a jump discontinuity at the point $(0, -1)$ as illustrated in Figure 3.4. The inflow boundary is composed of the bottom and right portions of $\partial\Omega$, $[x, -1] \cup [1, y]$, and the reduced problem solution $\hat{u}$ is constant along the streamlines

$$\left\{ (x, y) \,\Big|\, \tfrac{1}{2}y + \tfrac{\sqrt{3}}{2}x = \text{constant} \right\}, \tag{3.8}$$

with values determined by the inflow boundary condition. The discontinuity of the boundary condition causes $\hat{u}$ to be a discontinuous function with the value $\hat{u} = 0$ to the left of the streamline $y + \sqrt{3}x = -1$ and the value $\hat{u} = 1$ to the right. The diffusion term present in (3.1) causes this discontinuity to be smeared, producing an *internal layer* of width $O(\sqrt{\epsilon})$. There is also an exponential boundary layer near the top boundary $y = 1$, where the value of u drops rapidly from $u \approx 1$ to $u = 0$. Figure 3.4 shows contour and surface plots of the solution for $\epsilon = 1/200$.

An alternative characterization of streamlines may be obtained from the fact that in two dimensions, incompressibility via div $\vec{w} = 0$ implies that

$$\vec{w} = \left(-\frac{\partial\psi}{\partial y}, \frac{\partial\psi}{\partial x} \right)^{T}, \tag{3.9}$$

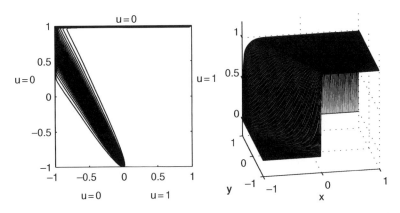

FIG. 3.4. Contour plot (left) and three-dimensional surface plot (right) of an accurate finite element solution of Example 3.1.3, for $\epsilon = 1/200$.

where $\psi(x,y)$ is an associated *stream function*. Defining the level curves of ψ as the set of points $(x,y) \in \Omega$ for which

$$\psi(x,y) = \text{constant}, \tag{3.10}$$

it can be shown that the streamlines in (3.8) are parallel to these level curves, see Problem 3.2. The upshot is that (3.10) can be taken as the definition of the streamlines for two-dimensional problems. This alternative characterization can also be applied in cases when the basic definition is not applicable. It is used, for example, in the following reference problem, which does not have an inflow boundary segment so that the corresponding reduced problem (3.3) does not have a uniquely defined solution.

3.1.4 Example: Zero source term, recirculating wind, characteristic boundary layers.

This example is known as the *double-glazing problem*: it is a simple model for the temperature distribution in a cavity with an external wall that is "hot". The wind $\vec{w} = (2y(1 - x^2), -2x(1 - y^2))$ determines a recirculating flow with streamlines

$$\{(x,y) \mid (1 - x^2)(1 - y^2) = \text{constant}\}.$$

All boundaries are of characteristic type. Dirichlet boundary conditions are imposed everywhere on $\partial\Omega$, and there are discontinuities at the two corners of the hot wall, $x = 1$, $y = \pm 1$. These discontinuities lead to boundary layers near these corners, as shown in Figure 3.5. Although these layers are comparable in width to those in Figure 3.3, it is emphasized that their structure is not accessible via asymptotic techniques, unlike the layers in the first three examples.

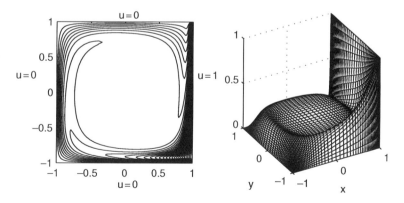

FIG. 3.5. Contour plot (left) and three-dimensional surface plot (right) of an accurate finite element solution of Example 3.1.4, for $\epsilon = 1/200$.

3.2   Weak formulation and the convection term

The standard weak formulation of the convection–diffusion equation (3.1) is determined essentially as in Chapter 1. For the boundary condition (3.2), the weak formulation is as follows:

Find $u \in H_E^1$ such that
$$\epsilon \int_\Omega \nabla u \cdot \nabla v + \int_\Omega (\vec{w} \cdot \nabla u)\, v = \int_\Omega fv + \epsilon \int_{\partial\Omega_N} g_N v \quad \text{for all } v \in H_{E_0}^1.$$

$$(3.11)$$

The presence of the convection term

$$c(u, v) = \int_\Omega (\vec{w} \cdot \nabla u)\, v \tag{3.12}$$

makes problem (3.11) very different from the analogue (1.17) derived from the Poisson equation. Let $a : \mathcal{H}^1(\Omega) \times \mathcal{H}^1(\Omega) \to \mathbb{R}$ denote the bilinear form on the left-hand side of (3.11) so that

$$a(u, v) := \epsilon \int_\Omega \nabla u \cdot \nabla v + \int_\Omega (\vec{w} \cdot \nabla u)\, v. \tag{3.13}$$

It is apparent that $a(u, v) \neq a(v, u)$, that is, $a(\cdot, \cdot)$ is *not* symmetric and so does not define an inner product on $H_{E_0}^1$; cf. Chapter 1.

One commonly held notion is that the convection term is skew-self-adjoint, that is, $c(u, v) = -c(v, u)$, so that after discretization it produces the skew-symmetric part of the matrix approximation of (3.1). It is instructive to investigate this notion here. Application of Green's theorem to (3.12) gives

$$c(u, v) = \int_\Omega (v\vec{w}) \cdot \nabla u = -\int_\Omega \text{div}\,(v\vec{w})\, u + \int_{\partial\Omega_N} uv\, \vec{w} \cdot \vec{n}$$

$$= -\int_\Omega [(v\,\text{div}\,\vec{w})\, u + (\vec{w} \cdot \nabla v)\, u] + \int_{\partial\Omega_N} uv\, \vec{w} \cdot \vec{n}$$

$$= -\int_\Omega (\vec{w} \cdot \nabla v)\, u + \int_{\partial\Omega_N} uv\, \vec{w} \cdot \vec{n},$$

where, for the last step, we have used the fact that $\text{div}\,\vec{w} = 0$. It follows that the first-order term is nearly skew-self-adjoint, but that there may be a nonzero self-adjoint perturbation given by

$$h(u, v) = c(u, v) + c(v, u) = \int_{\partial\Omega_N} uv\, \vec{w} \cdot \vec{n}, \tag{3.14}$$

which is associated with Neumann boundary conditions. Notice that the bilinear form

$$h(u, u) = \int_{\partial \Omega_N} u^2 \, \vec{w} \cdot \vec{n}$$

is positive if a Neumann boundary condition arises only on the outflow boundary; a Neumann condition on the inflow boundary typically makes a negative contribution to $h(u, u)$. We will return to this point later. For the Dirichlet problem ($\partial \Omega_N = \emptyset$), there is no boundary contribution and the convection term (3.12) is indeed skew-self-adjoint.

Using (3.13), and introducing the linear functional $\ell : \mathcal{H}^1(\Omega) \to \mathbb{R}$ so that

$$\ell(v) := \int_{\Omega} fv + \epsilon \int_{\partial \Omega_N} g_N v, \tag{3.15}$$

we can restate the weak formulation (3.11) in the following shorthand form:

Find $u \in H_E^1$ such that
$$a(u, v) = \ell(v) \quad \text{for all } v \in H_{E_0}^1 . \tag{3.16}$$

The following definitions introduce concepts used to establish the existence and uniqueness of the solution to (3.16). They are also used in developing the error analysis in Section 3.4. In the following, V denotes a Hilbert space with inner product $(\cdot, \cdot)_V$ and induced norm $\| \cdot \|_V$.

Definition 3.1 (Coercivity). A bilinear form $a(\cdot, \cdot)$ is said to be *coercive* (or *elliptic*) with respect to the norm $\| \cdot \|_V$ if there is a positive constant γ such that

$$a(u, u) \geq \gamma \|u\|_V^2 \quad \text{for all } u \in V.$$

Definition 3.2 (Continuity). A bilinear form $a(\cdot, \cdot)$ is *continuous* with respect to $\| \cdot \|_V$ if there is a positive constant Γ such that

$$a(u, v) \leq \Gamma \|u\|_V \|v\|_V \quad \text{for all } u, v \in V.$$

A linear functional $\ell(v)$ is *continuous* with respect to $\| \cdot \|_V$ if there is a constant Λ such that

$$\ell(v) \leq \Lambda \|v\|_V \quad \text{for all } v \in V.$$

If $a(\cdot,\cdot)$ and $\ell(\cdot)$ satisfy these criteria, then the *Lax–Milgram lemma* (see, e.g. Brenner & Scott [28, sect. 2.7], Quarteroni & Valli [152, sect. 5.1], or Renardi & Rogers [156]) implies that there exists a unique $u \in V$ for which

$$a(u,v) = \ell(v) \quad \text{for all } v \in V.$$

We will now use these tools to establish existence and uniqueness of the weak solution of the convection–diffusion equation by determining the constants γ, Γ and Λ in the definitions above. We assume that $\partial\Omega_D$, the portion of the boundary where Dirichlet conditions hold in (3.2), satisfies $\int_{\partial\Omega_D} ds \neq 0$, and for ease of exposition we assume homogeneous Dirichlet data $g_D = 0$ (otherwise, a more technical argument is needed). The latter condition implies that the spaces H_E^1 and $H_{E_0}^1$ given in (1.15)–(1.16) are the same, and in particular that (1.15) is closed under addition. Let V be taken to be this space, with the inner product $(u,v)_V = \int_\Omega \nabla u \cdot \nabla v$. The induced norm is therefore $\|u\|_V = \|\nabla u\|$.

From (3.14), it follows that the bilinear form (3.13) satisfies

$$a(u,u) = \epsilon \int_\Omega \nabla u \cdot \nabla u + \tfrac{1}{2} h(u,u).$$

If the Neumann condition holds only on the outflow or characteristic boundary, then $h(u,u) \geq 0$, and

$$a(u,u) \geq \epsilon \|\nabla u\|^2. \tag{3.17}$$

That is, $a(\cdot,\cdot)$ is coercive over $H_{E_0}^1$ with coercivity constant $\gamma = \epsilon$. Continuity over $H_{E_0}^1$ is established using

$$|a(u,v)| \leq \epsilon \left| \int_\Omega \nabla u \cdot \nabla v \right| + \left| \int_\Omega (\vec{w} \cdot \nabla u)\, v \right|.$$

The first-order term is bounded using several applications of the Cauchy–Schwarz inequality; we illustrate for the two-dimensional problem:

$$\left| \int_\Omega (\vec{w} \cdot \nabla u)\, v \right| \leq \int_\Omega (w_x^2 + w_y^2)^{1/2} \left(\left(\frac{\partial u}{\partial x} \right)^2 + \left(\frac{\partial u}{\partial y} \right)^2 \right)^{1/2} |v|$$

$$\leq \|\vec{w}\|_\infty \int_\Omega \left(\left(\frac{\partial u}{\partial x} \right)^2 + \left(\frac{\partial u}{\partial y} \right)^2 \right)^{1/2} |v|$$

$$\leq \|\vec{w}\|_\infty \|\nabla u\| \|v\|, \tag{3.18}$$

where $\|\vec{w}\|_\infty = \sup_x |\vec{w}(x)|$. Invoking the Cauchy–Schwarz inequality for the diffusion term, and applying Lemma 1.2 to the right-hand side of (3.18) then establishes the bound

$$|a(u,v)| \leq \Gamma_{\vec{w}} \|\nabla u\| \|\nabla v\|, \tag{3.19}$$

where the continuity constant is $\Gamma_{\vec{w}} = \epsilon + \|\vec{w}\|_\infty L$. The continuity of $\ell(v)$ of (3.15) may be established using the trace inequality of Lemma 1.5, and the norm equivalence in Proposition 1.6,

$$|\ell(v)| \leq \|f\| \|v\| + \epsilon \|g_N\|_{\partial\Omega_N} \|v\|_{\partial\Omega_N}$$
$$\leq \|f\| \|v\|_{1,\Omega} + \epsilon C_{\partial\Omega} \|g_N\|_{\partial\Omega_N} \|v\|_{1,\Omega}$$
$$\leq C_\Omega \left(\|f\| + \epsilon C_{\partial\Omega} \|g_N\|_{\partial\Omega_N} \right) \|\nabla v\|. \tag{3.20}$$

It follows that the weak formulation (3.16) has a unique solution.[1]

3.3   Approximation by finite elements

Nearly all of the formal aspects of finite element discretization of the convection–diffusion equation are the same as for the Poisson equation. A discrete weak formulation is defined using a discrete trial space $S_E^h \subset H_E^1$ and test space $S_0^h \subset H_{E_0}^1$, and implementation entails choosing bases for these spaces and constructing the finite element coefficient matrix and source vector. However, a significant difference for the convection–diffusion equation is that the presence of layers in the solution makes it more difficult to compute accurate discrete solutions. In this section, we discuss some things that can go wrong with the Galerkin discretization, and we motivate and examine the *streamline diffusion* method, which is designed to handle some of the difficulties.

3.3.1   *The Galerkin finite element method*

The Galerkin method for the convection–diffusion problem (3.16) is

Find $u_h \in S_E^h$ such that
$$a(u_h, v_h) = \ell(v_h) \quad \text{for all } v_h \in S_0^h. \tag{3.21}$$

Implementation issues are largely the same as those described in Chapter 1 for the Poisson equation, except that there are the additional computations required for the convection term (3.12). The difference here is that when $\mathcal{P} \gg 1$ the solution can have a steep gradient, and so appropriately fine grids are often needed to achieve accuracy.

Figures 3.6 and 3.7 illustrate Galerkin solutions to the first two reference problems using 16×16 square grids that are too coarse to resolve the layers in the case $\epsilon = 1/200$. These coarse grid solutions should be compared with the accurate ones shown in Figures 3.2 and 3.3. The discrete solution to Example 3.1.1 shown in Figure 3.6 is highly oscillatory at the boundary layer, and, in addition, these oscillations propagate into the domain along the streamlines. A similar difficulty arises for the characteristic layer in Example 3.1.2 shown in Figure 3.7. The main

---

[1] See Roos et al. [159, pp. 173ff] and references therein for a discussion of classical solutions.

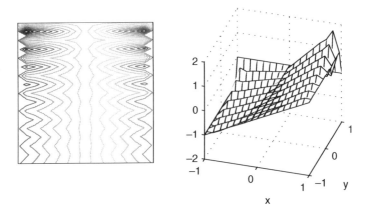

FIG. 3.6. Contour plot (left) and three-dimensional surface plot (right) of the Galerkin solution of Example 3.1.1, for $\epsilon = 1/200$, using $\boldsymbol{Q}_1$ approximation on a 16×16 square grid.

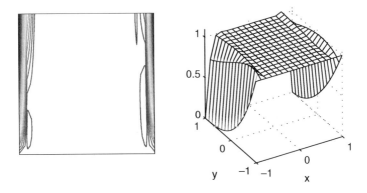

FIG. 3.7. Contour plot (left) and three-dimensional surface plot (right) of the Galerkin solution of Example 3.1.2, for $\epsilon = 1/200$, using $\boldsymbol{Q}_1$ approximation on a 16×16 square grid.

difference is that the oscillations are localized to the unresolved layer and do not propagate into the domain in this case.

To give some theoretical insight as to what is happening here, we note that (the proof is in the next section) the error in the Galerkin solution u_h satisfies

$$\|\nabla(u - u_h)\| \leq \frac{\Gamma_{\vec{w}}}{\epsilon} \inf_{v_h \in S_E^h} \|\nabla(u - v_h)\|, \qquad (3.22)$$

where $\Gamma_{\vec{w}}$ is the continuity constant in (3.19). This bound is a statement that the discrete solution is *quasi-optimal* in S_E^h, up to the constant

$$\frac{\Gamma_{\vec{w}}}{\epsilon} = 1 + \frac{\|\vec{w}\|_\infty L}{\epsilon} = 1 + \mathcal{P}, \tag{3.23}$$

where $\mathcal{P}$ is the Peclet number introduced in (3.6). While (3.22) implies that the approximation is good in the context of a diffusion-dominated problem — in the limit of $\mathcal{P} \to 0$ (3.22) recovers the best approximation bound (1.60) — it also suggests that the quality of the approximation can be poor for a convection-dominated problem. This lack of accuracy is typically manifested in the form of spurious oscillations. For example, for a constant wind, nonlocal oscillations (or *wiggles*) are generated if boundary layers are not resolved on the mesh, that is, if the *mesh Peclet number*

$$\mathcal{P}_h := \frac{\mathcal{P}}{2} \times \frac{h}{L} \tag{3.24}$$

is larger than unity. Extensive discussion and a more complete theoretical explanation of the issues involved will be given in Section 3.4. A good way to develop insight initially is to consider a vertical slice of the problem in Example 3.1.1. This avenue is explored in Problem 3.3.

The error bound (3.22) can be combined with the argument used to prove Theorem 1.10 in Section 1.5 to show that the accuracy of Galerkin discretization is certain to increase if uniform mesh refinement is used. For large Peclet numbers however, it is too costly to compute globally accurate solutions in this way. Instead, a more effective strategy is to refine the mesh *locally*, that is, only in regions where layers occur. In some cases it may be possible to do this without any a priori information about the solution. Using specially constructed fine meshes near outflow boundaries is an obvious tactic, since this is where exponential layers arise. An example of such a mesh is given in Figure 3.11 at the end of Section 3.4.1. Moreover, discontinuous boundary conditions on an inflow boundary inevitably give rise to an internal layer, so it is important to use a mesh such that the boundary discontinuity is commensurate with a generated layer of width $O(1/\sqrt{\mathcal{P}})$. (It makes little sense to try to solve the problem in Example 3.1.3 using uniform grids with h bigger than $1/20$, see Computational Exercise 3.1.) A similar consideration applies to characteristic layers, where the elements need to be small enough to resolve layers of width $O(1/\sqrt{\mathcal{P}})$ that typically arise. This issue is explored in Computational Exercise 3.2.

In general, of course, it is not possible to know about all non-smooth behavior of the solution, when, for example, the streamlines are complicated. In such cases, accurate solutions can be generated through the use of adaptive local mesh refinement: a discrete solution is computed on an initial grid, and then the error associated with this solution is estimated using techniques for a posteriori error approximation discussed in Section 3.4.2. This information is then used to refine the mesh in regions where the error is large so that an improved solution can be constructed on the refined mesh. For this strategy to be effective, it is important that errors do not propagate into regions where refinement is not needed. A similar issue arises for use of the multigrid algorithm for solving the discrete system, see Chapter 4. Even if the grid where the solution is computed (the "fine grid") provides suitable accuracy, the solution algorithm in this case requires a sequence of coarser grids, and it is important that the discretizations on these grids capture the character of the solution with a reasonable degree of accuracy. For these reasons, it is necessary to have available a discretization strategy that does not have the deficiencies exhibited by the Galerkin method. This issue is considered next.

3.3.2 *The streamline diffusion method*

The numerical results in the previous section show that the Galerkin discretization is inaccurate if the mesh is not fine enough to resolve steeply varying layers in the solution, and that inaccuracies may also pollute the discrete solution in regions of the domain where the exact solution is well-behaved. Such difficulties can be ameliorated by modifying the discretization to enhance its coercivity, thereby increasing stability. We now discuss an elegant approach (due to Hughes & Brookes [107]) for achieving this, the *streamline diffusion method*.

We will motivate this methodology by first considering a closely related approach (due to Brezzi et al. [35]) that, given a finite element mesh, enhances the quality of the discrete solution on that mesh by incorporating information interior to elements, where the Galerkin method provides no information. Loosely speaking, this is done by

- augmenting the space on which the solution is defined by adding basis functions with support contained entirely in element interiors, improving the quality of the discrete approximation in regions where resolution is lacking; and

- eliminating the unknowns associated with these new basis functions, producing an improved solution on the original grid.

Since the augmenting basis functions are local to elements and vanish on element boundaries, they are referred to as *bubble functions*. An assumption underlying this approach is that the problem produced on the given mesh contains as many unknowns as can be handled in the computational environment available. Therefore, the process is done implicitly, in the sense that no new unknowns are literally introduced into the problem. Instead, the weak formulation and corresponding algebraic system is modified in a way that reflects the introduction of new information. The procedure bears an analogy to *static condensation*, or, in algebraic terms, block Gaussian elimination, whereby the equations associated with unknowns on the element boundaries are modified by the elimination of unknowns associated with element interiors.

Let $\mathcal{T}_h$ denote a triangulation of Ω with individual elements $\{\triangle_k\}$. To keep this discussion simple, we assume we are solving a homogeneous problem; that is $g_D = 0$ and $g_N = 0$ in (1.2) cf. (1.18). Hence, the Galerkin discretization seeks $u_h \in S_0^h$ such that $a(u_h, v_h) = \ell(v_h)$ for all $v_h \in S_0^h$. Suppose S_0^h is augmented by another space $B_0^h \subset H_{E_0}^1$, producing a new, larger space

$$\hat{S}_0^h = S_0^h + B_0^h \subset H_{E_0}^1,$$

where we are assuming that functions in $\hat{S}_0^h$ are uniquely expressible as a sum of elements in the two spaces S_0^h, B_0^h, and that the restrictions of functions in $\hat{S}_0^h$ to individual elements $\triangle_k$ lie in $\mathcal{H}^2(\triangle_k)$. In principle, this expanded space could be used to construct a more accurate solution $\hat{u}_h \in \hat{S}_0^h$ such that

$$a(\hat{u}_h, \hat{v}_h) = \ell(\hat{v}_h) \quad \text{for all } \hat{v}_h \in \hat{S}_0^h. \tag{3.25}$$

With the assumption that the new space B_0^h contains only bubble functions, any $b_h \in B_0^h$ is uniquely expressible as

$$b_h = \sum_k b_h^{(k)},$$

where the support of $b_h^{(k)}$ lies in the interior of an individual element $\triangle_k \in \mathcal{T}_h$. Let us denote members of the augmented space as

$$\begin{aligned}
\hat{u}_h &= u_h + b_h, & u_h &\in S_0^h, & b_h &\in B_0^h, \\
\hat{v}_h &= v_h + c_h, & v_h &\in S_0^h, & c_h &\in B_0^h.
\end{aligned} \tag{3.26}$$

Since the relation (3.25) holds for all test functions $\hat{v}_h \in \hat{S}_0^h$, it is valid in particular when $\hat{v}_h$ is taken to be a member of the augmenting space B_0^h. That is, if $\hat{u}_h$ is expressed as in (3.26), then the augmenting function b_h must satisfy

$$a(b_h, c_h) = (f, c_h) - a(u_h, c_h) \quad \text{for all } c_h \in B_0^h. \tag{3.27}$$

If $c_h = c_h^{(k)}$, with support only in the interior of element $\triangle_k$, then

$$\ell(c_h) = \int_\Omega f c_h = \int_{\triangle_k} f c_h^{(k)},$$

and

$$\begin{aligned}
a(u_h, c_h) &= \epsilon \int_\Omega \nabla u_h \cdot \nabla c_h + \int_\Omega (\vec{w} \cdot \nabla u_h) c_h \\
&= \epsilon \int_{\triangle_k} \nabla u_h \cdot \nabla c_h^{(k)} + \int_{\triangle_k} (\vec{w} \cdot \nabla u_h) c_h^{(k)} \\
&= \int_{\triangle_k} (-\epsilon \nabla^2 u_h + \vec{w} \cdot \nabla u_h) c_h^{(k)},
\end{aligned}$$

where, in integrating by parts in the last step, we have used the fact that $c_h^{(k)} = 0$ on $\partial \triangle_k$. It follows that the expression on the right side of (3.27) is

$$\int_{\triangle_k} [f - (-\epsilon \nabla^2 u_h + \vec{w} \cdot \nabla u_h)] \, c_h^{(k)}.$$

Similarly, the left side of (3.27) is

$$a(b_h, c_h) = \int_{\triangle_k} (-\epsilon \nabla^2 b_h + \vec{w} \cdot \nabla b_h) \, c_h^{(k)}.$$

The fact that B_0^h is a suitably rich space then implies that $b_h^{(k)}$ satisfies a local version of (3.1):

$$-\epsilon \nabla^2 b_h^{(k)} + \vec{w} \cdot \nabla b_h^{(k)} = r_h^{(k)} \quad \text{in } \triangle_k, \tag{3.28}$$

where

$$r_h^{(k)} = f - (-\epsilon \nabla^2 u_h + \vec{w} \cdot \nabla u_h)$$

is the *element residual* associated with u_h, with $b_h^{(k)} = 0$ on $\partial \triangle_k$. This shows that, given $u_h \in S_0^h$, the function b_h can be viewed as a correction to u_h up to inter-element boundaries. Consequently the approach is often referred to as a *subgrid model*.

We will now introduce two simplifying assumptions, first, that the discretization S_0^h uses piecewise linear basis functions on $\mathcal{T}_h$, and second, that in the convection–diffusion equation (3.1), $\vec{w}$ and f are piecewise constant on $\mathcal{T}_h$. These assumptions imply that for any $u_h \in S_0^h$, the residual of u_h on $\triangle_k$, which now has the form $r_h^{(k)} = f - \vec{w} \cdot \nabla u_h$, is constant. In this case, the solution to equation (3.28) (and hence equation (3.27)) is given by

$$b_h^{(k)} = d_h^{(k)} r_h^{(k)}, \tag{3.29}$$

where $d_h^{(k)}$ satisfies

$$-\epsilon \nabla^2 d_h^{(k)} + \vec{w} \cdot \nabla d_h^{(k)} = 1 \text{ in } \triangle_k, \quad d_h^{(k)} = 0 \text{ on } \partial \triangle_k. \tag{3.30}$$

Notice that $d_h^{(k)}$ is a strictly positive function in $\triangle_k$ (see Problem 3.4).

Returning to relation (3.25), let the test function now be chosen as $\hat{v}_h = v_h \in S_0^h$. This leads to

$$a(u_h, v_h) = \ell(v_h) - a(b_h, v_h) = \ell(v_h) - \sum_k a(b_h^{(k)}, v_h). \tag{3.31}$$

In (3.31), the summand $a(b_h^{(k)}, v_h)$ satisfies

$$a(b_h^{(k)}, v_h) = \epsilon \int_{\triangle_k} \nabla b_h^{(k)} \cdot \nabla v_h + \int_{\triangle_k} (\vec{w} \cdot \nabla b_h^{(k)}) v_h$$

$$= -\epsilon \int_{\triangle_k} b_h^{(k)} \nabla^2 v_h - \int_{\triangle_k} \text{div} \, (v_h \vec{w}) \, b_h^{(k)}$$

$$= -\int_{\triangle_k} b_h^{(k)} (\vec{w} \cdot \nabla v_h), \tag{3.32}$$

where we have used the facts that v_h is piecewise linear (to eliminate the second-order term), and div $\vec{w} = 0$. Substitution of (3.29) into (3.32), together with the facts that $r_h^{(k)}$ and $\vec{w} \cdot \nabla v_h$ are piecewise constant on $\triangle_k$, leads to

$$a(b_h^{(k)}, v_h) = - \left(\int_{\triangle_k} d_h^{(k)} \right) r_h^{(k)} (\vec{w} \cdot \nabla v_h)$$

$$= - \left(\int_{\triangle_k} d_h^{(k)} \right) \frac{1}{|\triangle_k|} \int_{\triangle_k} (f - \vec{w} \cdot \nabla u_h)(\vec{w} \cdot \nabla v_h)$$

$$= -\delta_k \int_{\triangle_k} (f - \vec{w} \cdot \nabla u_h)(\vec{w} \cdot \nabla v_h), \tag{3.33}$$

where

$$\delta_k = \frac{1}{|\triangle_k|} \left(\int_{\triangle_k} d_h^{(k)} \right). \tag{3.34}$$

The characterization (3.33) and (3.34) means that (3.31) is equivalent to the condition

$$\epsilon \int_{\Omega} \nabla u_h \cdot \nabla v_h + \int_{\Omega} (\vec{w} \cdot \nabla u_h) v_h + \sum_k \delta_k \int_{\triangle_k} (\vec{w} \cdot \nabla u_h)(\vec{w} \cdot \nabla v_h)$$

$$= \ell(v_h) + \sum_k \delta_k \int_{\triangle_k} f \, \vec{w} \cdot \nabla v_h. \tag{3.35}$$

Imposing (3.35) for all $v_h \in S_0^h$ determines a new discrete system of equations with the same number of unknowns as the Galerkin system (3.21) but for which additional information associated with element interiors has been incorporated via (3.34). Moreover, setting $v_h = u_h$ shows that the bilinear form in (3.35) may be written as $a(u_h, u_h) + h(u_h, u_h)$ where

$$h(u_h, v_h) = \sum_k \delta_k \int_{\triangle_k} (\vec{w} \cdot \nabla u_h)(\vec{w} \cdot \nabla v_h).$$

The fact that $\delta_k > 0$ (left as an exercise, see Problem 3.4) implies that the coercivity of the bilinear form in (3.35) is stronger than that of $a(\cdot, \cdot)$.

The *streamline diffusion* (SD) method can be viewed as a parameterized weak formulation of the convection–diffusion equation associated with (3.35). For low order (linear or bilinear) elements, the SD method is defined by the requirement that (3.35) holds for all $v_h \in S_0^h$, where the parameters $\{\delta_k\}$, rather than being explicitly defined by (3.34), are instead "user-specified" quantities that can serve as tuning mechanisms to enhance the effectiveness of the approach. In the special case where both $\vec{w}$ or f are piecewise constant functions, the streamline diffusion discretization with parameter choice (3.34) corresponds to the addition and elimination of the bubble functions defined in (3.29)–(3.30) and thus offers the possibility of improved accuracy.

In practice, it is not feasible to compute $\{\delta_k\}$ by solving the convection–diffusion problem (3.30) in every element $\triangle_k$. A more practical alternative, explored later in this section, is to construct approximate solutions $d_*^{(k)}$ of (3.30), leading to the explicitly computable parameter

$$\delta_k^* = \frac{1}{|\triangle_k|} \left(\int_{\triangle_k} d_*^{(k)} \right). \tag{3.36}$$

Notice that only the mean value of $d_h^{(k)}$ is needed in (3.34), suggesting that even a crude approximate solution of (3.30) will lead to an effective method. Notice also that for the particular choice $\delta_k = 0$ for all k, the SD method reduces to the Galerkin discretization.

A more traditional derivation of the streamline diffusion discretization uses a specific *Petrov–Galerkin* formulation. This leads to a realization of the SD method that is defined for all types of elements and any order of approximation, with no element-oriented restrictions on the functions $\vec{w}$ or f. We now describe this approach. The basic idea of Petrov–Galerkin approximation is to specify a weak formulation in which the test space is taken to be different from the trial space. That is, given a problem $\mathcal{L}u = f$, the goal is to find $u \in S$ for some trial space S such that, for some (different) test space T and a suitable inner product,

$$(\mathcal{L}u, v) = (f, v) \quad \text{for all } v \in T. \tag{3.37}$$

In a streamline diffusion context, $\mathcal{L}$ is the convection–diffusion operator, the trial space is S_E^h, and the test space is taken to be the space spanned by functions of the form $v_h + \delta \vec{w} \cdot \nabla v_h$ where $v_h \in S_0^h$ and $\delta > 0$ is a constant parameter (locally defined parameter values will be introduced subsequently). With $u_h \in S_E^h$ denoting the trial function, the left-hand side of (3.37) is

$$(\mathcal{L}u, v) = \epsilon \int_\Omega \nabla u_h \cdot \nabla v_h - \epsilon \int_{\partial \Omega_N} v_h \frac{\partial u}{\partial n} + \int_\Omega (\vec{w} \cdot \nabla u_h) \, v_h$$
$$+ \delta \int_\Omega (\vec{w} \cdot \nabla u_h)(\vec{w} \cdot \nabla v_h) - \delta \epsilon \int_\Omega (\nabla^2 u_h)(\vec{w} \cdot \nabla v_h). \tag{3.38}$$

In the case $g_N = 0$, (3.38) simplifies since the boundary integral is zero. The tricky issue here is that since u_h is not required to have a second derivative, the last integral in (3.38) may not be well-defined. However, assuming that the restriction of functions in S_0^h to individual elements $\triangle_k$ lie in $\mathcal{H}^2(\triangle_k)$, a legitimate method can be constructed by replacing this integral by an element-wise sum, namely,

$$-\delta\epsilon \sum_k \int_{\triangle_k} (\nabla^2 u_h)(\vec{w} \cdot \nabla v_h). \tag{3.39}$$

This leads to the following general statement of the streamline diffusion discretization:

Find $u_h \in S_E^h$ such that
$$a_{sd}(u_h, v_h) = \ell_{sd}(v_h) \quad \text{for all } v_h \in S_0^h, \tag{3.40}$$

where

$$a_{sd}(u, v) := \epsilon \int_\Omega \nabla u \cdot \nabla v + \int_\Omega (\vec{w} \cdot \nabla u)v + \delta \int_\Omega (\vec{w} \cdot \nabla u)(\vec{w} \cdot \nabla v)$$
$$- \delta\epsilon \sum_k \int_{\triangle_k} (\nabla^2 u)(\vec{w} \cdot \nabla v), \tag{3.41}$$

$$\ell_{sd}(v) := \int_\Omega fv + \delta \int_\Omega f\,\vec{w} \cdot \nabla v,$$

The functional $\ell_{sd}(\cdot)$ is continuous over the space $\mathcal{H}_{E_0}^1$. In the case of linear or bilinear elements, the high order term (3.39) is zero, $a_{sd}(\cdot, \cdot)$ is continuous over the space $\mathcal{H}_{E_0}^1$, and (3.40) is identical to the bubble function formulation (3.35) with constant $\delta = \delta_k$ for all $\triangle_k$.

The SD methodology leads naturally to a different norm, the *streamline diffusion norm*

$$\|v\|_{sd} := \left(\epsilon \|\nabla v\|^2 + \delta \|\vec{w} \cdot \nabla v\|^2\right)^{1/2}, \tag{3.42}$$

defined on the space of trial functions $\mathcal{H}_{E_0}^1$. For large Peclet numbers, the character of the solution u is dominated by its behavior along the streamlines. Therefore, this norm, in which the streamline derivative plays a dominant role for large Peclet number, is a more meaningful measure than say, $\|\nabla v\|$. Moreover, in the case of linear or bilinear elements, the coercivity bound associated with the SD formulation (3.40), namely

$$a_{sd}(v_h, v_h) \geq \|v_h\|_{sd}^2 \tag{3.43}$$

is stronger than the corresponding Galerkin bound in the sense that coercivity does not degrade in the limit $\epsilon \to 0$; cf. (3.17). The importance of (3.43) will become apparent in Section 3.4. Ensuring that the problem (3.40) is uniformly

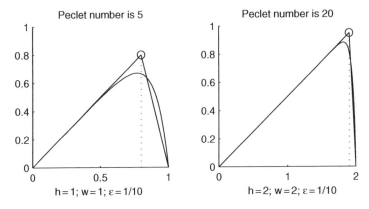

FIG. 3.8. Comparison of exact solution (curved line) of optimal bubble problem (3.45) with piecewise linear approximation, for two different Peclet numbers.

coercive over S_0^h with respect to $\|\cdot\|_{sd}$ is a more delicate issue when using P_2 or Q_2 approximation spaces, see Problem 3.5 for details.

To implement (3.40), we need a way to choose the parameter δ, or alternatively, to choose locally defined parameters δ_k^*, typically approximating δ_k in (3.34). A smart choice[2] is the following:

$$\delta_k^* = \begin{cases} \dfrac{h_k}{2|\vec{w}_k|}\left(1 - \dfrac{1}{\mathcal{P}_h^k}\right) & \text{if } \mathcal{P}_h^k > 1, \\[2mm] 0 & \text{if } \mathcal{P}_h^k \leq 1. \end{cases} \qquad (3.44)$$

Here, $|\vec{w}_k|$ is the ℓ_2 norm of the wind at the element centroid, h_k is a measure of the element length in the direction of the wind, and $\mathcal{P}_h^k := |\vec{w}_k| h_k/(2\epsilon)$ is the *element Peclet number*.

There are several ways of motivating the parameter choice embodied in (3.44). The most intuitive justification comes from looking at the one dimensional version of the optimal bubble problem (3.30):

$$-\epsilon u'' + wu' = 1 \quad \text{in } (0, h), \quad u(0) = u(h) = 0, \qquad (3.45)$$

see Problem 3.1. If $\mathcal{P}_h := hw/(2\epsilon)$ is larger than unity then the problem has a downstream boundary layer of width $2\epsilon/w$; see Figure 3.8. This means that the interior bubble solution $u(x)$ is closely approximated by a piecewise linear function $u_*(x)$ which agrees with the reduced problem solution $\hat{u} = x/w$ up to the transition point $x_* = h - h/\mathcal{P}_h$, $y_* = x_*/w$ (marked with a circle) and decreases linearly to zero in the boundary layer region $[x_*, h]$. The fact that the

---

[2]This has been extensively tested and is embodied in our IFISS software. On a rectangle with sides of length h_x, h_y in which the wind forms an angle $\theta = \arctan(|w_y/w_x|)$ at the centroid, h_k is given by $\min(h_x/\cos\theta, h_y/\sin\theta)$.

parameter choice in (3.44) is obtained by substituting the approximate bubble function solution u_* into (3.36) is left as an exercise, see Problem 3.6.

An important feature of the choice of δ_k^* is that the two limiting values are correctly defined. In particular, in the limit of $\mathcal{P}_h^k \to \infty$ we see that $\delta_k^* \to h_k/(2|\vec{w}_k|)$, which is the optimal choice in the convective limit, see for example Eriksson et al. [70, p. 463]. In the opposite extreme of a discretized problem where all the layers are resolved by the mesh, we see from (3.44) that the SD discretization reduces to the standard Galerkin method.

We conclude this section with one example demonstrating the effectiveness of the streamline diffusion method (3.40) as realized by solving problem (3.35) using the parameter choice (3.44). The plots in Figures 3.9 and 3.10 show the discrete solution of the problem in Example 3.1.3 with a mesh Peclet number $\mathcal{P}_h = 6.25$. Looking closely at the SD solution in Figure 3.10 it is evident that although there are small non-physical oscillations along the internal layer, the large amplitude oscillations evident in the Galerkin solution have been completely eliminated. For further insight into the relative behavior of Galerkin and SD approaches outside of exponential layers, see Problem 3.7.

The solution in Figure 3.10 should also be compared to the reference solution in Figure 3.4. This latter solution is also computed using the SD discretization, but it uses a much finer 128×128 grid, so that the mesh Peclet number is of order unity. The basic differences between these solutions are, first, that oscillations are imperceptible on the finer grid (since the interior layer is captured), and second, that the width of the boundary layer is significantly narrower in the finer grid case. The latter observation reinforces the point that resolution is always limited by the grid size; the bubble function stabilization underlying the SD method does not magically enable the resolution of features that have a scale finer than the

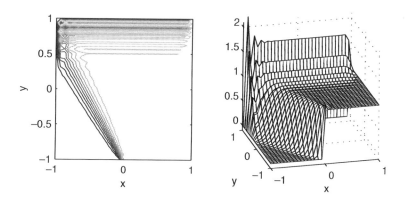

FIG. 3.9. Contour plot (left) and three-dimensional surface plot (right) of the Galerkin solution of Example 3.1.3 for $\epsilon = 1/200$, using $\boldsymbol{Q}_1$ approximation on a 32×32 square grid.

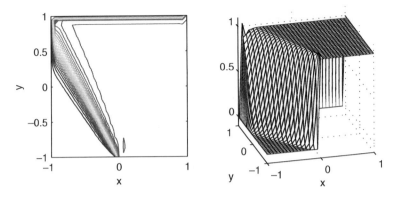

FIG. 3.10. Contour plot (left) and three-dimensional surface plot (right) of the streamline diffusion solution of Example 3.1.3, for $\epsilon = 1/200$, using Q_1 approximation on a 32×32 square grid.

basic grid used. The SD approach limits the damage caused by poor resolution of exponential layers and leads to a more accurate solution outside of the layers where the solution is not varying rapidly. These points will be explored in more detail in the next section.

3.4   Theory of errors

The tricky issue of finding tight bounds for the error associated with discretization of convection–diffusion problems is discussed in this section. We look at the issue from several different points of view. First, we discuss conventional a priori error bounds for the Galerkin and streamline diffusion discretizations. Second, we discuss a posteriori analysis of the error in a computed solution. The motivation is that associated local error bounds can readily be used to drive adaptive mesh refinement. This is particularly useful in the context of efficiently approximating solutions exhibiting layers.

3.4.1   *A priori error bounds*

The analysis of the Galerkin discretization begins with the defining equations (3.16) for the continuous problem and (3.21) for the discrete one, which together imply the Galerkin orthogonality property

$$a(u - u_h, w_h) = 0 \quad \text{for all } w_h \in S_0^h . \tag{3.46}$$

If v_h is an arbitrary function in S_E^h, then both u_h and v_h satisfy the essential boundary condition in (3.2), and hence $v_h - u_h \in S_0^h$. It follows that

$$a(u - u_h, u - u_h) = a(u - u_h, u - v_h) + a(u - u_h, v_h - u_h)$$
$$= a(u - u_h, u - v_h).$$

Application of the coercivity bound (3.17) and the continuity bound (3.19) then leads to the following estimate:

$$\epsilon \|\nabla(u - u_h)\|^2 \le a(u - u_h, u - u_h)$$
$$= a(u - u_h, u - v_h)$$
$$\le \Gamma_{\vec{w}} \|\nabla(u - u_h)\| \, \|\nabla(u - v_h)\|. \tag{3.47}$$

This establishes the quasi-optimality (3.22) of the Galerkin approximation. A formal statement is given below.

Theorem 3.3. $\|\nabla(u - u_h)\| \le \dfrac{\Gamma_{\vec{w}}}{\epsilon} \inf_{v_h \in S_E^h} \|\nabla(u - v_h)\|.$

As discussed in Section 3.3.1, the fact that the constant $\Gamma_{\vec{w}}/\epsilon$ blows up with the Peclet number suggests that the standard Galerkin discretization is likely to give a poor approximation in any case where problem (3.1)–(3.2) is convection-dominated. An alternative analysis that provides additional insight is given by the following error estimate.

Theorem 3.4. *If the variational problem (3.16) is solved using linear approximation on a shape regular triangular mesh, or bilinear approximation on a shape regular rectangular grid, then there exists a constant C, asymptotically proportional to $\mathcal{P}_h = h \|\vec{w}\|_\infty /\epsilon$, such that*

$$\|\nabla(u - u_h)\| \le Ch \|D^2 u\|, \tag{3.48}$$

where h is the length of the longest element edge and $\|D^2 u\|$ measures the $\mathcal{H}^2$-regularity of the target solution.

Proof The proof is clever. To start with, the right-hand side of (3.47) can be bounded as in (3.18):

$$\epsilon \|\nabla(u - u_h)\|^2 \le a(u - u_h, u - v_h)$$
$$= \epsilon \int_\Omega \nabla(u - u_h) \cdot \nabla(u - v_h) + \int_\Omega \vec{w} \cdot \nabla(u - u_h)(u - v_h)$$
$$\le \epsilon \|\nabla(u - u_h)\| \left[\|\nabla(u - v_h)\| + \frac{\|\vec{w}\|_\infty}{\epsilon} \|u - v_h\| \right]. \tag{3.49}$$

Assuming $\mathcal{H}^2$-regularity, the two terms in the square brackets of (3.49) can then be bounded separately using approximation theory as in Section 1.5.1. In particular, if we choose v_h to be the standard linear or bilinear interpolant, $\pi_h u$ say, then the interpolation estimates $\|\nabla(u - \pi_h u)\| \le C_1 h \|D^2 u\|$ from (1.76) and $\|u - \pi_h u\| \le C_0 h^2 \|D^2 u\|$ from Proposition 1.16, can be substituted into (3.49) to give the desired bound (3.48) with $C := \max\{C_0, C_1\}(1 + \mathcal{P}_h)$. $\qquad \square$

Remark 3.5. The statement of the bound (3.48) needs to be qualified in the vicinity of layers. For example, in the case of exponential boundary layer, looking at a typical solution (3.7), we see that streamline derivatives are inversely proportional to ϵ, and $\left\|D^2u\right\|_{\triangle_k}$ blows up as $\epsilon \to 0$ in elements $\triangle_k$ that lie within such a layer. If there are no layers in the solution, then the error bound in Theorem 3.4 is better than that in Theorem 3.3. Indeed, in the absence of layers (3.48) implies that the Galerkin approximation is optimal when $\mathcal{P}_h \le 1$.

We now turn our attention to the SD discretization (3.35) with parameter choice (3.44). To fix ideas, we restrict attention to Dirichlet boundary conditions and assume that $\vec{w}$ is constant, and is scaled so that $|\vec{w}| = 1$. We take a uniform grid of lowest order (P_1 or Q_1) elements on a uniform mesh of size h, so that δ_k in (3.44) is independent of k:

$$\delta := \max\left\{0, \frac{h}{2}\left(1 - \frac{1}{\mathcal{P}_h}\right)\right\}. \tag{3.50}$$

Assuming that $\mathcal{P}_h = h/(2\epsilon)$ is greater than unity implies that δ is a strictly positive constant. In this case, a representative error bound is the following.

Theorem 3.6. *If a constant coefficient Dirichlet problem (3.16) with $|\vec{w}| = 1$ is solved using the SD formulation (3.40) with δ given by (3.50) in conjunction with either linear or bilinear approximation on a uniform grid with $h > 2\epsilon$, then there exists a constant C, bounded independently of ϵ, such that*

$$\|u - u_h\|_{sd} \le Ch^{3/2}\|D^2u\|. \tag{3.51}$$

Here $\|v\|_{sd}$ is the streamline diffusion norm (3.42), and $\|D^2u\|$ measures the $\mathcal{H}^2$-regularity of the target solution.

The bound in Theorem 3.6 is essentially a special case of the general analysis given in Roos et al. [159, pp. 229ff]. A proof is outlined below.

The first step of the proof is to establish an orthogonality relation for the error $u - u_h$. Given that u is assumed to be in $\mathcal{H}^2(\Omega)$, u satisfies (3.1) in the sense that

$$\int_{\triangle_k}(-\epsilon\nabla^2u + \vec{w}\cdot\nabla u)q = \int_{\triangle_k} fq \quad \text{for all } q \in L_2(\triangle_k),$$

for all elements $\triangle_k \in \mathcal{T}_h$. Thus, for $v_h \in S_0^h$, we can choose $q = \vec{w}\cdot\nabla v_h$ in each element

$$\sum_k\int_{\triangle_k}(-\epsilon\nabla^2u + \vec{w}\cdot\nabla u)(\vec{w}\cdot\nabla v_h) = \sum_k\int_{\triangle_k} f\,\vec{w}\cdot\nabla v_h. \tag{3.52}$$

Given that $\partial\Omega_N = \emptyset$ and $v_h \in \mathcal{H}^1_{E_0}$, (3.11) combined with (3.52) implies that

$$\epsilon \int_\Omega \nabla u \cdot \nabla v_h + \int_\Omega (\vec{w} \cdot \nabla u) v_h + \delta \int_\Omega (\vec{w} \cdot \nabla u)(\vec{w} \cdot \nabla v_h)$$
$$- \delta\epsilon \sum_k \int_{\triangle_k} \nabla^2 u(\vec{w} \cdot \nabla v_h) = \int_\Omega f v_h + \delta \int_\Omega f\vec{w} \cdot \nabla v_h$$

for any $v_h \in S^h_0$. Then, in light of the definitions of $a_{sd}(\cdot, \cdot)$ and $\ell_{sd}(\cdot)$, see (3.41), it is evident that the SD formulation is *consistent* in the sense that

$$a_{sd}(u, v_h) = \ell_{sd}(v_h) \quad \text{for all } v_h \in S^h_0. \tag{3.53}$$

In addition, (3.40) may be subtracted from (3.53) to give the desired orthogonality condition:

$$a_{sd}(u - u_h, v_h) = 0 \quad \text{for all } v_h \in S^h_0. \tag{3.54}$$

The second step of the proof is to define $\pi_h u \in S^h_E$ to be the standard linear or bilinear interpolant of u at the nodal points of the discretization, and then to bound the error $\|u - u_h\|_{sd}$ in terms of the interpolation error $\|u - \pi_h u\|_{sd}$. Since $\pi_h u - u_h \in S^h_0$, the uniform coercivity bound (3.43) gives

$$\|\pi_h u - u_h\|^2_{sd} \leq a_{sd}(\pi_h u - u_h, \pi_h u - u_h)$$
$$= a_{sd}(\pi_h u - u, \pi_h u - u_h) + a_{sd}(u - u_h, \pi_h u - u_h)$$
$$= a_{sd}(u - \pi_h u, u_h - \pi_h u), \tag{3.55}$$

where the last step follows from (3.54). To simplify notation in the remainder of the proof, let

$$e_I := u - \pi_h u, \quad d_h := u_h - \pi_h u$$

denote the interpolation error and the difference between the interpolant and the SD solution, respectively. From the definition of $a_{sd}(\cdot, \cdot)$, (3.55) expands to

$$\|d_h\|^2_{sd} \leq \epsilon(\nabla e_I, \nabla d_h) + (\mathbf{w} \cdot \nabla e_I, d_h) + \delta(\mathbf{w} \cdot \nabla e_I, \mathbf{w} \cdot \nabla d_h)$$
$$- \delta\epsilon \sum_k (\nabla^2 e_I, \mathbf{w} \cdot \nabla d_h)_{\triangle_k}, \tag{3.56}$$

where $(u, v)_{\triangle_k}$ represents the element L_2 inner product. Then, using the trivial estimates

$$\epsilon^{1/2}\|\nabla v\| \leq \|v\|_{sd}, \quad \delta^{1/2}\|\mathbf{w} \cdot \nabla v\| \leq \|v\|_{sd},$$

the first three terms on the right-hand side of (3.56) are bounded as follows:

$$\epsilon|(\nabla e_I, \nabla d_h)| \le \epsilon^{1/2}\|\nabla e_I\|\,\epsilon^{1/2}\|\nabla d_h\| \le \epsilon^{1/2}\|\nabla e_I\|\,\|d_h\|_{sd}\,;$$

$$|(\mathbf{w}\cdot\nabla e_I, d_h)| = |(e_I, \mathbf{w}\cdot\nabla d_h)| \le \|e_I\|\,\|\mathbf{w}\cdot\nabla d_h\| \le \delta^{-1/2}\|e_I\|\,\|d_h\|_{sd}\,;$$

$$\delta|(\mathbf{w}\cdot\nabla e_I, \mathbf{w}\cdot\nabla d_h)| \le \delta^{1/2}\|\mathbf{w}\cdot\nabla e_I\|\,\delta^{1/2}\|\mathbf{w}\cdot\nabla d_h\| \le \|e_I\|_{sd}\,\|d_h\|_{sd}\,.$$

Notice that we explicitly use the fact that $e_I = 0$ on $\partial\Omega$ when integrating the second term by parts. The fourth term in the right-hand side of (3.56) can be bounded elementwise. A summation over the elements gives

$$\delta\left|\sum_k (\nabla^2 e_I, \mathbf{w}\cdot\nabla d_h)_{\triangle_k}\right| \le \delta\epsilon\sum_k \|\nabla^2 e_I\|_{\triangle_k}\,\|\mathbf{w}\cdot\nabla d_h\|_{\triangle_k},$$

and the discrete Cauchy–Schwarz inequality implies that

$$\delta\epsilon\left|\sum_k (\nabla^2 e_I, \mathbf{w}\cdot\nabla d_h)_{\triangle_k}\right| \le \delta\epsilon\left(\sum_k \|\nabla^2 e_I\|_{\triangle_k}^2\right)^{1/2}\left(\sum_k \|\mathbf{w}\cdot\nabla d_h\|_{\triangle_k}^2\right)^{1/2}$$

$$\le \delta^{1/2}\epsilon\left(\sum_k \|\nabla^2 e_I\|_{\triangle_k}^2\right)^{1/2}\delta^{1/2}\|\mathbf{w}\cdot\nabla d_h\|^2$$

$$\le \delta^{1/2}\epsilon\left(\sum_k \|\nabla^2 e_I\|_{\triangle_k}^2\right)^{1/2}\|d_h\|_{sd}\,.$$

Substitution of all these bounds into (3.56) leads to a bound for $\|d_h\|_{sd}$ in terms of norms of e_I. The triangle inequality gives the estimate

$$\|u - u_h\|_{sd} \le \|e_I\|_{sd} + \|d_h\|_{sd}$$

$$\le 2\|e_I\|_{sd} + \epsilon^{1/2}\|\nabla e_I\| + \delta^{-1/2}\|e_I\| + \delta^{1/2}\epsilon\left(\sum_k \|\nabla^2 e_I\|_{\triangle_k}^2\right)^{1/2}.$$

$$(3.57)$$

The final step in the proof is to bound the terms on the right-hand side of (3.57) using approximation theory. For example, since $|\vec{w}| = 1$, the first term can be bounded using the interpolation estimate $\|\nabla e_I\| \le C_1 h\,\|D^2 u\|$ of (1.76):

$$\|e_I\|_{sd} = \left(\epsilon\|\nabla e_I\|^2 + \delta\,\|\mathbf{w}\cdot\nabla e_I\|^2\right)^{1/2}$$

$$\le \epsilon^{1/2}\|\nabla e_I\| + \delta^{1/2}\,\|\mathbf{w}\cdot\nabla e_I\|$$

$$\le \left(\epsilon^{1/2} + \delta^{1/2}\right)\|\nabla e_I\|$$

$$\le C_1 h\left(\epsilon^{1/2} + \delta^{1/2}\right)\|D^2 u\|.$$

The third term is dealt with using the bound $\|e_I\| \le C_0 h^2 \|D^2 u\|$, see Proposition 1.16, and the last term is handled as follows:

$$\left(\sum_k \|\nabla^2 e_I\|_{\triangle_k}^2\right)^{1/2} = \left(\sum_k \|\nabla^2 u\|_{\triangle_k}^2\right)^{1/2} = \|\nabla^2 u\| \le \|D^2 u\|.$$

Finally, substitution of these inequalities into (3.57) and using the fact that $\epsilon < h$, gives

$$\|u - u_h\|_{sd} \le \left(3C_1 \delta^{1/2} h + 2C_1 h^{3/2} + C_0 \delta^{-1/2} h^2 + \delta^{1/2} h\right) \|D^2 u\|. \qquad (3.58)$$

The parameter $\delta = Ch$ thus nicely balances the approximation error and establishes the estimate (3.51). Notice that although the value of δ as determined by (3.50) depends on ϵ, the limiting value as $\epsilon \to 0$ is $\delta = h/2$. The upshot is that the constant in the parentheses on the right-hand side of (3.58) remains bounded in the limit $\epsilon \to 0$.

Remark 3.7. The significance of the SD error estimate (3.51) is the trivial bound

$$\delta^{1/2}\|\vec{w} \cdot \nabla(u - u_h)\| \le Ch^{3/2}\|D^2 u\|, \qquad (3.59)$$

implying that for smooth solutions (see Remark 3.5) the derivative of the error in the streamline direction is bounded independently of ϵ. It also represents an optimal approximation in terms of h (recall that $\delta = O(h)$).

Remark 3.8. It is interesting to compare the bound for the Galerkin error in Theorem 3.4 with the SD error bound in Theorem 3.6, particularly with respect to their dependence on the mesh Peclet number. Both bounds depend on $\|D^2 u\|$ and we will omit this factor from our deliberations here (recognizing however that this term depends on ϵ). We will use the notation $\alpha \lesssim \beta$ to mean $\alpha \le C\beta$ where C does not depend on β. The bound (3.48) implies that

$$\|\nabla(u - u_h)\| \lesssim \mathcal{P}_h h.$$

In contrast, with $\delta = Ch$, (3.59) shows that the streamline derivative satisfies

$$\|\vec{w} \cdot \nabla(u - u_h)\| \lesssim h.$$

Notice also that the bound $\epsilon^{1/2} \|\nabla(u - u_h)\| \lesssim h^{3/2}$ derived from the estimate (3.51) implies that the *cross-wind* derivative of the SD error satisfies

$$\|\vec{w}^\perp \cdot \nabla(u - u_h)\| \lesssim \mathcal{P}_h^{1/2} h.$$

The clear conclusion from this is that if $\mathcal{P}_h > 1$, then the SD approximation is more reliable than the Galerkin approximation in terms of both streamline and cross-wind components of the error.

On a grid that is manifestly too coarse to resolve an exponential boundary layer, the Galerkin solution u_h is oscillatory everywhere, see Figure 3.6. In contrast, the analogous SD solution u_h^* is often qualitatively accurate outside

Table 3.1 *Comparison of Galerkin and* SD *errors for Example 3.1.1 solved for different values of ϵ using a 16×16 uniform grid (grid level $\ell = 4$).*

ϵ	E_h	E_h^*	$h/(2\epsilon)$
1/16	1.185	1.185	1
1/64	4.917	4.006	4
1/256	1.255×10^1	8.948	16
1/1024	3.720×10^1	1.833×10^1	64
1/4096	1.347×10^2	3.688×10^1	256

of boundary layers even if the layers are not resolved. Computational experiments provide additional, quantitative, evidence that the SD discretization is more effective than standard Galerkin in such cases. As an illustration we take the analytic test problem in Example 3.1.1, and solve it for different values of ϵ using $\boldsymbol{Q}_1$ approximation on a fixed 16×16 grid of square elements. Computed values of the exact[3] Galerkin error $E_h := \|\nabla(u - u_h)\|$ (i.e. the left-hand side of (3.48)) and of $E_h^* := \|\nabla(u - u_h^*)\|$ where u_h^* is the corresponding SD solution, are presented in Table 3.1. The results are consistent with Remark 3.8: when $\mathcal{P}_h$ is increased from 1 by reducing ϵ while keeping h fixed, the error grows like $O(\mathcal{P}_h)$ for the Galerkin discretization and like $O(\mathcal{P}_h^{1/2})$ for the SD discretization. If ϵ becomes very small, then ultimately the boundary layer becomes "invisible" to the SD discretization and the solutions become independent of ϵ; see Computational Exercise 3.3.

It is also illuminating to compare the two discretization methods in the more realistic situation of solving the problem in Example 3.1.1 for a fixed value of ϵ, using successively finer meshes. Such a comparison is given in Table 3.2. While the SD errors E_h^* are smaller than the Galerkin counterparts E_h, what is rather striking here is the slow reduction in the global error as the grid is successively refined. With further grid refinement, it can be seen that the anticipated $O(h)$ error reduction is realized only after the exponential boundary layer is resolved (i.e. when $\mathcal{P}_h \leq 1$, so that the SD and Galerkin formulations coincide). Table 3.2 also shows the errors $R_h := \|\nabla(u - u_h)\|_{\Omega_\sqcup}$ and $R_h^* := \|\nabla(u - u_h^*)\|_{\Omega_\sqcup}$, computed on the subdomain $\Omega_\sqcup := (-1, 1) \times (-1, 3/4)$, which excludes the exponential layer. The tabulated values of R_h and R_h^* show the real advantage of the SD approximation, for which the solution is not corrupted by the unresolved exponential layer.

An additional, more accurate, set of results is given in Table 3.3. These results are obtained using grids that are specially designed to fit the boundary layer, see Figure 3.11. A strong theoretical foundation (originally developed by Shishkin [132]) motivates the use of such layer-adapted grids, see Computational Exercise 3.4. The art of constructing Shishkin grids lies in the definition of

---

[3]Computed with the `matlab` quadrature function `dblquad.m` with tolerance 10^{-13}.

Table 3.2 *Comparison of global errors for Example 3.1.1 for fixed $\epsilon = 1/64$, using uniform grids of width $h = 1/2^{l-1}$, where l is the grid refinement level (so the grid has dimensions $2^l \times 2^l$).*

l	E_h	E_h^*	R_h	R_h^*	$h/(2\epsilon)$
3	5.616	4.335	3.250	8.175×10^{-7}	8
4	4.917	4.006	1.483	1.642×10^{-5}	4
5	3.814	3.327	5.304×10^{-2}	1.112×10^{-5}	2
6	2.393	2.393	4.981×10^{-7}	4.981×10^{-7}	1

Table 3.3 *Errors for Example 3.1.1 for $\epsilon = 1/64$, using Shishkin grids of dimensions $2^l \times 2^l$.*

l	E_h	E_h^*
3	1.342	1.340
4	9.107×10^{-1}	9.106×10^{-1}
5	5.743×10^{-1}	5.743×10^{-1}
6	3.459×10^{-1}	3.459×10^{-1}

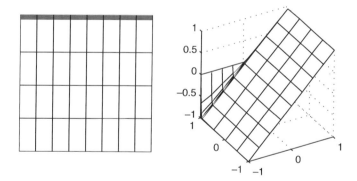

FIG. 3.11. 8×8 Shishkin grid (left) and surface plot (right) of the associated solution of Example 3.1.1 for $\epsilon = 1/64$.

the transition line $y = y_*$ which determines the boundary between the fine discretization that resolves the layer, and the coarse discretization outside of the layer. In Table 3.3, the specific point $y_* \equiv 1 - 2\epsilon \log_e n$ is used (see Roos et al. [159, p. 87]), and a uniform $n \times n/2$ grid of rectangular elements is taken on either side of the transition line. The outcome is the sought after $O(h)$ error reduction for both Galerkin and SD approximation. It is noteworthy that when the layer is resolved, addition of streamline diffusion makes an almost imperceptible change in the Galerkin solution — it is only added in the coarse grid region since $\mathcal{P}_h^k < 1$ for elements Δ_k in the fine grid.

3.4.2   A posteriori error bounds

The obvious motivation for developing a posteriori error estimation techniques is that convection–diffusion problems exhibit local features that make them ripe for adaptive refinement. Notwithstanding this goal, the fact that error estimation in a convection–diffusion context is a rapidly developing research area makes it difficult to survey all possible strategies. We thus have a more limited objective in this section, namely, to explore the effectiveness of the particular error estimator that is built into the IFISS software. This estimator requires the solution of local Poisson problems and is based on the approach developed in Section 1.5.2. Additionally, the performance of this local error estimator is assessed in the context of the reference problems of Section 3.1. Its effectiveness as a refinement indicator in a self-adaptive setting is also evaluated.

At the outset it must be stressed that the SD discretization has a crucial role to play in cases where exponential boundary layers are not resolved by the grid. The uniform 16×16 grid Galerkin solution illustrated in Figure 3.6 has large error in every element. In a self-adaptive setting, this would inevitably lead to a *global* grid refinement, which is costly and unnecessary. Use of the SD discretization instead ensures that the error is essentially localized to the unresolved boundary layer, see Computational Exercise 3.3. An effective self-adaptive refinement process would then produce local refinement that is confined to the layer of elements that delimit the unresolved exponential layer at the outflow boundary.

A particular error estimation strategy that has some attractive features (and is built into the IFISS software) is now described. Given some approximation, $u_h \in S_E^h$, to the solution $u \in \mathcal{H}_E^1$ of (3.16), the error $e = u - u_h \in \mathcal{H}_{E_0}^1$ satisfies

$$a(e, v) = \ell(v) - a(u_h, v) \quad \text{for all } v \in \mathcal{H}_{E_0}^1. \tag{3.60}$$

With the notation for element-wise forms $(u, v)_T := \int_T uv$ and $(\nabla u, \nabla v)_T := \int_T \nabla u \cdot \nabla v$ so that $a(u, v)_T := \epsilon(\nabla u, \nabla v)_T + (\vec{w} \cdot \nabla u, v)_T$, the error may be localized as in Section 1.5.2 to give the characterization

$$\sum_{T \in \mathcal{T}_h} a(e, v)_T = \sum_{T \in \mathcal{T}_h} \left[(f + \epsilon \nabla^2 u_h - \vec{w} \cdot \nabla u_h, v)_T - \frac{1}{2}\epsilon \sum_{E \in \mathcal{E}(T)} \left\langle \left[\!\!\left[\frac{\partial u_h}{\partial n} \right]\!\!\right], v \right\rangle_E \right], \tag{3.61}$$

where $\mathcal{E}(T)$ is the set of the four edges of element T. For the lowest order $\boldsymbol{P}_1$ or $\boldsymbol{Q}_1$ approximations over a triangular or rectangular element subdivision, respectively, $\nabla^2 u_h|_T = 0$, so that the *interior residual*

$$R_T := \{f - \vec{w} \cdot \nabla u_h\}|_T,$$

and the *flux jump* operator

$$R_E^* = \begin{cases} \dfrac{1}{2} \left[\!\left[\dfrac{\partial u_h}{\partial n} \right]\!\right] & E \in \mathcal{E}_{h,\Omega} \\ -\nabla u_h \cdot \vec{n}_{E,T} & E \in \mathcal{E}_{h,N} \\ 0 & E \in \mathcal{E}_{h,D} \end{cases}$$

can be defined. Then, following the path taken in Section 1.5.2, we write (3.61) as

$$\sum_{T \in \mathcal{T}_h} \{\epsilon(\nabla e, \nabla v)_T + (\vec{w} \cdot \nabla e, v)_T\}$$

$$= \sum_{T \in \mathcal{T}_h} \left[(R_T, v)_T - \epsilon \sum_{E \in \mathcal{E}(T)} \langle R_E^*, v \rangle_E \right] \quad \text{for } v \in \mathcal{H}_{E_0}^1. \quad (3.62)$$

A direct approximation of (3.62) suggests that a local convection–diffusion problem needs to be solved in order to compute a local error estimator. It turns out that an estimate of essentially the same quality (in a sense made precise later) is obtained by omitting the term $\|\vec{w} \cdot \nabla e\|_T$ in (3.62) and solving a local Poisson problem instead. This allows us to use the error estimation technique developed in Section 1.5.2. That is, recalling the correction space, $\mathcal{Q}_T$, of interior and edge bubble functions, see (1.96), we define the local error estimator

$$\eta_T := \|\nabla e_T\|_T, \quad (3.63)$$

where e_T solves the local Poisson problem: find $e_T \in \mathcal{Q}_T$ such that

$$\epsilon(\nabla e_T, \nabla v)_T = (R_T^0, v)_T - \epsilon \sum_{E \in \mathcal{E}(T)} \langle R_E^*, v \rangle_E, \quad (3.64)$$

for all $v \in \mathcal{Q}_T$, where R_T^0 represents a constant approximation of R_T. The upshot is that for each triangular (or rectangular) element a 4×4 (or 5×5) system of equations must be solved to compute e_T. The theoretical underpinning for the strategy (3.64) is the following result; this should be compared with Theorem 1.23.

Theorem 3.9. *If a variational problem* (3.16) *is solved using a grid of bilinear rectangular elements, and if the rectangle aspect ratio condition is satisfied with β_* given in Definition 1.18, then, the estimator η_T computed via* (3.64) *satisfies* *the upper bound property*

$$\|\nabla(u - u_h)\| \leq C(\beta_*) \left(\sum_{T \in \mathcal{T}_h} \eta_T^2 + \sum_{T \in \mathcal{T}_h} \left(\frac{h_T}{\epsilon} \|\vec{w}\|_T \right)^2 \left\| R_T - R_T^0 \right\|_T^2 \right)^{1/2}, \quad (3.65)$$

where C is independent of ϵ. The parameter h_T is the length of the longest edge of element T. The bound (3.65) holds for the Galerkin solution u_h satisfying (3.21) and the SD solution u_h satisfying (3.40) with δ given by (3.44).

Proof See Problem 3.8.   □

Remark 3.10. If f and $\vec{w}$ are both piecewise constant functions then the consistency error term $\left\| R_T - R_T^0 \right\|_T$ is identically zero. Otherwise, if f and $\vec{w}$ are smooth, this term represents a high-order perturbation. In any case the estimator η_T is reliable in the sense that the upper bound (3.65) is independent of h and ϵ.

Establishing that the estimated error η_T gives a lower bound on the local error is not possible. The difficulty is generic for any local error estimator: the local error is overestimated within exponential boundary layers wherever such layers are not resolved by the mesh. Thus, in contrast to the nice lower bound that holds when solving Poisson's equation, see Proposition 1.28, the tightest lower bound that can be established for the convection–diffusion equation is the following.

Proposition 3.11. *If the variational problem (3.16) with $\|\vec{w}\|_\infty = 1$ is solved via either the Galerkin formulation (3.21) or the SD formulation (3.40), using a grid of bilinear rectangular elements, and if the rectangle aspect ratio condition is satisfied, then the estimator η_T computed via (3.64) is a local lower bound for $e = u - u_h$ in the sense that*

$$\eta_T \leq C(\beta_{\omega_T}) \left(\|\nabla e\|_{\omega_T} + \sum_{T \subset \omega_T} \frac{h_T}{\epsilon} \|\vec{w} \cdot \nabla e\|_T + \sum_{T \subset \omega_T} \frac{h_T}{\epsilon} \left\| R_T - R_T^0 \right\|_T \right),$$
$$(3.66)$$

where C is independent of ϵ, and ω_T represents the patch of five elements that have at least one boundary edge E from the set $\mathcal{E}(T)$.

Proof The result is a generalization of Proposition 1.28. A proof can be found in Kay & Silvester [119]. The same estimate is obtained if local convection–diffusion problems are solved in place of (3.64), for details see Verfürth [205]. The restriction that $\|\vec{w}\|_\infty = 1$ can be removed; it is only included to simplify the proof.   □

From the bounds (3.65) and (3.66), the structure of the "optimality gap" term $\sum_{T \subset \omega_T} h_T/\epsilon \|\vec{w} \cdot \nabla e\|_T$, leads to the expectation that η_T will be an overestimate for $\|\nabla e\|_T$ in any element k where first, the element Peclet number, $h_T/2\epsilon$, is significantly bigger than unity, and second, the derivative of the error in the streamline direction, namely $\|\vec{w} \cdot \nabla e\|_T$, is commensurate with the derivative of the error in the cross-wind direction. Such a deterioration in performance is realized in the case of Example 3.1.1, if ϵ is decreased while keeping the grid fixed. As an example, estimated global errors $\eta = \left(\sum_{T \in \mathcal{T}_h} \eta_T^2 \right)^{1/2}$, and corresponding effectivity indices $X_\eta := \eta/E_h$, are presented in Table 3.4. For reference purposes, these results should be compared with the exact errors E_h given in Table 3.1. Notice that the global effectivity deteriorates like $O(\mathcal{P}_h)$ as $\mathcal{P}_h \to \infty$, suggesting that the bound (3.66) is tight in this instance.

Solving the problem using SD approximation in place of Galerkin approximation gives the estimated errors η^* and the associated effectivity indices $X_{\eta^*} = \eta^*/E_h^*$ presented in Table 3.5. These results are more encouraging in that the overestimation of the error is, in this case, limited to the set of elements that delimit the unresolved exponential layer. The upshot is that the global effectivity deteriorates like $O(\mathcal{P}_h^{1/2})$ as $\mathcal{P}_h \to \infty$. In considering the results

Table 3.4 *Estimated errors (η) and effectivity indices (X_η) for Example 3.1.1 solved with Galerkin approximation using a 16 × 16 uniform grid (grid level $\ell = 4$).*

ϵ	η	X_η	$h/(2\epsilon)$
1/16	1.515	1.279	1
1/64	1.578×10^1	3.210	4
1/256	1.649×10^2	1.314×10^1	16
1/1024	2.311×10^3	6.213×10^1	64
1/4096	3.627×10^4	2.693×10^2	256

Table 3.5 *Estimated errors (η^*) and effectivity indices (X_{η^*}) for Example 3.1.1 solved via SD approximation using a 16×16 uniform grid ($\ell = 4$).*

ϵ	η^*	X_{η^*}	$h/(2\epsilon)$
1/16	1.515	1.279	1
1/64	6.599	1.647	4
1/256	2.739×10^1	3.061	16
1/1024	1.108×10^2	6.045	64
1/4096	4.445×10^2	1.205×10^1	256

Table 3.6 *Comparison of estimated errors and effectivity indices for Example 3.1.1 and fixed $\epsilon = 1/64$, using Shishkin grids of dimension $2^\ell \times 2^\ell$ where ℓ is the grid refinement level.*

ℓ	Galerkin η	X_η	SD η^*	X_{η^*}
3	2.813	2.096	2.789	2.081
4	1.220	1.340	1.218	1.338
5	6.244×10^{-1}	1.087	6.243×10^{-1}	1.087
6	3.536×10^{-1}	1.022	3.536×10^{-1}	1.022

Table 3.7 *Estimated errors for Example 3.1.2 with $\epsilon = 1/200$, solved via SD approximation using sequences of uniform and stretched grids of dimension $2^\ell \times 2^\ell$ where ℓ is the grid refinement level.*

ℓ	Uniform η^*	Stretched η^*
3	5.998	3.476
4	2.695	1.085
5	1.294	4.392×10^{-1}
6	6.364×10^{-1}	1.883×10^{-1}

in Tables 3.4 and 3.5, we note that the improvement in performance of SD over Galerkin might be anticipated from (3.66) — in particular the better approximation of the streamline derivative suggests that the gap term will be smaller for SD in the limit as $\mathcal{P}_h \to \infty$.

The structure of the bound (3.66) suggests that η_T will have an effectivity close to unity if exponential layers are adequately resolved by the grid. To reinforce this point, estimated errors associated with the exact errors E_h and E_h^* for the Shishkin grid results in Table 3.3 are given in Table 3.6. The agreement between estimated and exact errors is quite impressive — especially since the elements in the fine discretization have large aspect ratio, for example, for the 8×8 grid it is 15:1. It is also rather satisfying that the global effectivity X_η seems to converge to unity as $h \to 0$.

The usefulness of the error estimation strategy in (3.64) is similarly apparent for problems with characteristic boundary layers like that in Example 3.1.2. Estimated errors for the specific problem illustrated in Figure 3.3 are given in Table 3.7. The error estimates for the uniform grid are commensurate with those shown in Table 3.6 for a Shishkin grid, suggesting that $\|\nabla(u - u_h^*)\|$ is decreasing

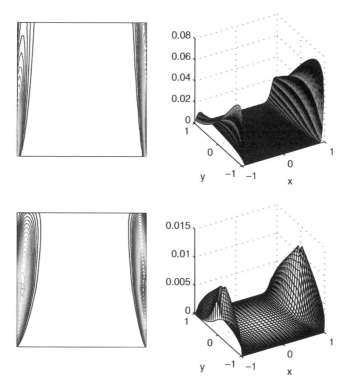

FIG. 3.12. Contour plot (left) and three-dimensional surface plot (right) of the estimated error η_T^* associated with the SD solution to Example 3.1.2 using a uniform 64×64 grid (top), and a stretched 64×64 grid (bottom).

like $O(h)$ as $h \to 0$. The distribution of the estimated error η_T^* on the uniform 64×64 grid, presented in the top of Figure 3.12, clearly suggests that use of a non-uniform set of points in the x-direction would yield a more even error distribution. This suggestion is reinforced by the results shown in Table 3.7 for a stretched grid.[4] Note that the error distribution for the stretched grid, shown at the bottom of Figure 3.12, is associated with the "reference" solution illustrated in Figure 3.3. The use of Shishkin grids for this problem is explored further in Computational Exercise 3.6.

Overestimation of the local error can still lead to some inefficiencies when using the estimator η_T to drive an adaptive refinement process. The problem in

---

[4]The grid is obtained by "stretching" near the characteristic boundaries. Starting from a point x_1 of distance d from $x = -1$, grid points near the left endpoint are constructed as $x_{k+1} = x_k + d\alpha^k$ where $\alpha > 1$ is a stretching factor. Points near the right endpoint $x = 1$ are defined in a similar manner, and an interior subinterval contains a uniform subgrid. This is implemented in the IFISS routine subint.

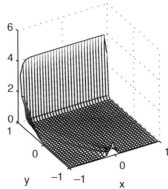

FIG. 3.13. Contour plot (left) and three-dimensional surface plot (right) of the
estimated error associated with the finite element solution to Example 3.1.3
given in Figure 3.10.

Example 3.1.3 is a perfect example of this. The issue is that the solution has both
an internal layer and an exponential boundary layer. Thus, computing a solution
on a coarse grid like that in Figure 3.10, the estimated error is excessively large
in elements that delimit the unresolved exponential layer. This is illustrated in
Figure 3.13, where the unresolved internal layer is barely visible on the surface
plot of the estimated error. This means that any adaptive refinement driven by
the estimated error will be localized to the exponential layer. However, if the
adaptive process is repeated, h_T/ϵ will eventually be close to unity in elements
defining the boundary layer, and subsequent refinement will home in on the
internal layer as required.

In the context of the incompressible fluid flow models discussed in later
chapters, exponential boundary layers only arise when downstream boundary
conditions are inappropriately specified. In particular, a "hard" Dirichlet boundary condition on an outflow boundary should never be imposed; a zero Neumann
condition is invariably more appropriate. Adopting this philosophy for the problem in Example 3.1.3 leads to the numerical solution in Figure 3.14. Notice that,
in this case, the estimator clearly "sees" the unresolved internal layer. This issue
is explored further in Computational Exercise 3.5.

3.5 Matrix properties

When solving a PDE numerically it is obvious that the stability of the discretization approach will determine the quality of the discrete solution. What is perhaps
less obvious (and also less generally appreciated) is that discretization profoundly
affects the properties of the algebraic systems obtained, and these issues in turn
have an impact on the performance and costs of solution algorithms for solving the systems. In this section, we discuss some properties of the algebraic

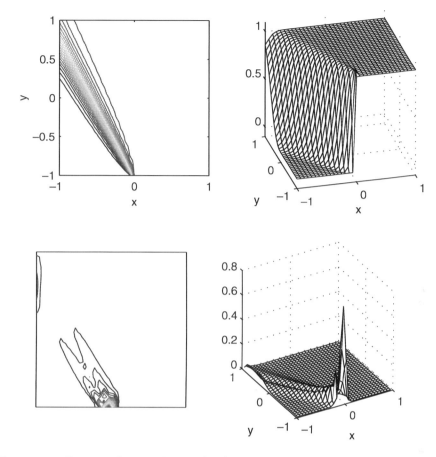

FIG. 3.14. Computed SD solution (top) and the associated estimated error (bottom) for Example 3.1.3 with a zero Neumann condition on the outflow boundary, for $\epsilon = 1/200$, and a 32×32 square grid. (These plots should be contrasted with Figures 3.10 and 3.13.)

systems generated by the finite element discretizations (3.21) and (3.40) of the convection–diffusion equation. Additional insight is then provided using Fourier analysis. This facilitates an understanding of the qualitative, oscillatory, form of discrete solutions that are a fact of life when living in a convection-dominated world. Algorithms for solving the associated matrix problems are considered in Chapter 4.

Let the basis for S_0^h be denoted by $\{\phi_j\}_{j=1}^n$ and let S_E^h be defined by augmenting this basis with $\{\phi_j\}_{j=n+1}^{n+n_\partial}$, so that the discrete solution u_h is expanded as $u_h = \sum_{j=1}^{n+n_\partial} \mathbf{u}_j \phi_j$, as in (1.18). The Galerkin system (3.21) is then equivalent to the problem

Find $\{\mathbf{u}_j\}_{j=1}^n$ such that
$$\sum_{j=1}^n a(\phi_j, \phi_i)\, \mathbf{u}_j = \hat{\ell}(\phi_i), \quad i = 1, \ldots, n,$$
where
$$\hat{\ell}(\phi_i) = \int_\Omega f\phi_i + \int_{\partial\Omega_N} g_N \phi_i$$
$$-\sum_{j=n+1}^{n+n_\partial} \left(\epsilon \int_\Omega \nabla\phi_j \cdot \nabla\phi_i + \int_\Omega (\vec{w} \cdot \nabla\phi_j)\, \phi_i \right) \mathbf{u}_j. \qquad (3.67)$$

This system can be written in matrix notation as
$$F\mathbf{u} = \mathbf{f}, \qquad (3.68)$$
where
$$F = [f_{ij}], \quad f_{ij} = a(\phi_j, \phi_i), \quad \mathbf{f}_i = \hat{\ell}(\phi_i).$$

The coefficient matrix F is the sum
$$F = \epsilon A + N, \qquad (3.69)$$
where
$$A = [a_{ij}], \quad a_{ij} = \int_\Omega \nabla\phi_j \cdot \nabla\phi_i,$$
$$N = [n_{ij}], \quad n_{ij} = \int_\Omega (\vec{w} \cdot \nabla\phi_j)\, \phi_i.$$

are discrete diffusion and convection operators, respectively. Consider the connection between these matrices and the bilinear forms discussed in Section 3.2. The form $a(u_h, u_h)$ satisfies
$$a(u_h, u_h) = \sum_{i=1}^n \left[\sum_{j=1}^n a(\phi_j, \phi_i)\mathbf{u}_j \right] \mathbf{u}_i$$
$$= \sum_{i=1}^n (F\mathbf{u})_i \mathbf{u}_i$$
$$= \langle F\mathbf{u}, \mathbf{u} \rangle.$$

Combining this with (3.69) gives
$$a(u_h, u_h) = \epsilon \langle A\mathbf{u}, \mathbf{u} \rangle + \langle N\mathbf{u}, \mathbf{u} \rangle.$$

For a better understanding of this relation, note that $\langle A\mathbf{u}, \mathbf{u}\rangle = \|\nabla u_h\|^2$ and recall (3.14), which shows the connection between the boundary conditions and the convection operator. In matrix notation, the discrete version of equation (3.14) has the form

$$H = N + N^T,$$

where

$$H = [h_{ij}], \quad h_{ij} = \int_{\partial\Omega_N} \phi_j \phi_i \vec{w} \cdot \vec{n}.$$

This leads to

$$\langle N\mathbf{u}, \mathbf{u}\rangle = -\langle N^T\mathbf{u}, \mathbf{u}\rangle + \langle H\mathbf{u}, \mathbf{u}\rangle = -\langle \mathbf{u}, N\mathbf{u}\rangle + \langle H\mathbf{u}, \mathbf{u}\rangle.$$

Since $\mathbf{u}$ is real, it follows that $\langle N\mathbf{u}, \mathbf{u}\rangle = \frac{1}{2}\langle H\mathbf{u}, \mathbf{u}\rangle$, and therefore

$$a(u_h, u_h) = \epsilon\langle A\mathbf{u}, \mathbf{u}\rangle + \frac{1}{2}\langle H\mathbf{u}, \mathbf{u}\rangle. \tag{3.70}$$

For Dirichlet problems, $H = 0$, and therefore (3.70) shows again that the coercivity constant is ϵ. More generally,

$$\frac{a(u_h, u_h)}{\|\nabla u_h\|^2} = \epsilon + \frac{\langle H\mathbf{u}, \mathbf{u}\rangle}{\langle A\mathbf{u}, \mathbf{u}\rangle}.$$

If there are Neumann conditions on part of $\partial\Omega$, then coercivity is degraded unless H is positive semi-definite, that is, unless the Neumann boundary conditions apply only on the outflow and characteristic boundaries. This restriction is also reasonable from a physical point of view; a Dirichlet condition at the inflow boundary corresponds to prescribed values of a quantity flowing into the domain.

For the streamline diffusion discretization, the discrete problem is

Find $\{\mathbf{u}_j\}_{j=1}^n$ such that

$$\sum_{j=1}^n a_{sd}(\phi_j, \phi_i)\, \mathbf{u}_j = \hat{\ell}_{sd}(\phi_i), \quad i = 1, \ldots, n,$$

where

$$\hat{\ell}_{sd}(\phi_i) = \hat{\ell}(\phi) - \sum_{j=n+1}^{n+n_\partial} \left(\sum_{k'} \delta_{k'} \int_{\triangle_{k'}} (\vec{w} \cdot \nabla\phi_j)(\vec{w} \cdot \nabla\phi_i) \right) \mathbf{u}_j.$$

Here $\hat{\ell}$ is as defined in (3.67), and the second sum is over elements $\triangle_{k'}$ with boundaries that intersect with $\partial\Omega_D$. In the case of low order (linear or bilinear) elements, the coefficient matrix for the streamline diffusion discretization has the form

$$F = \epsilon A + N + S. \tag{3.71}$$

The additional matrix S is given by

$$S = [s_{ij}], \quad s_{ij} = \sum_k \delta_k \int_{\triangle_k} (\vec{w} \cdot \nabla \phi_j)(\vec{w} \cdot \nabla \phi_i),$$

which can be viewed as a discrete diffusion operator associated with the streamline direction defined by $\vec{w}$. It is symmetric and positive semi-definite, so that it enhances coercivity.

3.5.1 Computational molecules and Fourier analysis

An illuminating exercise is to study the discrete operators obtained on uniform grids from constant coefficient problems ($\vec{w}$ constant in (3.1)) by viewing them as difference operators applied to the discrete vector of unknowns $\mathbf{u}$. A convenient notational device for this purpose is the *computational molecule* of the operator at an interior grid point. For example, the following picture of the matrix F

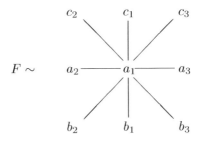

should be interpreted to mean that the value of $F\mathbf{u}$ at an interior mesh point with indices (i, j) is

$$[F\mathbf{u}]_{i,j} = c_2 \mathbf{u}_{i-1,j+1} + c_1 \mathbf{u}_{i,j+1} + c_3 \mathbf{u}_{i+1,j+1}$$
$$+ a_2 \mathbf{u}_{i-1,j} + a_1 \mathbf{u}_{i,j} + a_3 \mathbf{u}_{i+1,j}$$
$$+ b_2 \mathbf{u}_{i-1,j-1} + b_1 \mathbf{u}_{i,j-1} + b_3 \mathbf{u}_{i+1,j-1}.$$

Given a constant wind $\vec{w} = (w_x, w_y)$, the computational molecules for the matrices A (discrete diffusion), N (discrete convection), and S (streamline diffusion) defining F of (3.71) are shown in Figures 3.15 and 3.16. Figure 3.15 is for linear approximation on triangles with sides of length h, where the individual triangles are defined so that the hypotenuse is oriented along the northwest–southeast direction. Figure 3.16 shows the molecules for bilinear approximation on squares with sides of length h. Notice that the streamline diffusion parameter δ_k computed via (3.44) is the same in every element since h_k and $|\vec{w}_k|$ are independent of k. Computational molecules in the case of higher order P_2 and Q_2 approximation methods can be found in Morton [134, pp. 128ff].

To show the block structure of the coefficient matrices in this case, we further assume that the domain Ω is square, and that the boundary condition associated with our constant coefficient convection–diffusion problem is of Dirichlet

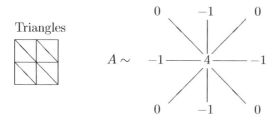

Triangles

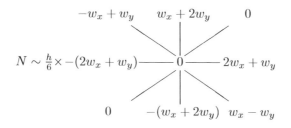

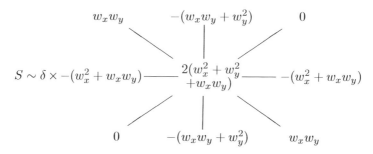

FIG. 3.15. Grid structure and computational molecules for discrete operators arising from P_1 approximation on triangles with sides of length h.

type. If the nodes associated with the P_1 or Q_1 approximation are labeled using a lexicographic ordering, see Figure 1.22, then the coefficient matrix F in (3.71) associated with the streamline diffusion discretization has a block tridiagonal form

$$F = \texttt{tridiag}(F_2, F_1, F_3) = \begin{pmatrix} F_1 & F_3 & & & 0 \\ F_2 & F_1 & F_3 & & \\ & \ddots & \ddots & \ddots & \\ & & F_2 & F_1 & F_3 \\ 0 & & & F_2 & F_1 \end{pmatrix}, \qquad (3.72)$$

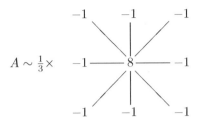

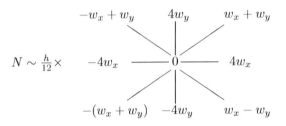

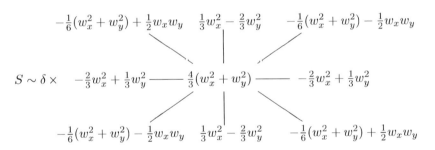

FIG. 3.16. Computational molecules for discrete operators arising from Q_1 approximation on squares with sides of length h.

where each block in (3.72) is itself a tridiagonal matrix, with constant entries along each of the diagonals. Indeed, the matrices $F_2 = \texttt{tridiag}(b_2, b_1, b_3)$, $F_1 = \texttt{tridiag}(a_2, a_1, a_3)$ and $F_3 = \texttt{tridiag}(c_2, c_1, c_3)$ are obtained by summing contributions from the molecules in Figures 3.15 or 3.16, corresponding to linear or bilinear approximation, respectively.

We observe that if the reference problem in Example 3.1.1 is discretized using Q_1 approximation over uniform grids of square elements, then the matrix F has additional structure in that the three matrices F_1, F_2 and F_3 are all *symmetric*. In this case the powerful tool of discrete Fourier analysis can be applied,[5] the basis for which is the following result.

Proposition 3.12. *The tridiagonal matrix* $T = \mathrm{tridiag}(\alpha_2, \alpha_1, \alpha_2)$ *of order* $k - 1$ *satisfies*

$$T\mathbf{v}_j = \lambda_j \mathbf{v}_j,$$

where for $1 \le j \le k - 1$, $\lambda_j = \alpha_1 + 2\alpha_2 \cos\left(\frac{j\pi}{k}\right)$, *and* $\mathbf{v}_j$ *is the discrete version of the function* $\sin(j\pi x)$ *on* $(0, 1)$,

$$\mathbf{v}_j = \sqrt{\tfrac{2}{k}} \left(\sin\left(j\tfrac{\pi}{k}\right), \sin\left(j\tfrac{2\pi}{k}\right), \dots, \sin\left(j\tfrac{(k-1)\pi}{k}\right) \right)^T.$$

Proof See Problem 3.9. □

Application of Proposition 3.12 to each of the three blocks of (3.72) gives

$$F_1 \mathbf{v}_j = \lambda_j \mathbf{v}_j, \quad \lambda_j = a_1 + 2a_2 \cos \frac{j\pi}{k},$$
$$F_2 \mathbf{v}_j = \theta_j \mathbf{v}_j, \quad \sigma_j = b_1 + 2b_2 \cos \frac{j\pi}{k}, \qquad (3.73)$$
$$F_3 \mathbf{v}_j = \sigma_j \mathbf{v}_j, \quad \theta_j = c_1 + 2c_2 \cos \frac{j\pi}{k}.$$

Let $V = [\mathbf{v}_1, \mathbf{v}_2, \dots, \mathbf{v}_{k-1}]$ denote the matrix of associated eigenvectors with $\mathbf{v}_j$ in its jth column, and let Λ, Σ and Θ denote the diagonal matrices with jth entries λ_j, σ_j and θ_j, respectively. Relation (3.73) can be written in terms of matrices as

$$F_1 V = V\Lambda, \quad F_2 V = V\Sigma, \quad F_3 V = V\Theta.$$

If $\mathcal{V}$ denotes the block diagonal matrix with diagonal entries V, it follows that

$$\mathcal{V}^T F \mathcal{V} = \begin{pmatrix} \Lambda & \Sigma & & & 0 \\ \Theta & \Lambda & \Sigma & & \\ & \ddots & \ddots & \ddots & \\ & & \Theta & \Lambda & \Sigma \\ 0 & & & \Theta & \Lambda \end{pmatrix}, \qquad (3.74)$$

so that each block is now a diagonal matrix. A simple reordering of the rows and columns of the matrix $\mathcal{V}^T F \mathcal{V}$ then gives the following result.

Proposition 3.13. *Let F be the block tridiagonal matrix* (3.72) *having symmetric tridigonal blocks of dimension* $k - 1$. *If* $\mathcal{V} = \mathrm{diag}(V, \dots, V)$ *where* V *is the*

---

[5]The technique is also applicable if a Neumann boundary condition holds on the outflow boundary.

associated matrix of discrete sine functions, then a permutation matrix $\tilde{P}$ can be constructed such that

$$\tilde{P}^T \mathcal{V}^T F \mathcal{V} \tilde{P} = \begin{pmatrix} \hat{T}_1 & & & 0 \\ & \hat{T}_2 & & \\ & & \ddots & \\ 0 & & & \hat{T}_{k-1} \end{pmatrix} =: \hat{T},$$

where $\hat{T}_j = \texttt{tridiag}(\theta_j, \lambda_j, \sigma_j)$ has coefficients (3.73) corresponding to the eigenvalues of F_2, F_1 and F_3 respectively.

Proof See Problem 3.10.   □

Fourier analysis, and, in particular, Proposition 3.13 will be used in the next section to explore why the Galerkin solution is prone to oscillations. It will also be useful in Chapter 4 when we study convergence properties of iterative solution algorithms.

3.5.2 Analysis of difference equations

The error analysis of Section 3.4 does not provide insight into the qualitative, oscillatory form displayed by discrete solutions of convection–diffusion problems. For one-dimensional problems, this phenomenon can be understood by viewing the discrete problem as a three-term recurrence relation; see Problems 3.3 and 3.7 for the details. This section outlines an analysis of the bilinear finite element discretization that generalizes this approach to higher dimensions. The methodology is limited to problems with constant wind $\vec{w}$ on uniform grids aligned with the flow, but for such problems it gives a clear picture of the oscillatory nature of the discrete Galerkin solution and insight into the corrective effects of streamline diffusion. It also provides an alternative motivation for the choice of parameter δ^* of (3.44).

The setting here then is the convection–diffusion equation (3.1) on the unit square $\Omega = \Omega_\square \equiv (0,1) \times (0,1)$, with a vertical wind $\vec{w} = (0,1)$, $f = 0$, and Dirichlet boundary conditions $u = g$. We also assume here that the boundary values along the characteristic boundaries, $g(0,y)$ on the left and $g(1,y)$ on the right (including the corners at $y = 0$ and $y = 1$), are constant and equal with value g_c. The discretization is on a uniform mesh of k^2 square elements, producing an algebraic system of equations $F\mathbf{u} = \mathbf{f}$ with $(k-1)^2$ unknowns associated with mesh points in the interior of $\Omega_\square$. As shown in Proposition 3.13, the linear system can be transformed via a similarity transformation into

$$\hat{T}\mathbf{y} = \hat{\mathbf{f}}, \tag{3.75}$$

where $\hat{T} = (\mathcal{V}\tilde{P})^T F (\mathcal{V}\tilde{P})$ is a block diagonal matrix with tridiagonal blocks, and $\hat{\mathbf{f}} = (\mathcal{V}\tilde{P})^T \mathbf{f}$.

The vector $\mathbf{y}$ can be viewed as consisting of a set of independent vertical slices, so that the transformed problem (3.75) can be analyzed with the tools used to study one-dimensional problems. There are $k - 1$ independent subproblems $\hat{T}_i \mathbf{y}_i = \hat{\mathbf{f}}_i$, each of which consists of a three-term recurrence (in m)

$$\theta_i y_{i,m-1} + \lambda_i y_{im} + \sigma_i y_{i,m+1} = \hat{f}_{im} . \tag{3.76}$$

The characteristic roots associated with this recurrence are then

$$\mu_1 = \mu_1(i) = \frac{-\lambda_i + \sqrt{\lambda_i^2 - 4\sigma_i\theta_i}}{2\sigma_i}, \quad \mu_2 = \mu_2(i) = \frac{-\lambda_i - \sqrt{\lambda_i^2 - 4\sigma_i\theta_i}}{2\sigma_i} . \tag{3.77}$$

Using these observations, it is possible to show that $\mathbf{y}$ has special structure. The derivation is rather technical and we omit it; details can be found in Elman & Ramage [67], [68].

Lemma 3.14. *The solution to the recurrence* (3.76) *has the form*

$$y_{im} = F_3(i) + [F_1(i) - F_3(i)] G_1(i, m) + [F_2(i) - F_3(i)] G_2(i, m), \tag{3.78}$$

where

$$G_1(i, m) = \frac{\mu_1^m - \mu_2^m}{\mu_1^k - \mu_2^k}, \quad G_2(i, m) = (1 - \mu_1^m) - (1 - \mu_1^k) \left(\frac{\mu_1^m - \mu_2^m}{\mu_1^k - \mu_2^k} \right), \tag{3.79}$$

and

$$F_1(i) = \sqrt{\frac{2}{k}} \sum_{p=1}^{k-1} g(x_p, 1) \sin \frac{ip\pi}{k}, \quad F_2(i) = g_c \sqrt{\frac{2}{k}} \sum_{p=1}^{k-1} \sin \frac{ip\pi}{k},$$

$$F_3(i) = \sqrt{\frac{2}{k}} \sum_{p=1}^{k-1} g(x_p, 0) \sin \frac{ip\pi}{k} . \tag{3.80}$$

The significance of this result is in showing the effect of boundary conditions on $\mathbf{y}$. The functions $F_1(i)$ and $F_3(i)$ derive from the outflow (top) and inflow (bottom) boundary conditions, respectively; except in the case where these are identical, the values of G_1 will have an impact on the structure of $\mathbf{y}$. Similarly, $F_2(i)$ derives from the boundary conditions along the characteristic (side) boundaries of $\Omega_\square$; except in the special case where the values at the inflow are identical to the constant value given on the sides, $F_2(i)$ will be different from $F_3(i)$, and G_2 will contribute to $\mathbf{y}$.

Next, we explore in detail the structure of the components G_1 and G_2 that determine $\mathbf{y}$. Consider the computational molecule for the $\boldsymbol{Q}_1$ Galerkin discretization:

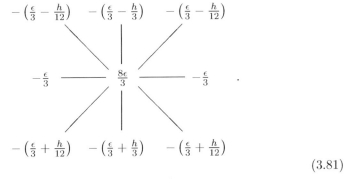

$$(3.81)$$

Substituting the quantities appearing here into (3.73) and (3.77) gives the characteristic roots

$$\mu_{1,2} = \frac{-\left(\dfrac{4-c_i}{2+c_i}\right)\dfrac{1}{\mathcal{P}_h} \pm \sqrt{1 + \dfrac{3(5+c_i)(1-c_i)}{(2+c_i)^2}\dfrac{1}{\mathcal{P}_h^2}}}{1 - \left(\dfrac{1+2c_i}{2+c_i}\right)\dfrac{1}{\mathcal{P}_h}}, \qquad (3.82)$$

where the mesh Peclet number is $\mathcal{P}_h = h/2\epsilon$, and $c_i = \cos(i\pi/k)$. This leads to statements about oscillations of the components of $\mathbf{y}$.

Lemma 3.15. *Suppose the convection–diffusion equation (3.1) with $\vec{w} = (0,1)$ and Dirichlet boundary conditions is discretized using $\mathbf{Q}_1$ Galerkin approximation on a uniform $k \times k$ grid. For any $\mathcal{P}_h$, if $\frac{2}{3}k < i < k-1$, then the functions $G_1(i,m)$ and $G_2(i,m)$ are oscillatory functions of m.*

Proof The characteristic roots satisfy $|\mu_1/\mu_2| < 1$, so that $G_1(i,m)$ can be expressed as

$$G_1(i,m) = \left(\frac{(\mu_1/\mu_2)^m - 1}{(\mu_1/\mu_2)^k - 1}\right)\mu_2^{m-k} = \chi\,\mu_2^m, \qquad (3.83)$$

where $\chi = \chi(i,m)$ has one sign for all i, m. This means that $G_1(i,m)$ alternates in sign as m varies if and only if μ_2 is negative. The numerator of μ_2 is always negative, so that oscillations will occur only when its denominator is positive. Let

$$\phi_i = \frac{1+2c_i}{2+c_i}, \qquad i = 1,\dots,k-1.$$

If $\mathcal{P}_h < \phi_i$, then $\mu_2 > 0$ and $G_1(i,m)$ is not oscillatory. On the other hand, if $\mathcal{P}_h > \phi_i$, then $\mu_2 < 0$ and $G_1(i,m)$ is oscillatory. But $\mathcal{P}_h > 0$ by definition, and $\phi_i < 0$ for $2/3\,k < i \le k-1$, so that $\mathcal{P}_h > \phi_i$ for these values of i independent of the value of $\mathcal{P}_h$. That is, $G_1(i,m)$ is an oscillatory function of m for any $\mathcal{P}_h$.

Next, let us turn our attention to G_2. Observe that (3.79) implies that $G_2(i, m) = (1 - \mu_1^m) - (1 - \mu_1^k)G_1(i, m)$. It can be shown that $0 < \mu_1 < 1$, so that $G_2(i, m)$ oscillates around the smooth function of m given by $1 - \mu_1^m$. $\square$

Note that $\phi_i < 1$ for all i, and the argument used in the proof shows that if $\mathcal{P}_h > 1$, then $G_1(i, m)$ is an oscillatory function of m for all i. It is also evident from (3.83) that the parity of the oscillations is independent of i. To use these observations to understand properties of the solution $\mathbf{u}$, we first consider an example with an exponential boundary layer.

Theorem 3.16. *If $\mathcal{P}_h > 1$, then for certain choices of boundary conditions on $\partial\Omega_\square$, the solution $\mathbf{u} = \{u_{jm}\}$ to the convection–diffusion problem of Lemma 3.15 is an oscillatory function of m.*

Proof Let the Dirichlet boundary conditions on $\partial\Omega_\square$ be given by $g(x, 1) = 1$ at the outflow boundary and $g = 0$ otherwise. This is a variant of the problem in Example 3.1.1, differing only in the boundary conditions along the vertical sides. With these boundary conditions, $F_2(i) = F_3(i) = 0$ in (3.80), so that only G_1 figures in (3.78). We have the relation $\mathbf{u} = \mathcal{V}\hat{P}\mathbf{y}$, or in more detail,

$$\mathbf{u}_{jm} = \sqrt{\frac{2}{k}} \sum_{i=1}^{k-1} \frac{ij\pi}{k} \mathbf{y}_{im}.$$

Combining this with (3.80), we can express $\mathbf{u}$ as

$$\mathbf{u}_{jm} = \sum_{i=1}^{k-1} d_{ij} G_1(i, m),$$

where

$$d_{ij} = \frac{2}{k} \sin\frac{ij\pi}{k} \sum_{p=1}^{k-1} \sin\frac{ip\pi}{k}.$$

The trigonometric identity (see Spiegel [182, 19.40])

$$\sum_{p=1}^{k-1} \sin\frac{ip\pi}{k} = \left(\frac{\sin(i\pi/2)}{\sin(i\pi/(2k))}\right) \sin\frac{(k-1)i\pi}{2k},$$

leads to the simplified expression for the coefficients d_{ij}

$$d_{ij} = \left(\sin^2\frac{i\pi}{2}\right)\left(\sin\frac{ij\pi}{k}\right)\left(\cot\frac{i\pi}{2k}\right).$$

This expression is particularly simple when $j = 1$: $d_{i1} = 2\cos^2(i\pi/2k)$. Thus, d_{i1} is positive when i is odd, and zero (hence non-negative) when i is even. Using this observation, we can easily understand the behavior of $\mathbf{u}$ near the left boundary of Ω. Since $\mathcal{P}_h > 1$, it follows that as a function of m, $G_1(i, m)$ oscillates about zero with the same parity for each i. Therefore, so does $\mathbf{u}_{1m}$.   □

Finally, we return to the exponential boundary layer and consider what this analysis says about the streamline diffusion discretization. The computational molecule is as shown below.

$$
\begin{array}{ccc}
-\left(\frac{\epsilon}{3} - \frac{h}{12} + \frac{\delta}{6}\right) & -\left(\frac{\epsilon}{3} - \frac{h}{3} + \frac{2\delta}{3}\right) & -\left(\frac{\epsilon}{3} - \frac{h}{12} + \frac{\delta}{6}\right) \\
& & \\
-\frac{\epsilon}{3} + \frac{\delta}{3} & \frac{8\epsilon}{3} + \frac{4\delta}{3} & -\frac{\epsilon}{3} + \frac{\delta}{3} \\
& & \\
-\left(\frac{\epsilon}{3} + \frac{h}{12} + \frac{\delta}{6}\right) & -\left(\frac{\epsilon}{3} + \frac{h}{3} + \frac{2\delta}{3}\right) & -\left(\frac{\epsilon}{3} + \frac{h}{12} + \frac{\delta}{6}\right)
\end{array}
$$

$$\tag{3.84}$$

Substitution into (3.73) and (3.77) gives the characteristic roots, in this case

$$
\mu_{1,2} = \frac{-2\frac{\delta}{h} - \left(\frac{4 - c_i}{2 + c_i}\right)\frac{1}{\mathcal{P}_h} \pm \sqrt{1 + \dfrac{12\frac{\delta}{h}(1 - c_i)}{(2 + c_i)}\frac{1}{\mathcal{P}_h} + \dfrac{3(5 + c_i)(1 - c_i)}{(2 + c_i)^2}\frac{1}{\mathcal{P}_h^2}}}{-2\frac{\delta}{h} + 1 - \left(\frac{1 + 2c_i}{2 + c_i}\right)\frac{1}{\mathcal{P}_h}}.
$$

$$\tag{3.85}$$

Note that this expression reduces to (3.82) when $\delta = 0$. Since oscillations occur when the denominator (of μ_2) turns positive, it is clear that the new term $-2\delta/h$ in the denominator of (3.85) helps prevent them. The analogue of Lemma 3.15 is given below. The proof is essentially identical.

Lemma 3.17. *For* SD *discretization of the problem of Lemma 3.15, if $\mathcal{P}_h > 1$, then for each value of $i = 1, \ldots, k - 1$ there exists a parameter*

$$
\delta_i^c = \frac{h}{2}\left(1 - \left(\frac{1 + 2c_i}{2 + c_i}\right)\frac{1}{\mathcal{P}_h}\right)
$$

$$\tag{3.86}$$

such that $\delta > \delta_i^c$ implies that $G_1(i, m)$ and $G_2(i, m)$ in (3.78) are non-oscillatory functions of m.

The two extremal values

$$
\underline{\delta} = \frac{h}{2}\left(1 - \frac{1}{\mathcal{P}_h}\right), \quad \overline{\delta} = \frac{h}{2}\left(1 + \frac{1}{\mathcal{P}_h}\right)
$$

$$\tag{3.87}$$

corresponding to the limits as $n \to \infty$ of δ_1^c and δ_{k-1}^c, respectively, serve as (strict) lower and upper bounds for δ_i^c for all i. This leads to a sufficient condition guaranteeing that the discrete SD solution u_h^* is not oscillatory.

Theorem 3.18. *For the convection–diffusion equation (3.1) with $\vec{w} = (0, 1)$ and Dirichlet boundary conditions, the $\boldsymbol{Q}_1$ streamline diffusion solution u_h^* does not exhibit oscillations in the vertical direction when $\delta \geq \bar{\delta}$.*

Let us consider the ramifications of these results. First, they confirm the conventional wisdom that $\mathcal{P}_h > 1$ leads to oscillatory solutions. It is also generally believed that the converse holds, that is, if $\mathcal{P}_h < 1$ then the discrete Galerkin solution is not oscillatory. However, this assertion is not true; for the problem defined in Theorem 3.16, there are examples in which $\mathcal{P}_h < 1$ but u_h is an oscillatory function. This is largely due to the result shown in Lemma 3.15, that the components of the "transformed" solution $\mathbf{y}$ are inherently oscillatory.

Finally, note that although the choice $\delta = \bar{\delta}$ results in smooth discrete solutions, these solutions tend to be overly diffuse, with boundary layers much wider than those displayed by the exact solution. This suggests that the restriction on δ given by Theorem 3.17 is too harsh. The value $\delta = \underline{\delta}$ is the same as the smart choice in (3.44). It produces solutions that are oscillatory, but in our experience the amplitudes of these oscillations are always small and the solutions are more accurate with respect to various measures of the error than those obtained using $\bar{\delta}$. This issue is explored in Computational Exercise 3.7. It is also clear from (3.87) that the two values have the same limit of $1/2$ as $\mathcal{P}_h \to \infty$.

Discussion and bibliographical notes

The convection–diffusion equation is the subject of three comprehensive monographs: Miller, O'Riordan & Shishkin [132], Morton [134], and Roos, Stynes & Tobiska [159]. The significance of this equation is noted by Morton, who states on the first page of his book that the models represented by it "are among the most widespread in all of science [and] engineering." We have attempted to touch upon the most important features of this problem, including the different character that the flows or boundary layers may have; the quality of finite element discretization; error analysis; and properties of the algebraic systems that arise from discretization. Many other aspects of the problem can be found in the three books cited above. An excellent concise introduction to convection–diffusion problems can be found in Eriksson et al. [70, chap. 18].

3.1. The reference problems are chosen to exhibit the main issues that arise in models of convection and diffusion and in the discretization of these models. These include:

- Boundary layers of exponential (Example 3.1.1) or characteristic (Examples 3.1.2, 3.1.3, 3.1.4) type, as well as internal layers (Example 3.1.3).

- Winds $\vec{w}$ of various types, including unidirectional (Examples 3.1.1, 3.1.2, 3.1.3) and recirculating (Example 3.1.4). Constant winds are primarily of use for analysis; the vertical wind of Example 3.1.1 is particularly simple to study in conjunction with grids aligned with the flow.

3.3. The oscillatory behavior of the discrete solutions of the convection–diffusion equation in cases of large mesh Peclet number is discussed by Gresho & Lee [93] and in Gresho & Sani [91, chap. 2]. The fact that boundary conditions play a role in the quality of the solution, and in particular, the fact that problems with characteristic layers or Neumann outflow conditions are not severely affected by large mesh Peclet numbers is discussed by Hedstrom & Osterheld [102] and Segal [170].

The streamline diffusion method originated with Hughes & Brooks [107]. (An earlier approach with a slightly different character was developed by Christie et al. [43].) There is an extensive literature on this methodology and variants of it. General descriptions can be found in the books [134], [159] as well as Johnson [111, chap. 9] and Quarteroni & Valli [152, chap. 8]; the latter reference also discusses generalizations called Galerkin-Least-Squares formulations. Most presentations of this idea use the Petrov–Galerkin formulation, where the test functions are allowed to be different from the trial functions by incorporating the streamline derivative. The reasons this leads to improved stability are not transparent. This point was clarified by the work of Brezzi, Marini & Russo [35], who showed the relation between streamline upwinding and the use of implicitly defined augmented finite element spaces, as presented in Section 3.3.2. The presentation given in this section follows Brezzi [31].

Shishkin grids constitute an alternative way to handle boundary layers, using grids specially tailored to the layer. This method was developed in Shishkin [173] and is discussed in [132] and [159, sect. 2.4]. Knowledge of the layer structure is needed to define the grid.

3.4. Pointwise a priori error analysis of the streamline diffusion finite element method is given in Johnson, Schatz & Wahlbin [114], and in Zhou & Rannacher [222]. Analysis of the residual free bubble formulation can be found in Brezzi, Marini & Suli [36] and in Brezzi et al. [34]. Conditions for uniform convergence with respect to ϵ are given in Stynes & Tobiska [188]. A posteriori error estimation for convection–diffusion problems was pioneered by Verfürth [205].

3.5. The discrete convection–diffusion equation has been used as an "archetypical" nonsymmetric system, in much the same way as the discrete Poisson equation is regarded as the "definitive" symmetric positive-definite system. We note in passing that some of the discussion is specialized to bilinear approximation, and that the matrices arising from linear approximation have slightly different properties (e.g. the Fourier analysis of Section 3.5.1 is not applicable). Our experience with solvers of the type discussed in the next chapter suggests that the performance characteristics are essentially independent of the approximation used.

Problems

3.1. Consider the one-dimensional convection–diffusion equation

$$-\epsilon u'' + wu' = 1 \text{ in } (0, L), \quad u(0) = u(L) = 0,$$

where ϵ and w are positive constants. Calculate the analytic solution $u(x)$, and compare it to the reduced problem solution $\hat{u} = x/w$ in the limit $2\epsilon/(wL) \to 0$. Show also that $\frac{du}{dx}(1) \to -\infty$ as $\epsilon \to 0$. Notice that $\frac{d\hat{u}}{dx}(1) = 1/w$.

3.2. Show that if the wind $\vec{w}$ is specified using the stream function as in (3.8) and (3.9), then the parameterized curves $\vec{c}(s)$ such that $\frac{d\vec{c}}{ds} = \vec{w}$ are parallel to the level curves of ψ.

3.3. Consider the one-dimensional convection–diffusion equation

$$-\epsilon u'' + u' = 0 \text{ in } (0, 1), \quad u(0) = 1, u(1) = 0.$$

Suppose this problem is discretized by Galerkin finite elements with linear elements on a uniform mesh of width h. Derive a closed form expression for the Galerkin solution, and use it to help explain the oscillatory nature of the solution when $h/(2\epsilon) > 1$.

3.4. By showing the positivity of the solution $d_h^{(k)}$ of (3.30), show that δ_k in (3.34) arising in the derivation of the streamline diffusion method must be a strictly positive parameter.

3.5. Assume that δ in (3.41) is chosen to ensure that $\delta \leq h_T^2/\epsilon C_k^2$, where C_k is the constant in the local inverse estimate

$$\left\| \nabla^2 u_k \right\|_{\triangle_k} \leq C_k h_T^{-1} \left\| \nabla u_k \right\|_{\triangle_k}, \tag{3.88}$$

with h_T and u_k defined on element $\triangle_k$ as in (1.104). Show that

$$\delta\epsilon \left| \sum_k \int_{\triangle_k} (\nabla^2 v_h)(\mathbf{w} \cdot \nabla v_h) \right| \leq \frac{1}{2} \|v_h\|_{sd}^2,$$

where $\|\cdot\|_{sd}$ is the streamline diffusion norm (3.42), and hence deduce the uniform ellipticity estimate

$$a_{sd}(v_h, v_h) \geq \frac{1}{2} \|v_h\|_{sd}^2 \quad \text{for all } v_h \in S_0^h. \tag{3.89}$$

3.6. Show that if $d_h(x)$ the solution of the one dimensional version of (3.30) is approximated by the piecewise linear function $d_*(x)$ illustrated in Figure 3.8, then the result is an SD discretization with parameter δ^* defined by (3.44). Show also that the "standard" parameter choice $\delta^* = h/2w$ is obtained by approximating d_h by the reduced problem solution $d_* = x/w$ (and ignoring the outflow boundary condition).

3.7. Let the convection–diffusion equation in Problem 3.3 be discretized by a streamline diffusion method with δ given by (3.44). Derive a closed form expression for the discrete solution in this case and compare it to the Galerkin solution.

3.8. Prove Theorem 3.9. Hint: start using (3.17) applied to $e := u - u_h$: that is, $\epsilon \|\nabla e\|^2 \le a(e, e)$. Then, follow the proof of Theorem 1.23. The streamline diffusion term can be bounded via

$$\delta_T \|\vec{w} \cdot e_h^*\|_T \le \frac{h_T}{2 \|\vec{w}\|_T} \|\vec{w} \cdot e_h^*\|_T \le h_T \|\nabla e\|_T .$$

3.9. Prove Proposition 3.12.

3.10. Prove Proposition 3.13.

3.11. Let matrices A and N be the discrete diffusion and convection operators, respectively, defined in Section 3.5. Prove that

$$|\langle N\mathbf{u}, \mathbf{v}\rangle| \le \|\vec{w}\|_\infty L \langle A\mathbf{u}, \mathbf{u}\rangle\langle A\mathbf{v}, \mathbf{v}\rangle,$$

where L is the Poincaré constant. Use this to prove that the largest singular value of the matrix $A^{-1/2}NA^{-1/2}$ is bounded by $\|\vec{w}\|_\infty L$. Here, $A^{1/2}$ is a matrix whose square is A.

Computational exercises

Numerical solutions to a convection–diffusion problem defined on the domain $\Omega_\square$ can be computed in two steps. First, running `square_cd` (or `ref_cd` to give the option of a Neumann condition on the top boundary) with wind vector $\vec{w}$ specified in function m-file `../convection/specific_wind`. Second, running `solve_cd` with boundary data specified in `../diffusion/specific_bc`. Running `cd_testproblem` automatically sets up the data files `specific_wind` and `specific_bc` associated with the problems in Examples 3.1.1–3.1.4.

3.1. Explore the solution to the variant of Example 3.1.3 in which the wind is vertical and the "hard" outflow boundary condition $u = 0$ at $y = 1$ is replaced by the condition $\partial u / \partial n = 0$.

3.2. Explore the solution to Example 3.1.2 using Galerkin and SD discretizations on a sequence of uniform grids. Find the coarsest grid for which a non-oscillatory layer solution is obtained in each case.

3.3. Write a function that postprocesses a Q_1 solution, u_h, and computes the global error $\|\nabla(u - u_h)\|$ for the analytic test problem in Example 3.1.1. Hence, verify the results given in Table 3.1. You might like to investigate the assertion that the SD solution becomes increasingly independent of ϵ in the limit $\epsilon \to 0$ (implying that the SD solution is intrinsically stable with respect to small changes in ϵ).

3.4. Generate a set of 16×16 Shishkin grids for the problem in Example 3.1.1, associated with the values of ϵ used in Table 3.1. Hence, use the function developed in Problem 3.3 to investigate the possibility that the SD error E_h^* can be bounded independently of ϵ using Shishkin grids.

3.5. Explore the convergence rate of the estimated error η_T, using a sequence of uniform grids to solve the variant of Example 3.1.3 with the "hard" outflow boundary condition $u = 0$ at $x = -1$ and $y = 1$, replaced by the condition $\partial u/\partial n = 0$.

3.6. Write a function to generate a Shishkin grid sequence that is appropriate for the characteristic layers for the problem illustrated in Figure 3.3. (Hint: define two transition points a distance $2\sqrt{\epsilon}\log_e N$ from the characteristic boundaries, and use a uniform subdivision in the y-direction.) Tabulate the estimated errors obtained using standard SD approximation on this grid sequence, and compare the results with those in Table 3.7.

3.7. Consider the effect of the streamline diffusion parameter on the accuracy of discrete solutions. In particular, use Example 3.1.1 with $\epsilon = 1/64$ and $1/200$, and choose several values of δ between 0 and $\bar{\delta}$ including $\underline{\delta}$. For each of these, compute the discrete solution on some grid for which $\mathcal{P}_h > 1$ and investigate the dependence of the L_∞-norm error on δ.

4

SOLUTION OF DISCRETE CONVECTION–DIFFUSION PROBLEMS

As shown in Chapter 3, the coefficient matrix arising from discretization of the convection–diffusion equation is nonsymmetric. To develop iterative solution algorithms for these problems, as well as those arising in other settings such as the Navier–Stokes equations, the algorithms discussed in Chapter 2 must be adapted to handle nonsymmetric systems of linear equations. In this chapter, we outline the strategies and issues associated with Krylov subspace iteration for general nonsymmetric systems, together with specific details for convection–diffusion systems associated with preconditioning and multigrid methods.

4.1 Krylov subspace methods

We are considering iterative methods for solving a system $F\mathbf{u} = \mathbf{f}$, where, for the moment (i.e. in this section), F represents an arbitrary nonsymmetric matrix of order n. Recall that for symmetric positive-definite systems, the conjugate gradient method has two properties that make it an effective iterative solution algorithm. It is *optimal*, in the sense that at the kth step, the energy norm of the error is minimized with respect to the k-dimensional Krylov space $\mathcal{K}_k(F, \mathbf{r}^{(0)})$. (Equivalently, the error is orthogonal to $\mathcal{K}_k(F, \mathbf{r}^{(0)})$ with respect to the energy inner product.) In addition, it is *inexpensive*: the number of arithmetic operations required at each step of the iteration is independent of the iteration count k. This also means that the storage requirements are fixed. Unfortunately, there are no generalizations of CG directly applicable to arbitrary nonsymmetric systems that have both of these properties. A Krylov subspace method for nonsymmetric systems of equations can display at most one of them: it can retain optimality but allow the cost per iteration to increase as the number of iterations grows, or it can require a fixed amount of computational work at each step but sacrifice optimality.

Before discussing what can be done, we note that one way to apply Krylov subspace methods to a nonsymmetric system is to simply create a symmetric positive definite one such as that defined by the normal equations $F^T F \mathbf{u} = F^T \mathbf{f}$. The problem could then be solved by applying the conjugate gradient method to the new system. This approach clearly inherits some of the favorable features of CG. However, the Krylov subspace generated is $\mathcal{K}_k(F^T F, F^T \mathbf{r}^{(0)})$ and therefore convergence will depend on properties of $F^T F$. For example, recall Theorem 2.4, which specifies a bound that depends on the condition number of the coefficient

166

matrix. Since the condition number of $F^T F$ is the square of that of F, this suggests that using CG in this way may be less effective than when it is applied directly to symmetric positive-definite systems. In our experience with problems arising in fluid mechanics such as the convection–diffusion equation, this is indeed the case; convergence of CG applied to the normal equations is slower than alternative approaches designed to be applied directly to nonsymmetric problems.

Let us consider instead iterative methods for systems with nonsymmetric coefficient matrices that generate a basis for $\mathcal{K}_k(F, \mathbf{r}^{(0)})$. Effective strategies are derived by exploiting the connection between algorithms for estimating eigenvalues of matrices (more precisely, for constructing nearly invariant subspaces of matrices) and those for solving systems. This connection was introduced in Section 2.4, where we established a relation between the conjugate gradient method and the Lanczos method for eigenvalues: the CG iterate is a linear combination of vectors generated by the Lanczos algorithm that constitute a basis for $\mathcal{K}_k(F, \mathbf{r}^{(0)})$. Here we will show how generalizations and variants of the Lanczos method for nonsymmetric matrices can be exploited in an analogous way.

4.1.1 GMRES

Our starting point is the *generalized minimum residual method* (GMRES), defined below. This algorithm, developed by Saad & Schultz [165], represents the standard approach for constructing iterates satisfying an optimality condition. It is derived by replacing the symmetric Lanczos recurrence (2.29) with the variant for nonsymmetric matrices known as the Arnoldi algorithm.

To show how this method works, we identify its relation to the Arnoldi method for eigenvalue computation. Starting with the initial vector $\mathbf{v}^{(1)}$, the main loop (on k) of Algorithm 4.1 constructs an orthonormal basis

$$\left\{ \mathbf{v}^{(1)}, \mathbf{v}^{(2)}, \ldots, \mathbf{v}^{(k)} \right\}$$

for the Krylov space $\mathcal{K}_k(F, \mathbf{v}^{(1)})$. To make $\mathbf{v}^{(k+1)}$ orthogonal to $\mathcal{K}_k(F, \mathbf{v}^{(1)})$, it is necessary to use all previously constructed vectors $\{\mathbf{v}^{(j)}\}_{j=1}^{k}$ in the computation. The construction in Algorithm 4.1 is analogous to the modified Gram–Schmidt process for generating an orthogonal basis. Let $V_k = [\mathbf{v}^{(1)}, \mathbf{v}^{(2)}, \ldots, \mathbf{v}^{(k)}]$ denote the matrix containing $\mathbf{v}^{(j)}$ in its jth column, for $j = 1, \ldots, k$, and let $H_k = [h_{ij}]$, $1 \leq i, j \leq k$, where entries of H_k not specified in the Algorithm are zero. Thus, H_k is an upper-Hessenberg matrix (i.e. $h_{ij} = 0$ for $j < i - 1$), and

$$\begin{aligned} FV_k &= V_k H_k + h_{k+1,k} \left[0, \ldots, 0, \mathbf{v}^{(k+1)} \right] \\ H_k &= V_k^T F V_k . \end{aligned} \tag{4.1}$$

The Arnoldi method for eigenvalues is to use the eigenvalues of H_k as estimates for those of F. This technique is a generalization of the Lanczos method that is applicable to nonsymmetric matrices. When F is symmetric, H_k reduces to the tridiagonal matrix produced by the Lanczos algorithm, and (4.1) is identical to (2.30).

> **Algorithm 4.1:** THE GMRES METHOD
> Choose $\mathbf{u}^{(0)}$, compute $\mathbf{r}^{(0)} = \mathbf{f} - F\mathbf{u}^{(0)}$, $\beta_0 = \|\mathbf{r}^{(0)}\|$, $\mathbf{v}^{(1)} = \mathbf{r}^{(0)}/\beta_0$
> for $k = 1, 2, \ldots$ until $\beta_k < \tau\beta_0$ do
> $\quad \mathbf{w}_0^{(k+1)} = F\mathbf{v}^{(k)}$
> $\quad$ for $l = 1$ to k do
> $\quad\quad h_{lk} = \langle \mathbf{w}_l^{(k+1)}, \mathbf{v}^{(l)} \rangle$
> $\quad\quad \mathbf{w}_{l+1}^{(k+1)} = \mathbf{w}_l^{(k+1)} - h_{lk}\mathbf{v}^{(l)}$
> $\quad$ enddo
> $\quad h_{k+1,k} = \|\mathbf{w}_{k+1}^{(k+1)}\|$
> $\quad \mathbf{v}^{(k+1)} = \mathbf{w}_{k+1}^{(k+1)}/h_{k+1,k}$
> $\quad$ Compute $\mathbf{y}^{(k)}$ such that $\beta_k = \left\|\beta_0\mathbf{e}_1 - \widehat{H}_k\mathbf{y}^{(k)}\right\|$ is minimized,
> $\quad\quad$ where $\widehat{H}_k = [h_{ij}]_{1 \le i \le k+1, 1 \le j \le k}$
> enddo
> $\mathbf{u}^{(k)} = \mathbf{u}^{(0)} + V_k\mathbf{y}^{(k)}$

To derive a Krylov subspace iteration for solving systems of equations, we let $\mathbf{u}^{(k)} \in \mathbf{u}^{(0)} + \mathcal{K}_k(F, \mathbf{r}^{(0)})$. For the choice $\mathbf{v}^{(1)} = \mathbf{r}^{(0)}/\beta_0$, with $\beta_0 = \|\mathbf{r}^{(0)}\|$ as in Algorithm 4.1, this is equivalent to

$$\mathbf{u}^{(k)} = \mathbf{u}^{(0)} + V_k\mathbf{y}^{(k)} \tag{4.2}$$

for some k-dimensional vector $\mathbf{y}^{(k)}$. But the first line of (4.1) can be rewritten as $FV_k = V_{k+1}\widehat{H}_k$, and this implies that the residual satisfies

$$\mathbf{r}^{(k)} = \mathbf{r}^{(0)} - AV_k\mathbf{y}^{(k)} = V_{k+1}\left(\beta_0\mathbf{e}_1 - \widehat{H}_k\mathbf{y}^{(k)}\right), \tag{4.3}$$

where $\mathbf{e}_1 = (1, 0, \ldots, 0)^T$ is the unit vector of size k. The vectors $\{\mathbf{v}^{(j)}\}$ are pairwise mutually orthogonal, so that

$$\|\mathbf{r}^{(k)}\| = \beta_k = \left\|\beta_0\mathbf{e}_1 - \widehat{H}_k\mathbf{y}^{(k)}\right\|. \tag{4.4}$$

In particular, the residual of the iterate (4.2) with smallest Euclidean norm is determined by the choice of $\mathbf{y}^{(k)}$ that minimizes the expression on the right side of (4.4).

This upper-Hessenberg least squares problem can be solved by transforming $\widehat{H}_k$ into upper triangular form $\begin{pmatrix} R_k \\ 0 \end{pmatrix}$, where R_k is upper triangular, using $k + 1$ plane rotations (which are also applied to $\beta_0\mathbf{e}_1$). Here, $\widehat{H}_k$ contains $\widehat{H}_{k-1}$ as a submatrix, so that in a practical implementation, R_k can be updated from R_{k-1}. Moreover, by an analysis similar to that leading to (2.37), it can be shown that $\|\mathbf{r}^{(k)}\|$ is available at essentially no cost. Hence, a step of the GMRES algorithm consists of constructing a new Arnoldi vector $\mathbf{v}^{(k+1)}$, determining the residual norm of the iterate $\mathbf{r}^{(k)}$ that would be obtained from $\mathcal{K}_k(F, \mathbf{r}^{(0)})$, and then either constructing $\mathbf{u}^{(k)}$ if the stopping criterion is satisfied, or proceeding to the next step otherwise.

By construction, the iterate $\mathbf{u}^{(k)}$ generated by the GMRES method is the member of the translated Krylov space

$$\mathbf{u}^{(0)} + \mathcal{K}_k(F, \mathbf{r}^{(0)})$$

for which the Euclidean norm of the residual vector is minimal. That is,

$$\|\mathbf{r}^{(k)}\| = \min_{p_k \in \Pi_k, \, p_k(0)=1} \|p_k(F)\mathbf{r}^{(0)}\|. \tag{4.5}$$

As in the analysis of the CG method, the Cayley–Hamilton theorem implies that the exact solution is obtained in at most n steps. Bounds on the norm of the residuals associated with the GMRES iterates are derived from the optimality condition.

Theorem 4.1. *Let $\mathbf{u}^{(k)}$ denote the iterate generated after k steps of GMRES iteration, with residual $\mathbf{r}^{(k)}$. If F is diagonalizable, that is, $F = V\Lambda V^{-1}$ where Λ is the diagonal matrix of eigenvalues of F, and V is the matrix whose columns are the eigenvectors, then*

$$\frac{\|\mathbf{r}^{(k)}\|}{\|\mathbf{r}^{(0)}\|} \leq \kappa(V) \min_{p_k \in \Pi_k, \, p_k(0)=1} \max_{\lambda_j} |p_k(\lambda_j)|, \tag{4.6}$$

where $\kappa(V) = \|V\| \, \|V^{-1}\|$ is the condition number of V. If, in addition, $\mathcal{E}$ is any set that contains the eigenvalues of F, then

$$\frac{\|\mathbf{r}^{(k)}\|}{\|\mathbf{r}^{(0)}\|} \leq \kappa(V) \min_{p_k \in \Pi_k, \, p_k(0)=1} \max_{\lambda \in \mathcal{E}} |p_k(\lambda)|. \tag{4.7}$$

Proof Assertion (4.6) is derived from the observations that, for any polynomial p_k,

$$\begin{aligned}
\|p_k(F)\mathbf{r}^{(0)}\| &= \|Vp_k(\Lambda)V^{-1}\mathbf{r}^{(0)}\| \\
&\leq \|V\| \, \|V^{-1}\| \, \|p_k(\Lambda)\| \, \|\mathbf{r}^{(0)}\| \\
&\leq \|V\| \, \|V^{-1}\| \max_{\lambda_j} |p_k(\lambda_j)| \, \|\mathbf{r}^{(0)}\|.
\end{aligned}$$

The bound (4.7) is an immediate consequence of (4.6). $\qquad\square$

These "minimax" bounds generalize the analogous results (2.11) and (2.12) for the conjugate gradient method. There are, however, two significant differences. First, there is the presence of the condition number $\kappa(V)$ of the matrix of eigenvectors. It is difficult to bound this quantity, but its presence is unavoidable for polynomial bounds entailing the eigenvalues of F. Second, it is more difficult to derive an error bound for the GMRES iterates in a form that is as clean as

Theorem 2.4 for CG. This is partly due to the factor $\kappa(V)$, but it also depends on the need for bounds from approximation theory for $\max_\lambda |p_k(\lambda)|$.

The results of Theorem 4.1 can be used to gain insight into the convergence behavior of GMRES by taking the kth root of (either of) the bounds. In particular,

$$
\left(\frac{\|r^{(k)}\|}{\|r^{(0)}\|}\right)^{1/k} \leq \kappa(V)^{1/k} \left(\min_{p_k \in \Pi_k,\, p_k(0)=1} \max_{\lambda_j} |p_k(\lambda_j)|\right)^{1/k}.
$$

$\kappa(V)$ does not depend on k, and it therefore follows that $\kappa(V)^{1/k} \to 1$ as k increases. This suggests consideration of the limit

$$
\rho := \lim_{k \to \infty} \left(\min_{p_k \in \Pi_k,\, p_k(0)=1} \max_{\lambda_j} |p_k(\lambda_j)|\right)^{1/k}. \tag{4.8}
$$

Since GMRES constructs the exact solution in a finite number of steps, this does not lead to a simple statement about the error at any given step of the computation. However, it does give insight into the asymptotic behavior for large enough k: as the iteration proceeds, it can be expected that the norm of the residual will be reduced by a factor roughly equal to ρ at each step. We refer to ρ of (4.8) as the *asymptotic convergence factor* of the GMRES iteration. It is an interesting fact that asymptotic estimates for large k are often descriptive of observed convergence behavior for $k \ll n$. It is rarely the case that n iterations are necessary for an accurate solution to be obtained.

For later analysis, we mention that for a simple (stationary) iteration (as in (2.26)) with iteration matrix $T = M^{-1}R$, the norms of successive error vectors will asymptotically (for large numbers of iterations) reduce by a factor which is simply the eigenvalue of T of maximum modulus. We will therefore denote by $\rho(T)$ the eigenvalue of T of maximum modulus since this reflects the ultimate rate of convergence for a simple iteration as does ρ defined above for GMRES iteration.

Returning to GMRES, a bound on ρ can be obtained using the fact that any polynomial $\chi_k \in \Pi_k$ with $\chi_k(0) = 1$ satisfies, for any set $\mathcal{E}$ that contains the eigenvalues of F,

$$
\min_{p_k \in \Pi_k,\, p_k(0)=1} \max_{\lambda \in \mathcal{E}} |p_k(\lambda)| \leq \max_{\lambda \in \mathcal{E}} |\chi_k(\lambda)|.
$$

This was used in Theorem 2.4 to construct a bound on the error for the CG method, where χ_k was taken to be a scaled and translated Chebyshev polynomial. The same approach can be used to derive a bound on the asymptotic convergence factor for the GMRES method when the enclosing set $\mathcal{E}$ is an ellipse in the complex plane.

Theorem 4.2. *Suppose F is diagonalizable and its eigenvalues all lie in an ellipse $\mathcal{E}$ with center c, foci $c \pm d$ and semi-major axis a, and $\mathcal{E}$ does not contain the origin. Then the asymptotic convergence factor for GMRES iteration is* *bounded as*

$$
\rho \leq \left[\frac{\left(\frac{a}{d} + \sqrt{\left(\frac{a}{d} \right)^2 - 1} \right)^k + \left(\frac{a}{d} + \sqrt{\left(\frac{a}{d} \right)^2 - 1} \right)^{-k}}{\left(\frac{c}{d} + \sqrt{\left(\frac{c}{d} \right)^2 - 1} \right)^k + \left(\frac{c}{d} + \sqrt{\left(\frac{c}{d} \right)^2 - 1} \right)^{-k}} \right]^{1/k} \approx \frac{a + \sqrt{a^2 - d^2}}{c + \sqrt{c^2 - d^2}}.
$$

Proof The proof uses the particular choice

$$
\chi_k(\lambda) = \frac{\tau_k \left(\frac{c-\lambda}{d} \right)}{\tau_k \left(\frac{c}{d} \right)},
$$

where $\tau_k(\zeta) = \cosh(k \cosh^{-1} \zeta)$ is the Chebyshev polynomial defined for complex ζ. The result is then obtained by bounding the maximum of $|\chi_k(\lambda)|$ for $\lambda \in \mathcal{E}$. A complete derivation of this bound can be found in Saad [164, pp. 188ff]. □

As we have observed, this analysis does not provide information on any specific step. The situation is somewhat better when the symmetric part of the coefficient matrix is positive-definite. Let F be written as $F = H + S$ where

$$
H = \frac{F + F^T}{2}, \quad S = \frac{F - F^T}{2}
$$

are the symmetric part and skew-symmetric part of F, respectively.[1] When H is positive-definite, it is possible to derive a more specific bound on the norm of the residual.

Proposition 4.3. *If the symmetric part of the coefficient matrix is positive-definite, then the residuals of the iterates generated by the* GMRES *algorithm satisfy*

$$
\frac{\|\mathbf{r}^{(k)}\|}{\|\mathbf{r}^{(0)}\|} \leq \left(1 - \frac{\lambda_{\min}(H)^2}{\lambda_{\min}(H)\lambda_{\max}(H) + \rho(S)^2} \right)^{k/2}, \tag{4.9}
$$

where $\rho(S)$ denotes the eigenvalue of S with maximum modulus.

Proof See Problem 4.2. □

The utility of the method derives in part from the fact that the solution is computed only after the stopping criterion is satisfied. This makes the computational work and storage less than what would be needed to update the solution whenever the Krylov subspace is augmented. (It also depends on the assumption that the stopping criterion uses the Euclidean norm of the residual.) As we have noted, however, there is a cost associated with the optimality property (4.5).

---

[1] An identical analysis also applies for complex matrices, where the Hermitian adjoint F^* is used in place of the transpose, and H represents the *Hermitian part* of F.

The work and storage requirements of GMRES grow like $O(kn)$, and these costs may become prohibitive for large k. One way to deal with this issue is to *restart* the GMRES algorithm when k becomes large. That is, some upper bound m on the dimension of the Krylov subspace is specified, and if the stopping criterion is not satisfied for $k \leq m$, then Algorithm 4.1 is stopped, $\mathbf{u}^{(m)}$ is constructed, and the iteration is restarted with $\mathbf{u}^{(m)}$ used in place of $\mathbf{u}^{(0)}$ as the new initial iterate. This approach is referred to as the GMRES(m) method. By design, it avoids the large costs of a full GMRES iteration, and the error bound of Proposition 4.3 still holds. However, convergence may be considerably slower than that of full GMRES. In particular, if the restart takes place before the asymptotic convergence behavior of the iteration is realized, then the restarted version may never display the asymptotic behavior achievable by GMRES. The next section is concerned with methods that do not satisfy an optimality condition but have fixed computational costs at each step.

4.1.2   *Biorthogonalization methods*

An alternative approach for developing Krylov subspace solvers is to force the computational costs at each step to be fixed. This can be done by starting from the variant of the Lanczos algorithm for nonsymmetric matrices that imposes biorthogonality conditions between members of two Krylov subspaces, one space associated with F and an auxiliary space associated with F^T. Suppose $\mathbf{v}^{(1)}$ and $\mathbf{w}^{(1)}$ are two vectors satisfying $\langle \mathbf{v}^{(1)}, \mathbf{w}^{(1)} \rangle = 1$. Let $\mathbf{v}^{(0)} = \mathbf{w}^{(0)} = 0$. The Lanczos algorithm for nonsymmetric matrices is given by

$$\tilde{\mathbf{v}}^{(j+1)} = F\mathbf{v}^{(j)} - \delta_j \mathbf{v}^{(j)} - \gamma_j \mathbf{v}^{(j-1)}$$
$$\tilde{\mathbf{w}}^{(j+1)} = F^T \mathbf{w}^{(j+1)} - \delta_j \mathbf{w}^{(j)} - \eta_j \mathbf{w}^{(j-1)}, \tag{4.10}$$
$$\mathbf{v}^{(j+1)} = \eta_{j+1} \tilde{\mathbf{v}}^{(j+1)}, \quad \mathbf{w}^{(j+1)} = \gamma_{j+1} \tilde{\mathbf{w}}^{(j+1)},$$

where $\delta_j = \langle \mathbf{w}^{(j)}, F\mathbf{v}^{(j)} \rangle$, and the scalars γ_{j+1} and η_{j+1} are chosen so that $\langle \mathbf{w}^{(j+1)}, \mathbf{v}^{(j+1)} \rangle = 1$. This procedure was developed as a method for estimating the eigenvalues of F using those of the tridiagonal matrix

$$G_k = \texttt{tridiag}\,[\gamma_j, \delta_j, \eta_{j+1}].$$

The two sets of vectors $\{\mathbf{v}^{(j)}\}$ and $\{\mathbf{w}^{(j)}\}$ are *biorthogonal*, that is they satisfy $\langle \mathbf{v}^{(i)}, \mathbf{w}^{(j)} \rangle = \delta_{ij}$. (See Problem 4.3.) If F is symmetric, then $\mathbf{v}^{(j)} = \mathbf{w}^{(j)}$, G_k is symmetric, and the process reduces to (2.29).

The Lanczos vectors can be used to construct approximate solutions to linear systems in a manner analogous to the way the Arnoldi basis is used in (4.2). Let $V_k = [\mathbf{v}^{(1)}, \mathbf{v}^{(2)}, \ldots, \mathbf{v}^{(k)}]$ and $W_k = [\mathbf{w}^{(1)}, \mathbf{w}^{(2)}, \ldots, \mathbf{w}^{(k)}]$. Given a starting vector $\mathbf{u}^{(0)}$, let $\mathbf{r}^{(0)} = \mathbf{f} - F\mathbf{u}^{(0)}$ denote the residual, and let $\mathbf{v}^{(1)} = \mathbf{r}^{(0)}$. Then an approximate solution is

$$\mathbf{u}^{(k)} = \mathbf{u}^{(0)} + V_k \mathbf{y}^{(k)}, \tag{4.11}$$

where $\mathbf{y}^{(k)}$ is computed so that the residual $\mathbf{r}^{(k)} = \mathbf{f} - F\mathbf{u}^{(k)}$ is orthogonal to the auxiliary set of vectors $\{\mathbf{w}^{(j)}\}$, $j = 1, \ldots, k$. Using

$$W_F^T \mathbf{r}^{(k)} = \mathbf{e}_1 - G_k \mathbf{y}^{(k)},$$

this can be done by computing the solution to the tridiagonal system of equations $G_k \mathbf{y}^{(k)} = \mathbf{e}_1$. The *Lanczos method* for nonsymmetric systems of equations constructs the two sets of biorthogonal vectors $\{\mathbf{v}^{(j)}\}$ and $\{\mathbf{w}^{(j)}\}$ using (4.10) and computes the solution using (4.11). The residual at step k satisfies

$$\mathbf{r}^{(k)} = \eta_{k+1}(\mathbf{e}_k^T \mathbf{y}^{(k)})\mathbf{v}^{(k+1)}, \tag{4.12}$$

so that the norm $\|\mathbf{r}^{(k)}\|$ can be monitored inexpensively using the norm of the term on the right side of this expression.

The recurrences in (4.10) are of fixed length and short. Thus, this approach contrasts with the GMRES algorithm in that it requires a fixed amount of computational work at each iteration, and it has fixed storage requirements. Since the computational costs are relatively low, there is also interest in variants of this approach in which the approximate solution is updated at each step. One such strategy is the *biconjugate gradient method* (BICG), developed by Fletcher [76]:

> **Algorithm 4.2:** THE BICONJUGATE GRADIENT METHOD
> Choose $\mathbf{u}^{(0)}$, compute $\mathbf{r}^{(0)} = \mathbf{f} - F\mathbf{u}^{(0)}$, set $\mathbf{p}^{(0)} = \mathbf{r}^{(0)}$
> Choose $\hat{\mathbf{r}}^{(0)}$ such that $\langle \mathbf{r}^{(0)}, \hat{\mathbf{r}}^{(0)} \rangle \neq 0$
> Set $\hat{\mathbf{p}}^{(0)} = \hat{\mathbf{r}}^{(0)}$
> for $k = 0$ until convergence do
> $\quad \alpha_k = \langle \hat{\mathbf{r}}^{(k)}, \mathbf{r}^{(k)} \rangle / \langle \hat{\mathbf{p}}^{(k)}, F\mathbf{p}^{(k)} \rangle$
> $\quad \mathbf{u}^{(k+1)} = \mathbf{u}^{(k)} + \alpha_k \mathbf{p}^{(k)}$
> $\quad \mathbf{r}^{(k+1)} = \mathbf{r}^{(k)} - \alpha_k F\mathbf{p}^{(k)}, \quad \hat{\mathbf{r}}^{(k+1)} = \hat{\mathbf{r}}^{(k)} - \alpha_k F^T \hat{\mathbf{p}}^{(k)}$
> $\quad \langle$Test for convergence$\rangle$
> $\quad \beta_k = \langle \hat{\mathbf{r}}^{(k+1)}, \mathbf{r}^{(k+1)} \rangle / \langle \hat{\mathbf{r}}^{(k)}, \mathbf{r}^{(k)} \rangle$
> $\quad \mathbf{p}^{(k+1)} = \mathbf{r}^{(k+1)} + \beta_k \mathbf{p}^{(k)}, \quad \hat{\mathbf{p}}^{(k+1)} = \hat{\mathbf{r}}^{(k+1)} + \beta_k \hat{\mathbf{p}}^{(k)}$
> enddo

This algorithm is related to the nonsymmetric Lanczos method in a manner analogous to the connection between CG and the symmetric Lanczos algorithm; see Section 2.4 and Problem 4.4. The benefit of the low cost per step of these methods is cancelled out to some extent by several drawbacks. In particular, the computation may *break down* and stop running for some reason before a solution is obtained. For example, in the BICG iteration, it is possible for either the numerator or the denominator of α_k to be zero, causing the iteration to stop making progress before finding the solution. In part, this is an inherent difficulty with the Lanczos process, in which it may happen that the vectors $\tilde{\mathbf{v}}^{(j+1)}$ and $\tilde{\mathbf{w}}^{(j+1)}$ are nonzero but orthogonal. In floating point arithmetic, exact orthogonality is unlikely to occur, but the vectors can become nearly orthogonal and this may cause the behavior of these methods to be erratic, that is, the error may be large during some stages of the iteration.[2] Besides these concerns about reliability, there are also issues of convenience of implementation. These methods require two matrix-vector products at each iteration, one by F and one by F^T. The latter matrix is not always easily available, and working with it, especially when preconditioning is involved, complicates the programming required for implementation. The extra matrix-vector product may also entail a nontrivial cost. Finally, there is also essentially no analysis of convergence of these solution strategies.

A considerable amount of research has been undertaken to address these deficiencies. We present a derivation of one such technique, the BICGSTAB algorithm, a *stabilized* variant of BICG developed by van der Vorst [198]. This discussion, which follows [198], is intended to give an idea of how the problem of improving the performance of Lanczos-based solvers is approached, and to specify an algorithm that is easy to implement. This method is one of many that have been proposed, and none of them is foolproof.

Since the BICG method is a Krylov subspace method, the residuals $\{\mathbf{r}^{(k)}\}$ and direction vectors $\{\mathbf{p}^{(k)}\}$ belong to the Krylov space $\mathcal{K}_k(F, \mathbf{r}^{(0)})$. This is equivalent to saying that

$$\mathbf{r}^{(k)} = \phi_k(F)\mathbf{r}^{(0)}, \quad \mathbf{p}^{(k)} = \psi_k(F)\mathbf{r}^{(0)},$$

where ϕ_k and ψ_k are polynomials of degree k. From the recurrences

$$\mathbf{r}^{(k)} = \mathbf{r}^{(k-1)} - \alpha_{k-1}F\mathbf{p}^{(k-1)}, \quad \mathbf{p}^{(k)} = \mathbf{r}^{(k)} + \beta_{k-1}\mathbf{p}^{(k-1)},$$

it follows that the polynomials satisfy the recurrences

$$\phi_k(\lambda) = \psi_{k-1}(\lambda) - \alpha_{k-1}\lambda\psi_{k-1}(\lambda), \quad \psi_k(\lambda) = \phi_k(\lambda) + \beta_{k-1}\psi_{k-1}(\lambda).$$

Let us now use this observation to construct an improved solution $\tilde{\mathbf{u}}^{(k)}$ with residual

$$\tilde{\mathbf{r}}^{(k)} = \eta_k(F)\phi_k(F)\mathbf{r}^{(0)} = \eta_k(F)\mathbf{r}^{(k)}, \tag{4.13}$$

where $\eta_k(\lambda)$ is some new polynomial to be determined. This idea is motivated by the observation that

$$\langle \hat{\mathbf{r}}^{(k)}, \mathbf{r}^{(k)} \rangle = \langle \phi_k(F^T)\hat{\mathbf{r}}^{(0)}, \phi_k(F)\mathbf{r}^{(0)} \rangle = \langle \hat{\mathbf{r}}^{(0)}, \phi_k(F)^2\mathbf{r}^{(0)} \rangle.$$

That is, the auxiliary vectors $\{\hat{\mathbf{r}}^{(k)}\}$ and $\{\hat{\mathbf{p}}^{(k)}\}$ in the BICG iteration can be viewed as producing information (implicit in the scalars $\{\alpha_k\}$ and $\{\beta_k\}$) associated with the higher degree polynomial $\phi_k(F)^2$.

---

[2]The BICG method is also somewhat more susceptible to breakdown than a method that constructs the iterate directly from the Lanczos vectors using (4.11) A similar statement applies to the conjugate gradient method. The correction of this difficulty for CG leads to the solution algorithms for symmetric indefinite systems considered in Chapter 2.

The BICGSTAB method constructs an alternative to $\phi_k(F)^2$ with the aim of producing a simple and more stable computation. Suppose the new polynomial is represented in factored form as

$$\eta_k(\lambda) = (1 - \omega_{k-1}\lambda) \cdots (1 - \omega_1\lambda)(1 - \omega_0\lambda).$$

In addition to the postulated improved residual $\tilde{\mathbf{r}}^{(k)}$ of (4.13), assume there is an alternative search direction vector

$$\tilde{\mathbf{p}}^{(k)} = \eta_k(F)\psi_k(F)\mathbf{r}^{(0)} = \eta_k(F)\mathbf{p}^{(k)}.$$

The new vectors $\tilde{\mathbf{r}}^{(k)}$ and $\tilde{\mathbf{p}}^{(k)}$ satisfy the recurrences

$$
\begin{aligned}
\tilde{\mathbf{r}}^{(k)} &= (1 - \omega_{k-1}F)\,\eta_{k-1}(F)\left[\phi_{k-1}(F) - \alpha_{k-1}F\psi_{k-1}(F)\right]\mathbf{r}^{(0)} \\
&= \left[\eta_{k-1}(F)\phi_{k-1}(F) - \alpha_{k-1}F\eta_{k-1}(F)\psi_{k-1}(F)\right]\mathbf{r}^{(0)} \\
&\quad - \omega_{k-1}F\left[\eta_{k-1}(F)\phi_{k-1}(F) - \alpha_{k-1}F\eta_{k-1}(F)\psi_{k-1}(F)\right]\mathbf{r}^{(0)} \\
&= \mathbf{s}^{(k-1)} - \omega_{k-1}F\mathbf{s}^{(k-1)}, \qquad\qquad\qquad\qquad\qquad (4.14)
\end{aligned}
$$

where $\mathbf{s}^{(k-1)} = \tilde{\mathbf{r}}^{(k-1)} - \alpha_{k-1}F\tilde{\mathbf{p}}^{(k-1)}$, and

$$
\begin{aligned}
\tilde{\mathbf{p}}^{(k)} &= (\eta_k(F)\phi_k(F) + \beta_{k-1}\psi_{k-1}(F))\,\mathbf{r}^{(0)} \\
&= \eta_k(F)\phi_k(F)\mathbf{r}^{(0)} + (1 - \omega_{k-1}F)\beta_{k-1}\eta_{k-1}(F)\psi_{k-1}(F)\mathbf{r}^{(0)} \\
&= \tilde{\mathbf{r}}^{(k)} + \beta_{k-1}\tilde{\mathbf{p}}^{(k-1)} - \omega_{k-1}F\tilde{\mathbf{p}}^{(k-1)}.
\end{aligned}
$$

Using $\mathbf{s}^{(k-1)}$, the recurrence for the residual vector $\tilde{\mathbf{r}}^{(k)}$ can be rewritten in the variant form

$$\tilde{\mathbf{r}}^{(k)} = \tilde{\mathbf{r}}^{(k-1)} - F\left(\alpha_{k-1}\tilde{\mathbf{p}}^{(k-1)} + \omega_{k-1}\mathbf{s}^{(k-1)}\right).$$

It then follows that if $\tilde{\mathbf{r}}^{(k-1)} = \mathbf{f} - F\tilde{\mathbf{u}}^{(k-1)}$ is the residual of some iterate $\tilde{\mathbf{u}}^{(k-1)}$, then the new iterate with residual $\tilde{\mathbf{r}}^{(k)}$ can be constructed as

$$\tilde{\mathbf{u}}^{(k)} = \tilde{\mathbf{u}}^{(k-1)} + \alpha_{k-1}\tilde{p}_{k-1} + \omega_{k-1}\mathbf{s}^{(k-1)}.$$

This derivation shows that the combination of the new set of scalar parameters $\{\omega_k\}$, which define $\eta_k(\lambda)$, together with the scalars $\{\alpha_k\}$ and $\{\beta_k\}$ from the BICG computation, provide a means for deriving a new iteration. The scalar ω_{k-1} of (4.14) is a free parameter, which can now be specified by analogy with the GMRES iteration to minimize the Euclidean norm of the residual. This gives

$$\omega_{k-1} = \langle F\mathbf{s}^{(k-1)}, \mathbf{s}^{(k-1)}\rangle / \langle F\mathbf{s}^{(k-1)}, F\mathbf{s}^{(k-1)}\rangle.$$

Finally, it can be shown that the BICG scalars can be constructed from quantities obtained directly from the new iteration, and the BICG iterates need not be computed at all; this result is left as an exercise (Problem 4.5). The resulting iteration is given below, where the "tilde" has now been eliminated from the notation.

> **Algorithm 4.3:** THE BICGSTAB METHOD
> Choose $\mathbf{u}^{(0)}$, compute $\mathbf{r}^{(0)} = \mathbf{f} - F\mathbf{u}^{(0)}$, set $\mathbf{p}^{(0)} = \mathbf{r}^{(0)}$
> Choose $\hat{\mathbf{r}}^{(0)}$ such that $\langle \mathbf{r}^{(0)}, \hat{\mathbf{r}}^{(0)} \rangle \neq 0$
> for $k = 0$ until convergence do
> $\quad \alpha_k = \langle \hat{\mathbf{r}}^{(0)}, \mathbf{r}^{(k)} \rangle / \langle \hat{\mathbf{r}}^{(0)}, F\mathbf{p}^{(k)} \rangle$
> $\quad \mathbf{s}^{(k)} = \mathbf{r}^{(k)} - \alpha_k F\mathbf{p}^{(k)}$
> $\quad \omega_k = \langle F\mathbf{s}^{(k)}, \mathbf{s}^{(k)} \rangle / \langle F\mathbf{s}^{(k)}, F\mathbf{s}^{(k)} \rangle$
> $\quad \mathbf{u}^{(k+1)} = \mathbf{u}^{(k)} + \alpha_k \mathbf{p}^{(k)} + \omega_k \mathbf{s}^{(k)}$
> $\quad \mathbf{r}^{(k+1)} = \mathbf{s}^{(k)} - \omega_k F\mathbf{s}^{(k)}$
> $\quad \langle$Test for convergence$\rangle$
> $\quad \beta_k = \left(\langle \hat{\mathbf{r}}^{(0)}, \mathbf{r}^{(k+1)} \rangle / \langle \hat{\mathbf{r}}^{(0)}, \mathbf{r}^{(k)} \rangle \right) (\alpha_k / \omega_k)$
> $\quad \mathbf{p}^{(k+1)} = \mathbf{r}^{(k+1)} + \beta_k (\mathbf{p}^{(k)} - \omega_k F\mathbf{p}^{(k)})$
> enddo

The convergence behavior of this method is typically superior to that of BICG. Like BICG, it requires a fixed amount of work and storage at each step. It also requires two matrix-vector products per iteration, but in this case only F (i.e. not F^T) is needed. We also reiterate that this idea improves upon but does not resolve the difficulties associated with algorithms having short recurrences. For example, it has been observed that BICGSTAB has trouble solving systems with eigenvalues that have large imaginary parts. The BICGSTAB(ℓ) method of Sleijpen & Fokkema [178] performs additional stabilization by minimizing over spaces of dimension ℓ greater than one. In practice, $\ell = 2$ is usually sufficient to resolve the difficulties caused by complex eigenvalues; see van der Vorst [199, chap. 9] for a description of BICGSTAB(2). There have also been alternative approaches devised incorporating a so-called "lookahead" strategy within the Lanczos computation, which are designed to minimize the possibility of breakdown occurring; see Freund & Nachtigal [80].

An implementation of BICGSTAB(ℓ) is included in the IFISS software. For an assessment of the relative merits of GMRES and BICGSTAB(2), see Computational Exercise 4.1.

4.2 Preconditioning methods and splitting operators

We next consider the issue of preconditioning the matrices F arising from discretization of the convection–diffusion equation, as in (3.68). The goal here is to devise preconditioning approaches that give rise to fast convergence of the Krylov subspace methods introduced in the previous section. Ideally the convergence rate of the preconditioned iterative method should be bounded independently of the problem parameters. There are two parameters in this case; the discretization parameter h associated with the subdivision, and the Peclet number, $\mathcal{P}$, associated with the underlying physical problem. To achieve convergence rates that are bounded independently of h, we build on the strategy in Chapter 2, that is, the use of a multigrid cycle as a preconditioner. This is done in Section 4.3. To realize convergence rates that do not deteriorate if the Peclet number is increased requires further insight. This is what we hope to provide within this section.

Let M denote a preconditioning operator for F. As we observed in Section 2.2, there are two requirements for effective preconditioners. First, M should in some sense represent a good approximation of F; for example, a desirable feature would be for the eigenvalues of FM^{-1} to be more tightly clustered than those of F. Second, the action of M^{-1} should be inexpensive to compute; that is, the solution of linear systems with M as the coefficient matrix should be cheap in terms of computational memory and operation counts. For nonsymmetric systems such (3.68), either of the preconditioned problems

$$[M^{-1}F]\,\mathbf{u} = M^{-1}\mathbf{f} \quad \text{or} \quad [FM^{-1}]\,[M\mathbf{u}] = \mathbf{f}, \tag{4.15}$$

can be solved using a Krylov subspace method. There is little difference in effectiveness between these *left-oriented* and *right-oriented* variants of the preconditioned problem. Our preference is for the right-oriented version — used in conjunction with GMRES, the norm that is minimized does not depend on the preconditioner.

As pointed out at the end of Section 2.2, there is a close connection between preconditioning and use of splittings to construct stationary iterative methods. Let the coefficient matrix F be decomposed into a splitting as

$$F = M - R.$$

Here M is also referred to as a splitting operator, and R can be viewed as the associated error matrix. Stationary methods compute a sequence of approximations to the solution $\mathbf{u}$ using the iteration

$$\mathbf{u}^{(k)} = M^{-1}(R\mathbf{u}^{(k-1)} + \mathbf{f}), \tag{4.16}$$

or, equivalently,

$$\mathbf{u}^{(k)} = \mathbf{u}^{(k-1)} + M^{-1}(\mathbf{f} - F\mathbf{u}^{(k-1)}). \tag{4.17}$$

It follows from (4.17) that the residuals $\mathbf{r}^{(k)} = \mathbf{f} - F\mathbf{u}^{(k)}$ satisfy

$$\mathbf{r}^{(k)} = p_k(FM^{-1})\,\mathbf{r}^{(0)},$$

so that stationary iteration is in fact a specialized example of a Krylov subspace method. (See Problem 4.6 for more on this.) Thus, the concepts of splitting and preconditioning are really two sides of the same coin: use of a Krylov subspace method to solve either of the systems of (4.15) is equivalent to *accelerating* the convergence of the stationary iteration.

We will discuss three fundamental issues in the following sections. First, the effectiveness of splitting operators for the discrete convection–diffusion problems and the effect of reordering rows and columns of the coefficient matrix; second, techniques for analyzing asymptotic rates of convergence of stationary iterations; and third, practical considerations. For the first two items, we will emphasize splittings derived from classical iterative methods based on relaxation, that is, Gauss–Seidel iteration. These can be very effective as preconditioners, and they are also amenable to analysis that provides insight into the requirements for efficient algorithms.

4.2.1 *Splitting operators for convection–diffusion systems*

To begin, we let $F = D - L - U$ where D denotes the block diagonal of F, $-L$ denotes the lower triangular matrix consisting of entries below the block diagonal D, and $-U$ is the analogous upper triangular matrix. For "point" versions of these ideas, D consists of the diagonal of F. For block versions, unknowns are grouped together and the blocks of D consist of the entries of the matrix corresponding to connections among all unknowns within the groups. A way to define the blocks of D for two-dimensional problems on uniform grids is to group lines of the grid together, as shown in Figure 4.1. The nonzero structure of the matrix F derived from Q_1 approximation on a uniform 6×6 grid is shown on the right, and thus the matrix D consists of a set of six tridiagonal matrices. Block matrices can also be defined for unstructured triangular meshes by partitioning the domain Ω into subdomains, and collecting together the unknowns within each subdomain. In such cases D will not have as regular a structure as the matrix of Figure 4.1, but it can always be defined to have a tightly banded form, which makes it easy to compute the solutions of systems involving D. Groupings of unknowns along lines (or "snakes") can be used to define block matrices when solving three-dimensional problems.

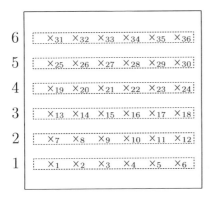
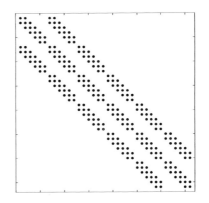

FIG. 4.1. Natural horizontal line ordering and nonzero structure of convection–diffusion matrix arising from Q_1 approximation.

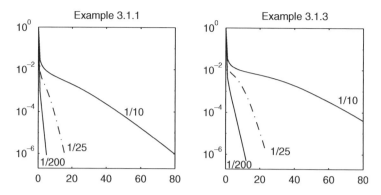

FIG. 4.2. Relative residuals $\|\mathbf{r}^{(k)}\|/\|\mathbf{f}\|$ vs. iterations for stationary line Gauss–Seidel solution of discrete versions of Examples 3.1.1 and 3.1.3, for $\boldsymbol{Q}_1$ approximation, 32×32 square grids (grid level $\ell = 5$) and SD discretization.

The classical stationary iterative methods are defined by the splittings[3]

$$\begin{aligned}
&\text{(block) Jacobi:} & M &= D, & R &= L + U, \\
&\text{(block) Gauss–Seidel:} & M &= D - L, & R &= U.
\end{aligned}$$

To give an idea of the utility of these techniques, we show in Figure 4.2 the behavior of the line Gauss–Seidel stationary iteration applied to Examples 3.1.1 and 3.1.3, where the blocks of D correspond to the lines of a uniform grid, as in Figure 4.1. The key point is the improved and very strong performance as the Peclet number $\mathcal{P}$ increases (i.e. $\epsilon \to 0$, since $\mathcal{P}$ is proportional to $1/\epsilon$). That is, although this simple iteration method is not especially effective as a solver for discrete Poisson problems, it is an effective method in the convection-dominated case.

This can be understood in large part by observing the connection between the physical and algebraic processes. The streamlines for these two problems are in the vertical (Example 3.1.1) or nearly vertical (Example 3.1.3) directions. The Gauss–Seidel iteration groups unknowns along horizontal lines and then "sweeps" across the domain from bottom to top in a direction parallel to the direction of flow, or nearly so for Example 3.1.3. Closer scrutiny of the two graphs in the figure reveals that performance is slightly better for Example 3.1.1. This can be explained by the fact that the sweeps in this case are exactly parallel to the flow whereas in Example 3.1.3 there is also a component of the wind in the horizontal direction.

---

[3]A generalization is the SOR splitting, $M = \frac{1}{\omega}(D - \omega L)$, $R = \frac{1}{\omega}[(1 - \omega)D + \omega U]$, where ω is a parameter. As we will see, the choice $\omega = 1$, giving the Gauss–Seidel iteration, is effective for the discrete convection–diffusion equation. In addition, a good choice of ω requires explicit knowledge of the largest eigenvalue of the Jacobi iteration matrix, which makes implementation more complicated. Therefore, we will not consider this splitting here.

We hasten to point out, however, that the effectiveness of these splitting operators is highly dependent on the ordering of the unknowns in the system, or more precisely, of the (block) rows and columns of the coefficient matrix. (The ordering of unknowns corresponds to an ordering of the columns of the coefficient matrix, which need not coincide with the ordering of the rows. In the following, however, we will assume that these two orderings are the same, in the sense that for any index i, the diagonal entry of the matrix in row i is the coefficient of the unknown with index i.) If a poor choice of ordering is used, then convergence may be much slower than that suggested by Figure 4.2. To demonstrate what can go wrong, we look again at Figure 4.1 and observe that there is an alternative "natural" ordering obtained by listing the lines from top to bottom rather than from bottom to top. In this case, for winds such as those of Examples 3.1.1 and 3.1.3, the resulting block relaxation would sweep across Ω in a direction opposite that of the flow. Another possibility is a *line red–black* ordering, where alternating lines of the grid are divided into two sets and ordered first within the sets; an example is shown in Figure 4.3. This strategy offers a potential advantage for parallel computation.

Figure 4.4 shows the performance of stationary solvers based on these two orderings as well as the original, flow-following one. (For the red–black ordering, lines with the same color follow the flow, as in Figure 4.3, although performance would be the same if they were labeled in the opposite direction.) It is evident that for both the red–black ordering and the natural ordering against the flow, convergence is slow for a large number of steps, although after the initial periods of slow convergence, the convergence rates are the same as for the iteration that follows the flow. For problems such as these with a constant flow, there is a rigorous analysis of this phenomenon; in particular, it can be shown that with red–black ordering on an $n \times n$ grid, there is a delay of about $n/2$ steps before

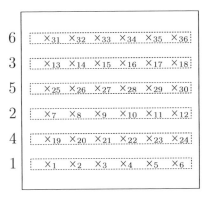
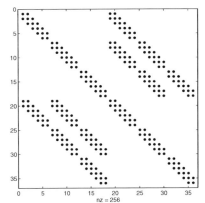

FIG. 4.3. Line red–black ordering and nonzero structure of Q_1 convection–diffusion matrix.

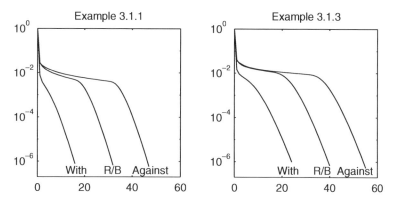

FIG. 4.4. Relative residuals $\|\mathbf{r}^{(k)}\|/\|\mathbf{f}\|$ vs. iterations for stationary line Gauss–Seidel solution of Examples 3.1.1 and 3.1.3, for $\epsilon = 1/25$, 32×32 square grids ($\ell = 5$) using SD discretization, and various orderings.

the asymptotic rate of convergence is exhibited, and with an ordering against the flow, the delay is approximately n steps. See Elman & Chernesky [62] for further discussion.

It is clear from these observations that ordering has a significant impact on performance. This will be equally important for problems with nonconstant wind and we will explore the practical consequences of this in Section 4.2.4. Before doing so, we turn our attention to the analysis of convergence properties of splitting methods. For constant coefficient problems, these results provide insight into the asymptotic convergence behavior exhibited in Figure 4.4.

4.2.2   Matrix analysis of convergence

One way to establish a bound on the asymptotic convergence factor associated with GMRES iteration would be to identify an ellipse containing the eigenvalues of the preconditioned operator FM^{-1} and then to use Theorem 4.2. In general of course, this is easier said than done. An alternative route is to derive a bound on $\rho(M^{-1}R)$. Since

$$FM^{-1} = I - RM^{-1}$$

and $\rho(M^{-1}R) = \rho(RM^{-1})$, this is the same thing as identifying a circle centered at $(1, 0)$ containing the spectrum of FM^{-1}. (Thus, $c = 1$ in Theorem 4.2, and we take the limit $d \to 0$ to give the bound $\rho \leq a$.) Bounding the radius of this circle is tantamount to understanding the asymptotic behavior of the stationary iteration (4.16).

For elaboration of this point, let $\mathbf{e}^{(s)} = \mathbf{u} - \mathbf{u}^{(s)}$ denote the error at the sth step of the stationary iteration. Then

$$\mathbf{e}^{(s)} = (M^{-1}R)^s \mathbf{e}^{(0)},$$

and

$$\|\mathbf{e}^{(s)}\| \leq \|(M^{-1}R)^s\| \, \|\mathbf{e}^{(0)}\|,$$

and therefore

$$\left(\frac{\|\mathbf{e}^{(s)}\|}{\|\mathbf{e}^{(0)}\|} \right)^{1/s} \leq \|(M^{-1}R)^s\|^{1/s}.$$

Analysis is based on the fact that

$$\lim_{s \to \infty} \|(M^{-1}R)^s\|^{1/s} = \rho(M^{-1}R). \tag{4.18}$$

The matrix $M^{-1}R$ is the *iteration matrix* associated with this algorithm, and the iteration is convergent if and only if the spectral radius $\rho(M^{-1}R)$ is less than one. Roughly speaking, for large enough s, the error decreases in magnitude by a factor of $\rho(M^{-1}R)$ at step s, and the spectral radius $\rho(M^{-1}R)$ is referred to as the *convergence factor* associated with the splitting. We are particularly interested in the case where the grouping of unknowns gives rise to a block consistently ordered matrix, see Young [220, p. 445], which implies that the spectral radii of the Jacobi and Gauss–Seidel iteration matrices are related by

$$\rho((D-L)^{-1}U) = \rho(D^{-1}R)^2, \tag{4.19}$$

where $R = L + U$. This property holds for the orderings discussed in Section 4.2.1. Thus, the key to the analysis of the Gauss–Seidel iteration is to bound $\rho(D^{-1}R)$. We consider two approaches for this. The first of these, described below, looks directly at the entries of the matrices D and R in the special case of grid aligned flow. The second approach, presented in the next section, exploits the relationship between the discrete and continuous problems.

We consider the specific example of (3.1) with $\vec{w} = (0, 1)$ on the square $\Omega = (0, 1) \times (0, 1)$, and a Dirichlet boundary condition. We take $\boldsymbol{Q}_1$ approximation on a uniform grid of k^2 elements, so that $h = 1/k$, giving the computational molecule in (3.81) (see also Figure 3.16). After scaling by $3/\epsilon$, we rewrite this as

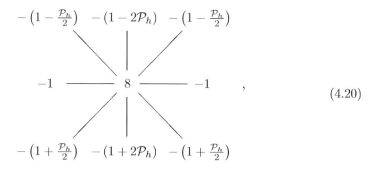

$$(4.20)$$

where $\mathcal{P}_h = h/(2\epsilon)$ is the mesh Peclet number. The block Jacobi operator D is determined by lines as in Figure 4.1, and it consists of a set of uncoupled tridiagonal matrices $\texttt{tridiag}[-1, 8, -1]$. For this problem, Fourier analysis can be used to construct bounds on convergence factors. The tools needed for this purpose have been developed in Section 3.5.1.

Suppose $\mathcal{S}$ is any nonsingular matrix, and let $\hat{F} = \mathcal{S}^{-1} F \mathcal{S}$. Consider the splitting

$$\hat{F} = \hat{D} - \hat{R},$$

where $\hat{D} = \mathcal{S}^{-1} D \mathcal{S}$, $\hat{R} = \mathcal{S}^{-1} R \mathcal{S}$. The eigenvalues of $D^{-1}R$ are the same as those of $\hat{D}^{-1}\hat{R}$. In Proposition 3.13 of Section 3.5.2, we used the choice $\mathcal{S} = V\tilde{P}$ to transform F into the tridiagonal matrix $\hat{T}$ of (3.75), where V derives from the matrix of eigenvectors of the tridiagonal blocks of F, and $\tilde{P}$ is a permutation matrix that reorders grid points from a row-oriented list to a column-oriented one. For this choice of $\mathcal{S}$, $\hat{D}$ is a diagonal matrix consisting of $k-1$ constant blocks of order $k-1$,

$$\hat{D} = \texttt{diag}(\hat{D}_1, \hat{D}_2, \ldots, \hat{D}_{k-1}), \quad \hat{D}_j = \lambda_j I, \quad \lambda_j = 8 - 2\cos\tfrac{j\pi}{k},$$

and $\hat{R} = \texttt{diag}(\hat{R}_1, \hat{R}_2, \ldots, \hat{R}_{k-1})$ is a block diagonal matrix whose jth block is the tridiagonal matrix

$$\hat{R}_j = \texttt{tridiag}\left(1 + 2\mathcal{P}_h + 2\left(1 + \tfrac{\mathcal{P}_h}{2}\right)\cos\tfrac{j\pi}{k}, 0, 1 - 2\mathcal{P}_h + 2\left(1 - \tfrac{\mathcal{P}_h}{2}\right)\cos\tfrac{j\pi}{k}\right).$$

The eigenvalues of $\hat{D}^{-1}\hat{R}$, and hence those of $D^{-1}R$, are then

$$\mu_{jm} = \frac{2\sqrt{(1 - 4\mathcal{P}_h^2) + 4(1 - \mathcal{P}_h^2)c_j + (4 - \mathcal{P}_h^2)c_j^2}\, c_m}{8 - 2c_j}, \quad 1 \le j,\ m \le k-1,$$

$$(4.21)$$

where $c_j = \cos\tfrac{j\pi}{k}$.

We seek the maximum value of this expression as a function of c_j and c_m, where the cosines take on discrete values in $(-1, 1)$. For simplicity, we will also allow the endpoints ± 1 to be considered. Since the focus here is on accurate discretization, we restrict attention to the case $\mathcal{P}_h \leq 1$. The following result establishes a bound on the spectral radius of the block Jacobi iteration matrix $\rho(D^{-1}R)$.

Theorem 4.4. *The spectral radius $\rho(D^{-1}R)$ of the iteration matrix derived from a block Jacobi splitting of the discrete convection–diffusion operator with computational molecule (4.20) is bounded as*

$$\rho(D^{-1}R) \leq \sqrt{1 - \mathcal{P}_h^2} \quad \text{for } \mathcal{P}_h \leq \sqrt{\tfrac{78}{101}} \approx 0.879,$$

and

$$\rho(D^{-1}R) \leq \tfrac{1}{15} \sqrt{23 \left(3 - \mathcal{P}_h^2\right)} \quad \text{for } \sqrt{\tfrac{78}{101}} < \mathcal{P}_h \leq 1.$$

Proof It is clear that the expression in (4.21) is bounded above using the choice $c_m = 1$. After squaring the result and simplifying, we require an upper bound with respect to c of

$$\sigma(c) = \frac{(1 - 4\mathcal{P}_h^2) + 4(1 - \mathcal{P}_h^2)c + (4 - \mathcal{P}_h^2)c^2}{(4 - c)^2}.$$

For $\mathcal{P}_h \leq 1/2$, $\sigma(c)$ is a strictly increasing function of $c \in [-1, 1]$, so that its maximal value, obtained at $c = 1$, is $1 - \mathcal{P}_h^2$. When $\mathcal{P}_h > 1/2$, the term $1 - 4\mathcal{P}_h^2$ in the numerator of $\sigma(c)$ is negative and the analysis is somewhat more technical. A straightforward derivation reveals that $\sigma(c)$ has a local minimum at

$$c^* = -\frac{1}{2}\left(\frac{3 - 4\mathcal{P}_h^2}{3 - \mathcal{P}_h^2}\right).$$

The spectral radius is then bounded by the maximum of

$$\{|\sigma(-1)|, |\sigma(c^*)|, |\sigma(1)|\}.$$

We can rule out $\sigma(-1) = \tfrac{1}{25}(1 - \mathcal{P}_h^2) < \sigma(1)$. Substitution of c^* into $\sigma(c)$ and some detailed algebra (where the fact that $(4 - \mathcal{P}_h^2)/(3 - \mathcal{P}_h^2) > 1$ is used to simplify the quadratic term) yields

$$\sigma(c^*) \geq -\frac{23}{9}\left(\frac{3 - \mathcal{P}_h^2}{(9 - 4\mathcal{P}_h^2)^2}\right) \geq -\frac{23}{225}(3 - \mathcal{P}_h^2).$$

This bound on $\sigma(c^*)$ is subsumed by $\sigma(1) = 1 - \mathcal{P}_h^2$ for $\mathcal{P}_h \leq \sqrt{\tfrac{78}{101}}$, but its absolute value is larger for larger values of $\mathcal{P}_h$. $\square$

A graph of this bound is shown in Figure 4.5. The figure also contains (dotted) a plot of the spectral radius for $\tfrac{78}{101} < \mathcal{P}_h \leq 1$. It is evident that the convergence

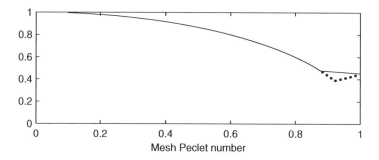

FIG. 4.5. Bound from Theorem 4.4 on the spectral radius of the line Jacobi
iteration matrix.

bound generally improves as $\mathcal{P}_h$ increases from zero, with a slight change of
character as $\mathcal{P}_h \to 1$.

Theorem 4.4 gives a good picture of the spectral radii in the case where $\mathcal{P}_h$
is not close to zero, but because of the decision to allow $c_j = 1$ and $c_m = 1$ in
the proof, the bounds are not tight as $h \to 0$. A more careful derivation, using
$c_1 = \cos \frac{\pi}{k}$ to produce the maxima with respect to both c_j and c_m, yields the
tighter asymptotic bound

$$\rho(D^{-1}R) \le \sqrt{1 - \left(2\pi^2 + \left(\tfrac{1}{2\epsilon}\right)^2\right)h^2} + o(h^2). \qquad (4.22)$$

The notation $\alpha(h) = o(\beta(h))$ means that $\alpha(h)/\beta(h) \to 0$ as $h \to 0$. To see the
difference between (4.22) and the result of Theorem 4.4, note that the square
root on the right side of (4.22) can be rewritten as $\sqrt{1 - \mathcal{P}_h^2 - 2\pi^2 h^2}$ for small
enough $\mathcal{P}_h$.

4.2.3   Asymptotic analysis of convergence

We next describe an alternative approach for analysis, due to Parter &
Steuerwalt [143, 145, 146], which derives a relation between certain eigenvalues associated with matrix splittings and those of the underlying differential
operator. We first take a general point of view with respect to the splitting
$F = M - R$. Then, as above, we look closely at the case $M = D$, where D is the
line Jacobi operator.

We are interested in the maximal eigenvalue of the matrix $M^{-1}R$, or,
equivalently, of the problem

$$\lambda M\mathbf{u} = R\mathbf{u}. \qquad (4.23)$$

Let us bring the coefficient matrix F into the discussion by subtracting $\lambda R u$ from both sides of this equation and dividing by λ. This leads to

$$F\mathbf{u} = \left(\frac{1-\lambda}{\lambda}\right) R\mathbf{u}, \tag{4.24}$$

which is a well-defined problem determining the nonzero eigenvalues of (4.23). Equivalently,

$$F\mathbf{u} = \mu_h \left(h^2 R\right) \mathbf{u} \tag{4.25}$$

with $\mu_h = (1 - \lambda)/(\lambda h^2)$ and h is the mesh parameter. It will be evident in a moment why we explicitly highlight the dependence of μ_h on the discretization; it is clear that λ also depends on it.

The eigenvalue problems (4.24)–(4.25) suggest a possible connection with a continuous eigenvalue problem

$$\mathcal{L}u = \mu \, \mathcal{R}u \ \text{ in } \Omega, \quad u = 0 \ \text{ on } \partial\Omega, \tag{4.26}$$

where $\mathcal{L} = -\epsilon\nabla^2 + \vec{w} \cdot \nabla$ is the convection–diffusion operator on the left side of (3.1) and $\mathcal{R}$ is to be determined. The key to the analysis is that for many examples of splittings, the matrix $h^2 R$ is approximately a weak multiplication operator. That is, there exists a smooth function $r(\vec{x})$ on Ω such that

$$h^2 \langle R\mathbf{u}, \mathbf{v} \rangle \approx (r u_h, v_h). \tag{4.27}$$

In this expression, which we will state more precisely below, u_h and v_h are the finite element functions determined by the vectors $\mathbf{u}$ and $\mathbf{v}$ (see Section 3.5). Thus, $\mathcal{R}u = r(\vec{x})u$. With $a(\cdot, \cdot)$ as in (3.13), the weak formulation of (4.26) is then

$$a(u, v) = \mu \, (r\, u, v) \quad \text{for all } v \in H^1_{E_0}. \tag{4.28}$$

The analysis proceeds under two fundamental assumptions, first, that (4.28) has a minimal real eigenvalue $\mu^{(0)} > 0$, and second, that the sequence of minimal eigenvalues of the discrete problems (4.25) derived from a sequence of mesh parameters h tending to 0 are convergent to $\mu^{(0)}$. The latter assumption is stated more precisely as

$$\mu_h^{(0)} = \mu^{(0)} + o(1) \quad \text{as } h \to 0. \tag{4.29}$$

(Recall that $\alpha(h) = o(1)$ means $\alpha(h) \to 0$ as $h \to 0$, see the comment following (4.22).) It will be seen below that (4.29) is a consequence of the weak multiplication property (4.27). The following result then establishes a connection between the maximal eigenvalue of $M^{-1}R$ and the minimal eigenvalue $\mu^{(0)}$ of the continuous problem.

Theorem 4.5. *Assume that the weak multiplication property* (4.27) *is valid, and suppose in addition that there is a maximal eigenvalue $\lambda_{\max}$ of $M^{-1}R$ that is* *real and positive, and, further, that $\lambda_{\max} = \rho(M^{-1}R) < 1$. Then the spectral radius satisfies*

$$\rho = 1 - \mu^{(0)}h^2 + o(h^2), \tag{4.30}$$

where $\mu^{(0)}$ is the minimal eigenvalue of the continuous problem (4.28).

Proof Since ρ is an eigenvalue of (4.23), it follows that $\mu_h = (1 - \rho)/(\rho h^2)$ is an eigenvalue of (4.25). But $\mu_h^{(0)}$ is minimal, so that

$$\frac{1 - \rho}{\rho h^2} \geq \mu_h^{(0)} = \mu^{(0)} + o(1).$$

This gives

$$\rho \leq \frac{1}{1 + (\mu^{(0)} + o(1))h^2} = 1 - \mu^{(0)}h^2 + o(h^2).$$

On the other hand, if μ_h is any eigenvalue of (4.25) for which $1 + \mu_h h^2 \neq 0$, then $\lambda = 1/(1 + \mu_h h^2)$ is an eigenvalue of (4.23), and by definition $\rho \geq |\lambda|$. For the particular choice $\mu_h = \mu_h^{(0)}$, the fact that $\lim_{h \to 0} \mu_h^{(0)} > 0$ implies that $1 + \mu_h^{(0)} h^2 > 0$ for small enough h. Consequently, for all small h, $|\lambda| = \lambda$ and

$$\rho \geq \lambda = \frac{1}{1 + (\mu^{(0)} + o(1))h^2} = 1 - \mu^{(0)}h^2 + o(h^2). \qquad \square$$

To derive a concrete bound from this result, it is necessary to have an understanding of the minimal eigenvalue of (4.28). This turns out to be straightforward for model problems in which Ω is square and the convection term is constant.

Lemma 4.6. *Let $\Omega = (0,1) \times (0,1)$. If $\vec{w} = (w_x, w_y)$ and $\mathcal{R} = r$ are constant functions, then the eigenfunctions and eigenvalues of (4.26) are*

$$u_{jm}(x, y) = e^{xw_x/(2\epsilon)} \sin(j\pi x)\, e^{yw_y/(2\epsilon)} \sin(m\pi y),$$

$$\Lambda_{jm} = \frac{\epsilon}{r}\left((j^2 + m^2)\pi^2 + \left(\frac{w_x}{2\epsilon}\right)^2 + \left(\frac{w_y}{2\epsilon}\right)^2\right).$$

The minimal eigenvalue $\mu^{(0)}$ corresponds to $j = m = 1$.

Proof The assertion can be checked by substitution into (4.26). $\qquad \square$

Substitution of $\mu^{(0)}$ from Lemma 4.6 into Theorem 4.5 gives the asymptotic value of $\rho(M^{-1}R)$, provided that the weak multiplication property (4.27) can be established for a constant function r. Before describing in detail how this is done, we first give an intuitive idea of what is involved. Consider Q_1 approximation on a uniform grid and take the block Jacobi splitting, where $M = D$ is determined by lines as in Figure 4.1. The properties of the "remainder term" of the splitting

$h^2 R$ are understood by viewing R as a difference operator and identifying an "average" value of Ru. That is, at the grid point with index (i, j),

$$
\begin{aligned}
[R\mathbf{u}]_{ij} &= \left(\frac{\epsilon}{3} + \frac{h}{12}(w_x + w_y)\right)\mathbf{u}_{i-1,j-1} + \left(\frac{\epsilon}{3} + \frac{h}{3}w_y\right)\mathbf{u}_{i,j-1} \\
&+ \left(\frac{\epsilon}{3} + \frac{h}{12}(w_x - w_y)\right)\mathbf{u}_{i+1,j-1} \\
&+ \left(\frac{\epsilon}{3} - \frac{h}{12}(-w_x + w_y)\right)\mathbf{u}_{i-1,j+1} + \left(\frac{\epsilon}{3} - \frac{h}{3}w_y\right)\mathbf{u}_{i,j+1} \\
&+ \left(\frac{\epsilon}{3} - \frac{h}{12}(w_x + w_y)\right)\mathbf{u}_{i+1,j+1}.
\end{aligned}
\tag{4.31}
$$

Inspection of (4.31) gives

$$
[R\mathbf{u}]_{ij} = 2\epsilon\,\mathbf{u}_{ij} + O(h),
\tag{4.32}
$$

so that $r = 2\epsilon$ in this example.

Thus, modulo (4.27), we have identified the asymptotic value of $\rho(M^{-1}R)$ in the limit as $h \to 0$. The power of this approach lies in the fact that the convergence factor can be identified essentially by inspection. We summarize formally in the following theorem.

Theorem 4.7. *Under the assumptions of Lemma 4.6, the spectral radius of the block Jacobi iteration matrix for Q_1 or P_1 approximation on a uniform grid of width h, has the asymptotic form*

$$
\rho(D^{-1}R) = 1 - \frac{1}{2}\left(2\pi^2 + \left(\frac{w_x}{2\epsilon}\right)^2 + \left(\frac{w_y}{2\epsilon}\right)^2\right)h^2 + o(h^2).
$$

Moreover, the spectral radius of the block Gauss–Seidel iteration matrix satisfies

$$
\rho((D - L)^{-1}U) = 1 - \left(2\pi^2 + \left(\frac{w_x}{2\epsilon}\right)^2 + \left(\frac{w_y}{2\epsilon}\right)^2\right)h^2 + o(h^2).
$$

Proof   The proof follows from Theorem 4.5, Lemma 4.6, and (4.32). $\square$

Returning to the issue of the weak multiplication property, recall that this whole approach depends on the assumption that $\mu_h^{(0)} \to \mu^{(0)}$, that is, that the minimal eigenvalue of the discrete problem (4.25) converges to the minimal eigenvalue of the continuous problem (4.28). This is reasonable from an intuitive point of view, although a full discussion of this point is beyond the scope of this book. A sufficient condition that can be used to establish its validity is obtained through the following statement of (4.27). For a proof, see Osborn [141], Parter [144], or Parter & Steuerwalt [146].

Theorem 4.8. *A sufficient condition ensuring that* $\mu_h^{(0)} \to \mu^{(0)}$ *is*

$$h^2 \langle R\mathbf{u}, \mathbf{v} \rangle = (ru_h, v_h) + e(u_h, v_h) \text{ for all } u_h, v_h \in S_0^h, \tag{4.33}$$

where

$$|e(u_h, v_h)| = \eta(h) \left(\|u_h\|_1^2 + \|v_h\|_1^2 + \|u_h\|_1 + \|v_h\|_1 \right) \tag{4.34}$$

and $\eta(h)$ *is a function that tends to 0 as* $h \to 0$.

The conditions (4.33)–(4.34) of Theorem 4.8 constitute a rigorous statement of the weak multiplication property. It remains to show how these conditions are established. For a sketch of this, we examine the part of R consisting of its action in just the vertical direction, that is, the two terms of (4.31) with indices $(i, j \pm 1)$:

$$[R^y \mathbf{u}]_{ij} = \frac{\epsilon}{3} \left(2\mathbf{u}_{ij} + [D_+^y \mathbf{u}]_{ij} - [D_-^y \mathbf{u}]_{ij} \right) - \frac{hw_y}{3} \left([D_+^y \mathbf{u}]_{ij} + [D_-^y \mathbf{u}]_{ij} \right),$$

where

$$[D_+^y \mathbf{u}]_{ij} = \mathbf{u}_{i,j+1} - \mathbf{u}_{ij}, \quad [D_-^y \mathbf{u}]_{ij} = \mathbf{u}_{ij} - \mathbf{u}_{i,j-1}$$

are classical difference operators. We also need two results whose proof is left as an exercise: first,

$$h^2 \langle \mathbf{u}, \mathbf{v} \rangle = (u_h, v_h) + e_1(u_h, v_h) \tag{4.35}$$

where $|e_1(u_h, v_h)| \leq h \left(\|u_h\|_1^2 + \|v_h\|_1^2 \right)$; and second,

$$\|D_+^y \mathbf{u}\| \leq c \|\nabla u_h\|, \quad \|D_-^y \mathbf{u}\| \leq c \|\nabla u_h\|, \tag{4.36}$$

where throughout this argument, c is a constant that does not depend on h. Taking the inner product of $h^2 R^y \mathbf{u}$ with $\mathbf{v}$ then gives

$$h^2 \langle R^y \mathbf{u}, \mathbf{v} \rangle = \frac{2\epsilon}{3} h^2 \langle \mathbf{u}, \mathbf{v} \rangle + \frac{\epsilon}{3} h^2 \left(\langle D_+^y \mathbf{u}, \mathbf{v} \rangle - \langle D_-^y \mathbf{u}, \mathbf{v} \rangle \right)$$
$$- \frac{w_y}{3} h^3 \left(\langle D_+^y \mathbf{u}, \mathbf{v} \rangle + \langle D_-^y \mathbf{u}, \mathbf{v} \rangle \right)$$
$$= \frac{2\epsilon}{3} (u_h, v_h) + \frac{2\epsilon}{3} e_1(u_h, v_h) + e_2(u_h, v_h),$$

where, from (4.35)–(4.36), a generic constant,

$$|e_2(u_h, v_h)| \leq ch^2 \left(\|D_+^y \mathbf{u}\| + \|D_-^y \mathbf{u}\| \right) \|\mathbf{v}\|$$
$$\leq ch \|\nabla u_h\| \|v_h\|_0$$
$$\leq ch \left(\|u_h\|_1^2 + \|v_h\|_1^2 \right).$$

This establishes the contribution to (4.34) coming from $R^y \mathbf{u}$. The other terms of $R\mathbf{u}$ are handled in a similar manner.

Remark 4.9. We assumed in this derivation (in Theorem 4.5) that there is a maximal eigenvalue of the iteration matrix that is real, positive and less than one (so that the iteration is convergent). For Q_1 approximation on square grids, this can be shown to hold using classic results in matrix analysis such as the Perron–Frobenius theory for non-negative matrices, see Axelsson [6, chap. 3], or Varga [202, chap. 2]. For other discretizations, including ones derived from P_1 approximation, a more technical argument must be used to establish (4.30), see Theorems 4.4 and 5.1 of [146].

Remark 4.10. We have applied these results to a model problem with constant wind, square domain and uniform grid. The restrictions are needed to derive a concrete bound such as that of Theorem 4.7. It is also shown in [146] that Theorem 4.5 holds for more general problems

4.2.4 *Practical considerations*

The theoretical results in Sections 4.2.2 and 4.2.3, particularly Theorems 4.4 and 4.7, provide analytic confirmation of the numerical results in Section 4.2.1, which suggested that the discrete convection–diffusion equation is easier to solve than the discrete Poisson equation. The theoretical bounds suggest that the spectral radii of the iteration matrix *decreases* as the Peclet number is increased from zero. (In Theorem 4.7, $(w_x/2\epsilon)^2 + (w_y/2\epsilon)^2 = \mathcal{P}^2$.) The analyses also complement each other. Theorem 4.7 depends on asymptotics with respect to the discretization, and it applies when the discretization is accurate in an asymptotic sense as $h \to 0$. Theorem 4.4 is less dependent on this issue and provides insight for a fixed mesh size where the Peclet number is allowed to vary within a range. The Fourier analysis approach can also be used to study the case of large $\mathcal{P}_h$ together with SD discretization — the details are worked out in Problem 4.7 and to show that a fast asymptotic convergence rate can be expected in this case. Notice that as $h \to 0$, the two analyses (using the tighter bound (4.22)) yield the same asymptotic bound for the spectral radii.

On the other hand, the theoretical results only provide information about performance in an asymptotic sense with respect to the iteration count k, that is, when k is such that $\|(M^{-1}R)^k\|^{1/k}$ is near its limiting value as in (4.18). The results show why convergence rates improve as convection becomes more dominant (Figure 4.2), and explain why the convergence rates for different orderings are eventually the same (Figure 4.4), but they do not touch upon the differences in the pre-asymptotic regime (i.e. with respect to k) for different orderings.

There remains the question of what to do in practical situations when the flow is more complex. For this, we can use what we have learned from simple flows to develop heuristics for constructing efficient methods. One approach is to try to order the unknowns in such a way that relaxation always follows the flow. This type of approach has attracted a lot of attention, but is inevitably complicated when the flow is not constant. In addition, it does not address the issue of recirculating flows of the type exhibited by Example 3.1.4.

We prefer an alternative strategy, based on building preconditioners that sweep in multiple directions. As we have seen, when the grid ordering is parallel to the underlying direction of flow, convergence of the Gauss–Seidel iteration is rapid. On the other hand, when the grid ordering is at odds with the flow, convergence is slow but relaxation does not do any harm. In particular, in Figure 4.4, the residuals do not increase during the initial, slowly convergent, stages of the iteration. These two observations motivate a stationary method consisting of multiple relaxation steps, each one of which attempts to follow the flow in part of the domain.

For two-dimensional problems, the algorithm performs four subsidiary steps, first sweeping from bottom to top in the domain as in the examples above, and then sweeping in succession from left to right, bottom to top and right to left. To be precise, assume F is the coefficient matrix derived from a natural left-to-right, bottom-to-top ordering of the grid as in Figure 4.1, and let $M^{(\uparrow)}$ denote the block Gauss–Seidel splitting that we have been studying. Similarly, let $F_{\rightarrow}$ denote the version of the discrete operator obtained by reordering the rows and columns of F so that points are first ordered from bottom to top, and then lines are ordered from left to right. Let $M_{F_{\rightarrow}}$ denote the block Gauss–Seidel splitting operator obtained by grouping points in vertical lines together in a manner analogous to the horizontal grouping used for $M^{(\uparrow)}$. If P is the permutation matrix that reorders grid points from a horizontal to a vertical ordering, then $M^{(\rightarrow)} = PM_{F_{\rightarrow}}P^{-1}$ is a splitting operator for F. Let $M^{(\downarrow)}$ and $M^{(\leftarrow)}$ be defined similarly, for top-to-bottom and right-to-left sweeps, respectively. One step of a stationary iteration using these four splittings starts from $\mathbf{u}^{(k)}$ and produces $\mathbf{u}^{(k+1)}$ as

$$
\begin{aligned}
\mathbf{u}^{(k+1/4)} &= \mathbf{u}^{(k)} + \left(M^{(\uparrow)}\right)^{-1}\left(\mathbf{f} - F\mathbf{u}^{(k)}\right) \\
\mathbf{u}^{(k+1/2)} &= \mathbf{u}^{(k)} + \left(M^{(\rightarrow)}\right)^{-1}\left(\mathbf{f} - F\mathbf{u}^{(k+1/4)}\right) \\
\mathbf{u}^{(k+3/4)} &= \mathbf{u}^{(k)} + \left(M^{(\downarrow)}\right)^{-1}\left(\mathbf{f} - F\mathbf{u}^{(k+1/2)}\right) \\
\mathbf{u}^{(k+1)} &= \mathbf{u}^{(k)} + \left(M^{(\leftarrow)}\right)^{-1}\left(\mathbf{f} - F\mathbf{u}^{(k+3/4)}\right).
\end{aligned}
\tag{4.37}
$$

For three-dimensional problems, there would be two additional sweeps derived from the third dimension, giving a total of six subsidiary steps.

The computation $\mathbf{u}^{(k)} \mapsto \mathbf{u}^{(k+1)}$ defines a linear operator M which can be viewed as a splitting operator defining a stationary iteration (4.16), or as a preconditioning operator. The idea is that for a general recirculating wind, at least one of these steps follows the flow in part of the domain, and none of them causes any damage. Moreover, if the general character of the flow is known, then a subset of the computations specified in (4.37) can be omitted, so that the cost per iteration decreases. For example, if the flow has no "right-to-left" component, then the step using $M^{(\leftarrow)}$ can be omitted.

Some experimental results demonstrating how this idea works are shown in Figure 4.6. Here, we compare the performance of the "standard" block

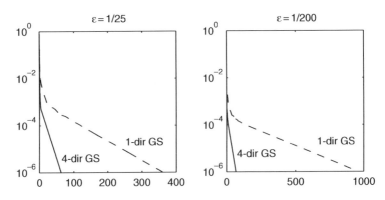

FIG. 4.6. Relative residuals $\|\mathbf{r}^{(k)}\|/\|\mathbf{f}\|$ vs. iterations for solution of SD discretization of Example 3.1.4 on a 32×32 grid ($\ell = 5$), using the block Gauss–Seidel (1-dir GS) and four-directional block Gauss–Seidel (4-dir GS) methods.

Gauss–Seidel stationary iteration to the "four-directional" variant defined by (4.37), for solving a problem with a recirculating wind, Example 3.1.4. For the first problem, with a relatively modest component of convection ($\epsilon = 1/25$, $\mathcal{P} = O(10)$), the iteration counts needed to satisfy the stopping criterion $\|\mathbf{r}^{(k)}\|/\|\mathbf{f}\| < 10^{-6}$ were 64 and 361, respectively, for the four-directional and one-directional variants. For the second problem, which is convection-dominated ($\epsilon = 1/200$, $\mathcal{P} = O(10^2)$), the counts were 68 and 952. The four-directional approach is four times as expensive, so that there is just a modest advantage for the first problem, but the improvement is dramatic in the second case.

The algebraic preconditioners based on incomplete factorization of F can also be viewed as having a character similar in spirit to this approach. For example, if F is derived from the natural left-to-right, bottom-to-top ordering of the grid, and $M = LU$ denotes an incomplete (ILU(0)) factorization of F, then the preconditioning operation

$$w = M^{-1}v = U^{-1}L^{-1}v$$

can be viewed as resembling two steps of the iteration (4.37), in which the forward solve (application of the action of L^{-1}) is a bottom-to-top sweep, and the backsolve (application of the action of U^{-1}) is a top-to-bottom sweep. It is also possible to define variants of this idea that generalize (4.37). For example, let $M^{(\uparrow)}$ represent the incomplete factorization of F, let $F_\rightarrow$ denote the reordered version of F defined above, and let $M^{(\rightarrow)} = PM_{F_\rightarrow}P^{-1}$, where $M_{F_\rightarrow}$ is the incomplete factorization of $F_\rightarrow$. Then

$$\begin{aligned}\mathbf{u}^{(k+1/2)} &= \mathbf{u}^{(k)} + (M^{(\uparrow)})^{-1}(\mathbf{f} - F\mathbf{u}^{(k)})\\ \mathbf{u}^{(k+1)} &= \mathbf{u}^{(k)} + (M^{(\rightarrow)})^{-1}(\mathbf{f} - F\mathbf{u}^{(k+1/2)})\end{aligned} \qquad (4.38)$$

defines a "two-directional" iteration based on ILU(0) factorization that is designed to handle multidirectional flows in two dimensions.

In summary, our experience of using iterative methods to solve discrete convection–diffusion problems is that it is vitally important to ensure that the matrix splitting relates to the underlying flow in a suitable way. (In particular, using a red–black ordering does not lead to fast convergence under any circumstances, see Computational Exercise 4.6.) We do not believe that it is possible to make a simple statement about any one superior method. Algebraic approaches based on incomplete factorization have the advantage of being general purpose methods. On the other hand, approaches based on Gauss–Seidel iteration are very simple to implement and can easily be adapted to handle strong convection.

Finally, it is emphasized that the effectiveness of either of these approaches can be enhanced by Krylov subspace iteration. As we have seen previously, simple iteration methods including multiple step versions such as (4.37) or (4.38) determine a linear operator from $\mathbf{u}^{(k)}$ to $\mathbf{u}^{(k+1)}$ that can be used as a preconditioner for the Krylov subspace methods described in Section 4.1. For simple flows such as those of Examples 3.1.1–3.1.3, the benefits of such an acceleration may only be marginal, see Computational Exercise 4.5. The gains are likely to be more significant for complex flows. By way of illustration, Figure 4.7 shows convergence histories for the GMRES algorithm using three general purpose preconditioners. Comparison with the corresponding results of Figure 4.6 shows that the accelerated solvers are much more efficient than the underlying simple iteration method in this case.

So far, we have avoided the issue of the behavior of preconditioned solvers with respect to the discretization parameter. As mentioned at the start

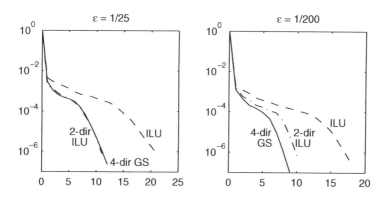

FIG. 4.7. Relative residuals $\|\mathbf{r}^{(k)}\|/\|\mathbf{f}\|$ vs. iterations for solution of SD discretization of Example 3.1.4 on a 32×32 grid ($\ell = 5$), using right-preconditioned GMRES with preconditioning by four-directional Gauss–Seidel (4-dir GS), ILU(0) (ILU), and two-directional ILU(0) (2-dir ILU).

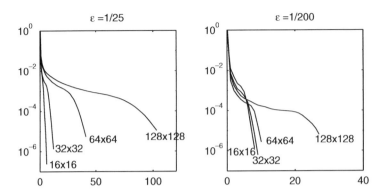

FIG. 4.8. Relative residuals $\|\mathbf{r}^{(k)}\|/\|\mathbf{f}\|$ vs. iterations for solution of SD discretization of Example 3.1.4 on uniform grids ($\ell = 4, 5, 6, 7$), using right-preconditioned GMRES with preconditioning by four-directional Gauss–Seidel.

of the section, an ideal preconditioner would exhibit a convergence rate that is independent of the mesh size. None of the preconditioning strategies that have been discussed thus far are ideal in this sense. An illustration is given in Figure 4.8, where convergence curves are plotted for systems arising from uniform grid discretization of Example 3.1.4 with two values of ϵ. Notice that the deterioration in performance with increasing refinement is more marked in the case $\epsilon = 1/25$. Indeed, the iteration counts appear to be inversely proportional to h in this case. The dependence on the grid is much less noticable in the case with dominant convection. For coarse grids, the streamline diffusion matrix represents a significant component of the algebraic system, and iteration counts are essentially independent of h. Once the grid becomes fine enough, however, the $O(h^{-1})$ behavior is again evident. The goal of grid-independent convergence (for any fixed value of ϵ) leads to consideration of multigrid algorithms, which is the topic of the next section.

4.3   Multigrid

As we showed in Chapter 2, multigrid methods can be used to solve the discrete Poisson equation with a rate of convergence independent of the discretization, leading to computational costs proportional to the size of the problem. This optimal efficiency is realized through the way multigrid handles components of the iteration error on two subspaces. On the fine grid subspace, a smoother derived from a splitting operator is used to eliminate oscillatory parts of the error. Once this is done, the iteration error is well-represented on the complementary coarse grid subspace, and a correction for this error can be obtained by solving a discrete problem on the coarser grid. Application of this strategy in a recursive manner leads to an optimal algorithm.

We will show here that the same approach can be used to solve the discrete convection–diffusion equation, again with computational costs proportional to the problem size. In contrast to the situation for the Poisson equation, however, analysis of multigrid methods in this setting is far from complete, and we make no claim to a definitive statement on this methodology. Our focus will be on empirical results, with the aim of developing an intuitive understanding of how to achieve good performance. We will emphasize two points:

1. CAREFUL CHOICES OF DISCRETIZATION. Suppose there is a (fine grid) discretization that resolves boundary layers and interior layers to an acceptable degree. The multigrid algorithm depends on a sequence of coarser grids that may not themselves be sufficiently fine to resolve the features of the problem. These grids are purely artifacts of the computational algorithm, but it is important that they be handled in an appropriate way. It turns out that using them with the streamline diffusion discretization strategy is a good choice.

2. SMOOTHING OPERATORS THAT TAKE ACCOUNT OF FLOWS. This point has already come up in Section 4.2 in the discussion of the general use of splitting operators and preconditioners, and the issues for multigrid are largely the same.

In discussing this methodology, we will assume we are seeking the solution to a problem $F\mathbf{u} = \mathbf{f}$ defined on a uniform fine grid with mesh width h, and that there is a set of coarser grids with mesh width $2^i h$, $i = 1, 2, \ldots, \ell - 1$. The approach is easily adapted to a set of hierarchical shape-regular triangulations.

4.3.1 Practical issues

We begin with a demonstration of the following "bottom-line" fact: if it is done correctly, multigrid is a robust and efficient solver for the convection–diffusion equation. What do we mean by "done correctly?" As shown in Chapter 3, streamline diffusion discretization of the convection–diffusion equation is superior to Galerkin discretization when the solution has steep layers that are not fully resolved by the grid. For large Peclet numbers, even if good resolution is achieved on the fine grid, it is inevitable that some of the coarser grids used by the multigrid solver will be too coarse to produce accurate solutions to the differential equation. Thus, it is natural to use SD for these coarse grids, and this dictates our strategy for discretizing at coarser levels. We use the SD parameter given by (3.44).

With regard to smoothing, we will build on the observations made in Section 4.2, namely, that effective stationary iterative methods come from splittings that are related to the underlying flow in the model. We illustrate this strategy using all four of the Example problems presented in Chapter 3. For Examples 3.1.1, 3.1.2 and 3.1.3, which have simple unidirectional flows, a smoothing step is defined to consist of one step of a block Gauss–Seidel iteration where the blocks correspond to lines derived from a left-to-right, bottom-to-top ordering of the grid as in Section 4.2.1. This is a "flow-following" smoothing operation. For Example 3.1.4, with a recirculating flow, a presmoothing step is specified to be one step of a four-directional block Gauss–Seidel iteration (4.37), and a postsmoothing step is a four-directional block Gauss–Seidel step with directions reversed. That is, instead of using (4.37), for postsmoothing, we apply $(M^{(\leftarrow)})^{-1}$ first, followed by $(M^{(\downarrow)})^{-1}$, etc. This gives the operations of presmoothing and postsmoothing a symmetric style.

There remains the question of how to transfer data between grids. Many authors have devised "operator-dependent" grid transfer operators, which incorporate properties of the underlying flow into the grid transfers; see Trottenberg et al. [196, section 7.7.5], Wesseling [214, section 5.4] and references therein. These are effective, but when coarse grid discretization is done in a stable manner as described above, it turns out that the "natural" grid transfer operators defined for the Poisson equation in Section 2.5 work perfectly well. That is, the coarse-to-fine grid transfer is achieved via natural inclusion (2.42), giving the matrix prolongation operator P; the corresponding restriction operator is $R = P^T$. These grid transfer operators have the advantage of being simpler to implement than operator-dependent ones.

Tables 4.1 and 4.2 show typical performance results for multigrid derived from this general approach, for the four Example problems described in Chapter 3. These experiments used V-cycle multigrid with one presmoothing step and one postsmoothing step at each level. The coarsest grid, on which a direct solve is performed, is of dimension 4×4. The results of Table 4.1 clearly demonstrate "textbook multigrid" convergence behavior, that is, iteration counts that are independent of the mesh size h. These results should be compared with those of Figure 4.8, which shows that ordinary preconditioned Krylov subspace iteration does depend on the mesh size.

Table 4.1 *Multigrid V-cycle iteration counts to satisfy* $\|\mathbf{r}^{(k)}\|/\|\mathbf{f}\| < 10^{-6}$, *starting with zero initial iterate, for two values of ϵ and various grids. Here, ℓ is the grid refinement level, $h = 2^{1-\ell}$, and the grid has dimensions $2^\ell \times 2^\ell$.*

	$\epsilon = 1/25$				$\epsilon = 1/200$			
	Example				Example			
ℓ	3.1.1	3.1.2	3.1.3	3.1.4	3.1.1	3.1.2	3.1.3	3.1.4
4	4	3	3	3	3	3	6	6
5	3	3	4	2	3	3	4	5
6	4	3	4	2	3	3	3	3
7	4	4	5	2	3	3	3	2

Suppose a less sophisticated strategy is used to define the multigrid algorithm. For example, Galerkin discretization could be used for the coarse grid discretizations, see Problem 4.8. Alternatively, the smoothing strategy could be derived from splitting operators that do not take into account properties of the flow. It should come as no surprise that with either of these choices the outcome is not as good.

Consider what happens if Galerkin discretization is used at every level instead of streamline diffusion, with no other modifications of the solution algorithm. We tested this approach for the benchmark problems used to generate Table 4.2. We only considered the first three values of ϵ in this table, $\epsilon = 1/25$, $1/50$ and $1/100$, so that reasonable accuracy is obtained on the fine grid. (The grid size is 128×128 ($h = 1/64$), and with these parameters, $\mathcal{P}_h < 1$ except when $\epsilon = 1/100$ for Examples 3.1.2 and 3.1.4, where the maximal value of $\mathcal{P}_h$ is approximately 1.5.) With this solution strategy, the multigrid algorithm is divergent for all cases. That is, Galerkin coarse grid discretization is completely ineffective.

What if instead the SD discretization is used wherever it is appropriate (i.e. wherever the mesh Peclet number is greater than one) but in combination with a less sophisticated smoother? To explore this, we replace the four-directional smoother for 3.1.4 with a single step of unidirectional Gauss–Seidel iteration. This leads to the performance depicted in Table 4.3. These results are not as bad: we are able to compute a solution for all parameter choices. However, in general, convergence is much slower than when four-directional smoothing is used, even if the higher cost per step of the latter approach is taken into account.

The empirical results of this section demonstrate the effectiveness of multigrid for solving the discrete convection–diffusion equation, when good choices are made for discretization and smoothing. Our recommended solution strategy

Table 4.2 *Multigrid V-cycle iteration counts to satisfy* $\|\mathbf{r}^{(k)}\|/\|\mathbf{f}\| < 10^{-6}$, *starting with zero initial iterate, for various ϵ on a 128×128 grid ($\ell = 7$).*

ϵ	Example			
	3.1.1	3.1.2	3.1.3	3.1.4
$1/25$	4	4	5	2
$1/50$	4	3	3	2
$1/100$	3	3	4	2
$1/200$	3	3	3	2
$1/400$	3	3	3	3
$1/800$	3	5	5	9

Table 4.3 *Multigrid V-cycle iteration counts to satisfy $\|\mathbf{r}^{(k)}\|/\|\mathbf{f}\| < 10^{-6}$, starting with zero initial iterate, for Example 3.1.4. Left: ℓ is the grid refinement level, $h = 2^{1-\ell}$, and the grid has dimensions $2^\ell \times 2^\ell$. Right: level $\ell = 7$.*

	ϵ		ϵ	Iters
ℓ	1/25	1/200	1/25	6
			1/50	8
4	9	18	1/100	29
5	15	23	1/200	40
6	8	31	1/400	52
7	6	40	1/800	74

consists of:

1. coarse grid operators defined by streamline diffusion discretization;
2. smoothing defined by flow-following stationary iteration, where complex flows are handled by multi-directional sweeps;
3. ordinary grid transfer operators defined by subspace inclusion and interpolation.

We will study the impact of these choices in more depth in Sections 4.3.2 and 4.3.3, as well as the reasons the alternative strategies have difficulties.

Remark 4.11. The first and third of these recommendations are unaffected by the structure of the underlying grid, but this is not true of the second one. In particular, for unstructured finite element meshes, there are no natural "lines" of the type we have used here. One way to handle this is to divide the domain into strips and then group unknowns according to their locations within the strips. Another, simpler, approach is to number grid points in lexicographical order according to their Cartesian coordinates, and then to use point relaxation. For a "horizontal" ordering, grid points are sorted in increasing order of their y-coordinates, and for each fixed value of y, in increasing order of their x-coordinates. Either strategy can be adapted to handle multi-directional sweeps. We have found the latter approach to be effective for unstructured meshes, see Wu & Elman [218].

Before concluding this discussion, we comment further on the accuracy and performance achieved in the various examples considered here. Throughout this section, we have been making the assumption that the fine grid discretization achieves "acceptable" accuracy. As we showed in Section 3.4, the streamline diffusion discretization achieves good accuracy outside of regions containing exponential boundary layers. For problems whose solutions contain only characteristic layers, the width of the layers is proportional to $\sqrt{\epsilon}$, and we can expect reasonable solutions on all of the domain Ω provided the mesh size h is of this magnitude or smaller. Among the benchmark problems considered here, Examples 3.1.2 and 3.1.4 have characteristic layers but not exponential layers, and the parameter combinations considered in the experiments described here produce qualitatively good solutions everywhere in Ω. For example, for the smallest value $\epsilon = 1/800$ used in Tables 4.2 and 4.3, $\sqrt{\epsilon} \approx 1/28$ and $h = 1/64$.

The other two problems, Examples 3.1.1 and 3.1.3, contain exponential boundary layers. If the mesh Peclet number is greater than one, then the global error is large, but, again, with SD discretization the errors are much smaller away from the boundary layers, see setion 3.4.1. In this situation, a posteriori error estimators (Section 3.4.2) can be used to identify regions where the error is large, and then local mesh refinement strategies can be used to improve the quality of discrete solutions.

Thus, for either type of boundary layer, it is important that solvers for the discrete convection–diffusion equation be reliable even when the mesh Peclet number becomes large. This is true for the multigrid strategies considered here, although it is worth noting that for some of the larger mesh Peclet numbers considered, performance degrades slightly, as for Example 3.1.4 with $\epsilon = 1/200$ on a 16×16 grid (Table 4.1) and $\epsilon = 1/800$ on a 128×128 grid (Table 4.2). It is difficult to give a complete explanation of this, but our suspicion is that it is caused by algebraic properties of the system rather than by the presence of errors in the discrete solution. When the mesh Peclet number becomes large, the character of the algebraic system changes: if $\mathcal{P}_h$ is small, then the coefficient matrix is diagonally dominant. This not the case when $\mathcal{P}_h > 1$, and this may affect the quality of smoothers. On the other hand, the exponential boundary layers of Examples 3.1.1 and 3.1.3 are not well resolved in many of the cases, but this in and of itself does not appear to have a significant impact on performance of the solvers.

Finally, we touch upon the question of what happens in the limiting case as $\epsilon \to 0$. Recall from (3.71) that the coefficient matrix has the form $F = \epsilon A + S + N$, so that if $\epsilon \to 0$ for h fixed, then F is independent of ϵ; moreover, as observed in Computational Exercise 3.3, the discrete solution also becomes independent of ϵ. Hence, this scenario is of limited practical interest. A more important question concerns what happens if resolution is also improved as the Peclet number is increased. This could be done by refining the mesh either globally or locally in the regions where layers appear. Although this is beyond the scope of this book, we have experience in [218] that indicates that the methods studied here are robust with respect to this question. If ϵ is reduced in conjunction with adaptive refinement of the mesh, there is a very slight degradation in performance of the multigrid solvers considered here.

4.3.2   *Tools of analysis: smoothing and approximation properties*

As explained in Chapter 2, at step i, the iteration errors for the two-grid algorithm with no postsmoothing satisfy

$$\mathbf{e}^{(i+1)} = \left[F^{-1} - P\bar{F}^{-1}P^T \right] \left[F(I - M^{-1}F)^k \right] \mathbf{e}^{(i)}, \tag{4.39}$$

where $\bar{F}$ is the coarse grid operator and k is the number of smoothing steps. The traditional analysis of multigrid depends on establishing a smoothing property

$$\|F(I - M^{-1}F)^k \mathbf{y}\| \leq \eta(k)\|\mathbf{y}\| \text{ with } \eta(k) \rightarrow 0 \text{ as } k \rightarrow \infty, \tag{4.40}$$

and an approximation property

$$\|(F^{-1} - P\bar{F}^{-1}P^T)\mathbf{y}\| \leq C\|\mathbf{y}\|,$$

for all vectors $\mathbf{y} \in \mathbb{R}^n$. At the time of this writing, there are no bounds of this type applicable to standard multigrid algorithms for the convection–diffusion equation. Nevertheless, insight into the multigrid process can be gained from empirical study of these quantities. In contrast to the Poisson equation, where it is natural to use the discrete energy norm, here we use the Euclidean norm in $\mathbb{R}^n$ and examine the two matrix norms

$$\|F(I - M^{-1}F)^k\| \quad \text{and} \quad \|F^{-1} - P\bar{F}^{-1}P^T\|.$$

We restrict our attention to the benchmark problems introduced in Chapter 3 that contain only characteristic layers, Examples 3.1.2 and 3.1.4. As we observed at the end of the previous section, for these two examples, accuracy of solutions is not degraded for moderately large values of $\mathcal{P}_h$. (It should also be noted that our experience with the other Example problems is similar to that described here.) There are two main points, which are highlighted by the data in Table 4.4 for the approximation property, and in Tables 4.5 and 4.6 for the smoothing property.

Table 4.4 addresses the approximation property, with computed values of $\|F^{-1} - P\bar{F}^{-1}P^T\|$ for both the SD and Galerkin discretizations. The goal here is for these quantities to be independent of the discretization mesh size h as $h \rightarrow 0$. This is true for the larger diffusion coefficient $\epsilon = 1/25$, but the trends are not clear for $\epsilon = 1/200$. For Example 3.1.4, the norm for the SD discretization is decreasing as h is reduced, whereas for Example 3.1.2 with both discretization strategies, it is increasing for the ranges of h in the table. What is evident, however, is that with respect to this measure, the Galerkin discretization is not dramatically different from the streamline diffusion discretization. Although Galerkin discretization does not represent an accurate approximation to the continuous operator when $\mathcal{P}_h$ is large, the coarse grid operator $\bar{F}$ is still "close" to the fine grid operator F in this measure.

Table 4.4 *Approximation property norms* $\|F^{-1} - P\bar{F}^{-1}P^T\|$ *for Examples 3.1.2 and 3.1.4, for $\epsilon = 1/25$ and $1/200$ and two discretization strategies. Here, ℓ is the grid refinement level, $h = 2^{1-\ell}$, and the grid has dimensions $2^\ell \times 2^\ell$.*

| | $\epsilon = 1/25$ | | | | $\epsilon = 1/200$ | | | |
| | Example 3.1.2 | | Example 3.1.4 | | Example 3.1.2 | | Example 3.1.4 | |
ℓ	SD	Galerkin	SD	Galerkin	SD	Galerkin	SD	Galerkin
4	13.1	24.7	12.6	12.9	62.1	88.0	301.0	104.7
5	32.2	34.5	12.5	12.5	92.6	185.5	192.5	113.5
6	40.0	40.4	12.6	12.6	117.8	342.3	119.3	110.4
7	42.6	43.0	12.7	12.7	330.0	568.4	108.7	110.4

Table 4.5 *Smoothing property norms* $\|F(I - M^{-1}F)^k\|$ *for Examples 3.1.2 and 3.1.4 with streamline-diffusion discretization. Here, ℓ is the grid refinement level, $h = 2^{1-\ell}$, and the grid has dimensions $2^\ell \times 2^\ell$.*

| | $\epsilon = 1/25$ | | | | $\epsilon = 1/200$ | | | |
| | Example 3.1.2 | | Example 3.1.4 | | Example 3.1.2 | | Example 3.1.4 | |
ℓ	$k=1$	$k=2$	$k=1$	$k=2$	$k=1$	$k=2$	$k=1$	$k=2$
4	0.031	0.006	0.008	0.002	0.004	0.0007	0.022	0.008
5	0.032	0.006	0.007	0.003	0.004	0.0008	0.006	0.002
6	0.028	0.006	0.006	0.004	0.004	0.0008	0.001	0.0002
7	0.040	0.010	0.006	0.003	0.004	0.0008	0.0008	0.0005

Table 4.6 *Smoothing property norms* $\|F(I - M^{-1}F)^k\|$ *for Examples 3.1.2 and 3.1.4 with Galerkin discretization. Here, ℓ is the grid refinement level, $h = 2^{1-\ell}$, and the grid has dimensions $2^\ell \times 2^\ell$.*

| | $\epsilon = 1/25$ | | | | $\epsilon = 1/200$ | | | |
| | Ex. 3.1.2 | | Ex. 3.1.4 | | Ex. 3.1.2 | | Ex. 3.1.4 | |
ℓ	$k=1$	$k=2$	$k=1$	$k=2$	$k=1$	$k=2$	$k=1$	$k=2$
4	119.8	2276.0	58.7	3×10^4	2×10^{14}	2×10^{17}	5×10^{27}	3×10^{56}
5	0.678	4.57	0.018	0.010	4×10^{16}	4×10^{18}	1×10^{52}	4×10^{105}
6	0.028	0.006	0.006	0.004	5×10^{19}	3×10^{22}	7×10^{71}	8×10^{145}
7	0.040	0.010	0.006	0.003	7×10^{34}	2×10^{37}	7×10^{69}	9×10^{141}

Tables 4.5 and 4.6 treat the smoothing property. Table 4.5 shows the norms $\|F(I - M^{-1}F)^k\|$ when the streamline diffusion method is used to discretize at all levels. The smoothers were those used in Section 4.3.1, one flow-following block Gauss–Seidel iteration for Example 3.1.2, and one four-directional block Gauss–Seidel iteration for Example 3.1.4. It is evident that this approach is effective, and the results suggest that the norm decreases in a manner required for the smoothing property (4.40) as the number of smoothing steps k increases. Table 4.6 shows analogous results when Galerkin discretization is used at all levels. Here, the contrast is dramatic: even though in principle the smoothers are tied to the underlying physics, they are completely ineffective when the Galerkin operators are inaccurate. The numbers in the table are too large to have real meaning, and we show them largely for dramatic effect, to emphasize how poorly these methods perform. Despite the intention of having the smoothers follow the flows, when the discrete operators do not represent these flows well, the smoothers are wildly ineffective.

Taken together, these results suggest that the difficulty that has for Galerkin discretizations come from the lack of an appropriate smoothing property. Because the approximation property is under control (or nearly so), this could in principle be overcome by devising alternative smoothing strategies. There is little need for this, though, since SD eliminates any such difficulties.

We also note an alternative interpretation of these data, namely, that these measures are not the best means of analyzing multigrid for this problem. This point of view has been taken by Olshanskii & Reusken [140], where an analysis based on a different factorization of the iteration matrix has been performed. Loosely speaking, in this analysis the splitting operator is incorporated into the approximation property. This leads to a nearly optimal convergence result for problems with grid aligned flow and outflow boundary conditions.

4.3.3   *Smoothing*

In this section we present a few graphical depictions of the effect of smoothing with the aim of increasing our intuitive understanding of the smoothing process for the convection–diffusion equation. Given a starting iterate $\mathbf{u}^{(0)}$, let $\tilde{\mathbf{u}}$ denote the result of performing some small number of steps of a stationary iteration (4.17). The goal of smoothing is to enable the resulting iteration error to be approximated well on a coarser grid. We emphasize here that it is the iteration error $\mathbf{u} - \tilde{\mathbf{u}}$ (to which there corresponds a functional error $u_h - \tilde{u}_h$) that is the concern here; this discussion does not have any bearing on the discretization error $u - u_h$.

For the purposes of illustration, we will use random initial data consisting of uniformly distributed random numbers in the interval $[-1, 1]$, which reveal the effects of smoothing more clearly than a smooth initial iterate such as zero. Figure 4.9 shows the initial error $\mathbf{u} - \mathbf{u}^{(0)}$ for Example 3.1.2, for $\epsilon = 1/200$ and streamline diffusion discretization on a 32×32 grid. The right side of this figure

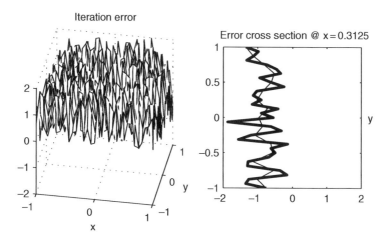

FIG. 4.9. Iteration error for Example 3.1.2 and cross-section of the iteration error along a selected vertical line, for $\epsilon = 1/200$, SD discretization on a 32×32 grid ($\ell = 5$), and a random initial iterate.

displays a cross-section of the initial error along a selected vertical line running across the domain at $x = .3125$. The cross-section shown here and throughout this section also contain thinner lines, corresponding to the restriction of the errors to a coarse 16×16 subgrid. It is obvious that the initial error is highly oscillatory. The most significant aspect of this from the point of view of multigrid is that, as seen in the cross-section, the fine grid error oscillates dramatically around its coarse grid restriction.

Now consider the smoothing effects of block Gauss–Seidel iteration. The plots at the top of Figure 4.10 show the result of performing one such smoothing step with sweeps that follow the flow, from bottom to top, starting with the initial data of Figure 4.9. As expected, this smoother does an excellent job of reducing all components, both oscillatory and smooth, of the error. Indeed, along the vertical cross-section depicted, the error and its restriction are indistinguishable. In contrast, the plots at the bottom of the figure show what happens when the Gauss–Seidel iteration sweeps go from top to bottom, opposite the direction of flow. In this case, we used *four* smoothing steps, and it can be seen that the error is smooth (and small) in the four grid lines processed at the end of the sweep, that is, at the bottom of the domain. But this iteration does not smooth the error in most of the domain.

Figure 4.11 examines similar issues for Example 3.1.4, which has a recirculating wind. The top part of the figure displays the errors obtained after one step of the four-directional block Gauss–Seidel iteration. These images are more typical of what multigrid achieves: the error is not as small as that occurring for the simpler Example 4.10, but it is *smooth* in all directions. This contrasts with

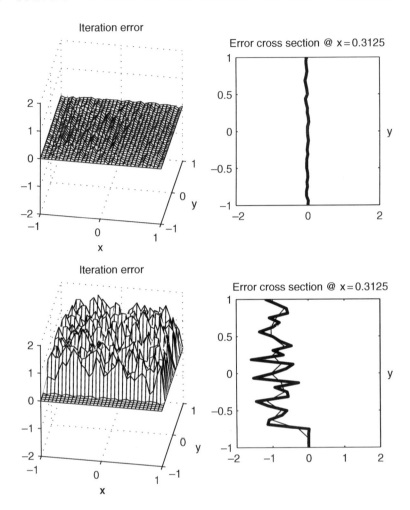

FIG. 4.10. Iteration error for Example 3.1.2 and cross-sections of the iteration error along a selected vertical line, for $\epsilon = 1/200$ and SD discretization on a 32×32 grid ($\ell = 5$). Top: after one flow-following block Gauss–Seidel iteration. Bottom: after four block Gauss–Seidel iterations against the flow.

the outcome obtained when a strategy that ignores the underlying physics is used, such as a unidirectional (bottom-to-top) line iteration. The bottom part of Figure 4.11 shows the results of this approach, where four steps of the bottom-to-top iteration are performed; this makes the computational effort essentially the same as for the four-directional smoother. It is evident that the unidirectional iteration fails to eliminate oscillations in part of the domain and is ineffective as a smoother. These observations are consistent with those made in Section 4.3.1

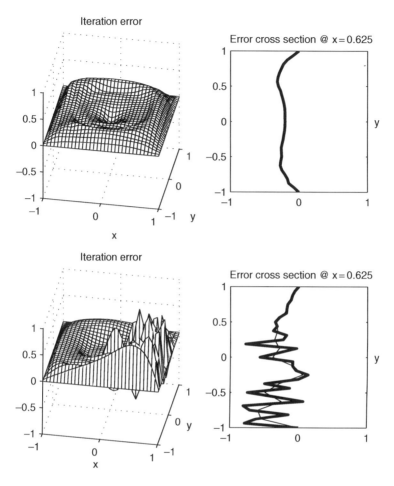

FIG. 4.11. Iteration error for Example 3.1.4 and cross-sections of the iteration
error along a selected vertical line, for $\epsilon = 1/200$ and SD discretization on
a 32×32 grid ($\ell = 5$). Top: after one four-directional block Gauss–Seidel
iteration. Bottom: after four bottom-to-top Gauss–Seidel iterations.

and they further demonstrate that a flow-following Gauss–Seidel iteration is a
good smoother for this problem.

4.3.4 *Analysis*

As we have noted, the analysis of multigrid methods for the convection–diffusion
equation is incomplete. In this section, we describe the main issue that makes
analysis difficult, and we outline some approaches that have been used to
circumvent it.

First, consider what goes wrong with the proof used in Section 2.5.1 to establish the approximation property for the Poisson equation in $\mathbb{R}^2$. We start with the convection–diffusion equation of (3.1) with forcing function f. Let u_h denote the discrete solution on a grid designated as the fine grid, and suppose both the fine and coarse grids are such that good resolution is achieved by Galerkin discretization. Let $F\mathbf{u} = \mathbf{y}$ denote the linear system of equations derived from the fine grid where $\mathbf{u} = F^{-1}\mathbf{y}$ contains the coefficients of u_h. Let $\bar{\mathbf{u}} = \bar{F}^{-1}R\mathbf{y}$ denote the analogous coefficients of the coarse grid solution u_{2h}. Exactly as in the proof of Theorem 2.9, the relation

$$\|(F^{-1} - P\bar{F}^{-1}R)\mathbf{y}\| = \|\mathbf{u} - P\bar{\mathbf{u}}\|$$

holds, where here the Euclidean norm is used instead of the discrete energy norm. The original argument can be continued by observing that

$$\begin{aligned}
\|\mathbf{u} - P\bar{\mathbf{u}}\| &\leq \left(\tfrac{1}{\lambda_{\min}(Q)}\right)^{1/2} \|u_h - u_{2h}\| \\
&\leq L\left(\tfrac{1}{\lambda_{\min}(Q)}\right)^{1/2} \|\nabla u_h - \nabla u_{2h}\| \\
&\leq L\left(\tfrac{1}{\lambda_{\min}(Q)}\right)^{1/2} (\|\nabla(u_h - u)\| + \|\nabla u_{2h} - \nabla u\|) \\
&\leq CL\left(\tfrac{1}{\lambda_{\min}(Q)}\right)^{1/2} h\|D^2 u\|,
\end{aligned}$$

where Q is the mass matrix, L is the Poincaré constant (see Lemma 1.2), and C is the constant from the a priori error bound in Theorem 3.4. Thus, noting that C is proportional to the mesh Peclet number $\mathcal{P}_h = h\|\vec{w}\|_\infty/\epsilon$, the fact that $\lambda_{\min}(Q)$ is proportional to h^2 in $\mathbb{R}^2$ gives the bound

$$\|(F^{-1} - P\bar{F}^{-1}R)\mathbf{y}\| \leq C\frac{h}{\epsilon}\|D^2 u\|,$$

where here, and in the remainder of this discussion, C represents a generic constant that does not depend on h or ϵ.

The proof of Theorem 2.9 is completed by invoking $\mathcal{H}^2$-regularity, which relates the second derivative of u to f. Unfortunately, if the solution of the convection–diffusion problem has layers, then this relation depends very strongly on ϵ:

$$\|D^2 u\| \leq \left(\frac{C}{\epsilon^\alpha}\right)\|f\|,$$

where $\alpha = 3/2$ if a Dirichlet condition holds on the outflow boundary and $\alpha = 1$ if a Neumann condition applies, see Roos et al. [159, p. 186]. Although mesh independence can be obtained through Lemma 2.8, the damage is done, as the constant that bounds

$$\frac{\|(F^{-1} - P\bar{F}^{-1}R)\mathbf{y}\|}{\|\mathbf{y}\|}$$

is of the form $1/\epsilon^{1+\alpha}$. This is too large to be handled by a compensating smoothing property bound.[4]

The weakness in this argument appears to lie in the decision to relate the approximation property to the discretization error. Recall, however, the observation made in Section 4.3.2, that in fact the approximation property nearly holds even if the discretization error is large. More precisely, the results of Table 4.4 suggest that the dependence on ϵ is closer to $1/\epsilon$ than what is obtained from regularity theory. Thus, it may still be possible to establish that two successive discrete problems are near enough to each other to enable multigrid to work well.

There is an extensive literature establishing rigorous bounds that demonstrate mesh-independent rates of convergence of multigrid for the discrete convection–diffusion equations in certain cases. To complete this discussion, we briefly discuss two such approaches.

1. PERTURBATION ARGUMENTS. The convection–diffusion operator can be written as

$$\mathcal{L}u = \mathcal{A}u + \mathcal{G}u,$$

where

$$\mathcal{A}u = -\epsilon\nabla^2 u + u, \quad \mathcal{G}u = \vec{w}\cdot\nabla u - u.$$

By viewing $\mathcal{G}$ as a perturbation of the self-adjoint operator $\mathcal{A}$, it is possible to generalize results for the self-adjoint problem to establish that multigrid methods display a convergence rate independent of the grid size or nearly so; see Bank [8], Bramble, Pasciak & Xu [24], Mandel [128], or Wang [209]. If $e_h^{(k)} = u_h - u_h^{(k)}$ is the (functional) iteration error for the kth multigrid step, then a bound requiring no assumptions on regularity (only that $u \in \mathcal{H}^1(\Omega)$) [209] is

$$\|e_h^{(k)}\|_1 \le \left(1 - \frac{1}{c(\epsilon)\log h^{-1}}\right)^k \|e_h^{(0)}\|_1,$$

where $c(\epsilon) = O(1/\epsilon)$. This leads to iteration counts on the order of $\epsilon^{-1}\log h^{-1}$. The extra factor of $\log h^{-1}$ makes this result marginally less than optimal and can be improved by making more stringent regularity assumptions. The dependence on the Peclet number is more dramatic than occurs in practice. These results require that the coarse grid width depends on ϵ.

---

[4]An alternative tactic for modifying the proof of Theorem 2.9 would involve bounding $\|u_h - u_{2h}\|$ directly using duality arguments, see Braess [19]. However, essentially the same conclusions would be reached.

2. A MATRIX-ANALYTIC APPROACH. Noting the difficulties associated with using regularity results, Reusken in [157] has developed a methodology based purely on linear algebraic arguments. This approach is for two-dimensional problems with constant coefficients and grid-aligned flow; the latter restrictions are typical of algebraic arguments and in this case allows the use of Fourier analysis to transform the coefficient matrix F to a set of tridiagonal matrices that are shown to have the structure of one-dimensional discrete convection–diffusion operators. (The same idea is used for a different purpose in Section 3.5.1.) The ideas are developed for a multigrid strategy that uses *semicoarsening*, in which coarsening of grids is performed in only one of the two dimensions, together with matrix-dependent grid-transfer operators. With these assumptions, a rigorous analysis establishes bounds on the Euclidian norms for the approximation property that are independent of both the discretization mesh size h and the diffusion coefficient ϵ. We also note that when algebraic multigrid methods are applied to the convection–diffusion equation, the coarse grids typically have a structure like that obtained from semicoarsening along streamlines.

Discussion and bibliographical notes

It has long been recognized that the combination of Krylov subspace methods with preconditioning is a good formula for development of efficient iterative methods. Preconditioners can be tailored to the problem being solved and in the best circumstances they enable convergence that is insensitive to fundamental aspects of the problem such as, in the present setting, the grid size or the Peclet number. Krylov subspace methods are of general applicability and enhance the effectiveness of preconditioners by reducing the effects of outliers (among eigenvalues) and conditioning. Thus, each of these components of the solution process has a dual role: preconditioners improve the performance of Krylov subspace methods by improving the conditioning (loosely defined) of the problem, or Krylov methods accelerate the convergence of iterative methods determined by preconditioning or splitting operators.

4.1. Early work on Krylov subspace methods for nonsymmetric systems of linear equations appears in the petroleum reservoir simulation literature, particularly Vinsome [206]. The major developments then follow two general trends, beginning with a flurry of activity from the late 1970s through the mid-1980s. These efforts were largely concerned with algorithms derived from optimality criteria, based on either minimization ("Ritz") or orthogonality ("Galerkin") conditions. (Theorem 2.2 shows that the conjugate gradient method satisfies both conditions.) Representative papers include Axelsson [5], Eisenstat, Elman & Schultz [55], Saad [162], and Young & Jea [221]. This work culminated with the GMRES algorithm developed by Saad & Schultz [165], in which the residual norm of the kth iterate is minimized over a Krylov space of dimension k. Optimality is achieved at a cost, however, in the sense that the complexity of the computation at the kth step of the iteration increases with the iteration counter k.

A collection of results from the late 1980s and 1990s emphasize suboptimal but efficient algorithms involving short-term recurrences. These are derived using biorthogonality conditions and can be viewed as generalizations of the biconjugate gradient algorithm originally proposed by Fletcher [76] in 1976. They include the BICG-SQUARED method developed by Sonneveld [180], the QUASI-MINUMUM RESIDUAL method of Freund & Nachtigal [80], van der Vorst's BICGSTAB method [198], and the BICGSTAB(ℓ) method developed by Sleijpen & Fokkema [178].

Two papers by Faber & Manteuffel [71, 72] demonstrate that there is no alternative to these choices; that is, for linear systems with general nonsymmetric coefficient matrices, there is no Krylov subspace method that is both optimal and uses short-term recurrences of fixed length. This question has also been addressed in the Russian literature; see Voevodin [207].

There is now an extensive literature on Krylov subspace methods. More detailed treatments and bibliographies can be found in the books by Axelsson [6], Greenbaum [89], Meurant [131], Saad [164], and Van der Vorst [199], and in the surveys by Eiermann & Ernst [54] and Freund, Golub & Nachtigal [79].

4.2. As we have observed, preconditioners tend to be more closely tailored to the problems to which they are applied. For the discrete convection–diffusion equation considered in this chapter, our emphasis has been on classical relaxation methods. These were the iterative methods of choice in the 1950s and their use predates that of computers, see discussions of this point in the books of Varga [202, p. 1] and Young [220, p. 456]. The techniques of analysis used in Sections 4.2.2 and 4.2.3 build on results on matrix splittings discussed in these references, although the utility of relaxation methods for convection–diffusion problems was not discussed there. This analysis can also be combined with matrix comparison theorems to establish convergence bounds for incomplete factorization preconditioners, see Elman & Golub [64].

The capability of relaxation methods to mimic the physics of the underlying process was pointed out by Chin & Manteuffel [40] and studied by Elman & Chernesky [62, 63]. It is also a well-established component of smoothing strategies used with multigrid for the convection–diffusion equation, see, for example, Hackbusch [97, section 10.4.3], Brandt & Yavneh [27]. This has also led to the development of *flow-directed* orderings of the underlying grid for problems without recirculations, as discussed in Han et al. [100], Farrell [73], Wang & Xu [208] as well as Bey & Wittum [15], Hackbusch & Probst [99] in the context of multigrid. The use of multi-directional sweeps essentially of the type considered in Section 4.2.4 to handle complex flows has also been considered in the multigrid literature, see Wesseling [214]. Insight into the impact of ordering on relaxation, in particular, the fact that the wrong ordering causes difficulties, can also be obtained by considering *pseudospectra* of the iteration matrix, see Trefethen [195].

Additional results on preconditioners can be found in many of the references listed above [6, 89, 131, 164, 199], as well as Hackbusch [98].

4.3. The standard means of demonstrating the effectiveness of multigrid is to allow the discretization mesh to become small and show that multigrid iteration counts are not affected by this. In the case of the convection–diffusion equation, a standard aspect of this (singularly perturbed) problem is to explore what happens when the diffusion coefficient becomes small. As has been seen in Chapter 3, it is not possible to separate the discretization from this process. Despite this, it is common to consider cases where ϵ is smaller than h, and because of this, the state of understanding of multigrid for this problem class is less crisp than for other topics treated in this book. Despite this difficulty, it is now a generally accepted fact that multigrid can be used to solve the discrete convection–diffusion equation efficiently. This message can be gleaned from standard references on multigrid that treat this subject, for example, Hackbusch [97, chapter 10], Trottenberg et al. [196, chapter 7], or Wesseling [214].

It is important that something be done to handle the instabilities associated with coarse grids, and that the properties of the flow be accounted for. The approach we have taken, streamline diffusion discretization coupled with suitable multidirectional relaxation, was shown to be effective in Ramage [154]. Our emphasis on streamline-diffusion for the first task comes from an effort to take a consistent point of view with respect to the discretization and solution strategies; in the context of finite element discretization, SD is straightforward and natural to use. Alternatives include matrix-dependent grid transfer operators, see Dendy [49], de Zeeuw [47]; or various upwinding strategies, as described in Bey & Wittum [15], Brandt & Yavneh [27], or de Zeeuw & van Asselt [48]. The latter approaches tend to be easy to use in the finite difference setting. Our choice of multidirectional relaxation as a smoother for complex flows stems from its simplicity. As noted above, for inflow-outflow problems, flow-directed orderings [15, 73, 99, 100, 208] constitute an alternative to this approach. Another strategy designed for flows with closed characteristics can be found in Yavneh, Venner & Brandt [219].

In addition to the methods of analysis discussed above, perturbation analysis and matrix oriented approaches, there has been a large amount of heuristic Fourier analysis of multigrid for the discrete convection–diffusion equation. This technique dates back to Brandt [25] and it is discussed at length for the convection–diffusion equation in Wesseling [214]. Although it falls short of providing rigorous bounds, it often leads to important insight. For example, Fourier analysis reveals the utility of so-called "alternating zebra" ordering strategies for this class of problems. These combine reordering strategies based on block red–black (or "white–black", whence zebra) permutations of the grid of the type considered in Section 4.2.1, with multiple directions of sweeps. Relaxation based on this strategy does not lead to an effective preconditioner for use with Krylov subspace methods, but it is effective as a smoother for multigrid and has the advantage, because of the red–black permutations, of being efficient to implement in parallel.

Problems

4.1. Identify the sequence of steps (Givens rotations) needed to transform the matrix $\hat{H}_k$ of (4.4) into upper triangular form, and show how to obtain the norm $\|\mathbf{r}^{(k)}\|$ as a byproduct of this computation.

4.2. Prove Proposition 4.3 using the following sequence of steps:

a. Show that the residual after the first GMRES iteration is given by $\mathbf{r}^{(1)} = \mathbf{r}^{(0)} - P\mathbf{r}^{(0)}$ where P is the orthogonal projector with respect to the Euclidean norm onto $\mathrm{span}\{F\mathbf{r}^{(0)}\}$. Use this to derive an expression for the scalar α_0 such that $\mathbf{r}^{(1)} = \mathbf{r}^{(0)} - \alpha_0 F\mathbf{r}^{(0)}$.

b. Show that

$$\left(\frac{\|\mathbf{r}^{(k)}\|}{\|\mathbf{r}^{(0)}\|}\right)^2 = 1 - \frac{\langle F\mathbf{r}^{(0)}, \mathbf{r}^{(0)}\rangle}{\langle \mathbf{r}^{(0)}, \mathbf{r}^{(0)}\rangle} \frac{\langle F\mathbf{r}^{(0)}, \mathbf{r}^{(0)}\rangle}{\langle F\mathbf{r}^{(0)}, F\mathbf{r}^{(0)}\rangle}.$$

c. Bound the quotients on the right side of the expression above as

$$\frac{\langle F\mathbf{r}^{(0)}, \mathbf{r}^{(0)}\rangle}{\langle \mathbf{r}^{(0)}, \mathbf{r}^{(0)}\rangle} \geq \lambda_{\min}(H)$$

$$\frac{\langle F\mathbf{r}^{(0)}, \mathbf{r}^{(0)}\rangle}{\langle F\mathbf{r}^{(0)}, F\mathbf{r}^{(0)}\rangle} \geq \frac{1}{\lambda_{\max}\left(\left(\frac{F^{-1}+F^{-T}}{2}\right)^{-1}\right)}.$$

d. Show that

$$\lambda_{\max}\left(\left(\frac{F^{-1}+F^{-T}}{2}\right)^{-1}\right) \leq \lambda_{\max}(H) + \rho(S)^2/\lambda_{\min}(H).$$

(Hint: Use the identity $\frac{X^{-1}+Y^{-1}}{2} = X^{-1}\left(\frac{X+Y}{2}\right)Y^{-1}$.)

4.3. a. Prove that the vectors $\{\mathbf{v}^{(j)}\}$ and $\{\mathbf{w}^{(j)}\}$ generated by the Lanczos process (4.10) satisfy $\langle \mathbf{v}^{(i)}, \mathbf{w}^{(j)}\rangle = \delta_{ij}$.
b. Prove that the residual generated by the nonsymmetric Lanczos method satisfies the relation (4.12).

4.4. Following the approach of Section 2.4, show that there is a connection between the BICG algorithm and the Lanczos construction (4.10).

4.5. Use the results of Problems 4.3a and 4.4 to prove that the scalars α_k and β_k given in the BICGSTAB method (Algorithm 4.3) are the same as those of the BICG method (Algorithm 4.2).

4.6. Show that the stationary iteration (4.17) produces a sequence of iterates $\{\mathbf{u}^{(k)}\}$ satisfying

$$\mathbf{r}^{(k)} = p_k(FM^{-1})\,\mathbf{r}^{(0)}$$
$$M^{-1}\mathbf{r}^{(k)} = p_k(M^{-1}F)\,(M^{-1}\mathbf{r}^{(0)}),$$

where $\mathbf{r}^{(k)} = \mathbf{f} - F\mathbf{u}^{(k)}$ and $p_k(0) = 1$. This implies that if F and M do not commute, then different residual polynomials will be generated by Krylov subspace solution of the left-oriented and right-oriented preconditioned systems (4.15).

4.7. Consider the convection–diffusion problem with grid-aligned flow ($\vec{w} = (0,1)$, $\Omega = (0,1) \times (0,1)$) together with a Dirichlet boundary condition. Let the problem be discretized using the SD formulation on a uniform grid, so that the computational molecule is given in (3.84). Use Fourier analysis to derive a bound analogous to that in Theorem 4.4 on the spectral radius of the line Jacobi iteration matrix.

4.8. Suppose F^h represents the discrete convection–diffusion operator obtained from Galerkin discretization (3.21). Show that the coarse grid version of this discrete operator satisfies $F^{2h} = P^T F^h P$ where $P = R^T$ is the prolongation operator derived from the natural inclusion of subspaces (2.42). Is the same statement true when F^h is derived from streamline-diffusion discretization (3.40)?

Computational exercises

Two iterative solver drivers are built into IFISS; a preconditioned Krylov subspace solver it_solve, and a multigrid solver mg_solve. Either of these may be called after having run the driver cd_testproblem. The function it_solve offers the possibility of using GMRES or BICGSTAB as a solver. In either case, the function calls the specialized matlab routines, gmres_r or bicgstab_l. Incomplete factorization preconditioning is achieved by calling the built-in matlab function luinc.

4.1. Use the function it_solve to compute discrete solutions of Example 3.1.2 in the cases $\epsilon = 1$, $1/10$ and $1/100$ using a 32×32 uniform grid. Compare the convergence curves obtained using GMRES, BICGSTAB, and BICGSTAB(2) without preconditioning. Which of these methods is the most cost-effective?

4.2. Use mg_solve to compare the effectiveness of Jacobi, point- and line-Gauss–Seidel and ILU as smoother for multigrid solution of Example 3.1.4. Can you explain what happens in the "bad" case?

4.3. In order to see the possibility of irregular convergence for BICGSTAB, run it_solve and select diagonal preconditioning. Then use BICGSTAB again with ILU and with multigrid preconditioning: do you notice any irregularity in the convergence with these better preconditioners?

4.4. Selecting the default values, run it_solve with multigrid preconditioning and then mg_solve for the same problem. Compare the performance with 16×16 and 32×32 grids for Examples 3.1.3 or 3.1.4.

4.5. Write a function to solve a discrete system using the stationary line Gauss–Seidel method, and hence verify the convergence plots given in Figure 4.2. Consider a 32×32 uniform grid SD discretization of Examples 3.1.2 and 3.1.4, and use your function and it_solve to assess the effectiveness of accelerating the convergence of the line Gauss–Seidel method using GMRES.

4.6. Write a function to postprocess the default grid data generated by IFISS so as to generate a red–black ordering of the unknowns as illustrated in Figure 4.3. Consider a 32×32 uniform grid SD discretization of Examples 3.1.2 and 3.1.4, and compare results using a conventional and red–black ordering of unknowns when using the line Gauss–Seidel splitting and ILU(0) factorization as preconditioners for GMRES.

5

THE STOKES EQUATIONS

The Stokes equation system

$$-\nabla^2 \vec{u} + \nabla p = \vec{0}$$
$$\nabla \cdot \vec{u} = 0 \tag{5.1}$$

is a fundamental model of viscous flow. The variable $\vec{u}$ is a vector-valued function representing the velocity of the fluid, and the scalar function p represents the pressure. Note that the Laplacian of a vector function is simply the vector obtained by taking the Laplacian of each component in turn. The first equation in (5.1) represents conservation of the momentum of the fluid (and so is the *momentum equation*), and the second equation enforces conservation of mass. This second equation is also referred to as the *incompressibility constraint*. The crucial modeling assumption made is that the flow is "low-speed", so that convection effects can be neglected. Such flows arise if the fluid is very viscous or else is tightly confined; examples are dust particles settling in engine oil, and the flow of blood in parts of the human body.

Of primary importance in this chapter is that the Stokes equations represent a limiting case of the more general Navier–Stokes equations, which are studied in Chapter 7. For both the Stokes and Navier–Stokes equations, the fact that the incompressibility constraint does not involve the pressure variable makes the construction of finite element approximations problematic. In particular, the discrete spaces used to approximate the velocity and pressure fields cannot be chosen independently of one another — there is a compatibility condition that needs to be satisfied if the resulting *mixed* approximation is to be effective. Unfortunately, as we will see, the simplest and most natural mixed approximation methods are unsuitable. The basic aim in this chapter is to identify the finite element spaces that are compatible, so as to define candidate approximation methods for the Navier–Stokes equations in Chapter 7.

The boundary value problem considered is equation (5.1) posed on a two- or three-dimensional domain Ω, together with boundary conditions on $\partial\Omega = \partial\Omega_D \cup \partial\Omega_N$ given by

$$\vec{u} = \vec{w} \text{ on } \partial\Omega_D, \quad \frac{\partial\vec{u}}{\partial n} - \vec{n}p = \vec{s} \text{ on } \partial\Omega_N, \tag{5.2}$$

where $\vec{n}$ is the outward-pointing normal to the boundary, and $\partial\vec{u}/\partial n$ denotes the directional derivative in the normal direction. The boundary conditions in (5.2) are analogous to the boundary conditions (1.14) for the Poisson problem in

Chapter 1. In order to ensure a unique velocity solution, it is assumed that the Dirichlet part of the boundary is nontrivial, that is $\int_{\partial\Omega_D} ds \neq 0$. Following the notation in Chapter 3, the Dirichlet boundary segment can be further subdivided according to its relation with the normal component of the imposed velocity:

$$\begin{aligned}
\partial\Omega_+ &= \{x \in \partial\Omega \,|\, \vec{w} \cdot \vec{n} > 0\}, &&\text{the } \textit{outflow boundary}, \\
\partial\Omega_0 &= \{x \in \partial\Omega \,|\, \vec{w} \cdot \vec{n} = 0\}, &&\text{the } \textit{characteristic boundary}, \\
\partial\Omega_- &= \{x \in \partial\Omega \,|\, \vec{w} \cdot \vec{n} < 0\}, &&\text{the } \textit{inflow boundary}.
\end{aligned}$$

If the velocity is specified everywhere on the boundary, that is, if $\partial\Omega_D \equiv \partial\Omega$, then the pressure solution to the Stokes problem (5.1) and (5.2) is only unique up to a constant (sometimes referred to as the *hydrostatic* pressure level). Moreover, integrating the incompressibility constraint over Ω using the divergence theorem gives

$$0 = \int_\Omega \nabla \cdot \vec{u} = \int_{\partial\Omega} \vec{u} \cdot \vec{n} = \int_{\partial\Omega} \vec{w} \cdot \vec{n}. \tag{5.3}$$

Thus, non-uniqueness of the pressure is associated with a *compatibility* condition on the boundary data

$$\int_{\partial\Omega_-} \vec{w} \cdot \vec{n} + \int_{\partial\Omega_+} \vec{w} \cdot \vec{n} = 0. \tag{5.4}$$

In simple terms, the volume of fluid entering the domain must be matched by the volume of fluid flowing out of the domain.

One important class of flow problems is called *enclosed flow*, characterized by specifying $\vec{w} \cdot \vec{n} = 0$ everywhere on $\partial\Omega$ (so that (5.4) is automatically satisfied). An example is given in the next section. To generate a nontrivial flow in such cases the imposed tangential velocity must be nonzero on some part of the boundary. If, on the other hand, the flow is not enclosed (if $\int_{\partial\Omega_-} \vec{w} \cdot \vec{n} \neq 0$), then the flow problem is of inflow/outflow type, and care must be taken to ensure that (5.4) is satisfied — otherwise (5.1) and (5.2) will have no solution.

One way of circumventing this difficulty is to replace any "hard" Dirichlet outflow condition by the Neumann outflow condition in (5.2), typically taking $\vec{s} = \vec{0}$. If such a "natural outflow" condition is specified, that is, if $\int_{\partial\Omega_N} ds \neq 0$, then $\vec{u} \cdot \vec{n}$ automatically adjusts itself on the outflow boundary to ensure that mass is conserved in the sense of (5.3). Moreover, the pressure solution to (5.1) and (5.2) is always unique in this case.

For a demonstration of this point, consider a two-dimensional flow out of a channel $\Omega = [0, L] \times [0, H]$, with $\partial\Omega_N = \{(L, y) \,|\, 0 < y < H\}$ and $\vec{w} = \vec{0}$ along the top and bottom components of $\partial\Omega$. With $\vec{u} = (u_x, u_y)^T$, the natural outflow condition is

$$\frac{\partial u_x}{\partial x} - p = s_x,$$

$$\frac{\partial u_y}{\partial x} = s_y. \tag{5.5}$$

Integrating the normal component over the outflow boundary gives

$$\int_0^H \frac{\partial u_x}{\partial x}\, dy - \int_0^H p\, dy = \int_0^H s_x\, dy. \tag{5.6}$$

The incompressibility constraint, together with the fact that $u_y(H) = 0 = u_y(0)$, then gives

$$\int_0^H p\, dy = -\int_0^H \frac{\partial u_y}{\partial y}\, dy - \int_0^H s_x\, dy = -\int_0^H s_x\, dy. \tag{5.7}$$

The constraint (5.7) automatically fixes the hydrostatic pressure level. For example, in the simplest case of $s_x = 0$ in (5.5), the average pressure at the outlet is set to zero by the condition (5.7). Throughout this and later chapters, the definition of the outflow boundary will be augmented to include any boundary part $\partial\Omega_N$ where Neumann conditions are specified.

Another important quantity in fluid dynamics is the *vorticity*: $\vec{\omega} = \nabla \times \vec{u}$. To see why, take the curl of the momentum equation in (5.1), giving

$$\vec{0} = \nabla \times \nabla^2 \vec{u} - \nabla \times \nabla p$$

$$= \nabla \times \nabla^2 \vec{u}$$

$$= \nabla \times (\nabla(\nabla \cdot \vec{u}) - \nabla \times \nabla \times \vec{u})$$

$$= -\nabla \times \nabla \times \vec{\omega}$$

$$= \nabla^2 \vec{\omega} - \nabla(\nabla \cdot \vec{\omega})$$

$$= \nabla^2 \vec{\omega}. \tag{5.8}$$

That is, in the Stokes flow limit, the vorticity is determined by a simple diffusion process; it satisfies Laplace's equation in Ω.

For two-dimensional Stokes flow, the vorticity–velocity system given by (5.8) and (5.1) can be simplified. Specifically, if the domain Ω is simply connected, then the incompressibility condition

$$\frac{\partial u_x}{\partial x} = -\frac{\partial u_y}{\partial y}$$

implies that there exists a scalar *stream function* $\psi : \Omega \to \mathbb{R}$, uniquely defined up to an additive constant, such that

$$u_x = \frac{\partial \psi}{\partial y}, \quad u_y = -\frac{\partial \psi}{\partial x}. \tag{5.9}$$

Using (5.9), we deduce that

$$-\nabla^2\psi = \frac{\partial u_y}{\partial x} - \frac{\partial u_x}{\partial y} = \omega, \tag{5.10}$$

where ω is the two-dimensional (scalar) vorticity variable that acts in a direction orthogonal to the x-y plane. The combination of (5.8) and (5.10) gives the *stream function–vorticity* formulation of the Stokes equations in two dimensions:

$$\nabla^2\omega = 0,$$
$$\nabla^2\psi + \omega = 0. \tag{5.11}$$

One issue that arises when solving this system instead of the velocity–pressure system (5.1) is the definition of a physical analogue of the boundary condition (5.2). Multiply-connected domains are also problematic, see Gresho & Sani [92, pp. 523–525] and Gunzburger [94, pp. 135–144] for a full discussion. Since there is no obvious generalization of (5.11) to three-dimensional flow, stream function-vorticity formulations are not discussed further in this book. Starting from a computed velocity solution however, with appropriate boundary conditions (5.10) may be solved numerically to generate a discrete stream function for flow visualization. Further details are given below.

5.1 Reference problems

Following the pattern of earlier chapters, examples of two-dimensional Stokes flow will be used to illustrate the behavior of finite element discretizations of (5.1) and (5.2). The quality and accuracy of such discretizations is the main topic in Sections 5.3 and 5.4, and special issues relating to solution algorithms are discussed in Chapter 6.

5.1.1 Example: Square domain $\Omega_\square = (-1,1)^2$, parabolic inflow boundary condition, natural outflow boundary condition, analytic solution.

The functions

$$u_x = 1 - y^2, \quad u_y = 0, \quad p = -2x + \text{constant}, \tag{5.12}$$

comprise a classical solution of the Stokes system (5.1), representing steady horizontal flow in a channel driven by a pressure difference between the two ends. More generally, (5.12) is a solution of the Navier–Stokes equations, called *Poiseuille flow*. A Poiseuille flow solution on $\Omega_\square$ can be computed numerically using the velocity solution in (5.12) to define a Dirichlet condition on the inflow boundary $x = -1$ and the characteristic boundary $y = -1$ and $y = 1$ (both fixed walls so $u_x = u_y = 0$). At the outflow boundary ($x = 1, -1 < y < 1$), the

Neumann condition

$$\frac{\partial u_x}{\partial x} - p = 0$$

$$\frac{\partial u_y}{\partial x} - 0$$

(5.13)

is satisfied. Notice that the Poiseuille flow solution satisfies (5.13) with the outflow pressure set to zero.

Having computed a finite element solution to the Stokes equation (5.1) subject to (5.2), a corresponding stream function solution can be computed by solving the Poisson equation (5.10) numerically. In general, the required boundary conditions for the stream function ψ are obtained by integrating the normal component of velocity along the boundary. In this case, since the normal velocity on a fixed wall is zero, ψ is constant on $y = -1$ and $y = 1$. Integrating along the inflow boundary $x = -1$ gives the Dirichlet condition

$$\psi(-1, y) = \int_{-1}^{y} u_x(-1, s) \, ds = \frac{2}{3} + y - \frac{y^3}{3}.$$

This condition implies that $\psi = 0$ on the bottom boundary, and $\psi = 4/3$ on the top boundary. Finally, taking the normal component of the outflow boundary condition (5.13) and integrating along the outflow gives the Neumann condition

$$\frac{\partial \psi}{\partial n}(1, y) = \int_{-1}^{y} p(1, s) \, ds,$$

where $p(1, s)$ is the computed finite element pressure on the outflow.

The stream function provides a useful mechanism for visualizing flow patterns. Specifically, as discussed earlier (3.10), streamlines are associated with level curves of the stream function, and thus they can be plotted using a contouring package. As an illustration, Figure 5.1 shows a contour plot of

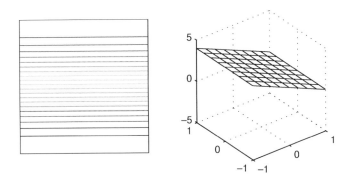

FIG. 5.1. Equally distributed streamline plot (left) and pressure plot (right) of a mixed finite element solution of Example 5.1.1.

streamlines at equally spaced levels computed from a finite element solution to a Poiseuille flow problem, together with a three-dimensional rendering of the corresponding pressure. The fact that streamlines also represent particle paths (in the case of steady flow) means that a marker introduced at the inflow boundary is simply carried along the channel and exits at the outflow boundary. Moreover, since the same volume of fluid flows between equally distributed streamlines, the fastest flow can be seen to be at the center of the channel.

5.1.2 Example: L-shaped domain $\Omega_{\bar{P}}$, parabolic inflow boundary condition, natural outflow boundary condition.

This example represents (slow) flow in a rectangular duct with a sudden expansion. (It is also called "flow over a step".) A Poiseuille flow profile is imposed on the inflow boundary ($x = -1$; $0 \le y \le 1$), and a no-flow (zero velocity) condition is imposed on the walls. The Neumann condition (5.13) is again applied at the outflow boundary ($x = 5$; $-1 < y < 1$) and automatically sets the mean outflow pressure to zero.

Figure 5.2 shows streamlines and a three-dimensional rendering of the pressure solution. The streamline plot suggests that the fluid moves fairly smoothly over the step. This is a little misleading however, since there is a singularity in this problem analogous to that in Example 1.1.2. Near the origin, the pressure behaves like $r^{-1/3}$, where r is the radial distance from the corner, and the solution is therefore unbounded! Once over the step, the fluid "recovers" and the horizontal velocity component tends to a parabolic profile as it moves towards the outflow boundary. The fact that the streamlines spread out means that the fluid is moving more slowly in the wider part — this must happen in order for the mass of the fluid entering at the inflow boundary to be conserved.

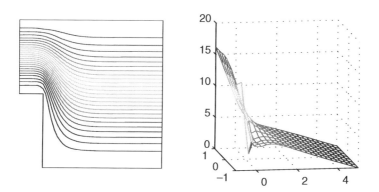

FIG. 5.2. Equally distributed streamline plot (left) and pressure plot (right) of a mixed finite element solution of Example 5.1.2.

5.1.3 Example: Square domain $\Omega_{\square}$, enclosed flow boundary condition.

This is a classic test problem used in fluid dynamics, known as *driven-cavity* flow. It is a model of the flow in a square cavity with the lid moving from left to right. Different choices of the nonzero horizontal velocity on the lid give rise to different computational models:

$$\{y = 1; -1 \leq x \leq 1 \mid u_x = 1\}, \qquad \text{a } \textit{leaky} \text{ cavity;}$$
$$\{y = 1; -1 < x < 1 \mid u_x = 1\}, \qquad \text{a } \textit{watertight} \text{ cavity;}$$
$$\{y = 1; -1 \leq x \leq 1 \mid u_x = 1 - x^4\}, \quad \text{a } \textit{regularized} \text{ cavity.}$$

These models give similar results as long as the approximation method is stable — this is discussed in more detail later in the chapter. A typical solution is illustrated in Figure 5.3. The streamlines are computed from the velocity solution by solving the Poisson equation (5.10) numerically subject to a zero Dirichlet boundary condition.

The Stokes driven-cavity flow pressure solution is anti-symmetric about the vertical centerline. The fact that the pressure is zero on this centerline can be

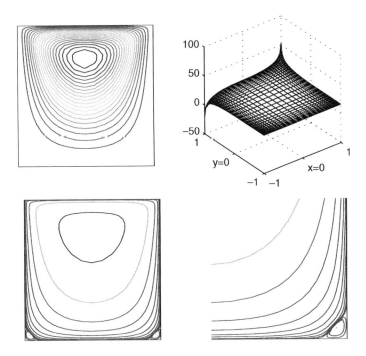

FIG. 5.3. Equally distributed streamline plot (top left), and pressure plot (top right) of a regularized lid-driven cavity problem, together with exponentially distributed streamlines and close up view of the recirculation in the bottom right corner.

used to check the convergence rate of numerical solutions. The flow essentially circulates around the cavity, but there are small counter-rotating recirculations (often called *Moffatt eddies*) in the bottom two corners. The strength of the secondary eddies is very small compared to the primary recirculation, so that stretched grids are needed to resolve them, and exponentially distributed contours of the stream function are needed to visualize them. In the cases of the leaky cavity and the watertight cavity the discontinuous horizontal velocity in the top corners generates a strong singularity in the pressure solution. Away from these corners the pressure is essentially constant.

5.1.4 Example: Square domain $\Omega_\square$, analytic solution.

This analytic test problem is associated with the following solution of the Stokes equation system:

$$u_x = 20xy^3; \quad u_y = 5x^4 - 5y^4; \quad p = 60x^2y - 20y^3 + \text{constant}. \quad (5.14)$$

It is a simple model of *colliding flow*, and a typical solution is illustrated in Figure 5.4. To solve this problem numerically, the finite element interpolant of the velocity in (5.14) is specified everywhere on $\partial\Omega_\square$. The Dirichlet boundary condition for the stream function calculation is the interpolant of the exact stream function: $\psi(x, y) = 5xy^4 - x^5$. We will return to this example when discussing error estimates in Section 5.4.1.

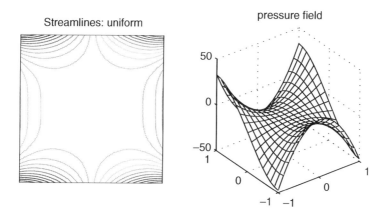

FIG. 5.4. Equally distributed streamline plot (left) and pressure plot (right) of a mixed finite element solution of Example 5.1.4.

5.2 Weak formulation

A weak formulation of the Stokes equation system (5.1) stems from the identities

$$\int_\Omega \vec{v} \cdot (-\nabla^2 \vec{u} + \nabla p) = 0, \tag{5.15}$$

$$\int_\Omega q \nabla \cdot \vec{u} = 0, \tag{5.16}$$

for all $\vec{v}$ and q in suitably chosen spaces of test functions. As in Chapter 1, the continuity requirements on the weak solution $(\vec{u}, p)$ can be reduced by "transferring" derivatives onto the test functions $\vec{v}$ and q:

$$\int_\Omega \vec{v} \cdot \nabla p = -\int_\Omega p \nabla \cdot \vec{v} + \int_\Omega \nabla \cdot (p\vec{v})$$

$$= -\int_\Omega p \nabla \cdot \vec{v} + \int_{\partial\Omega} p\vec{n} \cdot \vec{v}, \tag{5.17}$$

$$-\int_\Omega \vec{v} \cdot \nabla^2 \vec{u} = \int_\Omega \nabla\vec{u} : \nabla\vec{v} - \int_\Omega \nabla \cdot (\nabla\vec{u} \cdot \vec{v})$$

$$= \int_\Omega \nabla\vec{u} : \nabla\vec{v} - \int_{\partial\Omega} (\vec{n} \cdot \nabla\vec{u}) \cdot \vec{v}. \tag{5.18}$$

Here $\nabla\vec{u} : \nabla\vec{v}$ represents the componentwise scalar product, for example, in two dimensions $\nabla u_x \cdot \nabla v_x + \nabla u_y \cdot \nabla v_y$. Combining (5.15), (5.17) and (5.18) gives

$$\int_\Omega \vec{v} \cdot (-\nabla^2 \vec{u} + \nabla p) = \int_\Omega \nabla\vec{u} : \nabla\vec{v} - \int_\Omega p \nabla \cdot \vec{v} - \int_{\partial\Omega} \left(\frac{\partial \vec{u}}{\partial n} - p\vec{n} \right) \cdot \vec{v}, \tag{5.19}$$

for all $\vec{v}$ in a suitable set of test functions. The good news here is that the boundary term matches the Neumann condition in (5.2), and that the second term on the right-hand side of (5.19) mirrors the left-hand side of (5.16).

Equations (5.19) and (5.16) together with the boundary condition (5.2) lead to the following velocity solution and test spaces:

$$\mathbf{H}_E^1 := \{\vec{u} \in \mathcal{H}^1(\Omega)^d \,|\, \vec{u} = \vec{w} \text{ on } \partial\Omega_D\}, \tag{5.20}$$

$$\mathbf{H}_{E_0}^1 := \{\vec{v} \in \mathcal{H}^1(\Omega)^d \,|\, \vec{v} = \vec{0} \text{ on } \partial\Omega_D\}, \tag{5.21}$$

where $d = 2$ or $d = 3$ is the spatial dimension. (Since $\mathbf{H}_E^1$ is not closed, it is not a vector space unless $\vec{w} = \vec{0}$.) The fact that there are no pressure derivatives on the right-hand side of (5.19) means that $L_2(\Omega)$ is the appropriate space for p. Moreover, choosing the pressure test function q from $L_2(\Omega)$ ensures that the left-hand side of (5.16) is finite. The end product is thus the following weak formulation.

Find $\vec{u} \in \mathbf{H}_E^1$ and $p \in L_2(\Omega)$ such that

$$\int_\Omega \nabla \vec{u} : \nabla \vec{v} - \int_\Omega p \nabla \cdot \vec{v} = \int_{\partial \Omega_N} \vec{s} \cdot \vec{v} \quad \text{for all } \vec{v} \in \mathbf{H}_{E_0}^1, \tag{5.22}$$

$$\int_\Omega q \nabla \cdot \vec{u} = 0 \quad \text{for all } q \in L_2(\Omega). \tag{5.23}$$

The construction of the weak formulation implies that any solution of (5.1) and (5.2) satisfies (5.22)–(5.23). The equations (5.22) and (5.23) can also be identified with the optimality conditions of a *saddle-point* problem:

$$\inf_{\vec{v} \in \mathbf{H}_{E_0}^1} \sup_{q \in L_2(\Omega)} \int_\Omega |\nabla \vec{v}|^2 - \int_\Omega q \nabla \cdot \vec{v} - \int_{\partial \Omega_N} \vec{s} \cdot \vec{v}, \tag{5.24}$$

see Brezzi & Fortin [33, p. 14]. The fresh challenge here is to establish that weak solutions are uniquely defined (up to a constant pressure in the case of $\partial \Omega = \partial \Omega_D$). To this end, we consider the homogeneous version of (5.22)–(5.23): find $\vec{u} \in \mathbf{H}_{E_0}^1$ and $p \in L_2(\Omega)$ such that

$$\int_\Omega \nabla \vec{u} : \nabla \vec{v} - \int_\Omega p \nabla \cdot \vec{v} = 0 \quad \text{for all } \vec{v} \in \mathbf{H}_{E_0}^1, \tag{5.25}$$

$$\int_\Omega q \nabla \cdot \vec{u} = 0 \quad \text{for all } q \in L_2(\Omega), \tag{5.26}$$

and try to establish that $\vec{u} = \vec{0}$ and $p = $ constant. First, taking $\vec{v} = \vec{u}$ in (5.25) with $q = p$ in (5.26) we see that $\int_\Omega \nabla \vec{u} : \nabla \vec{u} = 0$. This, together with the fact that the bilinear form $a(\vec{u}, \vec{v}) := \int_\Omega \nabla \vec{u} : \nabla \vec{v}$ is coercive over $\mathbf{H}_{E_0}^1$ (see Problem 5.1) means that $\vec{u} = \vec{0}$ as required.

Turning to the pressure, substituting $\vec{u} = \vec{0}$ into (5.25) gives $\int_\Omega p \nabla \cdot \vec{v} = 0$ for all $\vec{v} \in \mathbf{H}_{E_0}^1$. It now remains to show that

$$\int_\Omega p \nabla \cdot \vec{v} = 0 \quad \text{for all } \vec{v} \in \mathbf{H}_{E_0}^1 \implies \begin{cases} p = \text{constant}, & \text{if } \partial \Omega = \partial \Omega_D; \\ p = 0, & \text{otherwise.} \end{cases} \tag{5.27}$$

Consider the case of an enclosed flow. Showing that $p = \text{constant}$ in (5.27) is done using the so-called *inf–sup* condition

$$\inf_{q \neq \text{constant}} \sup_{\vec{v} \neq \vec{0}} \frac{|(q, \nabla \cdot \vec{v})|}{\|\vec{v}\|_{1,\Omega} \|q\|_{0,\Omega}} \geq \gamma > 0, \tag{5.28}$$

where $\|\vec{v}\|_{1,\Omega} = (\int_{\Omega} \vec{v} \cdot \vec{v} + \nabla \vec{v} : \nabla \vec{v})^{1/2}$ is a norm for functions in $\mathbf{H}_{E_0}^1$, and $\|q\|_{0,\Omega} = \|q - (1/|\Omega|) \int_{\Omega} q\|$ is a *quotient space* norm. This condition can be viewed as an coercivity estimate for the bilinear form $b(\vec{v}, q) := \int_{\Omega} q \nabla \cdot \vec{v}$ on the spaces $L_2(\Omega)$ and $\mathbf{H}_{E_0}^1$, and it is used to establish the *existence* of a pressure solution.[1] To see how the inf–sup condition can be used to ensure (5.27), observe that (5.28) implies that for any nonconstant $q \in L_2(\Omega)$, there exists $\vec{v} = \vec{v}_q \in \mathbf{H}_{E_0}^1$ such that

$$\frac{(q, \nabla \cdot \vec{v})}{\|\vec{v}\|_{1,\Omega}} \geq \gamma \|q\|_{0,\Omega}.$$

For the particular choice $q = p$, the pressure solution, the numerator in this expression is zero for all functions $\vec{v}$. Therefore it must be that $\|p\|_{0,\Omega} = 0$. This means that $p = \bar{p}$, where $\bar{p}$ is the constant function $\bar{p} = (1/|\Omega|) \int_{\Omega} p$. In other words, the inf–sup condition is a sufficient condition for the pressure to be unique up to a constant in the case of enclosed flow. For other boundary conditions, the analysis is identical, except that the standard L_2-norm is used on the pressure space and the infimum in (5.28) includes all of $L_2(\Omega)$.

Ensuring that the *solvability* condition (5.27) holds when constructing finite element approximations of (5.22)–(5.23) is the main focus of the next section. The discrete version of the inf–sup condition (5.28) is key to the error analysis of finite element approximation methods for Stokes problems. This issue is discussed in detail in Section 5.4.

5.3   Approximation using mixed finite elements

The formal aspects of finite element discretization of the Stokes equations are the same as for the Poisson equation. A discrete weak formulation is defined using finite-dimensional spaces $\mathbf{X}_0^h \subset \mathbf{H}_{E_0}^1$ and $M^h \subset L_2(\Omega)$. The fact that these spaces are approximated independently leads to the nomenclature *mixed approximation*. Implementation entails defining appropriate bases for the chosen finite element spaces and construction of the associated finite element coefficient matrix. Specifically, given a velocity solution space $\mathbf{X}_E^h$, the discrete problem

---

[1] Establishing the inf–sup condition is technical and beyond the scope of this book; for details see [19, pp. 144–146], [28, pp. 241–242].

is: find $\vec{u}_h \in \mathbf{X}_E^h$ and $p_h \in M^h$ such that

$$\int_\Omega \nabla \vec{u}_h : \nabla \vec{v}_h - \int_\Omega p_h \nabla \cdot \vec{v}_h = \int_{\partial\Omega_N} \vec{s} \cdot \vec{v}_h \quad \text{for all } \vec{v}_h \in \mathbf{X}_0^h, \qquad (5.29)$$

$$\int_\Omega q_h \nabla \cdot \vec{u}_h = 0 \quad \text{for all } q_h \in M^h. \qquad (5.30)$$

To identify the corresponding linear algebra problem, we follow the procedure given in Section 1.3 and introduce a set of vector-valued (velocity) basis functions $\{\vec{\phi}_j\}$ such that,

$$\vec{u}_h = \sum_{j=1}^{n_u} \mathbf{u}_j \vec{\phi}_j + \sum_{j=n_u+1}^{n_u+n_\partial} \mathbf{u}_j \vec{\phi}_j, \qquad (5.31)$$

with $\sum_{j=1}^{n_u} \mathbf{u}_j \vec{\phi}_j \in \mathbf{X}_0^h$. We fix the coefficients $\mathbf{u}_j : j = n_u + 1, \ldots, n_u + n_\partial$ so that the second term interpolates the boundary data on $\partial\Omega_D$. Introducing a set of scalar (pressure) basis functions $\{\psi_k\}$, and setting

$$p_h = \sum_{k=1}^{n_p} \mathbf{p}_k \, \psi_k, \qquad (5.32)$$

we find that the discrete formulation (5.29)–(5.30) can be expressed as a system of linear equations

$$\begin{bmatrix} \mathbf{A} & B^T \\ B & O \end{bmatrix} \begin{bmatrix} \mathbf{u} \\ \mathbf{p} \end{bmatrix} = \begin{bmatrix} \mathbf{f} \\ \mathbf{g} \end{bmatrix}. \qquad (5.33)$$

The matrix $\mathbf{A}$ is called the *vector-Laplacian matrix,* and the matrix B is called the *divergence matrix.* The entries are given by

$$\mathbf{A} = [\mathbf{a}_{ij}], \quad \mathbf{a}_{ij} = \int_\Omega \nabla \vec{\phi}_i : \nabla \vec{\phi}_j, \qquad (5.34)$$

$$B = [b_{kj}], \quad b_{kj} = -\int_\Omega \psi_k \nabla \cdot \vec{\phi}_j, \qquad (5.35)$$

for i and $j = 1, \ldots, n_u$ and $k = 1, \ldots, n_p$. The entries of the right-hand side vector are

$$\mathbf{f} = [\mathbf{f}_i], \quad \mathbf{f}_i = \int_{\partial\Omega_N} \vec{s} \cdot \vec{\phi}_i - \sum_{j=n_u+1}^{n_u+n_\partial} \mathbf{u}_j \int_\Omega \nabla \vec{\phi}_i : \nabla \vec{\phi}_j, \qquad (5.36)$$

$$\mathbf{g} = [\mathbf{g}_k], \quad \mathbf{g}_k = \sum_{j=n_u+1}^{n_u+n_\partial} \mathbf{u}_j \int_\Omega \psi_k \nabla \cdot \vec{\phi}_j, \qquad (5.37)$$

and the function pair $(\vec{u}_h, p_h)$ obtained by substituting the solution vectors $\mathbf{u} \in \mathbb{R}^{n_u}$ and $\mathbf{p} \in \mathbb{R}^{n_p}$ into (5.31) and (5.32) is the *mixed* finite element solution. The system (5.33)–(5.37) is henceforth referred to as the *discrete Stokes problem.*

The spectral properties of the coefficient matrix in (5.33) are discussed in detail in Section 5.5.

In practice, the d components of velocity are always approximated using a single finite element space. In two dimensions, given a standard space of scalar finite element basis functions $\{\phi_j\}_{j=1}^n$, we set $n_u = 2n$ and define the velocity basis set

$$\{\vec{\phi}_1, \ldots, \vec{\phi}_{2n}\} := \{(\phi_1, 0)^T, \ldots, (\phi_n, 0)^T, (0, \phi_1)^T, \ldots, (0, \phi_n)^T\}. \qquad (5.38)$$

This componentwise splitting can be shown to induce (see Problem 5.2) a natural block partitioning of the Galerkin system (5.33). Specifically, with $\mathbf{u} := ([\mathbf{u}_x]_1, \ldots, [\mathbf{u}_x]_n, [\mathbf{u}_y]_1, \ldots, [\mathbf{u}_y]_n)^T$, (5.33) can be rewritten as

$$\begin{bmatrix} A & O & B_x^T \\ O & A & B_y^T \\ B_x & B_y & O \end{bmatrix} \begin{bmatrix} \mathbf{u}_x \\ \mathbf{u}_y \\ \mathbf{p} \end{bmatrix} = \begin{bmatrix} \mathbf{f}_x \\ \mathbf{f}_y \\ \mathbf{g} \end{bmatrix}, \qquad (5.39)$$

where the $n \times n$ matrix A is the scalar Laplacian matrix (discussed in detail in Section 1.6), and the $n_p \times n$ matrices B_x and B_y represent weak derivatives in the x and y directions:

$$A = [a_{ij}], \qquad a_{ij} = \int_\Omega \nabla\phi_i \cdot \nabla\phi_j, \qquad (5.40)$$

$$B_x = [b_{x,ki}], \qquad b_{x,ki} = -\int_\Omega \psi_k \frac{\partial\phi_i}{\partial x}, \qquad (5.41)$$

$$B_y = [b_{y,kj}], \qquad b_{y,kj} = -\int_\Omega \psi_k \frac{\partial\phi_j}{\partial y}. \qquad (5.42)$$

We now get to the central question: is the discrete Stokes problem (5.33)–(5.37) a faithful representation of the underlying continuous problem (5.1) and (5.2)? Simply looking at the matrix dimensions in (5.33) quickly shows why this is not automatic. If $n_p > n_u$ then the coefficient matrix in (5.33) is rank deficient by at least $n_p - n_u$! This means that choosing a high dimensional approximation for the pressure compared to that of the velocity is likely to lead to linear algebraic suicide.

As in the discussion of uniqueness in Section 5.2, the unique solvability of the matrix system (5.33) is determined by looking at the homogeneous system

$$\begin{aligned} \mathbf{A}\mathbf{u} + B^T\mathbf{p} &= \mathbf{0}, \\ B\mathbf{u} &= \mathbf{0}. \end{aligned} \qquad (5.43)$$

To start, premultiply the first equation by $\mathbf{u}^T$ and the second equation by $\mathbf{p}^T$. This implies that $\mathbf{u}^T\mathbf{A}\mathbf{u} = 0$, or using (5.39) that $\mathbf{u}_x^T A\mathbf{u}_x + \mathbf{u}_y^T A\mathbf{u}_y = 0$. Thus, since A is positive-definite it follows that $\mathbf{u} = \mathbf{0}$, implying unique solvability with respect to the velocity. On the other hand, unique solvability with respect to the pressure is problematic. Substituting $\mathbf{u} = \mathbf{0}$ into (5.43) gives $B^T\mathbf{p} = \mathbf{0}$, and implies that any pressure solution is only unique up to the null space of the matrix B^T. The bottom line is that if (5.33) is to properly represent a continuous Stokes problem, then the mixed approximation spaces need to be chosen carefully. Specifically, we have to ensure that $\mathtt{null}(B^T) = \{\mathbf{1}\}$ in the case of enclosed flow, and that $\mathtt{null}(B^T) = \{\mathbf{0}\}$, otherwise.

The conditions on the matrix B^T above are related to the solvability condition (5.27) introduced in the previous section. To make this connection concrete, we concentrate on the more challenging case of flow problems without a natural outflow boundary condition, $\partial\Omega = \partial\Omega_D$, and we use the fact that the null space of B^T is isomorphic to the constraint space

$$\mathbf{N}^h := \left\{ q_h \in M^h \, \bigg| \, \int_\Omega q_h \nabla \cdot \vec{v}_h = 0 \text{ for all } \vec{v}_h \in \mathbf{X}_0^h \right\}. \tag{5.44}$$

This can be seen by writing down the vector version of this space,

$$N^h := \{\mathbf{q} \in \mathbb{R}^m \, | \, \langle B^T \mathbf{q}, \mathbf{v} \rangle = 0 \quad \text{for all } \mathbf{v} \in \mathbb{R}^{n_u}\}. \tag{5.45}$$

Note that $p_h \in \mathbf{N}^h$ if and only if the condition (5.27) holds in the discrete case, that is,

$$\int_\Omega p_h \nabla \cdot \vec{v}_h = 0 \quad \text{for all } \vec{v}_h \in \mathbf{X}_0^h \implies p_h = \text{constant}. \tag{5.46}$$

In terms of the associated coefficient vector $\mathbf{p}$ of (5.32), this condition is

$$\langle B^T \mathbf{p}, \mathbf{v} \rangle = 0 \quad \text{for all } \mathbf{v} \in \mathbb{R}^{n_u} \implies \mathbf{p} = \text{constant}, \tag{5.47}$$

which is equivalent to requiring that $\mathtt{null}(B^T) = \{\mathbf{1}\}$. Any nonconstant functions $q_h \in \mathbf{N}^h$ are henceforth referred to as *spurious pressure modes*.

In the next section we will discuss specific mixed approximation methods that are compatible in the sense that (5.46) is satisfied on any grid. Before doing so, it must be emphasized that if the velocity is prescribed everywhere on the boundary, then $\mathbf{1} \in \mathtt{null}(B^T)$. This means that the discrete Stokes system (5.33) is *singular*, and that there is a compatibility condition on the right-hand side vector in (5.33) that must be satisfied in order for a discrete solution to exist. To see this, consider $p_h = p_h^* = \text{constant}$, and note that

$$\int_\Omega p_h \nabla \cdot \vec{v}_h = p_h^* \int_\Omega \nabla \cdot \vec{v}_h = p_h^* \int_{\partial\Omega} \vec{v}_h \cdot \vec{n} = 0$$

for all $\vec{v}_h \in \mathbf{X}_0^h$. This means that $p_h^* \in \mathbf{N}^h$ and implies that $\mathbf{1} \in \mathtt{null}(B^T)$. The associated solvability condition for the matrix system (5.33) is $\langle \mathbf{1}, \mathbf{g} \rangle = 0$; that is, the components of the right-hand side vector $\mathbf{g}$ must sum to zero. To study this, we start from the definition (5.37) and exploit the fact that the pressure basis functions form a partition of unity:

$$\sum_{k=1}^{n_p} g_k = \sum_{k=1}^{n_p} \sum_{j=n_u+1}^{n_u+n_\partial} \mathbf{u}_j \int_\Omega \psi_k \nabla \cdot \vec{\phi}_j$$

$$= \sum_j \mathbf{u}_j \int_\Omega \left(\sum_{k=1}^{n_p} \psi_k \right) \nabla \cdot \vec{\phi}_j$$

$$= \sum_j \mathbf{u}_j \int_\Omega \nabla \cdot \vec{\phi}_j$$

$$= \sum_j \mathbf{u}_j \int_{\partial\Omega} \vec{\phi}_j \cdot \vec{n}$$

$$= \int_{\partial\Omega} \underbrace{\sum_j \mathbf{u}_j \vec{\phi}_j \cdot \vec{n}}_{=: \vec{w} \cdot \vec{n}}.$$

This shows that the condition $\langle 1, \mathbf{g} \rangle = 0$ is just a discrete representation of the mass conservation condition (5.4). In simple terms, the specification of the velocity boundary data—that is, the values of $\mathbf{u}_j$ on the inflow and outflow boundaries—must be carefully chosen to ensure that the volume of fluid entering the computational domain is exactly matched by the volume of fluid flowing out. If an inflow/outflow problem is modeled, and the boundary velocity specification is such that

$$\int_{\partial\Omega_-} \sum_j \mathbf{u}_j \vec{\phi}_j \cdot \vec{n} + \int_{\partial\Omega_+} \sum_j \mathbf{u}_j \vec{\phi}_j \cdot \vec{n} \neq 0, \tag{5.48}$$

then the discrete Stokes problem (5.33) is not well posed! This illustrates why a Neumann condition should be applied on ouflow boundaries when modeling such problems.

We are now faced with the question of finding finite element approximation spaces for the Stokes problem. The discrete analogue of (5.28), referred to as *uniform inf–sup stability* or *the inf–sup condition*, requires the existence of a positive constant γ independent of h such that, for any conceivable grid,

$$\min_{q_h \neq \text{constant}} \max_{\vec{v}_h \neq \vec{0}} \frac{|(q_h, \nabla \cdot \vec{v}_h)|}{\|\vec{v}_h\|_{1,\Omega} \|q_h\|_{0,\Omega}} \geq \gamma. \tag{5.49}$$

This condition provides a guarantee that the discrete solvability condition (5.46) holds for any possible grid, and, as will be shown in Section 5.4, it is also crucial for establishing optimal a priori error bounds. It turns out, however, that choosing spaces for which the discrete inf–sup condition (5.49) holds is a delicate matter, and seemingly natural choices of velocity and pressure approximation do not work. For example, the simplest possible globally continuous approximation, bilinear approximation for both velocity components and pressure (the so-called $\boldsymbol{Q}_1$–$\boldsymbol{Q}_1$ approximation), is unstable. Indeed, for an enclosed flow problem on a square grid with an even number of elements, the null space of B^T is actually eight-dimensional! (The details of this particular case are worked out in Computational Exercise 5.1.) In general, care must be taken to make the velocity space rich enough compared to the pressure space, otherwise the discrete solution will be "over-constrained".

The rest of this section is devoted to inf–sup stability. Section 5.3.1 presents a methodology for establishing (5.49), which is illustrated using several choices of stable mixed approximation methods. Section 5.3.2 considers an alternative point of view, in which the inf–sup condition is circumvented by techniques of "stabilization".

5.3.1   Stable rectangular elements ($\boldsymbol{Q}_2$–$\boldsymbol{Q}_1$, $\boldsymbol{Q}_2$–$\boldsymbol{P}_{-1}$, $\boldsymbol{Q}_2$–$\boldsymbol{P}_0$)

One of the most elegant ways of establishing inf–sup stability is to use a *macroelement* construction. The idea is to consider "local enclosed-flow Stokes problems" posed on a subdomain $\mathcal{M}$ of Ω consisting of a small patch of elements arranged in a simple topology. This subdomain is referred to as a *macroelement*, and if the discrete solvability condition (5.46) is satisfied on it, it is called a *stable macroelement*. Theoretical results of Boland & Nicolaides [17] and Stenberg [183] show that for any grid that can be constructed by "patching together" such stable macroelements, the inf–sup condition (5.49) is satisfied. The method of analysis is to establish the discrete solvability condition (5.46) on macroelements, and then to establish a *connectivity condition* showing that (5.49) holds on the union of the macroelement patches. The details of the latter procedure are rather technical and it turns out to be more convenient to discuss them in conjunction with the error analysis. Therefore, we defer many of the details to Section 5.4. In this section, we illustrate the macroelement technique for establishing the local discrete solvability condition (5.46), and we use it to identify some stable mixed approximation methods for grids of rectangular elements.[2]

As we have observed, the $\boldsymbol{Q}_1$–$\boldsymbol{Q}_1$(bilinear) approximation is unstable. To get a stable method, the approximation of velocity needs to be enhanced relative to the pressure. The easiest way of doing this is to use biquadratic approximation for the velocity components rather than bilinear. The resulting mixed method is called $\boldsymbol{Q}_2$–$\boldsymbol{Q}_1$ approximation, but is often referred to in the literature as the *Taylor–Hood* method. The nodal positions are illustrated in Figure 5.5.

We first address the question of whether a macroelement consisting of a single rectangular element of this approximation is stable. Let $\mathcal{M} = \square_k$ denote such an element, with a local numbering as illustrated in Figure 5.6. For an enclosed flow problem posed over $\mathcal{M}$, there is one interior velocity node (9, marked with $\bullet$),

---

[2]The extension to isoparametrically mapped quadrilateral elements is also covered by Stenberg's analysis, see [183].

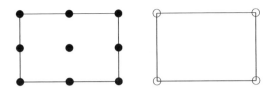

FIG. 5.5. Q_2–Q_1 element (● two velocity components; ○ pressure).

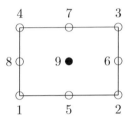

FIG. 5.6. Numbering for a single element macroelement.

and four pressure nodes, namely 1, 2, 3, and 4. To investigate the stability of the macroelement, we need to construct the matrix $B = [B_x \ B_y]$ using the definitions (5.41) and (5.42), and then, in light of (5.46)–(5.47), check to see if $\texttt{null}(B^T) = \{\mathbf{1}\}$. For this example, $\mathbf{v}$ has dimension 2 in (5.47), for the two velocity components defined on node 9, and $\mathbf{p}$ has dimension 4.

In this simple case, B_x^T and B_y^T are 1×4 row vectors,

$$B_x^T = [b_{x,j}], \quad b_{x,j} = -\int_{\square_k} \psi_j \frac{\partial \phi_9}{\partial x}, \tag{5.50}$$

$$B_y^T = [b_{y,j}], \quad b_{y,j} = -\int_{\square_k} \psi_j \frac{\partial \phi_9}{\partial y}, \tag{5.51}$$

where ϕ_9 is the standard biquadratic basis function discussed in Section 1.3.2. This function takes the value unity at node 9 and is zero at the other eight nodes. Similarly $\psi_j : j = 1, 2, 3, 4$ are the standard bilinear basis functions defined on $\square_k$. Noting that $\phi_9 = 0$ on the boundary of $\square_k$, we can evaluate (5.50) and (5.51) by first integrating by parts, and then using the tensor-product form of Simpson's rule:

$$\int_{\square_k} f(x, y) = \frac{|\square_k|}{36} (f_1 + f_2 + f_3 + f_4 + 4f_5 + 4f_6 + 4f_7 + 4f_8 + 16f_9),$$

with f_i representing $f(\vec{x}_i)$. (The rule is exact in this case, since we have a polynomial of degree three in each coordinate direction.) This gives

$$b_{x,j} = \int_{\square_k} \phi_9 \frac{\partial \psi_j}{\partial x} = \frac{4}{9} |\square_k| \frac{\partial \psi_j}{\partial x}(\vec{x}_9),$$

$$b_{y,j} = \int_{\square_k} \phi_9 \frac{\partial \psi_j}{\partial y} = \frac{4}{9} |\square_k| \frac{\partial \psi_j}{\partial y}(\vec{x}_9).$$

Evaluating the linear functions $\frac{\partial \psi_j}{\partial x}, \frac{\partial \psi_j}{\partial y} : j = 1, 2, 3, 4$ at the centroid, for example

$$\frac{\partial \psi_1}{\partial x}(\vec{x}_9) = -\frac{h_y}{2|\square_k|},$$

gives

$$B^T = \begin{bmatrix} B_x^T \\ B_y^T \end{bmatrix} = \begin{bmatrix} -2/_9 h_y & 2/_9 h_y & 2/_9 h_y & -2/_9 h_y \\ -2/_9 h_x & -2/_9 h_x & 2/_9 h_x & 2/_9 h_x \end{bmatrix}. \tag{5.52}$$

Note that the discrete operator B_x^T represents a scaled central difference approximation of the pressure x-derivative at the node $\vec{x}_9$:

$$0 = \frac{2}{9} h_y (-p_1 + p_2 + p_3 - p_4)$$

$$= \frac{4}{9} |\square_k| \left\{ \frac{(p_3 + p_2)/2 - (p_1 + p_4)/2)}{h_x} \right\} \approx \frac{4}{9} |\square_k| \frac{\partial p}{\partial x}(\vec{x}_9). \tag{5.53}$$

Finally, the linear system $B^T \mathbf{p} = \mathbf{0}$ is equivalent to

$$-p_1 + p_2 + p_3 - p_4 = 0,$$

$$-p_1 - p_2 + p_3 + p_4 = 0,$$

which is satisfied whenever $p_1 = p_3$ and $p_2 = p_4$. That is, B^T has a two-dimensional null space and the macroelement $\mathcal{M}$ in Figure 5.6 is not stable. Of course, we could have deduced this immediately by a simple constraint count. The system $B^T \mathbf{p} = \mathbf{0}$ contains two equations and four unknowns. Hence the dimension of $\texttt{null}(B^T)$ must be at least two.

This derivation shows that the macroelement test fails when it is applied to a single element of the Q_2–Q_1 discretization, so that the single element is not a stable macroelement. This does not mean that the discretization itself is unstable. Consider now the *two-element* patch $\mathcal{M} = \square_k \cup \square_m$, illustrated in Figure 5.7. For an enclosed flow problem posed over $\mathcal{M}$, there are three interior velocity nodes (7, 8, and 9), and six pressure nodes, namely 1, 2, 3, 4, 5 and 6. In this case, the matrix B^T is a 6×6 matrix, which offers the hope that the macroelement may be stable. To check this, we first note that the one-element patch matrix (5.52) gives the four rows of the stability system $B^T \mathbf{p} = 0$ that

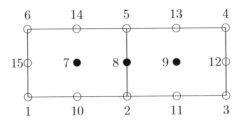

FIG. 5.7. Q_2–Q_1 macroelement numbering for two-element patch $\square_k \cup \square_m$.

correspond to the velocity nodes 7 and 9. Specifically, the node 7 equations imply that that $p_1 = p_5$ and $p_2 = p_6$, whereas the node 9 equations imply that $p_2 = p_4$ and $p_3 = p_5$. Combining these four equations implies that

$$p_1 = p_3 = p_5 = \xi,$$

$$p_2 = p_4 = p_6 = \eta.$$

To establish stability, we have to show that $\xi = \eta$ using one of the two equations associated with velocity node 8.

The vertical velocity component is the obvious one to consider first, since it represents a discrete approximation of $\partial p / \partial y = 0$ at the node $\vec{x}_8$. This suggests that the column entries corresponding to nodes 1, 3, 4 and 6 will be zero. To see this, consider the coefficient of p_1,

$$
\begin{aligned}
B^T_{y,k\,1} &= -\int_{\mathcal{M}} \psi_1 \frac{\partial \phi_8}{\partial y} \\
&= -\int_{\square_k} \psi_1 \frac{\partial \phi_8}{\partial y} \quad (\psi_1 = 0 \text{ in } \square_m) \\
&= \int_{\square_k} \phi_8 \frac{\partial \psi_1}{\partial y} \quad (\text{since } \phi_8 n_y = 0 \text{ on the boundary}) \\
&= \frac{1}{9} |\square_k| \frac{\partial \psi_1}{\partial y} (\vec{x}_8) \quad (\text{using Simpson's rule}) \\
&= 0,
\end{aligned}
$$

since $\psi_1 = 0$ on the vertical edge between nodes 2 and 5. By evaluating the other coefficients in the same way, it is easily shown that

$$(B^T_y)_k = \begin{bmatrix} 0 & -\frac{2}{9}h_x & 0 & 0 & \frac{2}{9}h_x & 0 \end{bmatrix}. \tag{5.54}$$

In the system $B^T \mathbf{p} = 0$, the equation corresponding to (5.54) implies that $p_2 = p_5$. This means that $\xi = \eta$, and $\mathtt{null}(B^T) = 1$ as required.

The analysis above shows that two-element patches of Q_2–Q_1 elements are stable. Our construction also shows that stability is independent of the sizes of the two adjoining rectangular elements, and that it does not depend on the orientation — if the macroelement in Figure 5.7 is rotated by $90°$, the horizontal velocity component on the inter-element edge gives an equation analogous to (5.54) connecting ξ and η. This geometric invariance is the key to establishing stability of mixed approximation on general grids. Once stability has been established on a reference macroelement, it holds for all patches of elements with the same topology. Moreover, any grid consisting of an even number of elements can be decomposed into two-element patches. This means that Q_2–Q_1 approximation is stable on *every* possible grid with an even number of elements.

To show the stability of Q_2–Q_1 approximation on grids containing an odd number of elements, three-element patches also need to be studied. This is quite straightforward however: appending one more element to the two-element patch in Figure 5.7 introduces two extra pressure nodes and two new interior velocity nodes. Taking the velocity node at the centroid of the new element gives two equations of the form (5.52), and macroelement stability then immediately follows using the fact that the original two-element patch is stable. Since all possible grids (other than a single-element grid) can be decomposed into two- and three-element patches, we deduce that the Q_2–Q_1 approximation is indeed uniformly stable.

We note in passing that the *serendipity* version of Q_2–Q_1 mixed approximation (obtained by excluding the centroid node from the velocity approximation space) is also uniformly stable. The details are worked out in Problem 5.3.

It is evident that the macroelement approach is especially easy to apply in cases where a one-element patch is stable. As we have just seen, this is not true for Q_2–Q_1 approximation, where it was necessary to examine (the slightly more complex) two-element and three-element patches. One-element patches suffice when certain globally *discontinuous* pressure approximations are used in place of the Q_1 pressure. Note that discontinuous pressure approximations are perfectly permissible in the framework of the weak formulation (5.22)–(5.23), since the only requirement on the pressure approximation space is that $M^h \subset L_2(\Omega)$. Unfortunately, using Q_2 velocity approximation together with a discontinuous Q_1 pressure approximation — so that there are four pressure degrees of freedom per element — is not a stable mixed method. However, a mixed approximation with Q_2 velocity and pressure defined *linearly* in each element (as illustrated in Figure 5.8) is perfectly stable. The notation P_{-1} is used to emphasize the fact that the approximation is discontinuous across inter-element boundaries.

In this case, since the pressure approximation p_h is discontinuous and has three degrees of freedom per element, (5.32) can be written in the form

$$p_h = \sum_{\Box_k \in \mathcal{T}_h} \sum_{j=1}^{3} p_j^k \psi_j := \sum_{\Box_k \in \mathcal{T}_h} p_h|_k, \qquad (5.55)$$

where $\{\psi_j\}$ are locally defined basis functions (zero outside of the element $\Box_k$), and p_j^k are the unknown local coefficients. Using the standard isoparametric

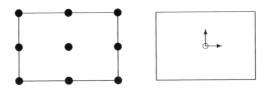

FIG. 5.8. Q_2–P_{-1} element (• velocity node; ○ pressure; $\xrightarrow{\uparrow}$ pressure derivative).

mapping (1.26)–(1.27), it is natural[3] to define the local linear function in terms of the reference element coordinates (ξ, η):

$$p_h|_k = p_1^k + p_2^k \xi + p_3^k \eta. \tag{5.56}$$

In the special case of a rectangular shaped element, mapping into the (x, y) coordinates using the isoparametric transformation (1.27), gives the locally defined pressure

$$p_h|_k = p_1^k \psi_1 + p_2^k \psi_2 + p_3^k \psi_3; \tag{5.57}$$

with basis functions given by

$$\psi_1(x, y) := 1; \quad \psi_2(x, y) := \frac{2}{h_x}(x - \bar{x}); \quad \psi_3(x, y) := \frac{2}{h_y}(y - \bar{y}); \tag{5.58}$$

where $(\bar{x}, \bar{y})$ are the coordinates of the centroid of the element. Note that (5.58) is a *Hermite* basis: the coefficients of the finite element solution vector satisfy

$$p_1^k = p_h|_{(\bar{x},\bar{y})}; \quad p_2^k = \frac{h_x}{2}\frac{\partial p_h}{\partial x}\bigg|_k; \quad p_3^k = \frac{h_y}{2}\frac{\partial p_h}{\partial y}\bigg|_k. \tag{5.59}$$

The uniform stability of the Q_2–P_{-1} approximation can now be established. Given the velocity node numbering in Figure 5.6, B_x^T and B_y^T are the 1×3 row vectors

$$B_x^T = [b_{x,j}], \quad b_{x,j} = -\int_{\Box_k} \psi_j \frac{\partial \phi_9}{\partial x} = \int_{\Box_k} \phi_9 \frac{\partial \psi_j}{\partial x} = \frac{\partial \psi_j}{\partial x}\int_{\Box_k} \phi_9, \tag{5.60}$$

$$B_y^T = [b_{y,j}], \quad b_{y,j} = -\int_{\Box_k} \psi_j \frac{\partial \phi_9}{\partial y} = \int_{\Box_k} \phi_9 \frac{\partial \psi_j}{\partial y} = \frac{\partial \psi_j}{\partial y}\int_{\Box_k} \phi_9. \tag{5.61}$$

Using Simpson's Rule $\int_{\Box_k} \phi_9 = \frac{4}{9}|\Box_k|$ and differentiating (5.58) gives

$$B^T = \begin{bmatrix} B_x^T \\ B_y^T \end{bmatrix} = \begin{bmatrix} 0 & 8/9h_y & 0 \\ 0 & 0 & 8/9h_x \end{bmatrix}. \tag{5.62}$$

---

[3]This is the approach that is implemented in the IFISS routine lshape.m.

Solving the linear system $B^T\mathbf{p}=\mathbf{0}$ associated with (5.62), we see that $p_2^k = p_3^k = 0$. Substituting into (5.57) gives $p_h = p_h|_k = p_1^k = $ constant, establishing that the one-element patch is stable. The uniform stability of $\mathbf{Q}_2$–$\mathbf{P}_{-1}$ mixed approximation is thus guaranteed using any grid of rectangular elements.

A simplification of the $\mathbf{Q}_2$–$\mathbf{P}_{-1}$ approximation can be obtained by setting $p_2^k = p_3^k = 0$ in (5.57) (leaving only the constant pressure component) and discarding two thirds of the discrete incompressibility constraints. The resulting method is referred to as the $\mathbf{Q}_2$–$\mathbf{P}_0$ approximation, and it is also uniformly stable by construction. However, as will be shown in Section 5.4.1, the simplification of $\mathbf{Q}_2$–$\mathbf{P}_{-1}$ is a false economy — with $\mathbf{Q}_2$–$\mathbf{P}_0$ the order of the velocity approximation is compromised by the reduced accuracy of the pressure approximation.

The attractive feature of such discontinuous pressure approximation methods is that mass is conserved at an element level. Taking $q_h = \psi_1$ in the weak incompressibility constraint equation (5.30) gives

$$0 = \int_\Omega \psi_1 \nabla \cdot \vec{u}_h = \int_{\square_k} \nabla \cdot \vec{u}_h = \int_{\partial \square_k} \vec{u}_h \cdot \vec{n}, \qquad (5.63)$$

showing that the flow into every element is balanced by the flow out. In contrast, with a continuous pressure approximation, mass is not conserved elementwise. Our computational experience is that discrete velocity solutions generated using discontinuous pressure elements are always more accurate than those generated using continuous pressure elements on the same grid. See however Exercise 5.6. This is especially noticable when modeling flow around obstacles, and in "low-flow regions" behind such obstacles, or close to corners of a flow domain. An example of such a flow, namely, high speed flow over a step, is one of the test problems in Chapter 7.

5.3.2 Stabilized rectangular elements ($\mathbf{Q}_1$–$\mathbf{P}_0$, $\mathbf{Q}_1$–$\mathbf{Q}_1$)

The results above show that the stability of mixed finite element methods boils down to properties of the null space of the matrix B^T. An approximation is unstable if $B^T\mathbf{p} = \mathbf{0}$ where $\mathbf{p}$ corresponds to some spurious pressure mode different from $\mathbf{1}$. In this section the concept of *stabilization* is described. Note that if $B^T\mathbf{p} = \mathbf{0}$, then $\begin{pmatrix}\mathbf{0}\\\mathbf{p}\end{pmatrix}$ is a null vector of the homogeneous system (5.43). The basic idea behind stabilization is to relax the incompressibility constraint in a special way so that this vector is no longer a null vector of the resulting coefficient matrix, and the discrete solutions $\vec{u}_h$ and p_h satisfy rigorous error bounds. Stabilization is applicable to any mixed approximation method. We restrict attention here to the two most important stabilized methods, $\mathbf{Q}_1$–$\mathbf{P}_0$ and $\mathbf{Q}_1$–$\mathbf{Q}_1$. The attraction of these methods is their simplicity.

The $\mathbf{Q}_1$–$\mathbf{P}_0$ mixed approximation is the lowest order conforming approximation method defined on a rectangular grid — it also happens to be the most famous example of an *unstable* mixed approximation method. It has the nodal positions illustrated in Figure 5.9. The constant pressure approximation is represented by a point value at the centroid.

The origin of the instability is the 2×2 four-element patch illustrated in Figure 5.10. Since the number of degrees of freedom is the same as for the single-element $\mathbf{Q}_2\text{–}\mathbf{Q}_1$ patch discussed in Section 5.3.1, it is immediately obvious that the 2×2 macroelement is not stable. Construction of the divergence matrix B in this case gives

$$B^T = \begin{bmatrix} B_x^T \\ B_y^T \end{bmatrix} = \begin{bmatrix} -\tfrac{1}{2}h_y & \tfrac{1}{2}h_y & \tfrac{1}{2}h_y & -\tfrac{1}{2}h_y \\ -\tfrac{1}{2}h_x & -\tfrac{1}{2}h_x & \tfrac{1}{2}h_x & \tfrac{1}{2}h_x \end{bmatrix}. \qquad (5.64)$$

This is a scalar multiple of the version of B^T for the single-element $\mathbf{Q}_2\text{–}\mathbf{Q}_1$ patch, see (5.52). The two rows are scaled central difference approximations of the pressure x-derivative and y-derivative at $\vec{x}_5$, as in (5.53). Solving $B^T\mathbf{p}=\mathbf{0}$ gives $p_1 = p_3$ and $p_2 = p_4$.

Now consider appending two additional elements onto the 2×2 patch to give the 2×3 six-element patch illustrated in Figure 5.11. This introduces an additional interior velocity node 6 and two new pressure nodes **5** and **6**. Clearly null(B^T) is still of dimension greater than one. In the system $B^T\mathbf{p}=\mathbf{0}$, the equation for node 5 implies that $p_1 = p_3$ and $p_2 = p_4$, and the equation for node 6 implies that $p_2 = p_6$ and $p_5 = p_3$. This means that the characteristic "checkerboard" pressure ("black" on the even numbered elements, "red" on the odd numbered ones) lives in the constraint space $\mathbf{N}^h$ defined in (5.44). This is an example of a *spurious pressure mode*.

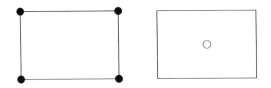

FIG. 5.9. $\mathbf{Q}_1\text{–}\mathbf{P}_0$ element ($\bullet$ two velocity components; $\circ$ pressure).

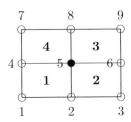

FIG. 5.10. Numbering for a 2×2 $\mathbf{Q}_1\text{–}\mathbf{P}_0$ macroelement.

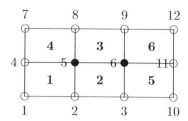

FIG. 5.11. Numbering for a 2×3 $\boldsymbol{Q}_1$–$\boldsymbol{P}_0$ macroelement.

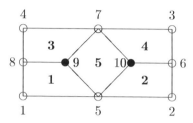

FIG. 5.12. Numbering for the stable five-element $\boldsymbol{Q}_1$–$\boldsymbol{P}_0$ macroelement.

The construction above shows why the checkerboard mode ± 1 is such a persistent feature. Generalizing the above argument shows that $\pm 1 \in \mathtt{null}(B^T)$ for all possible rectangular $n \times m$ macroelements. (However, if $n \geq 3$ and $m \geq 3$ then $n_u \geq n_p - 1$ and so the ill-posedness cannot be deduced by a simple constraint count.) An immediate consequence is that if a Stokes problem is solved using $\boldsymbol{Q}_1$–$\boldsymbol{P}_0$ approximation on a rectangular grid[4] with the velocity imposed everywhere on the boundary, then *two* compatibility conditions on the right-hand side vector in (5.33) must be satisfied in order for a discrete solution to exist. First, as discussed above, $\langle \mathbf{1}, \mathbf{g} \rangle = 0$ so that discrete mass conservation holds, and second, the "spurious" condition $\langle \pm\mathbf{1}, \mathbf{g} \rangle = 0$. A full discussion of this issue can be found in Gresho & Sani [92, pp. 572–578].

It should also be mentioned however that $\boldsymbol{Q}_1$–$\boldsymbol{P}_0$ approximation is often stable if *non-rectangular* grids are used. As an example, it can be shown (see Problem 5.4) that the five-element patch illustrated in Figure 5.12 is a stable macroelement. This macroelement could be used as a building block for constructing stable $\boldsymbol{Q}_1$–$\boldsymbol{P}_0$ approximation on grids of quadrilateral elements.

The motivation for stabilizing $\boldsymbol{Q}_1$–$\boldsymbol{P}_0$ approximation on rectangular grids is the computational convenience. In the rest of this section, it is shown that macroelements provide a useful framework for deriving stabilized methods. Starting from an unstable macroelement, the role of stabilization is to "relax" the

---

[4] As used by the IFISS software for discretizing the problems in Examples 5.1.1–5.1.4.

incompressibility constraint on the discrete velocity in the hope of obtaining a well-posed local Stokes problem defined on the macroelement. If this can be achieved, then standard macroelement theory can then be used to show that a discrete problem defined on a grid constructed by patching together "stabilized macroelements" is well posed. From a linear algebra perspective, the stabilization provides a consistent regularization of the singular matrix that arises in the unstabilized case.

For a concrete example, consider the 2×2 macroelement in Figure 5.10. Constructing the discrete Stokes system (5.33), it is easily shown that the pressure coefficient vector $\mathbf{p}$ solves the 4×4 *pressure Schur complement* system

$$BA^{-1}B^{T}\mathbf{p} = BA^{-1}\mathbf{f} - \mathbf{g}, \tag{5.65}$$

where B^T is the 2×4 matrix in (5.64), and $\mathbf{A}$ is the 2×2 diagonal matrix, defined from (5.39) and (5.40), with entries

$$\mathbf{a}_{11} = \mathbf{a}_{22} = \frac{4}{3}\left(\frac{h_y}{h_x} + \frac{h_x}{h_y}\right). \tag{5.66}$$

The local Stokes problem is well posed by definition if, and only if, the symmetric positive semi-definite matrix $S := BA^{-1}B^{T}$ in (5.65) satisfies the pressure solvability condition

$$\texttt{null}(S) = \texttt{null}(B^{T}) = 1. \tag{5.67}$$

Computing the pressure Schur complement matrix from (5.64) and (5.66) gives

$$S = \frac{1}{4\mathbf{a}_{11}}\begin{bmatrix} h_x^2 + h_y^2 & h_x^2 - h_y^2 & -h_x^2 - h_y^2 & -h_x^2 + h_y^2 \\ h_x^2 - h_y^2 & h_x^2 + h_y^2 & -h_x^2 + h_y^2 & -h_x^2 - h_y^2 \\ -h_x^2 - h_y^2 & -h_x^2 + h_y^2 & h_x^2 + h_y^2 & h_x^2 - h_y^2 \\ -h_x^2 + h_y^2 & -h_x^2 - h_y^2 & h_x^2 - h_y^2 & h_x^2 + h_y^2 \end{bmatrix}. \tag{5.68}$$

The eigenvalues and eigenvectors of S are given by

$$\left\{0, \begin{bmatrix} 1 \\ 1 \\ 1 \\ 1 \end{bmatrix}\right\}, \left\{0, \begin{bmatrix} 1 \\ -1 \\ 1 \\ -1 \end{bmatrix}\right\}, \left\{\frac{3h_x h_y}{4(1+\alpha^2)}, \begin{bmatrix} 1 \\ 1 \\ -1 \\ -1 \end{bmatrix}\right\}, \left\{\frac{3h_x h_y \alpha^2}{4(1+\alpha^2)}, \begin{bmatrix} 1 \\ -1 \\ -1 \\ 1 \end{bmatrix}\right\}, \tag{5.69}$$

$$\underbrace{}_{\mathbf{q_1}} \quad \underbrace{}_{\mathbf{q_2}} \quad \underbrace{}_{\mathbf{q_3}} \quad \underbrace{}_{\mathbf{q_4}}$$

where $\alpha = h_y/h_x$ is the element aspect ratio. The source of the macroelement instability ($\mathbf{q_2} = \pm 1$) is again evident.

Now consider relaxing the incompressibility constraint in the discrete formulation. In linear algebra terms, this introduces a relaxation parameter $\beta > 0$

together with a 4×4 stabilization matrix C in place of the zero block in the discrete system (5.33):

$$\begin{bmatrix} \mathbf{A} & B^T \\ B & -\beta C \end{bmatrix} \begin{bmatrix} \mathbf{u} \\ \mathbf{p} \end{bmatrix} = \begin{bmatrix} \mathbf{f} \\ \mathbf{g} \end{bmatrix}. \tag{5.70}$$

The stabilized pressure vector $\mathbf{p}$ thus satisfies

$$(B\mathbf{A}^{-1}B^T + \beta C)\mathbf{p} = B\mathbf{A}^{-1}\mathbf{f} - \mathbf{g}. \tag{5.71}$$

By analogy with (5.67), the discrete problem (5.71) is well posed if the stabilized pressure matrix $S_\beta := B\mathbf{A}^{-1}B^T + \beta C$ is a symmetric positive semi-definite matrix (suggesting that C itself should be a symmetric positive semi-definite matrix) with the following stability condition satisfied,

$$\mathtt{null}(S_\beta) = 1. \tag{5.72}$$

Alternatively, the condition (5.72) is equivalent to requiring that the macroelement stabilization matrix C satisfy two properties; first, to ensure *consistency* we require that $1 \in \mathtt{null}(C)$ (this precludes the use of inconsistent "penalty methods"), and second, to provide *stabilization* we require that $\mathbf{p}_*^T C \mathbf{p}_* > 0$ for all spurious pressure modes $\mathbf{p}_* \neq 1$ in $\mathtt{null}(B^T)$.

Having identified the rogue eigenvector, $\mathbf{p}_* = \mathbf{q_2}$ in (5.69), we can easily identify one choice of a stabilization matrix: $C_* = \mathbf{q_2 q_2}^T$. This has the explicit form

$$C_* = \begin{bmatrix} 1 & -1 & 1 & -1 \\ -1 & 1 & -1 & 1 \\ 1 & -1 & 1 & -1 \\ -1 & 1 & -1 & 1 \end{bmatrix}. \tag{5.73}$$

The matrix $(1/4)C_*$ is an orthogonal projection matrix. It shares the eigenvector $\mathbf{q_2}$ with S, and the associated eigenvalue is 4. Also, since $\mathbf{q_2}$ is orthogonal to the other eigenvectors of S, C_* shares those with S as well, with eigenvalues equal to 0. It follows that the eigenvalues and eigenvectors of S_β are

$$\{0, \mathbf{q_1}\}, \quad \{4\beta, \mathbf{q_2}\}, \quad \left\{\frac{3|\square|}{4(1+\alpha^2)}, \mathbf{q_3}\right\}, \quad \left\{\frac{3\alpha^2|\square|}{4(1+\alpha^2)}, \mathbf{q_4}\right\}, \tag{5.74}$$

where $|\square| = h_x h_y$ is the element area.

From (5.74), we see that the macroelement compatibility condition (5.72) is satisfied for any value of $\beta > 0$. There remains the question of what is a good value to use for β. It is clear that β should not be too small, since the point of stabilization is remove the checkerboard mode $\mathbf{q_2}$ from the null space of the Schur complement. On the other hand, β should also not be too large. To see why, observe that the right-hand side of (5.71) is a linear combination of the three eigenvectors $\mathbf{q_2}$, $\mathbf{q_3}$ and $\mathbf{q_4}$. The solution $\mathbf{p}$ is determined by multiplying each of these components by the inverse of the associated eigenvalue. If β is very large, then $\mathbf{q_2}$ will not figure prominently in the solution, and accuracy will be compromised. Thus, choosing β requires a balance between stability and accuracy. The expression (5.74) suggests that a natural choice is $\beta_* = \frac{1}{4}|\square|$, which normalizes the local component of the unstable mode to the element area,

$$\beta_* C_* \mathbf{q_2} = |\square|\mathbf{q_2}. \tag{5.75}$$

Let $Q = |\square| I$ be the mass matrix (1.114) associated with the P_0 approximation. From the construction of C_*, we see that if $\beta \leq \beta_*$, then

$$\beta \frac{\mathbf{p}^T C_* \mathbf{p}}{\mathbf{p}^T Q \mathbf{p}} \leq 1 \quad \text{for all } \mathbf{p} \in \mathbb{R}^{n_p}, \tag{5.76}$$

and β_* is the largest value for which the boundedness condition (5.76) holds. This condition will be explored further in Section 5.5.2, and we will refer to β_* as the *ideal* parameter value. For this choice, the unstable mode $\mathbf{p}_{cb} = \pm 1$ is "moved" from a zero eigenvalue (instability) to a unit eigenvalue (stability without loss of accuracy):

$$\frac{\mathbf{p}_{cb}^T S \mathbf{p}_{cb}}{\mathbf{p}_{cb}^T Q \mathbf{p}_{cb}} = 0, \quad \frac{\mathbf{p}_{cb}^T S_{\beta_*} \mathbf{p}_{cb}}{\mathbf{p}_{cb}^T Q \mathbf{p}_{cb}} = 1. \tag{5.77}$$

To apply this stabilization strategy on general grids of rectangular elements, we need to subdivide the grid into 2×2 patches and then construct a block-diagonal "grid" stabilization matrix $\mathbf{C} = \mathtt{diag}[C_*, \ldots, C_*]$, with C_* given in (5.73). The stabilized Stokes coefficient matrix, K_β, is then given by

$$K_\beta := \begin{bmatrix} \mathbf{A} & B^T \\ B & -\beta\mathbf{C} \end{bmatrix}. \tag{5.78}$$

The stabilization matrix C_* is an example of a *minimal* stabilization operator. The only part of the pressure approximation space M_h affected by the stabilization is the component corresponding to the rogue eigenvector $\mathbf{q_2}$ in (5.69). Although the concept of minimal stabilization can be generalized to arbitrary macroelements, this approach is limited in general by the requirement that all spurious pressure modes need to be known explicitly. An alternative approach that does not have this limitation is known as *pressure jump stabilization*. This strategy provides a different macroelement stabilization matrix C^* that is symmetric and consistent ($\mathbf{1} \in \mathtt{null}(C^*)$).

To see the construction of C^*, consider the 2×2 macroelement patch of rectangular elements in Figure 5.10, and observe that the area-weighted inter-element pressure jumps satisfy

$$h_x h_y \left[(p_1 - p_2)^2 + (p_2 - p_3)^2 + (p_3 - p_4)^2 + (p_4 - p_1)^2 \right] = \mathbf{p}^T C^* \mathbf{p}, \tag{5.79}$$

with C^*, the macroelement jump stabilization matrix, given by

$$C^* = h_x h_y \begin{bmatrix} 2 & -1 & 0 & -1 \\ -1 & 2 & -1 & 0 \\ 0 & -1 & 2 & -1 \\ -1 & 0 & -1 & 2 \end{bmatrix}. \tag{5.80}$$

The eigenvalues and eigenvectors of C^* are

$$\{0, \mathbf{q_1}\}, \quad \{4|\square|, \mathbf{q_2}\}, \quad \{2|\square|, \mathbf{q_3}\}, \quad \{2|\square|, \mathbf{q_4}\}. \tag{5.81}$$

Since $\mathbf{p}^T C^* \mathbf{p} > 0$ for all $\mathbf{p}$ not proportional to $\mathbf{1}$, all nonconstant macroelement pressure vectors are affected by the jump stabilization. By combining (5.74) and (5.81), it can be seen that the eigenvalues and eigenvectors of S_β are

$$\{0, \mathbf{q_1}\}, \{4\beta|\square|, \mathbf{q_2}\}, \left\{ \frac{3|\square|}{4(1+\alpha^2)} + 2\beta|\square|, \mathbf{q_3} \right\}, \left\{ \frac{3\alpha^2|\square|}{4(1+\alpha^2)} + 2\beta|\square|, \mathbf{q_4} \right\},$$
$$\tag{5.82}$$

showing that the macroelement compatibility condition (5.72) is satisfied for all $\beta > 0$. Notice that the *ideal* value of the stabilization parameter (associated with the upper bound (5.76)) is given by $\beta := \beta^* = 1/4$. It turns out that this "ideal" choice also ensures stability independently of the rectangle aspect ratio. Further details are given in Problem 5.5.

The generalization of the jump stabilization strategy to any rectangular $n \times m$ macroelement $\mathcal{M}$ is straightforward. In terms of the underlying discrete formulation, the original incompressibility constraint (5.30) is augmented by a jump stabilization term

$$-\int_{\mathcal{M}} q_h \nabla \cdot \vec{u}_h - \frac{1}{4}|M| \sum_{e \in \Gamma_{\mathcal{M}}} \langle [\![p_h]\!]_e, [\![q_h]\!]_e \rangle_{\bar{E}} = 0 \quad \text{for all } q_h \in M^h. \tag{5.83}$$

Here, $|M|$ is the mean element area within the macroelement, the set $\Gamma_{\mathcal{M}}$ consists of interior element edges in $\mathcal{M}$, $[\![\cdot]\!]_e$ is the jump across edge e, and $\langle p, q \rangle_{\bar{E}} = (1/|E|) \int_E pq$. (The stabilization term in (5.83) is the natural generalization of the operator $\beta^* C^*$, see Problem 5.6.) Independent of the dimensions n and m, the fact that $\mathbf{p}^T C^* \mathbf{p} > 0$ for all $\mathbf{p} \neq \mathbf{1}$ ensures that the solvability condition (5.72) is satisfied.

To apply the jump stabilization strategy on general grids, we need to be able to construct the grid from locally stabilized patches. We can then define a block diagonal "grid" stabilization matrix, $\mathbf{C}$, which has the contributing macroelement stabilization matrices as its diagonal blocks. For example, if a uniform grid of square elements is decomposed into 2×2 macroelements, then the stabilized Stokes coefficient matrix, K_{β^*}, is given by

$$K_{\beta^*} := \begin{bmatrix} \mathbf{A} & B^T \\ B & -\mathbf{C} \end{bmatrix}, \tag{5.84}$$

where $\mathbf{C} = \frac{1}{4}\mathrm{diag}[C^*, \ldots, C^*]$, with C^* given in (5.80). The fact that (5.72) is satisfied on every macroelement implies that the stabilized Stokes problem is well posed. For a flow problem with velocity prescribed everywhere on the boundary, $\mathrm{null}(K_{\beta^*}) = [\mathbf{0}^T \quad \mathbf{1}^T]^T$, and if a natural outflow boundary condition holds, the stabilized Stokes coefficient matrix is guaranteed to be non-singular, that is, $\mathrm{null}(K_{\beta^*}) = \mathbf{0}$. Moreover, it is easily shown, see Problem 5.6, that the *ideal* value $\beta^* = 1/4$ results in a "perfect" stabilization for grids of 2×2 macroelements, in that

$$\frac{\mathbf{p}_{cb}^T S_{\beta^*} \mathbf{p}_{cb}}{\mathbf{p}_{cb}^T Q \mathbf{p}_{cb}} = 1. \tag{5.85}$$

Another property of the jump stabilization strategy that needs to be emphasized is that $\vec{u}_h$, the velocity solution in (5.83), satisfies

$$0 = \int_{\mathcal{M}} \nabla \cdot \vec{u}_h = \int_{\partial \mathcal{M}} \vec{u}_h \cdot \vec{n}. \tag{5.86}$$

Thus, although elementwise mass conservation is lost in the stabilized formulation, local incompressibility is retained at the macroelement level. Establishing (5.86) is left as an exercise; see Problem 5.7.

We conclude the discussion of Q_1–P_0 approximation with an example demonstrating the importance of stabilization in practice. The plots in Figure 5.13 show stabilized and unstabilized pressure solutions for the expansion flow problem in Example 5.1.2. The grid is uniform with square elements of width $h = 1/8$, and is referred to as the 16×48 grid.[5] Because of the natural outflow boundary condition, the Stokes system matrix (5.84) is non-singular in both cases. (If the velocity on the outflow boundary were imposed instead, then the Stokes system matrix would be rank-deficient by one for $\beta \neq 0$, and by two otherwise.) Although the pressure is smooth near the outflow boundary in both cases, the singularity

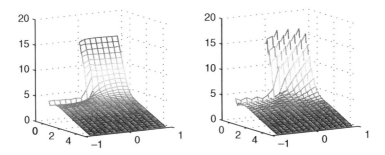

FIG. 5.13. Pressure solutions corresponding to a stabilized (left, $\beta = \beta^*$) and unstabilized (right, $\beta = 0$) Q_1–P_0 mixed approximation of Example 5.1.2.

[5]In the IFISS software the corresponding grid parameter is $\ell = 4$.

in the pressure at the corner excites spurious oscillations in the unstabilized case. This pressure instability does not seem to harm the velocity solution however — the streamlines corresponding to the stabilized and unstabilized solutions are indistinguishable to plotting accuracy.

Now consider the Q_1–Q_1 approximation method, which has velocity and pressure degrees of freedom defined at the same set of grid points. This regular arrangement of data is very appealing from the point of view of ease of programming and computational efficiency. For a uniform grid of square elements of size h, Q_1–Q_1 approximation can be stabilized by perturbing the incompressibility constraint (5.30) with a pressure Laplacian term,

$$-\int_\Omega q_h \nabla \cdot \vec{u}_h - \beta h^2 (\nabla p_h, \nabla q_h) = 0 \quad \text{for all } q_h \in M^h, \tag{5.87}$$

where $\beta > 0$. The attraction of this formulation is that the corresponding system of linear equations is extremely easy to construct. It takes the form

$$\begin{bmatrix} A & B^T \\ B & -\beta h^2 A \end{bmatrix} \begin{bmatrix} \mathbf{u} \\ \mathbf{p} \end{bmatrix} = \begin{bmatrix} \mathbf{f} \\ \mathbf{g} \end{bmatrix}, \tag{5.88}$$

where the stabilization matrix A is the (singular) bilinear stiffness matrix associated with a Neumann boundary condition. This stabilization strategy (5.87) has two big limitations, however. The first is that the Laplacian stabilization term is only appropriate in the case of uniform square grids. If a rectangular grid is used instead, the pressure Laplacian operator in (5.87) must be replaced by a local non-isotropic diffusion operator weighted by the edge lengths. Without this fix, symptoms of instability are increasingly evident as the element aspect ratio is increased. More importantly, this strategy requires a "good" choice of stabilization parameter β. It is very easy to over-stabilize by using a parameter value that is too large. This issue is explored in Computational Exercise 5.2.

A much better way to stabilize Q_1–Q_1 is with a novel approach suggested by Dohrmann & Bochev [51]. Motivated by the rigorous error analysis of mixed methods based on continuous pressure approximation,[6] the deficiency of Q_1–Q_1 approximation can be associated with the mismatch between the discrete divergence of the velocity field (a subspace of the space P_0 of discontinuous piecewise constant functions) and the actual discrete pressure space Q_1. To get into the "right space", a suitable pressure stabilization operator is needed, namely

$$C(p_h, q_h) = (p_h - \Pi_0 p_h, q_h - \Pi_0 q_h), \tag{5.89}$$

where Π_0 is the L^2 projection from M_h into the space P_0. Note that this projection is defined locally: $\Pi_0 p_h$ is a constant function in each element $\square_k \in \mathcal{T}_h$.

---

[6]The technical details are discussed in Section 5.4.1.

It is determined simply by local averaging,

$$\Pi_0 p_h|_{\square_k} = \frac{1}{|\square_k|} \int_{\square_k} p_h \quad \text{for all } \square_k \in \mathcal{T}_h. \tag{5.90}$$

In the case of a rectangular grid, given the standard bilinear basis functions $\{\psi_j\}_{j=1}^4$, and substituting the local expansion

$$p_h|_{\square_k} = p_1^k \psi_1 + p_2^k \psi_2 + p_3^k \psi_3 + p_4^k \psi_4 \tag{5.91}$$

into (5.90), we see that

$$\Pi_0 p_h|_{\square_k} = \frac{(p_1^k + p_2^k + p_3^k + p_4^k)}{4},$$

that is, the projected pressure is the mean of the vertex values. Given that

$$C(p_h, q_h) = \sum_{\square_k \in \mathcal{T}_h} \int_{\square_k} (p_h - \Pi_0 p_h)|_{\square_k} (q_h - \Pi_0 q_h)|_{\square_k}, \tag{5.92}$$

it is clear that the associated grid stabilization matrix C can be assembled from element contribution matrices in the same way as a standard Galerkin matrix. Indeed, using (5.90), it can easily be shown that the 4×4 contribution matrix C_k is given by

$$C_k = Q_k - \mathbf{q}\mathbf{q}^T |\square_k|, \tag{5.93}$$

where Q_k is the 4×4 element mass matrix for the bilinear discretization, and $\mathbf{q} = [\frac{1}{4}, \frac{1}{4}, \frac{1}{4}, \frac{1}{4}]^T$ is the local averaging operator. The null space of C_k consists of constant vectors. The details are left as an exercise, see Problem 5.8.

The fact that $\mathbf{1} \in \text{null}(C_k)$ means that the assembled stabilization operator is consistent, that is, $\mathbf{1} \in \text{null}(C)$. There is, however, a fundamental difference between the stabilization matrix C associated with (5.89) and the $\mathbf{P}_0$ stabilization matrix $\mathbf{C}$ in (5.84). By construction, all macroelement constant functions are in the null space of $\mathbf{C}$, and so they are "invisible" to the stabilization. In contrast, the only functions from M^h lying in the null space of the $\mathbf{Q}_1$ stabilization matrix C are constant functions over Ω, that is, $\text{null}(C) = \text{span}\{\mathbf{1}\}$. The key point is that the $\mathbf{P}_0$ stabilization is *local* (by construction, to ensure local incompressibility), whereas the $\mathbf{Q}_1$ stabilization is *global*.

The really appealing feature of the $\mathbf{Q}_1$–$\mathbf{Q}_1$ stabilization strategy (5.89) is that there is no (explicit) stabilization parameter. The unparameterized method is to assemble the grid stabilization matrix C from the element matrices (5.93), and then to solve the system

$$\begin{bmatrix} A & B^T \\ B & -C \end{bmatrix} \begin{bmatrix} \mathbf{u} \\ \mathbf{p} \end{bmatrix} = \begin{bmatrix} \mathbf{f} \\ \mathbf{g} \end{bmatrix}. \tag{5.94}$$

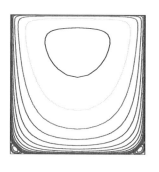 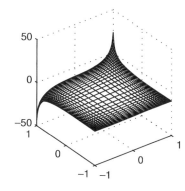

FIG. 5.14. Exponentially distributed streamline plot (left) and pressure plot of the stabilized Q_1–Q_1 approximation of Example 5.1.3 using a 32×32 stretched grid.

This approach gives remarkably accurate solutions even in the case of strongly stretched grids. An illustration is given in Figure 5.14. The appropriateness of the implicit parameter value $\beta = 1$ in (5.94) should not be a big surprise, however, since the stabilization matrix C clearly inherits the "correct" mass matrix scaling from (5.93). This implies the ideal scaling property discussed earlier, see (5.76):

$$\frac{\mathbf{p}^T C \mathbf{p}}{\mathbf{p}^T Q \mathbf{p}} \leq 1 \quad \text{for all } \mathbf{p} \in \mathbb{R}^{n_p}. \tag{5.95}$$

5.3.3 Triangular elements

In this section we identify some stable mixed approximation methods for unstructured meshes of triangular elements. Unsurprisingly, the lowest order mixed approximation methods based on globally continuous linear approximation for both velocity components, together with either a discontinuous constant pressure $(P_1$–$P_0)$, or a continuous linear pressure $(P_1$–$P_1)$, are both unstable. In the P_1–P_0 case, the approximation can be readily stabilized using the pressure jump stabilization in (5.83) together with an appropriate macroelement subdivision. The pressure projection stabilization strategy (5.89) is the recommended way of restoring stability in the P_1–P_1 case.

Instead of stabilizing the P_1–P_1 method, the velocity approximation can be enhanced relative to the pressure to produce a uniformly stable method. Specifically, the discussion in Section 5.3.1 suggests using quadratic approximation for the velocity components rather than linear. The resulting mixed method is called P_2–P_1 approximation. It has the nodal positions illustrated in Figure 5.15.

The macroelement technique discussed in Section 5.3.1 can also be used to establish the uniform stability of the mixed approximation in this case. The starting point is the two-element patch $\mathcal{M} = \triangle_k \cup \triangle_m$ illustrated in Figure 5.16.

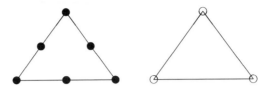

FIG. 5.15. P_2–P_1 element ($\bullet$ two velocity components; $\circ$ pressure).

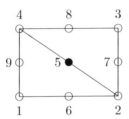

FIG. 5.16. P_2–P_1 macroelement numbering for two-element patch $\triangle_k \cup \triangle_m$.

In this simple case, B_x^T and B_y^T are 1×4 row vectors,

$$B_x^T = [b_{x,j}], \quad b_{x,j} = -\int_{\triangle_k} \psi_j \frac{\partial \phi_5}{\partial x} - \int_{\triangle_m} \psi_j \frac{\partial \phi_5}{\partial x}, \qquad (5.96)$$

$$B_y^T = [b_{y,j}], \quad b_{y,j} = -\int_{\triangle_k} \psi_j \frac{\partial \phi_5}{\partial y} - \int_{\triangle_m} \psi_j \frac{\partial \phi_5}{\partial y}, \qquad (5.97)$$

where ϕ_5 is the quadratic bubble function that takes the value unity at node 5 and is zero at the other eight nodes. Similarly ψ_j: $j = 1, 2, 3, 4$ are the standard linear basis functions defined on $\triangle_k \cup \triangle_m$. Evaluating the integrals gives

$$B^T = \begin{bmatrix} B_x^T \\ B_y^T \end{bmatrix} = \begin{bmatrix} -1/6 h_y & 1/6 h_y & 1/6 h_y & -1/6 h_y \\ -1/6 h_x & -1/6 h_x & 1/6 h_x & 1/6 h_x \end{bmatrix}, \qquad (5.98)$$

and writing out the associated linear system $B^T \mathbf{p} = \mathbf{0}$ gives

$$-p_1 + p_2 + p_3 - p_4 = 0,$$

$$-p_1 - p_2 + p_3 + p_4 = 0.$$

A useful interpretation is obtained by combining these two equations. Subtracting the equations gives a discrete approximation to $\partial p/\partial s = 0$ at the node $\vec{x}_5$ where s is tangential to the inter-element edge. This implies that the pressures p_2 and p_4 at the vertices at each end of the edge are equal. Summing the equations

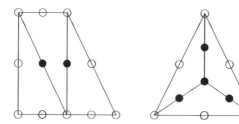

FIG. 5.17. P_2–P_1 three-element patches.

gives a discrete approximation to $\partial p/\partial n = 0$ at the node $\vec{x}_5$ where n is normal to the inter-element edge. This implies that the pressures p_1 and p_3 at the two "off-edge" vertices are equal. Notice that this result mirrors that obtained for Q_2–Q_1 approximation (Section 5.3.1) in the case of a single element macroelement, so the two-element patch does not establish stability. This conclusion is completely general with respect to triangle structure—the equality of the inter-element edge pressures and of the off-edge pressures is independent of the shape and orientation of the two triangles forming the macroelement.

Since two-element P_2–P_1 patches are not stable, we need to consider three-element patches. There are two generic cases to consider, as illustrated in Figure 5.17. Both of these macroelement configurations are clearly stable. For example, appending another element to the two-element patch in Figure 5.16 introduces an extra pressure node and one new interior velocity node as illustrated in Figure 5.17. The additional velocity node adds two rows to the matrix B^T of (5.98). The condition $B^T\mathbf{p} = \mathbf{1}$ boils down to an equation relating the pressures p_2 and p_3 in Figure 5.16, and an equation relating p_4 to the newly introduced pressure node. Thus macroelement stability is assured. It is also clear that if an additional element is appended to any stable P_2–P_1 patch then the new macroelement is also stable. Thus, since all possible grids (other than trivial single-element or two-element grids) can be decomposed into stable three-, four- or five-element patches, we deduce that the P_2–P_1 approximation is indeed uniformly stable.

Turning to discontinuous pressure approximation, the triangular P_2–P_{-1} (quadratic velocity with discontinuous linear pressure) analogue of the Q_2–P_{-1} mixed approximation method is *not* stable. There is no interior velocity degree of freedom in the triangle case, and so the nonconstant pressure components cannot be stabilized on a one-element patch. To get a stable method the quadratic velocity approximation in each element must be enhanced by adding a cubic bubble function to each component. The resulting mixed method is uniformly stable. The resulting mixed method is called P_{2*}–P_{-1} approximation, but is also referred to in the literature as the *Crouzeix–Raviart* method. It has the degrees of freedom illustrated in Figure 5.18.

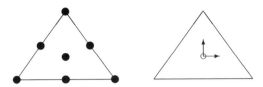

FIG. 5.18. P_{2*}–P_{-1} element (• two velocity components; ○ pressure; $\stackrel{\uparrow}{\rightarrow}$ pressure derivative).

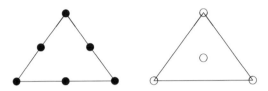

FIG. 5.19. P_2–P_{-1*} element (• two velocity components; ○ pressure).

Instead of adding velocity degrees of freedom, the alternative route to stability is to use a piecewise constant pressure approximation in place of the piecewise linear pressure. This gives the uniformly stable P_2–P_0 mixed approximation. Unfortunately, as in the Q_2–P_0 case, the order of the velocity approximation is compromised by the reduced accuracy of the pressure approximation. A third possibility is to construct a *hybrid* pressure approximation by combining the continuous linear pressure approximation with the discontinuous constant pressure approximation.[7] The resulting mixed method is referred to as the P_2–P_{-1*} approximation and enjoys the best of both worlds; it has locally incompressibility in the sense of (5.63), and yet it does not have its accuracy compromised by the lower order pressure. The degrees of freedom are illustrated in Figure 5.19. Perhaps surprisingly, the P_2–P_{-1*} approximation is also uniformly stable.

5.3.4   *Brick and tetrahedral elements*

There are no real surprises in going from mixed approximation defined on rectangles in $\mathbb{R}^2$ to mixed approximation defined on bricks in $\mathbb{R}^3$. Indeed, the uniform stability of the brick counterparts of the Q_2–Q_1 and Q_2–P_{-1} rectangle elements may be readily established using the macroelement technique of Section 5.3.1. (The specific case of Q_2–Q_1 approximation is explored in Problem 5.10.)

---

[7]Since this gives two different ways of representing constant functions, the matrix B^T has a two-dimensional null space in the case of enclosed flow boundary conditions.

Stabilized versions of the equal order Q_1–Q_1 mixed method, and the lowest order Q_1–P_0 mixed method are obvious extensions of the corresponding rectangular element case. In particular, the jump stabilized Q_1–P_0 brick approximation method has a modified incompressibility constraint defined on an $m \times n \times k$ macroelement $\mathcal{M}$ exactly as in (5.83),

$$-\int_{\mathcal{M}} q_h \nabla \cdot \vec{u}_h - \beta \, |M| \sum_{e \in \Gamma_{\mathcal{M}}} \langle [\![p_h]\!]_e, [\![q_h]\!]_e \rangle_{\bar{E}} = 0 \quad \text{for all } q_h \in M^h. \tag{5.99}$$

The only difference in three dimensions is that $|M|$ represents the mean element *volume*, and the set $\Gamma_{\mathcal{M}}$ consists of the interior *faces* within the macroelement. Note that, $\langle p, q \rangle_{\bar{E}} := (1/|E|) \int_E pq$, as in (5.83). The final point that needs to be emphasized is that the smallest macroelement configuration allowed in three dimensions is the $2 \times 2 \times 2$ element bundle.

We note in passing that the Q_1–P_0 brick element generates a system of linear algebraic equations whose matrix has a much richer structure than that of the Q_1–P_0 rectangle. Indeed, because of all the intrinsic two-dimensional symmetries, the dimension of the null space of B^T is actually proportional to k in the case of an $k \times k \times k$ subdivision of square bricks! See Brezzi & Fortin [33, p. 244] for further discussion.

In complete contrast, the construction of stable mixed approximation methods for tetrahedral elements mimicking stable mixed methods on triangular elements is not automatic. The reason for this is that mid-face nodes play the role of connecting inter-element pressures in three dimensions, whereas mid-edge nodes do this job in two dimensions. What makes life awkward is that standard quadratic approximation on a tetrahedra does not involve mid-face nodes. Triquadratic approximation on a brick, on the other hand, does have a node on every mid-face, see Figures 1.13 and 1.12. As a consequence, the tetrahedral version of the P_{2*}–P_{-1} approximation supplements the standard quadratic approximation for each velocity component with four additional face bubble functions, as well as an interior bubble function, in order to ensure the stability of the mixed approximation. See [33, p. 217] for further details. Despite these difficulties, macroelement techniques can nevertheless be used to show that the continuous pressure P_2–P_1 mixed approximation is uniformly stable in the case of tetrahedral elements.

5.4 Theory of errors

We start from the weak formulation of the Stokes problem. If we define the continuous bilinear forms $a : \mathbf{H}^1 \times \mathbf{H}^1 \to \mathbb{R}$, and $b : \mathbf{H}^1 \times L_2(\Omega) \to \mathbb{R}$, so that

$$a(\vec{u}, \vec{v}) := \int_{\Omega} \nabla \vec{u} : \nabla \vec{v}, \quad b(\vec{v}, q) := -\int_{\Omega} q \nabla \cdot \vec{v}, \tag{5.100}$$

and the continuous functional $\ell : \mathbf{H}^1 \to \mathbb{R}$, so that

$$\ell(\vec{v}) := \int_{\partial \Omega_N} \vec{s} \cdot \vec{v}; \qquad (5.101)$$

then the weak formulation (5.22)–(5.23) can be concisely restated:

Find $\vec{u} \in \mathbf{H}_E^1$ and $p \in L_2(\Omega)$ such that
$$a(\vec{u}, \vec{v}) + b(\vec{v}, p) = \ell(\vec{v}) \quad \text{for all } \vec{v} \in \mathbf{H}_{E_0}^1, \qquad (5.102)$$
$$b(\vec{u}, q) = 0 \quad \text{for all } q \in L_2(\Omega).$$

With a conforming mixed approximation, the corresponding discrete problem (5.29)–(5.30) is given by:

Find $\vec{u}_h \in \mathbf{X}_E^h$ and $p_h \in M^h$ such that
$$a(\vec{u}_h, \vec{v}_h) + b(\vec{v}_h, p_h) = \ell(\vec{v}_h) \quad \text{for all } \vec{v}_h \in \mathbf{X}_0^h, \qquad (5.103)$$
$$b(\vec{u}_h, q_h) = 0 \quad \text{for all } q_h \in M^h.$$

Our task here is to estimate the quality of the mixed approximation, that is, to bound $\|\vec{u} - \vec{u}_h\|_X$ and $\|p - p_h\|_M$ with respect to appropriate norms $\|\cdot\|_X$ and $\|\cdot\|_M$. We outline the derivation of a priori error bounds in the next section, and go on to discuss computable error bounds in Section 5.4.2.

5.4.1 A priori error bounds

For clarity of exposition, we assume throughout this section that we are solving an enclosed flow problem. In this case, the natural norms are the energy norm for the velocity $\|\vec{v}\|_X := \|\nabla \vec{v}\|$, and the quotient norm for the pressure $\|p\|_M := \|p\|_{0,\Omega}$ defined earlier. For a flow problem that has a natural boundary condition, the standard L_2 norm for the pressure can be used instead.

The basic mixed approximation error estimate is a consequence of the coercivity and inf–sup conditions that were introduced in the discussion of the uniqueness of the weak solution at the end of Section 5.2. We summarize the result in the following theorem.

Theorem 5.1. *Consider the discrete formulation (5.103). Assume that there exist constants, $\alpha^* > 0$ and $\gamma^* > 0$ independent of h, such that the following conditions hold:*

(I) $\mathbf{X}_0^h$-*coercivity*

$$a(\vec{v}_h, \vec{v}_h) \geq \alpha^* \|\vec{v}_h\|_X^2 \quad \text{for all } \vec{v}_h \in \mathbf{X}_0^h,$$

(II) discrete inf–sup stability

$$\max_{\vec{v}_h \in \mathbf{X}_0^h} \frac{b(\vec{v}_h, q_h)}{\|\vec{v}_h\|_X} \geq \gamma^* \|q_h\|_M \quad \text{for all } q_h \in M^h.$$

Then a discrete solution $(\vec{u}_h, p_h)$ exists and is unique. Moreover, if $(\vec{u}, p)$ is the solution of (5.102), then there exists a constant $C > 0$ such that

$$\|\vec{u} - \vec{u}_h\|_X + \|p - p_h\|_M \leq C \left(\inf_{\vec{v}_h \in \mathbf{X}_E^h} \|\vec{u} - \vec{v}_h\|_X + \inf_{q_h \in M^h} \|p - q_h\|_M \right). \quad (5.104)$$

The conditions *I)* and *II)* are also referred to as the Babuška–Brezzi stability conditions. The derivation of (5.104) is outlined below. The first step is to define the "big" symmetric bilinear form

$$B((\vec{u}, p); (\vec{v}, q)) = a(\vec{u}, \vec{v}) + b(\vec{u}, q) + b(\vec{v}, p), \quad (5.105)$$

and the corresponding functional $F((\vec{v}, q)) = \ell(\vec{v})$. Choosing the successive test vectors $(\vec{v}, 0)$ and $(\vec{0}, q)$ shows that the Stokes problem (5.102) can be rewritten in the form: find $(\vec{u}, p) \in \mathbf{H}_E^1 \times L_2(\Omega)$ such that

$$B((\vec{u}, p); (\vec{v}, q)) = F((\vec{v}, q)) \quad \text{for all } (\vec{v}, q) \in \mathbf{H}_{E_0}^1 \times L_2(\Omega). \quad (5.106)$$

The discrete problem (5.103) is similarly given by: find $(\vec{u}_h, p_h) \in \mathbf{X}_E^h \times M^h$ such that

$$B((\vec{u}_h, p_h); (\vec{v}_h, q_h)) = F((\vec{v}_h, q_h)) \quad \text{for all } (\vec{v}_h, q_h) \in \mathbf{X}_0^h \times M^h. \quad (5.107)$$

The Babuška theory is based on the following property, referred to as *B-stability*: there exists a constant $\gamma > 0$ such that

$$\max_{(\vec{v}_h, q_h) \in \mathbf{X}_0^h \times M^h} \frac{B((\vec{w}_h, s_h); (\vec{v}_h, q_h))}{\|(\vec{v}_h, q_h)\|_{X \times M}} \geq \gamma \|(\vec{w}_h, s_h)\|_{X \times M} \quad (5.108)$$

for all $(\vec{w}_h, s_h) \in \mathbf{X}_0^h \times M^h$. The product norm is defined by $\|(\vec{v}, q)\|_{X \times M} = \|\vec{v}\|_X + \|q\|_M$. Using (5.108), the error bound in Theorem 5.1 is established via a two-stage process. The first is to show that the Babuška–Brezzi stability conditions imply *B*-stability. The second is to show that (5.108) implies the desired error estimate (5.104). Each of the stages is summarized below in the form of a lemma.

Lemma 5.2. *Under the assumptions of Theorem 5.1, the stability inequality (5.108) is satisfied.*

Proof We begin by observing that if, for any pair $(\vec{w}_h, s_h) \in \mathbf{X}_0^h \times M^h$, there exists a pair $(\vec{v}_h, q_h) \in \mathbf{X}_0^h \times M^h$ such that

$$B((\vec{w}_h, s_h); (\vec{v}_h, q_h)) \geq \gamma_1 \|(\vec{w}_h, s_h)\|_{X \times M}^2 \quad (5.109)$$

and

$$\|(\vec{v}_h, q_h)\|_{X \times M} \leq \gamma_2 \|(\vec{w}_h, s_h)\|_{X \times M}, \tag{5.110}$$

then (5.108) is satisfied with $\gamma = \gamma_1 / \gamma_2$.

To make progress, note that the inf–sup stability condition implies that for any $s_h \in M^h$, there exists a corresponding vector $\vec{z}_h \in \mathbf{X}_0^h$ such that

$$\|\vec{z}_h\|_X = \|s_h\|_M, \quad -b(\vec{z}_h, s_h) \geq \gamma^* \|s_h\|_M^2. \tag{5.111}$$

Also, if we define the operator norm

$$\|a\| = \sup_{\substack{\|\vec{u}\|_X = 1 \\ \|\vec{v}\|_X = 1}} |a(\vec{u}, \vec{v})|,$$

we immediately have the following bound

$$|a(\vec{u}_h, \vec{v}_h)| \leq \|a\| \, \|\vec{u}_h\|_X \, \|\vec{v}_h\|_X \quad \text{for all } \vec{u}_h, \vec{v}_h \in \mathbf{X}_0^h. \tag{5.112}$$

Given $(\vec{w}_h, s_h) \in \mathbf{X}_0^h \times M^h$, we then choose $\vec{v}_h = \vec{w}_h - \delta \vec{z}_h$ and $q_h = -s_h$ with the specific constant $\delta = \alpha^* \gamma^* / \|a\|^2$. Using (5.111) and (5.112) together with the Babuška–Brezzi conditions then gives

$$
\begin{aligned}
B((\vec{w}_h, s_h); (\vec{v}_h, q_h)) &= B((\vec{w}_h, s_h); (\vec{w}_h - \delta \vec{z}_h, -s_h)) \\
&= a(\vec{w}_h, \vec{w}_h - \delta \vec{z}_h) + b(\vec{w}_h, -s_h) + b(\vec{w}_h - \delta \vec{z}_h, s_h) \\
&= a(\vec{w}_h, \vec{w}_h) - \delta a(\vec{w}_h, \vec{z}_h) - \delta b(\vec{z}_h, s_h) \\
&\geq \alpha^* \|\vec{w}_h\|_X^2 - \delta a(\vec{w}_h, \vec{z}_h) - \delta b(\vec{z}_h, s_h) \\
&\geq \alpha^* \|\vec{w}_h\|_X^2 - \delta \|a\| \, \|\vec{w}_h\|_X \, \|\vec{z}_h\|_X - \delta b(\vec{z}_h, s_h) \\
&\geq \alpha^* \|\vec{w}_h\|_X^2 - \delta \|a\| \, \|\vec{w}_h\|_X \, \|s_h\|_M + \delta \gamma^* \|s_h\|_M^2 \\
&= \frac{\alpha^*}{2} \|\vec{w}_h\|_X^2 + \delta \gamma^* \|s_h\|_M^2 \\
&\quad + \alpha^* \left(\frac{1}{2} \|\vec{w}_h\|_X^2 - \delta \frac{\|a\|}{\alpha^*} \|\vec{w}_h\|_X \, \|s_h\|_M \right) \\
&\geq \frac{\alpha^*}{2} \|\vec{w}_h\|_X^2 + \delta \gamma^* \|s_h\|_M^2 - \frac{\delta^2}{2\alpha^*} \|a\|^2 \|s_h\|_M^2 \\
&\geq \frac{\alpha^*}{2} \|\vec{w}_h\|_X^2 + \frac{\delta \gamma^*}{2} \left(2 - \frac{\delta \|a\|^2}{\alpha^* \gamma^*} \right) \|s_h\|_M^2 \\
&\geq \frac{\alpha^*}{2} \|\vec{w}_h\|_X^2 + \frac{\delta \gamma^*}{2} \|s_h\|_M^2 \\
&\geq 2\gamma_1 (\|\vec{w}_h\|_X^2 + \|s_h\|_M^2) \\
&\geq \gamma_1 (\|\vec{w}_h\|_X + \|s_h\|_M)^2.
\end{aligned}
$$

This establishes (5.109) with $\gamma_1 = \min(\alpha^*/4, \delta\gamma^*/4)$. Moreover,

$$
\begin{aligned}
\|(\vec{v}_h, q_h)\|_{X \times M} &= \|\vec{v}_h\|_X + \|q_h\|_M \\
&= \|\vec{w}_h - \delta\vec{z}_h\|_X + \|-s_h\|_M \\
&\leq \|\vec{w}_h\|_X + \delta \|\vec{z}_h\|_X + \|s_h\|_M \\
&\leq \|\vec{w}_h\|_X + \delta \|s_h\|_M + \|s_h\|_M \\
&\leq \gamma_2(\|\vec{w}_h\|_X + \|s_h\|_M),
\end{aligned}
$$

thus establishing (5.110) with $\gamma_2 = 1 + \delta$.   □

Lemma 5.3. *Given a B-stable approximation* (5.107) *of the weak formulation* (5.106)*, the error satisfies the quasi-optimal bound* (5.104)*.*

Proof Since $\mathbf{X}_0^h \times M^h \subset \mathbf{H}_{E_0}^1 \times L_2(\Omega)$, we subtract (5.107) from (5.106) to give the Galerkin orthogonality property:

$$
B((\vec{u} - \vec{u}_h, p - p_h); (\vec{v}_h, q_h)) = 0 \quad \text{for all } (\vec{v}_h, q_h) \in \mathbf{X}_0^h \times M^h. \tag{5.113}
$$

Given a pair $(\vec{v}_h, q_h) \in \mathbf{X}_0^h \times M^h$, define $\vec{w}_h = \vec{u}_h - \vec{v}_h \in \mathbf{X}_0^h$ and $r_h = p_h - q_h \in M^h$. Then, for all $(\vec{z}_h, s_h) \in \mathbf{X}_0^h \times M^h$,

$$
\begin{aligned}
B((\vec{w}_h, r_h); (\vec{z}_h, s_h)) &= B((\vec{u}_h - \vec{v}_h, p_h - q_h); (\vec{z}_h, s_h)) \\
&= B((\vec{u}_h - \vec{u} + \vec{u} - \vec{v}_h, p_h - p + p - q_h); (\vec{z}_h, s_h)) \\
&= B((\vec{u} - \vec{v}_h, p - q_h) + (\vec{u} - \vec{u}_h, p - p_h); (\vec{z}_h, s_h)) \\
&= B((\vec{u} - \vec{v}_h, p - q_h); (\vec{z}_h, s_h)) \tag{5.114}
\end{aligned}
$$

making use of (5.113) in the last step. Since $a(\cdot, \cdot)$ and $b(\cdot, \cdot)$ are continuous with respect to the spaces $\mathbf{H}_{E_0}^1$ and $L_2(\Omega)$, the form $B(\cdot, \cdot)$ must also be continuous with respect to $\mathbf{H}_{E_0}^1 \times L_2(\Omega)$. Therefore

$$
B((\vec{u} - \vec{v}_h, p - q_h); (\vec{z}_h, s_h)) \leq \|B\| \, \|(\vec{u} - \vec{v}_h, p - q_h)\|_{X \times M} \, \|(\vec{z}_h, s_h)\|_{X \times M},
$$

for all $(\vec{z}_h, s_h) \in \mathbf{X}_0^h \times M^h$. Using (5.114) and rearranging then gives the bound

$$
\max_{(\vec{z}_h, s_h) \in \mathbf{X}_0^h \times M^h} \frac{B((\vec{u}_h - \vec{v}_h, p_h - q_h); (\vec{z}_h, s_h))}{\|(\vec{z}_h, s_h)\|_{X \times M}} \leq \|B\| \, \|(\vec{u} - \vec{v}_h, p - q_h)\|_{X \times M}. \tag{5.115}
$$

Combining the bound (5.115) with the B-stability bound (5.108) yields the estimate

$$
\|\vec{u}_h - \vec{v}_h\|_X + \|p_h - q_h\|_M \leq \frac{\|B\|}{\gamma} (\|\vec{u} - \vec{v}_h\|_X + \|p - q_h\|_M). \tag{5.116}
$$

Finally, the triangle inequality and (5.116) gives

$$\begin{aligned}
\|\vec{u} - \vec{u}_h\|_X + \|p - p_h\|_M &= \|\vec{u} - \vec{v}_h + \vec{v}_h - \vec{u}_h\|_X + \|p - q_h + q_h - p_h\|_M \\
&\leq \|\vec{u} - \vec{v}_h\|_X + \|p - q_h\|_M \\
&\quad + \|\vec{u}_h - \vec{v}_h\|_X + \|p_h - q_h\|_M \\
&\leq C\left(\|\vec{u} - \vec{v}_h\|_X + \|p - q_h\|_M\right),
\end{aligned}$$

which establishes the desired error bound (5.104) with $C := 1 + \|B\|/\gamma$. $\square$

Remark 5.4. The $\mathbf{X}_0^h$-coercivity condition in Theorem 5.1 is automatically satisfied for a well-chosen norm $\|\cdot\|_X$. If $\|\vec{v}\|_X \equiv \|\nabla\vec{v}\|$, then $\alpha^* = 1$. Consequently, the discrete inf–sup stability condition in Theorem 5.1 is frequently referred to as the "Babuška–Brezzi stability condition".

In contrast, the discrete inf–sup stability condition in Theorem 5.1 is *not* automatically satisfied. Specifically, if the discrete problem is not uniquely solvable because of the existence of one or more *spurious* pressure modes $p_h^* \in \mathbf{N}^h$, then, from (5.44), we have that $b(\vec{v}_h, p_h^*) = 0$ for all $\vec{v}_h \in \mathbf{X}_0^h$. This means that it is not possible to find $\gamma^* > 0$ in Theorem 5.1. If $\gamma^* = 0$, then scrutiny of the proof of this theorem reveals that (5.108) only holds with $\gamma = 0$ (since $\gamma_1 = 0$ in Lemma 5.2), and then the right-hand side of the estimate (5.104) is not bounded! Another possibility is that the inf–sup constant might depend on the mesh parameter, for example $\gamma^* = O(h)$. In this scenario, the discrete Stokes problem is uniquely solvable on any given grid, but since the constant in the estimate (5.104) looks like $C = 1 + O(h^{-2})$, it is likely that the rate of convergence will be slower than desired. An h-dependent inf–sup constant will also have an impact on the performance of iterative solvers applied to the discretized matrix systems. (This degradation in efficiency is what makes stabilization of Q_1–P_0 so essential in $\mathbb{R}^3$.) We will return to this issue in Section 5.5.

The macroelement viewpoint introduced in Section 5.3.1 holds the key here. In particular, it can be shown (see Lemma 5.8) that the discrete inf–sup condition is automatically satisfied if the underlying grid can be subdivided into a partition of macroelements that are all stable (i.e. for a local Stokes problem over each macroelement, the null space of the matrix B^T is one-dimensional), as long as each boundary between adjoining macroelements has at least one "interior" velocity node. This latter requirement is known as the *macroelement connectivity condition*. If the approximation is inf–sup stable, then the constant C in (5.104) is bounded away from infinity independently of h, so that, if the underlying Stokes problem is sufficiently regular (see below), an optimal rate of convergence is achieved.

The extension of the concept of $\mathcal{H}^2$ regularity used in deriving approximation error bounds in Section 1.5.1 is the following.

Definition 5.5 (Stokes problem regularity). The variational problem (5.102) is said to be $\mathcal{H}^2$-*regular* if there exists a constant C_Ω such that for every $\vec{f} \in \mathbf{L}_2(\Omega)$ and associated functional $\ell(\vec{v}) := \int_\Omega \vec{f} \cdot \vec{v}$, there is a solution $(\vec{u}, p) \in \mathbf{H}^1_{E_0} \times L_2(\Omega)$ that is also in $\mathbf{H}^2(\Omega) \times \mathcal{H}^1(\Omega)$ such that

$$\|\vec{u}\|_{2,\Omega} + \|p\|_{1,\Omega} \le C_\Omega \|\vec{f}\|.$$

More generally, a necessary condition for (5.102) to be $\mathcal{H}^3$ regular is that $\|\vec{u}\|_{3,\Omega} + \|p\|_{2,\Omega} < \infty$.

The specific case of two-dimensional $\boldsymbol{Q}_2$–$\boldsymbol{Q}_1$ approximation will be used to illustrate the derivation of an optimal a priori error estimate. Notice that the mid-edge node built into the $\boldsymbol{Q}_2$ approximation ensures that the macroelement connectivity condition is always satisfied. We state the error bound in the form of a theorem.

Theorem 5.6. *If the variational problem* (5.102) *is solved using* $\boldsymbol{Q}_2$–$\boldsymbol{Q}_1$ *approximation on a grid* $\mathcal{T}_h$ *of rectangular elements that satisfies the aspect ratio condition* (*see* Definition 1.18), *then there exists a constant* C_2 *such that*

$$\|\nabla(\vec{u} - \vec{u}_h)\| + \|p - p_h\|_{0,\Omega} \le C_2\, h^2 \left(\|D^3\vec{u}\| + \|D^2p\| \right), \tag{5.117}$$

where $\|D^3\vec{u}\|$ *and* $\|D^2p\|$ *measures the* $\mathcal{H}^3$-*regularity of the target solution, and* h *is the length of the longest edge in* $\mathcal{T}_h$.

Notice that the error estimate (5.117) immediately follows from combining the stable approximation bound (5.104) with the biquadratic and bilinear interpolation error bounds

$$\left\|\nabla(u - \pi_h^2 u)\right\|^2_{\square_k} \le Ch_k^4 \left\|D^3 u\right\|^2_{\square_k}, \tag{5.118}$$

$$\left\|p - \pi_h^1 p\right\|^2_{\square_k} \le Ch_k^4 \left\|D^2 p\right\|^2_{\square_k}, \tag{5.119}$$

which are the rectangular element analogues of the bounds in Proposition 1.22 and Proposition 1.16, respectively. Proving Theorem 5.6 is thus reduced to an exercise in inf–sup stability theory — we need to use macroelement stability to rigorously establish that the $\boldsymbol{Q}_2$–$\boldsymbol{Q}_1$ approximation approach is inf–sup stable for any grid of rectangular elements. In the following description, M_0^h is used to represent the pressure space with the constant pressure mode removed, that is, $M_0^h := \left\{ p_h \in M^h, \int_\Omega p_h = 0 \right\}$.

The essence of the macroelement theory can be condensed into two lemmas. The first gives a theoretical underpinning for the stable macroelements that were introduced in Section 5.3.1. Given a macroelement $\mathcal{M}$, let the spaces $\mathbf{X}_0^{\mathcal{M}}$ and $M^{\mathcal{M}}$ represent macroelement versions of the spaces $\mathbf{X}_0^h$ and M_0^h, respectively. For example, $\mathbf{X}_0^{\mathcal{M}}$ contains functions from $\mathbf{X}_0^h$ that are restricted to $\mathcal{M}$, and have the value zero on the boundary of $\mathcal{M}$. Note that any collection of elements that can be continuously mapped onto a master element forms an equivalence class.

Lemma 5.7. *Let $\mathcal{E}_{M_i}$ be a class of macroelements equivalent to a reference macroelement $\widehat{\mathcal{M}}_i$. If all the macroelements $\mathcal{M} \in \mathcal{E}_{M_i}$ are stable, then there exists a constant $\gamma_{M_i} > 0$ such that*

$$\max_{\vec{v}_h \in \mathbf{X}_0^{\mathcal{M}}} \frac{b(\vec{v}_h, q_h)}{\|\nabla \vec{v}_h\|} \geq \gamma_{M_i} \|q_h\| \quad \text{for all } q_h \in M_0^{\mathcal{M}},$$

holds for every $\mathcal{M} \in \mathcal{E}_{M_i}$.

Proof See Stenberg [183, Lemma 3.1]. □

In the context of $\boldsymbol{Q}_2$–$\boldsymbol{Q}_1$ approximation, two equivalence classes are known to be stable: the two-element patch shown in Figure 5.7, and any three-element patch. The associated inf–sup constants are denoted by γ_{M_2} and γ_{M_3}. The key thing here is that any conceivable grid of rectangular elements can be partitioned into a combination of two- and three-element patches.

Given such a macroelement partitioning, say $\mathcal{T}_{\mathcal{M}}$, let Π_h represent the L_2 projection operator mapping M_0^h onto the subspace

$$Q_h := \left\{ \mu_h \in M_0^h, \mu_h|_{\mathcal{M}} \text{ is constant for all } \mathcal{M} \in \mathcal{T}_{\mathcal{M}} \right\}.$$

Notice that for every $p_h \in M_0^h$, the function $(I - \Pi_h)p_h \in M_0^{\mathcal{M}}$ for all macroelements $\mathcal{M} \in \mathcal{T}_{\mathcal{M}}$. Lemma 5.7 then implies that for every macroelement $\mathcal{M}$, there exists a locally defined function $\vec{v}_{\mathcal{M}} \in \mathbf{X}_0^{\mathcal{M}}$ such that

$$b(\vec{v}_{\mathcal{M}}, (I - \Pi_h)p_h) \geq \gamma_M \|(I - \Pi_h)p_h\|^2_{\mathcal{M}} \tag{5.120}$$

and

$$\|\nabla \vec{v}_{\mathcal{M}}\|_{\mathcal{M}} \leq \|(I - \Pi_h)p_h\|_{\mathcal{M}}, \tag{5.121}$$

where $\gamma_M := \min(\gamma_{M_2}, \gamma_{M_3})$. Let us now define the global function $\vec{v}_h$ so that

$$\vec{v}_h|_{\mathcal{M}} = \vec{v}_{\mathcal{M}} \quad \text{for all } \mathcal{M} \in \mathcal{T}_{\mathcal{M}}.$$

Since $\vec{v}_h = 0$ on $\partial \mathcal{M}$ for every macroelement, we have that $\vec{v}_h \in \mathbf{X}_0^h$, and also

$$b(\vec{v}_h, \Pi_h p_h) = \int_\Omega \Pi_h \, p_h \nabla \cdot \vec{v}_h = 0. \tag{5.122}$$

Moreover, the local bounds (5.120) and (5.121), combined with (5.122), imply that for any $p_h \in M_0^h$ a function $\vec{v}_h \in \mathbf{X}_0^h$ exists such that

$$b(\vec{v}_h, p_h) = b(\vec{v}_h, (I - \Pi_h)p_h) \geq \gamma_M \|(I - \Pi_h)p_h\|^2 \tag{5.123}$$

and

$$\|\nabla \vec{v}_h\| \leq \|(I - \Pi_h)p_h\|. \tag{5.124}$$

In simple terms, the bounds (5.123) and (5.124) imply that the $\boldsymbol{Q}_2$–$\boldsymbol{Q}_1$ mixed approximation method is stable up to constant functions defined on macroelements. The only remaining question is whether or not functions in the space Q_h provide a stable mixed approximation — here the macroelement connectivity condition is the key. This is the subject of the second lemma.

Lemma 5.8. *In the case of $\boldsymbol{Q}_2$ velocity approximation, for every $p_h \in M_0^h$ there exists a corresponding velocity vector $\vec{w}_h \in \mathbf{X}_0^h$, and a constant C_2 such that*

$$b(\vec{w}_h, \Pi_h p_h) = \|\Pi_h p_h\|^2, \quad \|\nabla \vec{w}_h\| \leq C_2 \|\Pi_h p_h\|. \tag{5.125}$$

Proof See Stenberg [183, Lemma 3.3]. □

The point here is that for the projected pressure $\Pi_h p_h$, which is constant on macroelements, it is possible to use the macroelement connectivity condition to construct a velocity vector $\vec{w}_h$ satisfying the bound required for the inf–sup condition; see [183] for details.

Finally, to show discrete inf–sup stability, we can follow the construction used in the proof of Lemma 5.2. Given an arbitrary $p_h \in M_0^h$, we take $\vec{v}_h$ defined in (5.123)–(5.124) and $\vec{w}_h$ defined in (5.125), and choose $\vec{z}_h \in \mathbf{X}_0^h$ so that $\vec{z}_h = \vec{v}_h + \delta \vec{w}_h$ with $\delta = 2\gamma_M/(1 + C_2^2)$. We then have that

$$
\begin{aligned}
b(\vec{z}_h, p_h) &= b(\vec{v}_h, p_h) + \delta\, b(\vec{w}_h, p_h) \\
&\geq \gamma_M \|(I - \Pi_h)p_h\|^2 + \delta\, b(\vec{w}_h, \Pi_h p_h) + \delta\, b(\vec{w}_h, (I - \Pi_h)p_h) \\
&\geq \gamma_M \|(I - \Pi_h)p_h\|^2 + \delta \|\Pi_h p_h\|^2 + \delta\, b(\vec{w}_h, (I - \Pi_h)p_h) \\
&\geq \gamma_M \|(I - \Pi_h)p_h\|^2 + \delta \|\Pi_h p_h\|^2 - \delta \|\nabla \cdot \vec{w}_h\| \|(I - \Pi_h)p_h)\| \\
&\geq \gamma_M \|(I - \Pi_h)p_h\|^2 + \delta \|\Pi_h p_h\|^2 - \delta \|\nabla \vec{w}_h\| \|(I - \Pi_h)p_h)\| \\
&\geq \gamma_M \|(I - \Pi_h)p_h\|^2 + \delta \|\Pi_h p_h\|^2 - \delta\, C_2 \|\Pi_h p_h\| \|(I - \Pi_h)p_h)\| \\
&= \gamma_M \|(I - \Pi_h)p_h\|^2 + \frac{\delta}{2} \|\Pi_h p_h\|^2 \\
&\quad + \delta \left(\frac{1}{2} \|\Pi_h p_h\|^2 - C_2 \|\Pi_h p_h\| \|(I - \Pi_h)p_h)\| \right) \\
&\geq \gamma_M \|(I - \Pi_h)p_h\|^2 + \frac{\delta}{2} \|\Pi_h p_h\|^2 - \frac{\delta}{2} C_2^2 \|(I - \Pi_h)p_h\|^2 \\
&\geq \frac{\delta}{2} \|(I - \Pi_h)p_h\|^2 + \frac{\delta}{2} \|\Pi_h p_h\|^2 \\
&\geq \frac{\delta}{2} \|p_h\|^2. \tag{5.126}
\end{aligned}
$$

Moreover,

$$\|\nabla \vec{z}_h\| \leq \|\nabla \vec{v}_h\| + \delta \|\nabla \vec{w}_h\|$$
$$\leq \|(I - \Pi_h)p_h\| + \delta C_2 \|\Pi_h p_h\| \leq C \|p_h\|. \tag{5.127}$$

It follows from (5.126) and (5.127) that

$$\max_{\vec{v}_h \in \mathbf{X}_0^h} \frac{b(\vec{v}_h, p_h)}{\|\nabla \vec{v}_h\|} \geq \frac{b(\vec{z}_h, p_h)}{\|\nabla \vec{z}_h\|} \geq \frac{\delta}{2C} \|p_h\| \quad \text{for all } p_h \in M_0^h, \tag{5.128}$$

implying discrete inf–sup stability of the $\mathbf{Q}_2$–$\mathbf{Q}_1$ approximation with a constant $\gamma^* := \gamma_M/C(1 + C_2^2)$. As noted earlier, the optimal error estimate (5.117) follows.

Remark 5.9. In the simple case of the Poiseuille flow (5.12), the right-hand side of (5.117) is identically zero. This means that, independent of the grid used, a computed $\mathbf{Q}_2$–$\mathbf{Q}_1$ solution of Example 5.1.1 should always be identical to the exact solution. This issue is explored in Computational Exercise 5.3.

To illustrate the tightness of the error bound in Theorem 5.6, some errors $\|\nabla(\vec{u} - \vec{u}_h)\|$ and $\|p - p_h\|_{0,\Omega}$ corresponding to $\mathbf{Q}_2$–$\mathbf{Q}_1$ approximation of the colliding flow in Example 5.1.4 are given in Table 5.1. Both of the error measures can be seen to decrease by a factor of four with successive refinement — consistent with the bound (5.117). This is to expected since the exact solution (5.14) is perfectly regular.

Using the macroelement argument above, the $\mathbf{Q}_2$–$\mathbf{P}_{-1}$ and $\mathbf{Q}_2$–$\mathbf{P}_0$ methods of Section 5.3.1 can also be shown to be inf–sup stable for any grid of rectangular elements. Some errors that result using these alternatives to $\mathbf{Q}_2$–$\mathbf{Q}_1$ mixed approximation are given in Table 5.2. Notice that the $\mathbf{Q}_2$–$\mathbf{P}_{-1}$ results mirror the $\mathbf{Q}_2$–$\mathbf{Q}_1$ results since the velocity errors decrease by a factor of four with successive refinement. In contrast, the $\mathbf{Q}_2$–$\mathbf{P}_0$ velocity errors seem to be decreasing by a factor of two with successive refinement. Although the $\mathbf{Q}_2$–$\mathbf{P}_0$ approximation is perfectly stable, the difference here is that the approximation bound associated with piecewise constant approximation of the pressure, namely

$$\left\|p - \pi_h^0 p\right\|_{\square_k}^2 \leq C h_k^2 \left\|D^1 p\right\|_{\square_k}^2, \tag{5.129}$$

Table 5.1 *Errors for Example 5.1.4 using $\mathbf{Q}_2$–$\mathbf{Q}_1$ approximation: ℓ is the grid refinement level and $h = 2^{2-\ell}$.*

ℓ	$\|\nabla(\vec{u} - \vec{u}_h)\|$	$\|p - p_h\|_{0,\Omega}$	n_u
3	2.264×10^0	1.896×10^0	98
4	5.618×10^{-1}	4.605×10^{-1}	450
5	1.399×10^{-1}	1.144×10^{-1}	1922

Table 5.2 *Velocity errors* $\|\nabla(\vec{u} - \vec{u}_h)\|$ *for Example 5.1.4:* ℓ *is the grid refinement level and* $h = 2^{2-\ell}$

ℓ	$Q_2\text{-}P_{-1}$	$Q_2\text{-}P_0$	n_u
3	2.371×10^0	1.035×10^1	98
4	5.716×10^{-1}	6.071×10^0	450
5	1.407×10^{-1}	3.249×10^0	1922

is of lower order than the biquadratic velocity approximation bound (5.118). The right-hand side of the quasi-optimal error bound (5.104) is thus reduced to $O(h)$, consistent with the $Q_2\text{-}P_0$ results.

Remark 5.10. The $\mathcal{H}^3$-regularity of the target solution is essential if the $O(h^2)$ convergence rate in Theorem 5.6 is to be realized. In contrast, since the problem in Example 5.1.2 is not even $\mathcal{H}^2$ regular, the rate of convergence when solving this problem on a uniform grid sequence is slower than $O(h)$—independently of the order of the mixed approximation!

The lack of regularity of practical flow problems (such as those in Example 5.1.2 and Example 5.1.3) provides the motivation for stabilizing the lowest order methods, as discussed in Section 5.3.2 and Section 5.3.3. The pressure jump stabilized methods—that is the $Q_1\text{-}P_0$ and $P_1\text{-}P_0$ approximation methods in two or three dimensions—are discussed first. The following analogue of Theorem 5.6 is well known.

Theorem 5.11. *Suppose that an* $\mathcal{H}^2$*-regular problem* (5.102) *is solved on a grid* $\mathcal{T}_h$ *of rectangular elements that satisfies the aspect ratio condition, using an ideally stabilized* $Q_1\text{-}P_0$ *approximation. That is, with* $\beta = \beta^* = 1/4$ *fixed, the discrete incompressibility constraint in* (5.103) *is replaced by*

$$b(\vec{u}_h, q_h) - \frac{1}{4} \sum_{\mathcal{M} \in \mathcal{T}_M} |\mathcal{M}| \sum_{e \in \Gamma_M} \langle [\![p_h]\!]_e, [\![q_h]\!]_e \rangle_E = 0 \quad \text{for all } q_h \in M^h, \quad (5.130)$$

see (5.83), *where the common boundary between the macroelements* $\mathcal{M} \in \mathcal{T}_M$ *has at least one interior velocity node. Then there exists a constant* C_1 *such that*

$$\|\nabla(\vec{u} - \vec{u}_h)\| + \|p - p_h\|_{0,\Omega} \le C_1 \, h \, \left(\|D^2 \vec{u}\| + \|D^1 p\| \right), \quad (5.131)$$

where h *is the length of the longest edge in* $\mathcal{T}_h$.

Proof See Kechkar & Silvester [120]. (The parameter value β_0 in the proof of Theorem 3.1 in [120] can be identified with β^*.) ☐

The proof of Theorem 5.11 is an elegant and simple generalization of the macroelement theory described above. The first step of the proof is to define the bilinear form corresponding to the left-hand side of the stabilized discrete formulation, namely,

$$\mathcal{B}_h((\vec{u}_h, p_h); (\vec{v}_h, q_h)) = a(\vec{u}_h, \vec{v}_h) + b(\vec{u}_h, q_h) + b(\vec{v}_h, p_h) - c_h(p_h, q_h), \quad (5.132)$$

which is defined on the space $\mathbf{X}_0^h \times M^h$, with

$$c_h(p_h, q_h) := \frac{1}{4} \sum_{M \in \mathcal{T}_M} |M| \sum_{e \in \Gamma_M} \langle [\![p_h]\!]_e, [\![q_h]\!]_e \rangle_E. \quad (5.133)$$

As in the stable case, the quasi-optimal error bound (5.131) follows from the analogue of Lemma 5.2 (i.e. $\mathcal{B}_h$-stability in the sense of (5.109) and (5.110)), together with a generalization of Lemma 5.3. Moreover, $\mathcal{B}_h$-stability can also be established here using a macroelement argument, see Problem 5.11. There are two components to this; first, the macroelement connectivity condition in the statement of Theorem 5.11 ensures that the $\mathbf{Q}_1$–P_0 method is stable up to constant functions defined on macroelements. Second, the stabilization term controls the nonconstant macroelement pressure component. This makes up for the lack of local stability in the sense of Lemma 5.7. Indeed, the analogue of Lemma 5.7 in the locally stabilized case is the following result.

Lemma 5.12. *Let $\mathcal{E}_{M_i}$ be a class of macroelements equivalent to a reference macroelement $\widehat{\mathcal{M}}_i$. There exists a constant $\gamma_{M_i} > 0$ such that*

$$|M| \sum_{e \in \Gamma_M} \langle [\![q_h]\!]_e, [\![q_h]\!]_e \rangle_{\bar{E}} \geq \gamma_{M_i} \|q_h\|^2 \quad \text{for all } q_h \in M_0^{\mathcal{M}},$$

and all $\mathcal{M} \in \mathcal{E}_{M_i}$.

Proof See Kechkar & Silvester [120, Lemma 3.1]. □

Remark 5.13. The right-hand side of the $\mathbf{Q}_1$–P_0 error bound (5.131) is not zero in the case of Poiseuille flow (5.12). This means that, in contrast with the case of higher order approximation, computed $\mathbf{Q}_1$–P_0 solutions of Example 5.1.1 do not coincide with the exact solution. This issue is explored in Computational Exercise 5.4.

Finally, to complete our discussion of stabilized methods, we return to the pressure projection stabilization (5.89) of the $\mathbf{Q}_1$–$\mathbf{Q}_1$ approximation. The analogue of Theorem 5.11 is the following error estimate.

Theorem 5.14. *Suppose that an $\mathcal{H}^2$-regular problem (5.102) is solved on a grid $\mathcal{T}_h$ of rectangular elements that satisfies the aspect ratio condition, using stabilized $\mathbf{Q}_1$–$\mathbf{Q}_1$ approximation. That is, the discrete incompressibility constraint in (5.103) is replaced by*

$$b(\vec{u}_h, q_h) - (p_h - \Pi_0 p_h, q_h - \Pi_0 q_h) = 0 \quad \text{for all } q_h \in M^h, \quad (5.134)$$

where Π_0 is the L^2 projection from M_h into the space $\mathbf{P}_0$. Then there exists a constant C_1, such that

$$\|\nabla(\vec{u} - \vec{u}_h)\| + \|p - p_h\|_{0,\Omega} \leq C_1 \, h \, \left(\left\|D^2\vec{u}\right\| + \left\|D^1p\right\|\right), \tag{5.135}$$

where h is the length of the longest edge in $\mathcal{T}_h$.

The proof has a key ingredient, so called "weak inf–sup stability". We outline the construction below; further details can be found in Bochev et al. [16, Theorem 5.1]. We assume, for simplicity, that we have a uniform grid of square elements of size h. The following (sub-optimal) bound is the first component of the proof.

Lemma 5.15. *Given a uniform square grid of size h. If the approximation spaces are given by $\mathbf{X}_0^h \subset \mathbf{H}_{E_0}^1$ and $M^h \subset \mathcal{H}^1(\Omega)$, then there exist positive constants c_1 and c_2 such that*

$$\max_{\vec{v}_h \in \mathbf{X}_0^h} \frac{b(\vec{v}_h, q_h)}{\|\nabla\vec{v}_h\|} \geq c_1 \, \|q_h\| - c_2 h \, \|\nabla q_h\| \quad \text{for all } q_h \in M^h. \tag{5.136}$$

Proof The bound is a direct result of the inf–sup stability of the underlying weak formulation, and applies to any mixed method based on continuous pressure approximation. See Bochev et al. [16, Lemma 2.1], or Brezzi & Fortin [33, p. 55] for details. □

Notice that the term $h\,\|\nabla q_h\|$ on the right-hand side of (5.136) quantifies the inf–sup "deficiency" of $\mathbf{Q}_1$–$\mathbf{Q}_1$ approximation, and this is the motivation for the simple pressure Laplacian stabilization strategy embodied in (5.87). Returning to Theorem 5.14, the second component of the proof is to establish that this term is controlled by the pressure projection stabilization operator in (5.134). The following result is just what is needed.

Lemma 5.16. *There exists a constant C such that*

$$h\,\|\nabla p_h\| \leq C\,\|p_h - \Pi_0 p_h\| \quad \text{for all } p_h \in M^h. \tag{5.137}$$

Proof Since $\Pi_0 p_h$ is piecewise constant, we have that $\nabla(\Pi_0 p_h)|_{\square_k} = 0$ for all $\square_k \in \mathcal{T}_h$. Using the local inverse estimate $\|\nabla q_h\|_{\square_k} \leq C_I h^{-1} \|q_h\|_{\square_k}$ (see Lemma 1.26), then gives

$$h^2\,\|\nabla p_h\|^2 \leq \sum_{\square_k} h^2\,\|\nabla p_h\|_{\square_k}^2 = \sum_{\square_k} h^2\,\|\nabla(p_h - \Pi_0 p_h)\|_{\square_k}^2$$

$$\leq \sum_{\square_k} C_I\,\|p_h - \Pi_0 p_h\|_{\square_k}^2 = C_I\,\|p_h - \Pi_0 p_h\|^2,$$

as required. □

Table 5.3 *Errors for Example 5.1.4 using stabilized mixed approximation methods: ℓ is the grid refinement level and $h = 2^{1-\ell}$.*

	Q_1–P_0 $(\beta = \beta^*)$		Q_1–Q_1		
ℓ	$\|\nabla(\vec{u} - \vec{u}_h)\|$	$\|p - p_h\|_{0,\Omega}$	$\|\nabla(\vec{u} - \vec{u}_h)\|$	$\|p - p_h\|_{0,\Omega}$	n_u
3	8.889×10^0	9.840×10^0	8.542×10^0	7.940×10^0	98
4	4.455×10^0	4.398×10^0	4.124×10^0	2.500×10^0	450
5	2.224×10^0	2.066×10^0	2.021×10^0	7.533×10^{-1}	1922
6	1.111×10^0	1.005×10^0	1.001×10^0	2.248×10^{-1}	7938

By combining (5.136) with (5.137), we see that there exist positive constants c_1 and c_3 such that

$$\max_{\vec{v}_h \in \mathbf{X}_0^h} \frac{b(\vec{v}_h, q_h)}{\|\nabla \vec{v}_h\|} \geq c_1 \|q_h\| - c_3 \|q_h - \Pi_0 q_h\| \quad \text{for all } q_h \in M^h. \tag{5.138}$$

The final step of the proof of Theorem 5.14 is to establish $\mathcal{B}_h$-stability using the construction in Lemma 5.2. The details are left as an exercise, see Problem 5.12. The quasi-optimal error bound (5.135) follows from a generalization of Lemma 5.3 exactly as in the stabilized Q_1–P_0 case.

A set of errors computed for the ideally stabilized approximations of the colliding flow problem are given in Table 5.3. With Q_1–P_0 approximation both error measures can be seen to decrease by a factor of two with successive refinement — consistent with the bound (5.131). Using Q_1–Q_1 the velocity error is consistent with (5.135), whereas the pressure error is decreasing faster than might be expected.

Remark 5.17. The bound (5.136) provides an alternative route to establishing the inf–sup stability of mixed methods involving continuous pressure approximation like P_2–P_1. The mechanism for controlling the term $h \|\nabla q_h\|$ in such cases is a bound expressing the stability of the mixed approximation with respect to the "wrong norms":

$$\max_{\vec{v}_h \in \mathbf{X}_0^h} \frac{b(\vec{v}_h, q_h)}{\|\vec{v}_h\|} \geq \gamma \|\nabla q_h\| \quad \text{for all } q_h \in M^h. \tag{5.139}$$

This alternative stability bound will feature in later chapters on iterative solvers. We defer further discussion until Section 5.5.

5.4.2 A posteriori error bounds

In this section we build on the foundation laid in Section 1.5.2 and describe an efficient a posteriori estimation strategy (due to Ainsworth & Oden [3]) for the Stokes equation. The error estimator is computed by solving a local Poisson problem for each component of velocity, and this makes it easy to generalize the estimation approach to the linearized Navier–Stokes equations, see Section 7.4.2.

It also avoids the awkward complication of having to ensure inf–sup stability of mixed approximations when solving local Stokes problems.

The starting point for the theoretical analysis is the weak formulation (5.106). Given some approximation, $(\vec{u}_h, p_h) \in \mathbf{X}_E^h \times M^h$, to the solution $(\vec{u}, p)$ satisfying (5.106), we have that

$$
\begin{aligned}
B((\vec{u} - \vec{u}_h, p - p_h); (\vec{v}, q)) &= B((\vec{u}, p) - (\vec{u}_h, p_h); (\vec{v}, q)) \\
&= B((\vec{u}, p); (\vec{v}, q)) - B((\vec{u}_h, p_h); (\vec{v}, q)) \\
&= F((\vec{v}, q)) - B((\vec{u}_h, p_h); (\vec{v}, q)) \\
&= \ell(\vec{v}) - a(\vec{u}_h, \vec{v}) - b(\vec{v}, p_h) - b(\vec{u}_h, q), \\
&= \ell(\vec{v}) - \int_\Omega \nabla \vec{u}_h : \nabla \vec{v} + \int_\Omega p_h \nabla \cdot \vec{v} + \int_\Omega q \nabla \cdot \vec{u}_h
\end{aligned}
$$

for all $(\vec{v}, q) \in \mathbf{H}_{E_0}^1 \times L_2(\Omega)$, with $\ell(\vec{v}) = \int_{\partial \Omega_N} \vec{s} \cdot \vec{v}$. We can now follow the path taken in Section 1.5.2. That is, after breaking the integrals into element contributions, the second and third terms on the right-hand side can be integrated by parts elementwise to give

$$
\begin{aligned}
&- \int_T \nabla \vec{u}_h : \nabla \vec{v} + \int_T p_h \nabla \cdot \vec{v} \\
&= \int_T \vec{v} \cdot (\nabla^2 \vec{u}_h - \nabla p_h) - \sum_{E \in \mathcal{E}(T)} \left\langle \frac{\partial \vec{u}_h}{\partial n_{E,T}} - p_h \vec{n}_{E,T}, \vec{v} \right\rangle_E,
\end{aligned} \qquad (5.140)
$$

where $\mathcal{E}(T)$ is the set of the edges ($\mathbb{R}^2$) or faces ($\mathbb{R}^3$) of element T, and $\vec{n}_{E,T}$ is the outward pointing normal. The velocity approximation typically has a discontinuous normal derivative across inter-element boundaries. Thus it is convenient to generalize the flux jump in (1.92) to give a *stress jump* across edge or face E adjoining elements T and S:

$$
[\![\nabla \vec{u}_h - p_h \vec{\mathbf{I}}]\!] := ((\nabla \vec{u}_h - p_h \vec{\mathbf{I}})|_T - (\nabla \vec{u}_h - p_h \vec{\mathbf{I}})|_S) \vec{n}_{E,T}.
$$

(Note that if a C^0 pressure approximation is used, then the jump in $p_h \vec{\mathbf{I}}$ is zero.) Furthermore, if $\mathcal{E}_h = \mathcal{E}_{h,\Omega} \cup \mathcal{E}_{h,D} \cup \mathcal{E}_{h,N}$ denotes the splitting into interior, Dirichlet & Neumann boundary edges (or faces), we define the *equidistributed stress jump operator* by

$$
\vec{R}_E^* := \begin{cases} \frac{1}{2}[\![\nabla \vec{u}_h - p_h \vec{\mathbf{I}}]\!] & E \in \mathcal{E}_{h,\Omega}, \\ \vec{s} - \left(\frac{\partial \vec{u}_h}{\partial n_{E,T}} - p_h \vec{n}_{E,T}\right) & E \in \mathcal{E}_{h,N}, \\ 0 & E \in \mathcal{E}_{h,D}. \end{cases} \qquad (5.141)
$$

We also define the elementwise interior residuals by $\vec{R}_T := \{\nabla^2 \vec{u}_h - \nabla p_h\}|_T$ and $R_T := \{\nabla \cdot \vec{u}_h\}|_T$, respectively. This gives

$$B((\vec{u} - \vec{u}_h, p - p_h); (\vec{v}, q)) = \sum_{T \in \mathcal{T}_h} \left[(\vec{R}_T, \vec{v})_T - \sum_{E \in \mathcal{E}(T)} \left\langle \vec{R}_E^*, \vec{v} \right\rangle_E + (R_T, q)_T \right],$$

for all $(\vec{v}, q) \in \mathbf{H}_{E_0}^1 \times L_2(\Omega)$. Finally, if we rewrite this in component form, then the errors $\vec{e} := \vec{u} - \vec{u}_h \in \mathbf{H}_{E_0}^1$ and $\epsilon := p - p_h \in L_2(\Omega)$ can be seen to be characterized by a set of *localized Stokes problems*

$$\sum_{T \in \mathcal{T}_h} \{(\nabla \vec{e}, \nabla \vec{v})_T - (\epsilon, \nabla \cdot \vec{v})_T\} = \sum_{T \in \mathcal{T}_h} \left[(\vec{R}_T, \vec{v})_T - \sum_{E \in \mathcal{E}(T)} \left\langle \vec{R}_E^*, \vec{v} \right\rangle_E \right]$$

$$- \sum_{T \in \mathcal{T}_h} (q, \nabla \cdot \vec{e})_T = \sum_{T \in \mathcal{T}_h} (R_T, q)_T \qquad (5.142)$$

for all $(\vec{v}, q) \in \mathbf{H}_{E_0}^1 \times L_2(\Omega)$.

The representation (5.142) opens the way to deriving a posteriori error estimates for a general mixed approximation. We concentrate on the stabilized $\boldsymbol{Q}_1$–$\boldsymbol{P}_0$ or $\boldsymbol{P}_1$–$\boldsymbol{P}_0$ approximation methods for the remainder of the section. We only consider two-dimensional $\boldsymbol{Q}_1$–$\boldsymbol{P}_0$ approximation in detail but note that the extension to three dimensions and to $\boldsymbol{Q}_1$–$\boldsymbol{Q}_1$ approximation are completely straightforward.[8] Notice that the interior residual terms in (5.142) are trivial to compute using either $\boldsymbol{Q}_1$–$\boldsymbol{P}_0$ or $\boldsymbol{P}_1$–$\boldsymbol{P}_0$. In either case, R_T is piecewise constant and $\vec{R}_T$ is identically zero. Introducing appropriate higher order spaces, in this case $\vec{\mathcal{Q}}_T := (\mathcal{Q}_T)^2$ for the two velocity components (see (1.97)) and $\boldsymbol{P}_1(T)$ for the pressure, we follow the Ainsworth & Oden strategy (see [4, Section 9.2]) and compute element functions $\vec{e}_T \in \vec{\mathcal{Q}}_T$ and $\epsilon_T \in \boldsymbol{P}_1(T)$ such that

$$(\nabla \vec{e}_T, \nabla \vec{v})_T = - \sum_{E \in \mathcal{E}(T)} \left\langle \vec{R}_E^*, \vec{v} \right\rangle_E \quad \text{for all } \vec{v} \in \vec{\mathcal{Q}}_T \qquad (5.143)$$

$$(\epsilon_T, q)_T = (R_T, q)_T \quad \text{for all } q \in \boldsymbol{P}_1(T). \qquad (5.144)$$

The system (5.143) and (5.144) represents an attractive alternative to the local Stokes problem embodied in (5.142). First, (5.143) decouples into a pair of local Poisson problems, and, as the stress jump term $\vec{R}_E^*$ is piecewise linear, it may be computed exactly by sampling it at the mid-point of the edge. Second, since $R_T \in \boldsymbol{P}_1(T)$, the solution of (5.144) is immediate: $\epsilon_T = \nabla \cdot \vec{u}_h$. The local error estimator is then simply the "energy norm" of the element error, that is

$$\eta_T^2 := \|\nabla \vec{e}_T\|_T^2 + \|\epsilon_T\|_T^2 = \|\nabla \vec{e}_T\|_T^2 + \|\nabla \cdot \vec{u}_h\|_T^2, \qquad (5.145)$$

and the global error estimator is $\eta := \left(\sum_{T \in \mathcal{T}_h} \eta_T^2 \right)^{1/2}$.

[8]The a posteriori error estimation built into IFISS is restricted to the lowest order $\boldsymbol{Q}_1$–$\boldsymbol{P}_0$ and $\boldsymbol{Q}_1$–$\boldsymbol{Q}_1$ methods.

Table 5.4 *Estimated errors for Example 5.1.4 using Q_1–P_0 approximation with $\beta = \beta^*$: ℓ is the grid refinement level and $h = 2^{1-\ell}$.*

ℓ	ξ	$\|\nabla \cdot \vec{u}_h\|_\Omega$	η
3	8.124×10^0	6.157×10^{-1}	8.148×10^0
4	4.597×10^0	1.658×10^{-1}	4.560×10^0
5	2.401×10^0	4.256×10^{-2}	2.402×10^0
6	1.219×10^0	1.074×10^{-2}	1.219×10^0

To illustrate the effectiveness of this error estimator, some computed values of η in the case of the analytic test problem in Example 5.1.4 are given in Table 5.4. The estimated velocity errors, $\xi := (\sum_{T \in \mathcal{T}_h} \|\nabla \vec{e}_T\|_T^2)^{1/2}$, are stunningly close to the corresponding errors $\|\nabla(\vec{u} - \vec{u}_h)\|$ given in Table 5.3. Notice that the velocity divergence error $\|\nabla \cdot \vec{u}_h\|_\Omega$ given in Table 5.4 is clearly converging like $O(h^2)$, that is, it is *super-convergent*—and as a result we see that η converges to ξ as $h \to 0$. Such numerical results strongly suggest that the error estimation strategy provides a reliable measure of the global error. This assertion is made precise in the following theorem.

Theorem 5.18. *If the variational problem (5.102) is solved using a grid of Q_1–P_0 rectangular elements with a pressure jump stabilization term (5.130), and if the rectangle aspect ratio condition is satisfied, then the estimator η_T in (5.145) computed via (5.143)–(5.144) generates the upper bound*

$$\|\nabla(\vec{u} - \vec{u}_h)\| + \|p - p_h\|_{0,\Omega} \leq \frac{C}{\gamma_\Omega} \left(\sum_{T \in \mathcal{T}_h} \eta_T^2 \right)^{1/2}, \qquad (5.146)$$

where C depends only on the aspect-ratio constant given in Definition 1.18, and γ_Ω is the so-called continuous B-stability constant. The estimator η_T also provides a local lower bound, that is

$$\eta_T \leq C \left(\|\nabla(\vec{u} - \vec{u}_h)\|_{\omega_T} + \|p - p_h\|_{0,\omega_T} \right), \qquad (5.147)$$

where ω_T represents the patch of five elements that have at least one boundary edge E from the set $\mathcal{E}(T)$.

An outline of the derivation of the upper bound is given below. All the details can be found in Kay & Silvester [118]. The key to handling the stabilization term is an error orthogonality property. This is established in the following proposition.

Proposition 5.19. *The solution $(\vec{u}_h, p_h)$ satisfying the jump-stabilized formulation $\mathcal{B}_h((\vec{u}_h, p_h); (\vec{v}_h, q_h)) = F((\vec{v}_h, q_h))$ for all $(\vec{v}_h, q_h) \in \mathbf{X}_0^h \times M^h$, is consistent* with the underlying variational formulation (5.106) in the sense that

$$B((\vec{u} - \vec{u}_h, p - p_h); (\vec{v}_h, 0)) = 0 \quad \text{for all } \vec{v}_h \in \mathbf{X}_0^h. \tag{5.148}$$

Proof See Problem 5.13.   □

The same technique used to derive the analogous result (1.101) in the case of Poisson's equation is applicable here. Specifically, choosing $\vec{v}_h$ in (5.148) to be the quasi-interpolant $\vec{v}_h^*$ as in Lemma 1.25 gives

$$B((\vec{u} - \vec{u}_h, p - p_h); (\vec{v}, q)) = B((\vec{u} - \vec{u}_h, p - p_h); (\vec{v} - \vec{v}_h^*, q))$$

for all $(\vec{v}, q) \in \mathbf{H}_{E_0}^1 \times L_2(\Omega)$. Then, localizing and integrating by parts exactly as in the derivation of (5.142) gives

$$B((\vec{u} - \vec{u}_h, p - p_h); (\vec{v}, q)) = \sum_{T \in \mathcal{T}_h} \left[(R_T, q)_T - \sum_{E \in \mathcal{E}(T)} \left\langle \vec{R}_E^*, \vec{v} - \vec{v}_h^* \right\rangle_E \right].$$

Hence, using the property (1.103) of the quasi-interpolant, together with the following continuous B-stability bound

$$\sup_{(\vec{v},q) \in \mathbf{H}_{E_0}^1 \times L_2(\Omega)} \frac{B((\vec{w}, s); (\vec{v}, q))}{(\|\nabla \vec{v}\|^2 + \|q\|_{0,\Omega}^2)^{1/2}} \geq \gamma_\Omega (\|\nabla \vec{w}\|^2 + \|s\|_{0,\Omega}^2)^{1/2},$$

which is valid for all $(\vec{w}, s) \in \mathbf{H}_{E_0}^1 \times L_2(\Omega)$, leads to the following *residual estimator* bound

$$\|\nabla(\vec{u} - \vec{u}_h)\| + \|p - p_h\|_{0,\Omega} \leq \frac{C}{\gamma_\Omega} \left(\sum_{T \subset \mathcal{T}_h} \left\{ \|R_T\|_T^2 + \sum_{E \in \mathcal{E}(T)} h_E \left\| \vec{R}_E^* \right\|_E^2 \right\} \right)^{1/2}. \tag{5.149}$$

Finally (5.149) can be extended to give the upper bound (5.146) exactly as in Section 1.5.2. For further details see Verfürth [203] and Kay & Silvester [118].

Remark 5.20. The estimated error can be shown to be a strict upper bound on the exact error if a stress jump equilibration technique is combined with the strategy of solving a higher order system of the form (5.143)–(5.144). See Ainsworth & Oden [4, Section 9.2] for further details.

Some plots of the estimated error distribution associated with computed solutions to the problems in Examples 5.1.2 and 5.1.3 are presented in Figures 5.20 and 5.21, respectively. The structure of the error can be seen to be quite different in these two cases. In the expansion flow the singularity at the step seems to really dominate. The estimated error is also concentrated in the top corners in the the cavity flow but the regularization makes it much smoother. The components of the estimated error are given in Table 5.5. These results, taken together with the bound (5.146), imply that the finite element solution is

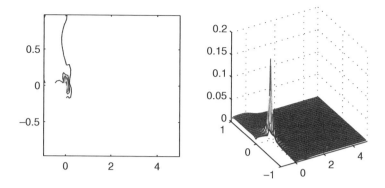

FIG. 5.20. Estimated error η_T associated with a 32×96 square grid ($\ell = 5$) $\boldsymbol{Q}_1$–$\boldsymbol{P}_0$ solution to the expansion flow problem in Example 5.1.2: $\beta = \beta^*$.

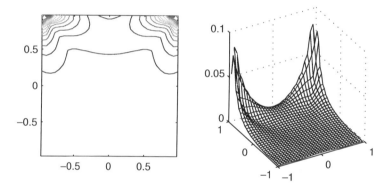

FIG. 5.21. Estimated error η_T associated with a 32×32 square grid ($\ell = 5$) $\boldsymbol{Q}_1$–$\boldsymbol{P}_0$ solution to the regularized driven cavity problem in Example 5.1.3: $\beta = \beta^*$.

Table 5.5 *Estimated errors for regularized driven cavity flow using $\boldsymbol{Q}_1$–$\boldsymbol{P}_0$ approximation with $\beta = \beta^*$: ℓ is the grid refinement level and $h = 2^{1-\ell}$.*

ℓ	ξ	$\|\nabla \cdot \vec{u}_h\|_\Omega$	η
3	8.694×10^{-1}	6.999×10^{-2}	8.722×10^{-1}
4	6.050×10^{-1}	2.333×10^{-2}	6.054×10^{-1}
5	3.780×10^{-1}	7.160×10^{-3}	3.781×10^{-1}
6	2.221×10^{-1}	2.089×10^{-3}	2.221×10^{-1}

converging to the exact solution in this case, albeit at a sub-optimal rate. Similar (slow) convergence rates are obtained in the expansion flow example. The case of the watertight cavity is explored in Computational Exercise 5.5.

5.5   Matrix properties

The spectral properties of the matrices arising from mixed approximation of the Stokes equation are analyzed in this section. The results will be used in the next chapter to predict the behavior of iterative solvers applied to the discrete Stokes problem (5.33), that is,

$$\begin{bmatrix} \mathbf{A} & B^T \\ B & O \end{bmatrix} \begin{bmatrix} \mathbf{u} \\ \mathbf{p} \end{bmatrix} = \begin{bmatrix} \mathbf{f} \\ \mathbf{g} \end{bmatrix}. \tag{5.150}$$

The immediate task is to define matrices representing norms of functions $\vec{v}_h \in \mathbf{X}_0^h$ and $q_h \in M^h$. To do this, we follow the construction of Section 1.6, and define the norms

$$\|\nabla \vec{v}_h\| := \langle \mathbf{A}\mathbf{v}, \mathbf{v} \rangle^{1/2}, \tag{5.151}$$

$$\|q_h\| := \langle Q\mathbf{q}, \mathbf{q} \rangle^{1/2}, \tag{5.152}$$

where $\mathbf{v}$ and $\mathbf{q}$ are vectors of the coefficients associated with the velocity and pressure basis sets $\{\vec{\phi}_j\}_{j=1}^{n_u}$ and $\{\psi_k\}_{k=1}^{n_p}$. The matrix $\mathbf{A}$ is the discrete vector-Laplacian

$$\mathbf{A} = [\mathbf{a}_{ij}], \quad \mathbf{a}_{ij} = \int_\Omega \nabla \vec{\phi}_i : \nabla \vec{\phi}_j, \quad i, j = 1, \ldots, n_u; \tag{5.153}$$

and the matrix Q is the *pressure mass matrix*

$$Q = [q_{kl}], \quad q_{kl} = \int_\Omega \psi_k \psi_l, \quad k, l = 1, \ldots, n_p. \tag{5.154}$$

Using a standard componentwise vector basis for a two-dimensional problem as in (5.39) gives

$$\langle \mathbf{A}\mathbf{v}, \mathbf{v} \rangle \equiv \langle A\mathbf{v}_x, \mathbf{v}_x \rangle + \langle A\mathbf{v}_y, \mathbf{v}_y \rangle, \tag{5.155}$$

where $n_u = 2n$, and A is the $n \times n$ discrete (scalar-) Laplacian matrix. The spectral properties of A are analyzed in Section 1.6. Combining (5.155) with (1.119) and Remark 1.33 leads to the following characterization.

Theorem 5.21. *With P_1, P_2, Q_1 or Q_2 approximation on a shape regular, quasi-uniform subdivision of $\mathbb{R}^2$, the matrix $\mathbf{A}$ in (5.150) satisfies*

$$ch^2 \le \frac{\langle \mathbf{A}\mathbf{v}, \mathbf{v} \rangle}{\langle \mathbf{v}, \mathbf{v} \rangle} \le C \quad \text{for all } \mathbf{v} \in \mathbb{R}^{n_u}, \tag{5.156}$$

where h is the length of the longest edge in the mesh or grid, and c and C are positive constants that are independent of h.

The (rectangular) divergence matrix B in (5.150) needs to be looked at from a different perspective. Specifically, we need to introduce the idea of *generalized singular values*. These are real numbers σ associated with the following generalized eigenvalue problem:

$$\begin{bmatrix} O & B^T \\ B & O \end{bmatrix} \begin{bmatrix} \mathbf{v} \\ \mathbf{q} \end{bmatrix} = \sigma \begin{bmatrix} \mathbf{A} & O \\ O & Q \end{bmatrix} \begin{bmatrix} \mathbf{v} \\ \mathbf{q} \end{bmatrix}. \tag{5.157}$$

The null space of B is readily identified with zero eigenvalues: if $\sigma = 0$, then $B^T\mathbf{q} = 0$ and $B\mathbf{v} = 0$. The nonzero eigenvalues σ are associated with nonzero pressures (and velocity vectors $\mathbf{v} \notin \mathtt{null}(B)$). To see this, take the inner product of $(\mathbf{v}^T, -\mathbf{q}^T)^T$ with (5.157) to give

$$\langle \mathbf{v}, B^T\mathbf{q} \rangle - \langle \mathbf{q}, B\mathbf{v} \rangle = 0 = \sigma \left(\langle \mathbf{v}, \mathbf{A}\mathbf{v} \rangle - \langle \mathbf{q}, Q\mathbf{q} \rangle \right), \tag{5.158}$$

and so if $\sigma \neq 0$ we have that $\langle \mathbf{A}\mathbf{v}, \mathbf{v} \rangle = \langle Q\mathbf{q}, \mathbf{q} \rangle$ (or equivalently, $\|\nabla \vec{v}_h\| = \|q_h\|$). In this case, $\mathbf{v}$ and $\mathbf{q}$ can be independently eliminated from (5.157). This leads to a Rayleigh-quotient characterization of the nonzero eigenvalues σ in terms of the remaining variable:

$$\frac{\langle B\mathbf{A}^{-1}B^T\mathbf{q}, \mathbf{q} \rangle}{\langle Q\mathbf{q}, \mathbf{q} \rangle} = \sigma^2 = \frac{\langle B^T Q^{-1}B\mathbf{v}, \mathbf{v} \rangle}{\langle \mathbf{A}\mathbf{v}, \mathbf{v} \rangle}. \tag{5.159}$$

In the special case of $\mathtt{null}(B^T) = \mathbf{0}$, the matrix B has n_p positive singular values $0 < \sigma_1 \leq \sigma_2 \leq \cdots \leq \sigma_{n_p}$, characterized by (5.159).

The singular value construction in (5.159) can be used to give an algebraic interpretation of the discrete inf–sup stability condition (5.49):

$$\gamma \leq \min_{q_h \neq \text{constant}} \max_{\vec{v}_h \neq \vec{0}} \frac{|(q_h, \nabla \cdot \vec{v}_h)|}{\|\nabla \vec{v}_h\| \|q_h\|}. \tag{5.160}$$

In particular, using the matrix norms in (5.151) and (5.152), and given that $|(q_h, \nabla \cdot \vec{v}_h)| = |\langle \mathbf{q}, B\mathbf{v} \rangle|$, we have

$$\gamma \leq \min_{\mathbf{q} \neq 1} \max_{\mathbf{v} \neq 0} \frac{|\langle \mathbf{q}, B\mathbf{v} \rangle|}{\langle \mathbf{A}\mathbf{v}, \mathbf{v} \rangle^{1/2} \langle Q\mathbf{q}, \mathbf{q} \rangle^{1/2}}$$

$$= \min_{\mathbf{q} \neq 1} \frac{1}{\langle Q\mathbf{q}, \mathbf{q} \rangle^{1/2}} \max_{\mathbf{w} = \mathbf{A}^{1/2}\mathbf{v} \neq 0} \frac{|\langle \mathbf{q}, B\mathbf{A}^{-1/2}\mathbf{w} \rangle|}{\langle \mathbf{w}, \mathbf{w} \rangle^{1/2}}$$

$$= \min_{\mathbf{q} \neq 1} \frac{1}{\langle Q\mathbf{q}, \mathbf{q} \rangle^{1/2}} \max_{\mathbf{w} \neq 0} \frac{|\langle \mathbf{A}^{-1/2}B^T\mathbf{q}, \mathbf{w} \rangle|}{\langle \mathbf{w}, \mathbf{w} \rangle^{1/2}}$$

$$= \min_{\mathbf{q} \neq 1} \frac{\langle \mathbf{A}^{-1/2}B^T\mathbf{q}, \mathbf{A}^{-1/2}B^T\mathbf{q} \rangle^{1/2}}{\langle Q\mathbf{q}, \mathbf{q} \rangle^{1/2}}$$

$$= \min_{\mathbf{q} \neq 1} \frac{\langle B\mathbf{A}^{-1}B^T\mathbf{q}, \mathbf{q} \rangle^{1/2}}{\langle Q\mathbf{q}, \mathbf{q} \rangle^{1/2}},$$

since the maximum is attained when $\mathbf{w} = \pm \mathbf{A}^{-1/2} B^T \mathbf{q}$. Thus, we see that

$$\gamma^2 = \min_{\mathbf{q} \neq 1} \frac{\langle B\mathbf{A}^{-1}B^T \mathbf{q}, \mathbf{q} \rangle}{\langle Q\mathbf{q}, \mathbf{q} \rangle}, \tag{5.161}$$

and comparing with (5.159), we see that $\gamma \equiv \sigma_{\min}$. The matrix $B\mathbf{A}^{-1}B^T$ is called the *pressure Schur complement*. Incidentally, (5.159) also provides an alternative characterization of the inf–sup constant, namely

$$\gamma^2 = \min_{\{\mathbf{v} \in \mathbb{R}^{n_u} \,|\, \langle \mathbf{A}\mathbf{v}, \mathbf{u} \rangle = 0, \, \mathbf{u} \in \texttt{null}(B)\}} \frac{\langle B^T Q^{-1} B\mathbf{v}, \mathbf{v} \rangle}{\langle \mathbf{A}\mathbf{v}, \mathbf{v} \rangle}. \tag{5.162}$$

Looking ahead to Chapter 6, it turns out that establishing uniform bounds for the generalized singular values of B is the key to designing fast iterative methods for solving the discrete Stokes problems. Moreover, if an unstable mixed approximation is used, so that $\gamma \to 0$ as $h \to 0$, then iterative solvers do not perform optimally — this is the major motivation for stabilizing $\boldsymbol{Q}_1$–$\boldsymbol{P}_0$. We return to this point after discussing the more straightforward case of stable mixed approximation.

5.5.1 Stable mixed approximation

Bounds for the eigenvalues of the mass matrix scaled pressure Schur complement $Q^{-1}B\mathbf{A}^{-1}B^T$ are discussed in this section. The first case we consider is that of flow problems with velocity specified everywhere on the boundary.

Theorem 5.22. *For any flow problem with $\partial\Omega = \partial\Omega_D$, discretized using a uniformly stable mixed approximation on a shape regular, quasi-uniform subdivision of $\mathbb{R}^2$, the pressure Schur complement matrix $B\mathbf{A}^{-1}B^T$ is spectrally equivalent to the pressure mass matrix Q:*

$$\gamma^2 \leq \frac{\langle B\mathbf{A}^{-1}B^T \mathbf{q}, \mathbf{q} \rangle}{\langle Q\mathbf{q}, \mathbf{q} \rangle} \leq 1 \quad \text{for all } \mathbf{q} \in \mathbb{R}^{n_p} \text{ such that } \mathbf{q} \neq 1. \tag{5.163}$$

The inf–sup constant, γ, is bounded away from zero independently of h, and the "effective" condition number satisfies $\kappa(B\mathbf{A}^{-1}B^T) \leq C/(c\gamma^2)$, where c and C are the constants given by (1.116):

$$ch^2 \leq \frac{\langle Q\mathbf{q}, \mathbf{q} \rangle}{\langle \mathbf{q}, \mathbf{q} \rangle} \leq Ch^2 \quad \text{for all } \mathbf{q} \in \mathbb{R}^{n_p}. \tag{5.164}$$

Proof The lower bound follows immediately from (5.161). The upper bound is a consequence of properties of the underlying differential operators. First, using the Cauchy–Schwarz inequality, we have that

$$|(q_h, \nabla \cdot \vec{v}_h)| \leq \|q_h\| \, \|\nabla \cdot \vec{v}_h\| . \tag{5.165}$$

Second, noting that $\vec{v}_h \in \mathbf{H}_{E_0}^1 \equiv \mathbf{H}_0^1$ (since $\partial\Omega = \partial\Omega_D$) and substituting into the identity

$$\int_\Omega \nabla\vec{w} : \nabla\vec{v} \equiv \int_\Omega (\nabla \cdot \vec{w})(\nabla \cdot \vec{v}) + \int_\Omega (\nabla \times \vec{w}) : (\nabla \times \vec{v}), \qquad (5.166)$$

which is valid for all functions $\vec{w}$, $\vec{v}$ in the Sobolev space

$$\mathbf{H}_0^1 = \left\{ \vec{u} \in \mathcal{H}^1(\Omega)^2 \big| \vec{u} = \vec{0} \text{ on } \partial\Omega \right\},$$

we find that

$$\|\nabla\vec{v}_h\|^2 = \|\nabla \cdot \vec{v}_h\|^2 + \|\nabla \times \vec{v}_h\|^2 \geq \|\nabla \cdot \vec{v}_h\|^2. \qquad (5.167)$$

Finally, combining (5.165) with (5.167) gives

$$\frac{|(q_h, \nabla \cdot \vec{v}_h)|}{\|\nabla\vec{v}_h\| \, \|q_h\|} \leq 1.$$

Expressing this in terms of matrices, as in the derivation of (5.161), gives the upper bound in (5.163).

The condition number bound follows from writing

$$\gamma^2 \frac{\langle Q\mathbf{q}, \mathbf{q}\rangle}{\langle \mathbf{q}, \mathbf{q}\rangle} \leq \frac{\langle B\mathbf{A}^{-1}B^T\mathbf{q}, \mathbf{q}\rangle}{\langle \mathbf{q}, \mathbf{q}\rangle} \leq \frac{\langle Q\mathbf{q}, \mathbf{q}\rangle}{\langle \mathbf{q}, \mathbf{q}\rangle}$$

and using the mass matrix bound (5.164). □

Remark 5.23. The bounds in (5.163) also hold in the case of stable mixed approximations using tetrahedral or brick elements on a quasi-uniform discretization of a domain in $\mathbb{R}^3$. The mass matrix bounds are given by (1.117).

To illustrate the tightness of the bounds in (5.163), computed extremal eigenvalues for $\mathbf{Q}_2$–$\mathbf{Q}_1$ and $\mathbf{Q}_2$–$\mathbf{P}_{-1}$ mixed approximations of an enclosed flow problem on a sequence of uniform square grids are given in Table 5.6. The upper

Table 5.6 *Extremal eigenvalues of the matrix $Q^{-1}B\mathbf{A}^{-1}B^T$ for a sequence of square grids defined on $\Omega_\square = [-1, 1] \times [-1, 1]$: ℓ is the grid refinement level and $h = 2^{2-\ell}$.*

ℓ	$\mathbf{Q}_2$–$\mathbf{Q}_1$		$\mathbf{Q}_2$–$\mathbf{P}_{-1}$	
	λ_2	λ_{n_p}	λ_2	λ_{n_p}
3	0.2254	0.9951	0.2563	0.9977
4	0.2140	0.9997	0.2352	0.9999
5	0.2074	1.0000	0.2223	1.0000
6	0.2027	1.0000	0.2137	1.0000

bound of unity is clearly tight in both cases. The inf–sup eigenvalue λ_2 seems to be converging to an asymptotic value that is bigger than $1/5$ in either case.

The analogous eigenvalue bounds for flow problems with a Neumann condition applied on part of the boundary are given in Proposition 5.24.

Proposition 5.24. *For any flow problem with $\int_{\partial\Omega_N} ds \neq 0$, discretized using a uniformly stable mixed approximation on a shape regular, quasi-uniform subdivision of $\mathbb{R}^2$, the pressure Schur complement matrix satisfies*

$$\gamma_N^2 \leq \frac{\langle B\mathbf{A}^{-1}B^T\mathbf{q}, \mathbf{q}\rangle}{\langle Q\mathbf{q}, \mathbf{q}\rangle} \leq 2 \quad \text{for all } \mathbf{q} \in \mathbb{R}^{n_p}. \qquad (5.168)$$

The constant γ_N is bounded away from zero independently of h.

Proof See Problem 5.9.   □

Computed eigenvalues of the scaled Schur complement for the discrete matrices arising from $\boldsymbol{Q}_2$–$\boldsymbol{Q}_1$ approximation of Examples 5.1.3 and 5.1.2 are illustrated in Figure 5.22. For the enclosed flow, all the eigenvalues λ_2 through λ_{n_p} live in the interval $[0.2, 1]$. For the expansion flow with a natural outflow, the eigenvalues λ_2 through λ_{n_p} live in the interval $[0.09, 1.5]$. The eigenvalue λ_1 is bounded away from zero, taking the values of 0.0251, 0.0249, 0.0248 and 0.0248 on the four grids illustrated. Note that the number of eigenvalues satisfying $\lambda > 1$ is proportional to the number of elements on the outflow boundary.

The $\boldsymbol{Q}_2$–$\boldsymbol{Q}_1$ mixed approximation is also known to be stable with respect to the "wrong norms" (i.e. measuring velocity in $\mathbf{L}_2(\Omega)$ and pressure in $\mathcal{H}^1(\Omega)$, see (5.139)). This implies that there exists an inf–sup constant δ bounded away from zero such that

$$\delta \leq \min_{q_h \neq \text{constant}} \max_{\vec{v}_h \neq \vec{0}} \frac{|(q_h, \nabla \cdot \vec{v}_h)|}{\|\vec{v}_h\| \, \|\nabla q_h\|}. \qquad (5.169)$$

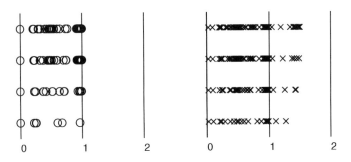

FIG. 5.22. Computed eigenvalues of $Q^{-1}B\mathbf{A}^{-1}B^T$ for $\boldsymbol{Q}_2$–$\boldsymbol{Q}_1$ approximation with uniformly refined square grids ($\ell = 2$ bottom to $\ell = 5$ top); ○ square domain and enclosed flow boundary condition (Example 5.1.3); × step domain with natural outflow boundary condition (Example 5.1.2).

Looking ahead to the development of fast iterative solvers for the Navier–Stokes equations in Chapter 8, it is useful here to express the alternative inf–sup constant δ in terms of finite element matrices. To do this, we have to define the *velocity mass matrix* $\mathbf{Q}$,

$$\mathbf{Q} = [\mathbf{q}_{ij}], \quad \mathbf{q}_{ij} = \int_\Omega \vec{\phi}_i \cdot \vec{\phi}_j, \quad i, j = 1, \ldots, n_u; \tag{5.170}$$

and the *pressure Laplacian* matrix A_p,

$$A_p = [a_{kl}], \quad a_{kl} = \int_\Omega \nabla \psi_k \cdot \nabla \psi_l, \quad k, l = 1, \ldots, n_p. \tag{5.171}$$

Then, following the derivation of (5.157), the condition (5.169) can be rewritten as

$$\delta^2 = \min_{\mathbf{q} \neq 1} \frac{\langle B\mathbf{Q}^{-1}B^T \mathbf{q}, \mathbf{q} \rangle}{\langle A_p \mathbf{q}, \mathbf{q} \rangle}. \tag{5.172}$$

The matrix $B\mathbf{Q}^{-1}B^T$ in (5.172) is commonly referred to as the "consistent pressure-Poisson matrix", see Gresho & Sani [92, pp. 640–644]. The matrix is singular for enclosed flow boundary conditions, but non-singular otherwise. Computing extremal eigenvalues for $\mathbf{Q}_2$–$\mathbf{Q}_1$ approximation we observe that

$$\frac{1}{2} < \frac{\langle B\mathbf{Q}^{-1}B^T \mathbf{q}, \mathbf{q} \rangle}{\langle A_p \mathbf{q}, \mathbf{q} \rangle} \leq 1 \quad \text{for all } \mathbf{q} \in \mathbb{R}^{n_p}, \mathbf{q} \neq 1, \tag{5.173}$$

on uniform grids, and for an enclosed flow boundary condition. Further details are given in Computational Exercise 5.8.

Remark 5.25. The pressure Laplacian A_p in (5.171) is only defined for continuous pressure approximation spaces, whereas the consistent pressure-Poisson matrix $B\mathbf{Q}^{-1}B^T$ is also defined for discontinuous pressure approximation. The pressure Laplacian in (5.171) represents a discrete *Neumann problem* and is singular: $\mathtt{null}(A_p) = 1$. (This is because the pressure basis functions form a partition of unity, see the discussion of (1.24) in Chapter 1.) This is independent of the boundary conditions that are associated with the flow problem.

5.5.2 *Stabilized mixed approximation*

In this section we consider linear algebra aspects of the stabilized system that arises in the case of $\mathbf{Q}_1$–$\mathbf{P}_0$ and $\mathbf{Q}_1$–$\mathbf{Q}_1$ mixed approximation:

$$\begin{bmatrix} \mathbf{A} & B^T \\ B & -C \end{bmatrix} \begin{bmatrix} \mathbf{u} \\ \mathbf{p} \end{bmatrix} = \begin{bmatrix} \mathbf{f} \\ \mathbf{g} \end{bmatrix}, \tag{5.174}$$

see (5.84) and (5.94), respectively. The main task here is to identify a suitable generalization of the inf–sup constant γ. Having done this, we proceed to derive Rayleigh quotient bounds analogous to the ones established for inf–sup stable approximations in the previous section.

The discrete system (5.174) is associated with the following discrete formulation: find $\vec{u}_h \in \mathbf{X}_E^h$ and $p_h \in M^h$ such that

$$\int_\Omega \nabla \vec{u}_h : \nabla \vec{v}_h - \int_\Omega p_h \nabla \cdot \vec{v}_h = \int_{\partial \Omega_N} \vec{s} \cdot \vec{v}_h \quad \text{for all } \vec{v}_h \in \mathbf{X}_0^h, \tag{5.175}$$

$$-\int_\Omega q_h \nabla \cdot \vec{u}_h - c(p_h, q_h) = 0 \quad \text{for all } q_h \in M^h, \tag{5.176}$$

where $c(\cdot, \cdot) : M^h \times M^h \to \mathbb{R}$ is the stabilization operator that generates the matrix C in (5.174). Let us now introduce the following functional, $s : M^h \to \mathbb{R}$, such that

$$s(q_h) := \max_{\vec{v}_h \neq \vec{0}} \frac{(q_h, \nabla \cdot \vec{v}_h)}{\|\nabla \vec{v}_h\|} + c(q_h, q_h)^{1/2}. \tag{5.177}$$

This functional is the key to the stability of the formulation (5.175)–(5.176).

Definition 5.26 (Uniform stabilization). The discrete problem (5.175)–(5.176) is said to be *uniformly stabilized* if there exists a constant δ independent of h such that

$$s(q_h) \geq \delta \|q_h\| \quad \text{for all } q_h \in M^h. \tag{5.178}$$

The value $\delta = \min_{q_h \neq 0}\{s(q_h)/\|q_h\|\}$ is called the *generalized* inf–sup constant. Note that, our terminology is consistent in that, if the stabilization term is omitted, (5.178) simply reduces to the standard definition of γ in (5.160). The notion of uniform stabilization provides an alternative point of view to the framework of $\mathcal{B}_h$-stability discussed in Section 5.4.1.

Remark 5.27. Establishing (5.178) for the case of Q_1–Q_1 approximation can be done very simply using "Verfürth's trick". The idea is to combine the weak inf–sup stability inequality (5.138):

$$s(q_h) \geq \max_{\vec{v}_h \in \mathbf{X}_0^h} \frac{b(\vec{v}_h, q_h)}{\|\nabla \vec{v}_h\|} \geq c_1 \|q_h\| - c_3 \|q_h - \Pi_0 q_h\|, \tag{5.179}$$

with the stabilization property

$$s(q_h) \geq c(q_h, q_h)^{1/2} = \|q_h - \Pi_0 q_h\|. \tag{5.180}$$

This is easily done by multiplying (5.180) by c_3 and then adding to (5.179), giving

$$(1 + c_3)s(q_h) \geq c_1 \|q_h\| \quad \text{for all } q_h \in M^h, \tag{5.181}$$

that is, (5.178) is satisfied with $\delta = c_1/(1 + c_3)$.

To get back to linear algebra, expressing the uniform stabilization condition (5.178) in the matrix notation of the previous section leads to the following characterization:

$$\langle BA^{-1}B^T\mathbf{q}, \mathbf{q}\rangle^{1/2} + \langle C\mathbf{q}, \mathbf{q}\rangle^{1/2} \geq \delta \langle Q\mathbf{q}, \mathbf{q}\rangle^{1/2} \quad \text{for all } \mathbf{q} \in \mathbb{R}^{n_p}. \qquad (5.182)$$

This condition can be simplified by removing the square roots (using the fact that $a + b \leq (\sqrt{a} + \sqrt{b})^2 \leq 2a + 2b$ for all positive numbers a and b). The upshot is that uniform stabilization via (5.178) is equivalent to the condition that

$$\langle BA^{-1}B^T\mathbf{q}, \mathbf{q}\rangle + \langle C\mathbf{q}, \mathbf{q}\rangle \geq \delta^2 \langle Q\mathbf{q}, \mathbf{q}\rangle \quad \text{for all } \mathbf{q} \in \mathbb{R}^{n_p}. \qquad (5.183)$$

The generalized inf–sup constant is thus given by

$$\delta^2 = \min_{\mathbf{q} \neq 1} \frac{\langle (BA^{-1}B^T + C)\mathbf{q}, \mathbf{q}\rangle}{\langle Q\mathbf{q}, \mathbf{q}\rangle}. \qquad (5.184)$$

This is the obvious generalization of the unstabilized case (5.161): stability is determined by the smallest nonzero eigenvalue of the scaled Schur complement $Q^{-1}(BA^{-1}B^T + C)$ associated with (5.174). Note however, that the alternative characterization (5.162) cannot be generalized to the system (5.174).

To determine upper bounds on the eigenvalues of $Q^{-1}(BA^{-1}B^T + C)$, the ideal stabilization property in (5.76) or (5.95),

$$\frac{\langle C\mathbf{p}, \mathbf{p}\rangle}{\langle Q\mathbf{p}, \mathbf{p}\rangle} \leq 1 \quad \text{for all } \mathbf{p} \in \mathbb{R}^{n_p}, \qquad (5.185)$$

can be combined with the upper bounds derived in Section 5.5.1. The result is the following bound:

$$\frac{\langle (BA^{-1}B^T + C)\mathbf{p}, \mathbf{p}\rangle}{\langle \mathbf{p}, Q\mathbf{p}\rangle} \leq \begin{cases} 2, & \text{if } \partial\Omega = \partial\Omega_D, \\ 3, & \text{otherwise.} \end{cases} \qquad (5.186)$$

The issue of the *optimal* choice of $\boldsymbol{Q}_1$–$\boldsymbol{P}_0$ stabilization parameter, β in (5.70), can now be resolved. If β is not "ideally chosen" then the Schur complement bounds are less attractive. Specifically, using (5.85), we observe that

$$\frac{\langle (BA^{-1}B^T + \beta C)\mathbf{p}_{cb}, \mathbf{p}_{cb}\rangle}{\langle Q\mathbf{p}_{cb}, \mathbf{p}_{cb}\rangle} = \beta \frac{\langle C\mathbf{p}_{cb}, \mathbf{p}_{cb}\rangle}{\langle Q\mathbf{p}_{cb}, \mathbf{p}_{cb}\rangle} = 4\beta,$$

which means that if β is smaller than a critical value $\beta_0 < 1/4$, then the smallest eigenvalue of the Schur complement δ^2 is proportional to β. Conversely, if $\beta > \beta^* = 1/4$, then the largest eigenvalue of the Schur complement is proportional to β. The choice $\beta = \beta^*$ is thus the largest $\boldsymbol{Q}_1$–$\boldsymbol{P}_0$ parameter that ensures that the effective condition number (i.e. the ratio of extremal nonzero eigenvalues) of the stabilized pressure Schur complement $Q^{-1}(BA^{-1}B^T + \beta C)$ is independent of β. This β-dependent behavior is generic and is illustrated in Figure 5.23, where the effective condition number of the stabilized Schur complement is plotted for a sample set of values of β. Notice that for both $\boldsymbol{Q}_1$–$\boldsymbol{P}_0$ approximation or $\boldsymbol{Q}_1$–$\boldsymbol{Q}_1$ approximation, the "ideal" parameter value

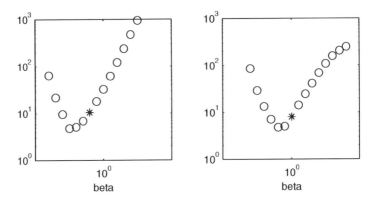

FIG. 5.23. Computed condition numbers of $Q^{-1}(B\mathbf{A}^{-1}B^T + \beta C)$ with $\beta = 2^k$, for an enclosed flow boundary condition (Example 5.1.3) using $\mathbf{Q}_1$–$\mathbf{P}_0$ approximation (left) and $\mathbf{Q}_1$–$\mathbf{Q}_1$ approximation (right) and a 16×16 square grid ($\ell = 4$).

$\beta = \beta^*$ (corresponding to the symbol $*$ on the plot) is slightly to the right of the parameter value that minimizes the condition number.

We summarize our ideal stabilization bounds in the following theorem.

Theorem 5.28. *For any flow problem with $\partial \Omega = \partial \Omega_D$, discretized using a ideally stabilized mixed approximation on a shape regular subdivision of $\mathbb{R}^2$, the pressure Schur complement matrix satisfies*

$$\delta^2 \leq \frac{\langle (B\mathbf{A}^{-1}B^T + C)\mathbf{q}, \mathbf{q} \rangle}{\langle Q\mathbf{q}, \mathbf{q} \rangle} \leq 2 \quad \textit{for all } \{\mathbf{q} \in \mathbb{R}^{n_p}; \mathbf{q} \neq \mathbf{1}\}. \tag{5.187}$$

The generalized inf–sup constant, δ, is bounded away from zero independently of h. For a flow problem with $\int_{\partial \Omega_N} \mathrm{d}s \neq 0$ the corresponding bound is

$$\delta_N^2 \leq \frac{\langle (B\mathbf{A}^{-1}B^T + C)\mathbf{q}, \mathbf{q} \rangle}{\langle Q\mathbf{q}, \mathbf{q} \rangle} \leq 3 \quad \textit{for all } \mathbf{q} \in \mathbb{R}^{n_p}, \tag{5.188}$$

with δ_N bounded away from zero independently of h.

Remark 5.29. The bounds in (5.187) and (5.188) also hold in the case of ideally stabilized mixed approximations using tetrahedral or brick elements on a quasi-uniform discretization of a domain in $\mathbb{R}^3$.

To illustrate the tightness of the bounds in (5.187), computed extremal eigenvalues for stabilized $\mathbf{Q}_1$–$\mathbf{Q}_1$ and $\mathbf{Q}_1$–$\mathbf{P}_0$ approximation of an enclosed flow problem on a sequence of uniform square grids are given in Table 5.7. (The case of inflow/outflow boundary conditions is the topic of Computational Exercise 5.9.) The upper bound in (5.187) is clearly an overestimate in both cases. The inf–sup eigenvalue λ_2 seems to be converging to an asymptotic value that is bigger

Table 5.7 *Extremal eigenvalues of the matrix* $Q^{-1}(B\mathbf{A}^{-1}B^T + C)$ *for a sequence of square grids defined on* $\Omega_\square = [-1,1] \times [-1,1]$: ℓ *is the grid refinement level and* $h = 2^{1-\ell}$.

	Q_1–Q_1		Q_1–P_0	
ℓ	λ_2	λ_{n_p}	λ_2	λ_{n_p}
3	0.2668	1.2146	0.2809	1.7238
4	0.2423	1.2399	0.2522	1.7441
5	0.2275	1.2470	0.2339	1.7486

than $1/5$ in either case. The effective condition numbers, λ_{n_p}/λ_2, can be seen to be very similar to those of the stable mixed approximations in Table 5.6.

Remark 5.30. The extremal eigenvalues of the *unstabilized* Q_1–P_0 Schur complement matrix $Q^{-1}B\mathbf{A}^{-1}B^T$ behave very differently. For the problem in Table 5.7, eigenvalue λ_2 is zero because of the spurious checkerboard mode. Moreover, the smallest nonzero eigenvalues $\lambda_3, \lambda_4, \ldots$ are unstable: they all decrease like $O(h)$ as the grid is refined, and so the effective condition number λ_{n_p}/λ_3 blows up as $h \to 0$! These unstable eigenvalues also arise in the case of non-enclosed flow boundary conditions, see Computational Exercise 5.9. Indeed, the associated eigenvectors are the origin of the unphysical oscillations that are visible in the unstabilized Q_1–P_0 pressure solution in Figure 5.13.

Discussion and bibliographical notes

The Stokes equations also model the displacement of an incompressible elastic solid, rubber for example, when subject to external forces. Denoting the displacement vector by $\vec{u}$, the strain rate tensor by $\vec{\epsilon} := [\nabla\vec{u} + (\nabla\vec{u})^T]/2$ and the total stress tensor by $\vec{\sigma} := 2\mu\vec{\epsilon} - p\mathbf{I}$ (where μ is the constant viscosity), the basic model for incompressible isotropic elasticity is the *stress-divergence* form of the Stokes equations

$$\nabla \cdot \vec{\sigma} = \vec{0},$$

$$\nabla \cdot \vec{u} = 0.$$

The identity $\nabla \cdot \vec{\epsilon} \equiv [\nabla^2\vec{u} + \nabla(\nabla \cdot \vec{u})]/2$ implies that this system is equivalent to the Stokes equation system (5.1). The discretized versions of the two models are quite different however, see Gresho & Sani [92, Section 3.8]. The natural boundary condition $\vec{n} \cdot \vec{\sigma} = \vec{s}$ associated with the stress-divergence formulation corresponds to specifying a physical force or traction on $\partial\Omega_N$. Such a condition is not nearly as useful as the outflow condition in (5.2) when modeling fluid flow.

A slightly compressible elastic solid may be modeled by replacing the divergence-free constraint above by the relation

$$\nabla \cdot \vec{u} + \frac{p}{\lambda} = 0,$$

where λ is the so-called Lamé parameter. (This parameter becomes unbounded in the incompressible limit.) If λ is finite, then the "pressure" p can be eliminated from the system to give the compressible constitutive equation: $\vec{\sigma} := 2\mu\vec{\epsilon} + \lambda\nabla \cdot \vec{u}\vec{\mathbf{I}}$. In a fluid flow context, the introduction of the (artificial) parameter λ and the subsequent elimination of the pressure is called a penalty method. This idea was popularized by Hughes et al. in the classic paper [106]. The attraction of this approach is that the resulting linear algebra system is symmetric and positive definite. Discretization of the penalty formulation is tricky however. Typically, different quadrature rules must be applied to the μ and λ terms. If the same quadrature is used for both terms, the penalty method is equivalent to using equal-order mixed interpolation, which implies instability in the incompressible limit. For a detailed discussion, see Hughes [108, Section 4.4].

There is little loss of generality in the fact that there is no forcing term for the momentum equation in (5.1). A conservative body force (e.g. that due to gravity when the fluid is supported below) is the gradient of a scalar field and thus (in the absence of outflow boundaries) can be incorporated into the system by simply redefining the pressure.

5.1. Poiseuille flow is also a classical solution of the Stokes (and Navier–Stokes) equations in a cylindrical coordinate system. Flow in a straight pipe of circular cross-section is axisymmetric with a parabolic profile in the radial direction, and having a linear pressure variation in the axial direction. For further discussion, see for example, Acheson [1].

Driven flows exhibit almost all the phenomena that can arise in incompressible flows; eddies, chaotic particle motions, transition and turbulence. The beautiful review of Shankar & Deshpande [171] gives details. The theory of Moffatt eddies is developed in his classic paper [133].

5.3. The issue of incompatible discrete approximations was first realized in the 1960s by the finite difference community. The first discretization method to give good results was the remarkable MAC (Marker and Cell) scheme devised by Harlow & Welch [101]. A rigorous proof of the MAC scheme's stability was established later, see Nicolaides [137].

We have restricted our attention to low order approximations — so called h-finite element methods — throughout the chapter. Useful mixed approximation methods in a hp finite element setting are families of mixed approximations which are uniformly inf–sup stable with respect to both the mesh parameter h and the polynomial degree p. The best known example of such a family is $\boldsymbol{Q}_k$–$\boldsymbol{P}_{-(k-1)}$ for $k \geq 2$. A proof of the hp uniform stability is given by Bernardi & Maday [14]. There are also families of mixed methods which are uniformly stable with respect to h, but which are not uniformly stable with respect to p. The best known examples are the Taylor–Hood families Q_k–Q_{k-1} and P_k–P_{k-1} for $k \geq 2$; see Stenberg & Suri [184] for the details.

The instability of Q_1–P_0 approximation was exposed by Fortin [77]. The linear algebra issues underlying the checkerboard mode were explained by Sani et al. [166]. The existence of the more vicious pressure modes associated with a h-dependent inf–sup constant was established by Boland & Nicolaides [18]. A quantitative description of these so-called "pesky modes" can be found in Gresho & Sani [92, pp. 686–691].

Some discretizations of the Stokes equations, notably spectral approximation methods, give rise to a nonsymmetric linear equation system of the form

$$\begin{bmatrix} \mathbf{A} & G \\ B & O \end{bmatrix} \begin{bmatrix} \mathbf{u} \\ \mathbf{p} \end{bmatrix} = \begin{bmatrix} \mathbf{f} \\ \mathbf{g} \end{bmatrix},$$

where $\mathbf{A} = \mathbf{A}^T$ but $G \neq B^T$. Solvability of the underlying formulation in such cases requires two inf–sup conditions, one each for the discrete gradient G and the discrete divergence B; see Nicolaides [136].

5.4. The origin of the Babuška–Brezzi condition lies in two classic papers; [7] and [30]. A different way of establishing that a mixed approximation strategy is inf–sup stable is to use Fortin's method, see [77]. A posteriori error estimation for Stokes equations was pioneered by Verfürth [203] and Bank & Welfert [10].

If highly stretched grids are used, then the dependence of the inf–sup constant on the aspect ratio of the element is likely to be an important issue. The best mixed approximation methods have an inf–sup constant which is independent of the aspect ratio. Schötzau & Schwab [167] introduce a macroelement framework for establishing that a given mixed approximation method is inf–sup stable on stretched grids. It turns out that both the stabilized Q_1–P_0 and the Q_2–Q_1 mixed methods are uniformly inf–sup stable on "edge macroelements", which typically arise in resolving shear layers. The Q_2–P_{-1} mixed approximation is not uniformly stable with respect to grid stretching however; see Ainsworth & Coggins [2] for details.

The Brezzi saddle-point theory underlying Theorem 5.1 only requires that the bilinear form $a(\cdot, \cdot)$ be coercive over the subspace

$$\mathbf{Z}^h = \{\vec{v}_h \in \mathbf{X}_0^h \,|\, b(\vec{v}_h, q_h) = 0 \text{ for all } q_h \in M^h\}.$$

The Stokes problem is an atypical example of a saddle-point problem in that coercivity holds over the whole space $\mathbf{X}_0^h$. A more typical example of saddle-point theory is that of potential flow in which case the bilinear form $a(\cdot, \cdot)$ corresponds to a velocity mass matrix $\mathbf{Q}$ and is only coercive on the subspace $\mathbf{Z}^h$ (with respect to a weaker norm). This is a source of difficulty when coupling a potential (or Darcy) ground-water flow model with a Stokes flow. The theory underlying such coupled problems was recently developed by Mardal et al. [129]. Finding velocity–pressure pairs that are simultaneously stable for both flow regimes is not easy!

The approximation theory motivating the use of *hp* finite element methods for Stokes problems has been developed over the last decade, see Schwab [169]. Optimal methods involve using fine mesh low order approximations in the neighborhood of corners, with a geometrically increasing degree of approximation moving towards the interior of the domain where the Stokes solution is analytic. Further details can be found in Schötzau et al. [168] and Toselli & Schwab [194].

5.5. An accessible discussion the inf–sup condition — written from the perspective of ensuring the well-posedness of the discrete Stokes system — is contained in Brezzi [32]. Useful spectral properties of the Stokes system may also be obtained by exploiting the discrete analogue of the so-called Crouzeix–Velte decomposition of the space $\mathbf{H}^1_{E_0}$. See Stoyan [185, 186] for further details.

Problems

5.1. Verify that the bilinear form $a(\vec{u}, \vec{v}) := \int_\Omega \nabla \vec{u} : \nabla \vec{v}$ is coercive (see Definition 3.1), and continuous (see Definition 3.2) with respect to the norm $\|\cdot\|_{1,\Omega}$ over the space $\mathbf{H}^1_{E_0}$ defined in (5.21).

5.2. Verify that the specific choice of basis functions (5.38) applied to the Stokes system (5.33)–(5.37) leads to the matrix partitioning (5.39).

5.3. Consider the serendipity version of $\boldsymbol{Q}_2$–$\boldsymbol{Q}_1$ mixed approximation obtained by excluding the centroid node as illustrated in the figure below.

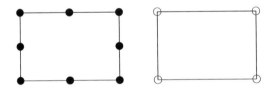

Show that a two-element patch macroelement is *not* stable in this case, and then determine the "smallest" patch that is stable.

5.4. Verify that the five-element patch of $\boldsymbol{Q}_1$–$\boldsymbol{P}_0$ elements illustrated in Figure 5.12 is a stable macroelement. Show further that if an additional element is appended to the stable patch then the new macroelement is also stable.

5.5. Show that, independent of the aspect ratio h_y/h_x, the eigenvalues (5.82) of the 2×2 grid local-jump stabilized matrix $1/(h_x h_y) S_{\beta^*}$ lie in the interval $[1/2, 5/4]$.

5.6. Verify that in the case of a 2×2 patch of rectangular elements, the local jump stabilization matrix $C^*_{2\times2}$ associated with the relaxed incompressibility constraint (5.83) is given by (5.80). Hence, show that the corresponding matrix

$C_{2\times3}^*$ for the 2×3 patch of Q_1–P_0 rectangles illustrated in Figure 5.11 is given by

$$C_{2\times3}^* = h_x h_y \begin{bmatrix} 2 & -1 & 0 & -1 & 0 & 0 \\ -1 & 3 & -1 & 0 & -1 & 0 \\ 0 & -1 & 3 & -1 & 0 & -1 \\ -1 & 0 & -1 & 2 & 0 & 0 \\ 0 & -1 & 0 & 0 & 2 & -1 \\ 0 & 0 & -1 & 0 & -1 & 2 \end{bmatrix}.$$

Making the parameter choice $\beta^* = 1/4$ gives a grid stabilization matrix $\mathbf{C} = \beta^* \mathtt{diag}\,[C_{2\times2}^*, \ldots, C_{2\times3}^*]$ associated with grids that are composed of locally stabilized 2×2 and 2×3 macroelement patches. Show that, for this choice of $\mathbf{C}$, the uniform stabilization condition

$$\frac{\mathbf{p}_{cb}^T S_{\beta_*} \mathbf{p}_{cb}}{\mathbf{p}_{cb}^T \mathcal{Q} \mathbf{p}_{cb}} \geq 1,$$

is satisfied, that is, the *ideal* choice $\beta^* = 1/4$ ensures uniform stabilization for any rectangular grid that can be decomposed into such macroelements.

5.7. Suppose we have a consistent local stabilization, that is, where $\mathbf{1} \in \mathtt{null}(C^*)$ for every macroelement $\mathcal{M}$ in a grid. Show that the discrete velocity $\vec{u}_h$ associated with $\mathbf{u}$, the solution of

$$\begin{bmatrix} \mathbf{A} & \mathbf{B}^T \\ \mathbf{B} & -\mathbf{C} \end{bmatrix} \begin{bmatrix} \mathbf{u} \\ \mathbf{p} \end{bmatrix} = \begin{bmatrix} \mathbf{f} \\ \mathbf{g} \end{bmatrix}$$

with $\mathbf{C} = \mathtt{diag}\,[\frac{1}{4}C^*, \ldots, \frac{1}{4}C^*]$, also satisfies

$$0 = \int_{\mathcal{M}} \nabla \cdot \vec{u}_h, \quad \text{for all } \mathcal{M}.$$

5.8. Verify that in the case of a rectangular element $\square_k$, with edge lengths h_x and h_y and area $|\square| = h_x \times h_y$, the local stabilization matrix C associated with the Q_1 pressure projection operator (5.92), is given by

$$C_k = Q_k - \mathbf{q}\mathbf{q}^T |\square|,$$

where Q_k is the 4×4 element mass matrix, and $\mathbf{q}^T = [\frac{1}{4}, \frac{1}{4}, \frac{1}{4}, \frac{1}{4}]$. Show, therefore, that C_k is a symmetric and positive semi-definite matrix, and that it has a one-dimensional null space spanned by the constant vector $\mathbf{q}$.

5.9. Prove Proposition 5.24. (Hint: for the lower bound, consider the cases of $\mathbf{q} = \mathbf{1}$ and $\{\mathbf{q} \in \mathbb{R}^m; \mathbf{q} \neq \mathbf{1}\}$ separately.)

5.10. Consider the Q_2–Q_1 mixed approximation in three dimensions. Find the simplest patch of bricks that determines a stable macroelement. Show further that if an additional element is appended to the stable patch then the new macroelement is also stable.

5.11. Given that any grid can be partitioned into macroelements such that nonconstant pressure components are stabilized in the sense of Lemma 5.12, establish the global bound

$$c_h(p_h, p_h) \geq \gamma_M \left\| (I - \Pi_h) p_h \right\|^2 \quad \text{for all } p_h \in M_0^h,$$

where Π_h represents the L_2 projection from M_0^h onto the macroelement constant pressure space Q_h. Then, given an arbitrary pair $(\vec{v}_h, q_h) \in \mathbf{X}_0^h \times M^h$, and assuming that (5.125) is valid, take $\vec{z}_h = \vec{v}_h - \delta \vec{w}_h$ and $r_h = -q_h$ with an appropriately chosen constant δ, so as to establish $\mathcal{B}_h$-stability:

$$\mathcal{B}_h((\vec{v}_h, q_h); (\vec{z}_h, r_h)) \geq \kappa_1 \left(\|\nabla \vec{v}_h\| + \|q_h\| \right)^2,$$

and

$$\|\nabla \vec{z}_h\| + \|r_h\| \leq \kappa_2 \left(\|\nabla \vec{v}_h\| + \|q_h\| \right).$$

(Hint: look at the proof of Lemma 5.2.)

5.12. Use the construction in Lemma 5.2 to establish $\mathcal{B}_h$-stability in the case of stabilized $\mathbf{Q}_1$–$\mathbf{Q}_1$ approximation: for any $(\vec{w}_h, s_h) \in \mathbf{X}_0^h \times M^h$ there exists $(\vec{v}_h, q_h) \in \mathbf{X}_0^h \times M^h$ such that

$$\mathcal{B}_h((\vec{w}_h, s_h); (\vec{v}_h, q_h)) \geq \kappa_1 \left(\|\nabla \vec{w}_h\| + \|s_h\| \right)^2,$$
$$\|\nabla \vec{v}_h\| + \|q_h\| \leq \kappa_2 \left(\|\nabla \vec{w}_h\| + \|s_h\| \right),$$

where

$$\mathcal{B}_h((\vec{u}_h, p_h); (\vec{v}_h, q_h))$$
$$= (\nabla \vec{u}_h, \nabla \vec{v}_h) + b(\vec{u}_h, q_h) + b(\vec{v}_h, p_h) - (p_h - \Pi_0 p_h, q_h - \Pi_0 q_h).$$

(Hint: the condition (5.138) implies that for any $s_h \in M^h$, there exists a corresponding vector $\vec{z}_h \in \mathbf{X}_0^h$ such that

$$\|\nabla \vec{z}_h\| = \|s_h\|, \quad -b(\vec{z}_h, s_h) \geq c_1 \|s_h\|^2 - c_3 \|(I - \Pi_0) s_h\|^2.$$

Thus, given an arbitrary pair $(\vec{w}_h, s_h)$ take $\vec{v}_h = \vec{w}_h - \delta \vec{z}_h$ and $q_h = -s_h$ with $\delta := \min(c_1, 1/c_3)$.)

5.13. Prove Proposition 5.19.

Computational exercises

Two specific flow domains are built into IFISS; the square $\Omega_\square := (-1, 1) \times (-1, 1)$ and an expanding channel $\Omega_{\sqcup} := \{(-1, 0) \times (0, 1)\} \cup \{(0, 5) \times (-1, 1)\}$. Numerical solutions to a Stokes flow problem defined on one of these domains can be computed in two steps. For the square domain, running `square_stokes` (or `pipe_stokes` to give the option of a Neumann condition on the right-hand boundary) sets up the requisite finite element matrices. For the expanding domain, running `step_stokes` does the same job. Then running either `solve_stokes` or `solve_step_stokes` generates a flow solution associated with the velocity boundary data in the M-file `../stokes_flow/specific_flow`. The associated stream function solution is computed by solving a Poisson problem with the boundary data in the M-file `../diffusion/specific_bc`. Running the driver `stokes_testproblem` automatically sets up the data files `specific_flow` and `specific_bc` associated with the problems in Examples 5.1.1–5.1.4.

5.1. Compute the divergence matrix B in the case of an enclosed flow problem, as in Example 5.1.4, discretized using unstabilized $\boldsymbol{Q}_1$–$\boldsymbol{Q}_1$ approximation. Take a sequence of 4×4, 8×8, and 16×16 square grids, and use the matlab `svd` function to show that the null space of B^T is eight-dimensional. Then, repeat the same exercise with a natural outflow condition on one of the boundary segments as in Example 5.1.1. Is the unstabilized $\boldsymbol{Q}_1$–$\boldsymbol{Q}_1$ approximation method viable in this case?

5.2. Write a function that computes the stabilized $\boldsymbol{Q}_1$–$\boldsymbol{Q}_1$ solution satisfying the system (5.88). Use your function to solve the regularized driven cavity problem in Example 5.1.3 on a 32×32 square grid. Take values of β in the interval $[10^{-3}, 10^3]$ and compare the solutions obtained with a reference solution computed using $\boldsymbol{Q}_2$–$\boldsymbol{P}_{-1}$ approximation. Having determined a "best" parameter value, solve the cavity flow problem using the 16×16, 32×32 and 64×64 stretched grid sequence that is generated by IFISS. Compare results with those obtained using the pressure projection stabilized $\boldsymbol{Q}_1$–$\boldsymbol{Q}_1$ approximation and the same grid sequence.

5.3. Verify that the $\boldsymbol{Q}_2$–$\boldsymbol{Q}_1$ solution of Example 5.1.1 is identical to the *exact* flow solution (5.12). Take a sequence of uniform square grids. Repeat this experiment using the $\boldsymbol{Q}_2$–$\boldsymbol{P}_{-1}$ and $\boldsymbol{Q}_2$–$\boldsymbol{P}_0$ methods instead. Can you explain the difference in the observed behavior?

5.4. Consider solving the Poiseuille flow problem in Example 5.1.1 using the default stabilized $\boldsymbol{Q}_1$–$\boldsymbol{P}_0$ and $\boldsymbol{Q}_1$–$\boldsymbol{Q}_1$ methods. Take a sequence of uniform square grids, and compute the exact velocity error in each case. Repeat this experiment using the unstabilized $\boldsymbol{Q}_1$–$\boldsymbol{P}_0$ method (i.e. set the stabilization parameter β to zero). Is there any difference in the behavior?

5.5. Consider solving the watertight driven cavity flow problem in Example 5.1.3 using the default stabilized $\boldsymbol{Q}_1$–$\boldsymbol{P}_0$ method. Tabulate the estimated error η for a sequence of uniform square grids, and hence estimate the rate of convergence. Compare your results with those in Table 5.5. Can you explain the difference in the observed convergence rates?

5.6. This is an important exercise. It explores the issue of local mass conservation using the step flow problem in Example 5.1.2. The function `flowvolume` postprocesses a flow solution and computes the volume of fluid crossing a specified vertical grid line. Take the default 16×48 grid and use `flowvolume` to generate horizontal velocity profiles at $x = -1$, $x = 0$ and $x = 5$ for the default stabilized $\boldsymbol{Q}_1$–$\boldsymbol{P}_0$ and $\boldsymbol{Q}_1$–$\boldsymbol{Q}_1$ methods. Compare the results with those obtained using the higher order $\boldsymbol{Q}_2$–$\boldsymbol{P}_{-1}$ and $\boldsymbol{Q}_2$–$\boldsymbol{Q}_1$ methods.

5.7. This exercise builds on Exercise 5.6. Consider a flow problem defined on the contracting channel domain featured in Exercise 1.5. A Poiseuille inflow profile and a natural outflow can be realized by running `quad_stokes` with boundary conditions given in functions `quad_flow` and `quad_bc`. Using `flowvolume`, compare the computed horizontal velocity profiles at $x = 0$, $x = 1$ and $x = 4$ using $\boldsymbol{Q}_2$–$\boldsymbol{P}_{-1}$ approximation and the default quadrilateral element grid. Contrast the results with those obtained using the nonconservative $\boldsymbol{Q}_2$–$\boldsymbol{Q}_1$ approximation method.

5.8. Write a function that computes the eigenvalues satisfying the *alternative* inf–sup eigenvalue problem,

$$B \mathbf{Q}^{-1} B^T \mathbf{q} = \sigma A_p \mathbf{q},$$

see (5.172), in the context of $\boldsymbol{Q}_2$–$\boldsymbol{Q}_1$ approximation. Tabulate the extremal eigenvalues arising in the case of the cavity flow problem in Example 5.1.3 on a sequence of uniformly refined grids. Then, repeat the exercise using unstabilized $\boldsymbol{Q}_1$–$\boldsymbol{Q}_1$ approximation and compare the results.

5.9. Write a function that computes the extremal eigenvalues of the inf–sup eigenvalue problem,

$$(B \mathbf{A}^{-1} B^T + C) \mathbf{q} = \sigma Q \mathbf{q},$$

see (5.183). Tabulate the extremal eigenvalues arising from $\boldsymbol{Q}_1$–$\boldsymbol{P}_0$ approximation of the flow problem in Example 5.1.2 on a sequence of uniformly refined grids. Then, take the unstabilized approximation and compare results using the same grid sequence.

6

SOLUTION OF DISCRETE STOKES PROBLEMS

The *inf–sup* stability condition associated with mixed approximation of the Stokes problem is of central importance when it comes to finding fast and reliable iterative solution methods. Indeed, although it is not always necessary to use stable approximations to compute an accurate velocity field, it is *crucial* for the construction of fast convergent iterative methods. This is the subject of this chapter.

Whatever solution strategy is employed, the key issue for the discrete Stokes problem is indefiniteness. The Galerkin matrix

$$K = \begin{bmatrix} \mathbf{A} & B^T \\ B & -C \end{bmatrix} \tag{6.1}$$

(see (5.150) or (5.174)) is symmetric, reflecting the self-adjointness of the continuous Stokes operator, but always *indefinite*: it has both positive and negative eigenvalues. To see this, apply the Sylvester Law of Inertia [87, p. 403] to the *congruence transform*[1] of the Galerkin matrix (6.1)

$$\begin{bmatrix} \mathbf{A} & B^T \\ B & -C \end{bmatrix} = \begin{bmatrix} I & 0 \\ B\mathbf{A}^{-1} & I \end{bmatrix} \begin{bmatrix} \mathbf{A} & 0 \\ 0 & -(B\mathbf{A}^{-1}B^T + C) \end{bmatrix} \begin{bmatrix} I & \mathbf{A}^{-1}B^T \\ 0 & I \end{bmatrix} \tag{6.2}$$

to reveal that K has n_u positive eigenvalues and at least $n_p - 1$ negative eigenvalues for both a uniformly stable ($C = 0$) or a stabilized method. For an enclosed flow, the hydrostatic pressure $\mathbf{u} = \mathbf{0}$, $\mathbf{p} = \mathbf{1}$ lies in the kernel of K and so there will be $n_p - 1$ negative eigenvalues of the negative Schur complement $-(B\mathbf{A}^{-1}B^T + C)$; for a non-enclosed flow, there will be n_p negative eigenvalues, since then $-(B\mathbf{A}^{-1}B^T + C)$ is negative-definite (see (5.188)). Here, as in Chapter 5, n_u is the total number of velocity and n_p the total number of pressure degrees of freedom.

The indefiniteness of the Stokes system is fundamental. If the mesh size h is reduced and the discrete problem size correspondingly increased, then both the number of positive eigenvalues and the number of negative eigenvalues increase; some authors refer to such systems as being highly (or strongly) indefinite.

Our concern in this chapter is preconditioned Krylov subspace methods for the discrete Stokes equations, with preconditioners derived from efficient solution to certain subsidiary problems on the velocity and pressure spaces. Although

---

[1]Symmetric matrices R and S are congruent if $R = XSX^T$ for any non-singular matrix X. The Sylvester Law of Inertia says that congruent matrices have the same number of positive, zero and negative eigenvalues.

indefiniteness can cause the CG method to fail, there are good alternatives for symmetric indefinite matrices. The MINRES method (Algorithm 2.4) is robust and is the method of choice. Even though CG may work for some symmetric indefinite systems, it is not robust in general and the two extra vector operations that MINRES requires in an iteration is a very small price to pay for a robust algorithm. MINRES is a built-in algorithm in `matlab`.

For multigrid methods applied directly to (6.1), there is the immediate difficulty that simple iterations such as the Jacobi and Gauss–Seidel methods are not well-defined because of zero diagonals. For certain mixed approximations, incomplete factorizations can be employed as smoothers, but the power of multigrid can be used in a different way. The simple multigrid algorithm for the Poisson problem, Algorithm 2.7 in Chapter 2, can form part of a very effective preconditioner for the discrete Stokes problem.

6.1   The preconditioned MINRES method

In the generic context of solving a symmetric and indefinite matrix system

$$K\mathbf{x} = \mathbf{b}, \tag{6.3}$$

MINRES is characterized as follows (see Section 2.4 and especially (2.38)). The kth iterate $\mathbf{x}^{(k)}$ lies in the translated Krylov subspace

$$\mathbf{x}^{(0)} + \text{span}\{\mathbf{r}^{(0)}, K\mathbf{r}^{(0)}, K^2\mathbf{r}^{(0)}, \ldots, K^{k-1}\mathbf{r}^{(0)}\},$$

where $\mathbf{r}^{(0)} = \mathbf{b} - K\mathbf{x}^{(0)}$ is the initial residual. Moreover, the kth residual vector satisfies

$$\mathbf{r}^{(k)} \in \mathbf{r}^{(0)} + \text{span}\{K\mathbf{r}^{(0)}, K^2\mathbf{r}^{(0)}, \ldots, K^k\mathbf{r}^{(0)}\}, \tag{6.4}$$

the defining condition being that the Euclidean norm $\|\mathbf{r}^{(k)}\|$ is minimal over all vectors from this space. As in Chapter 2 with CG (see (2.9)) we describe vectors in the shifted (or affine) Krylov subspace (6.4) in terms of polynomials in K operating on the initial residual $\mathbf{r}^{(0)}$. If Π_k is the set of real polynomials of degree less than or equal to k, then the residual vectors defined by the MINRES iterates satisfy $\mathbf{r}^{(k)} = p_k(K)\mathbf{r}^{(0)}$, where $p_k \in \Pi_k$, $p_k(0) = 1$, and p_k is optimal in the sense that $\|\mathbf{r}^{(k)}\|$ is minimal. Expanding in terms of the eigenvectors $\mathbf{w}_j$ of K gives

$$\mathbf{r}^{(0)} = \sum_j \alpha_j \mathbf{w}_j, \quad K\mathbf{w}_j = \lambda_j \mathbf{w}_j$$

we have

$$\mathbf{r}^{(k)} = p_k(K) \sum_j \alpha_j \mathbf{w}_j = \sum_j \alpha_j p_k(\lambda_j) \mathbf{w}_j$$

so that

$$\|\mathbf{r}^{(k)}\| = \min_{p_k \in \Pi_k, p_k(0)=1} \left\| \sum \alpha_j p_k(\lambda_i) \mathbf{w}_j \right\|$$

$$= \min_{p_k \in \Pi_k, p_k(0)=1} \left(\sum \alpha_j^2 p_k(\lambda_j)^2 \langle \mathbf{w}_j, \mathbf{w}_j \rangle \right)^{1/2}$$

$$\leq \min_{p_k \in \Pi_k, p_k(0)=1} \max_j |p_k(\lambda_j)| \left(\sum \alpha_j^2 \langle \mathbf{w}_j, \mathbf{w}_j \rangle \right)^{1/2},$$

or

$$\|\mathbf{r}^{(k)}\| \leq \min_{p_k \in \Pi_k, p_k(0)=1} \max_j |p_k(\lambda_j)| \, \|\mathbf{r}^{(0)}\|. \tag{6.5}$$

Note the similarity with the convergence estimate (2.11) for CG: symmetry of the matrix ensures orthogonality of the eigenvectors in both cases, with the consequence that convergence depends only on the eigenvalues. This is the important property that is lost with nonsymmetric matrices. It is also apparent that for MINRES it is the Euclidean norm of the residual that is reduced in an optimal manner rather than the **A**-norm, which is not defined here (see Theorem 2.2 and (2.10)).

The key difference here, however, is that indefinite matrices have both positive and negative real eigenvalues $\{\lambda_j\}$. Thus the polynomial that achieves the minimum in (6.5) has to take the value of unity at the origin and yet be small at points on the real axis on either side. Even if an inclusion set $[-a, -b] \cup [c, d]$, $a, b, c, d > 0$ can be found for the eigenvalues, the optimal polynomial is not as accessible here as in the positive-definite case. We will return to convergence analysis in Section 6.2.4, but it is sufficient to realize at this stage that the number of iterations required for convergence for an indefinite systems must be more than for a positive definite system with the same positive eigenvalues. Thus, in order to achieve rapid convergence, preconditioning will be at least as important here as in the symmetric positive-definite case.

It is desirable to ensure that any preconditioner does not destroy the symmetry of the discrete Stokes problem; otherwise iterative methods for nonsymmetric systems would have to be employed as for discrete convection–diffusion problems. To preserve symmetry in the preconditioned system, a symmetric and positive-definite preconditioner $M = HH^T$ is required.

The symmetric system

$$H^{-1}KH^{-T}\mathbf{y} = H^{-1}\mathbf{b}, \quad \mathbf{y} = H^T\mathbf{x}. \tag{6.6}$$

clearly has the same solution as (6.3) and since $H^{-1}KH^{-T}$ is a congruence transform on K, the coefficient matrix has the same number of positive, zero and negative eigenvalues as K by the Sylvester Law of Inertia.

For any approximation $\mathbf{x}^{(k)}$ to the solution to (6.3), the corresponding residual for the preconditioned system is

$$H^{-1}(\mathbf{b} - K\mathbf{x}^{(k)}) = H^{-1}\mathbf{r}^{(k)} = H^T\mathbf{z}^{(k)},$$

where $\mathbf{r}^{(k)}$ is the residual for the original system (6.3) and $\mathbf{z}^{(k)} = M^{-1}\mathbf{r}^{(k)}$ is the preconditioned residual. Thus, it is

$$\|H^{-1}\mathbf{r}^{(k)}\| = \|\mathbf{r}^{(k)}\|_{M^{-1}}$$

that is minimized over

$$H^{-1}\left(\mathbf{r}^{(0)} + \mathrm{span}\{KM^{-1}\mathbf{r}^{(0)}, (KM^{-1})^2\mathbf{r}^{(0)}, \dots, (KM^{-1})^k\mathbf{r}^{(0)}\}\right)$$

when the positive-definite preconditioner M is employed with the MINRES method.

The preconditioned MINRES convergence estimate becomes

$$\frac{\|\mathbf{r}_k\|_{M^{-1}}}{\|\mathbf{r}_0\|_{M^{-1}}} \leq \min_{p_k \in \Pi_k, p_k(0)=1} \max_\lambda |p_k(\lambda)|, \tag{6.7}$$

where the maximum is over eigenvalues λ of $M^{-1}K$. The role of preconditioning in this case is to cluster both the positive and the negative eigenvalues so that the polynomial approximation error in (6.7) is acceptably small after a few iterations. In view of the similarity transformation $H^{-T}(H^{-1}KH^{-T})H^T = M^{-1}K$, it is evident that, exactly as in the discussion of CG (see Section 2.2), reference to H is an artifact of the analysis. In practical computation, only M, or, more precisely, only a procedure for evaluating the action of M^{-1} is required.

Note also that a positive-definite preconditioner is needed in order for $\|\cdot\|_{M^{-1}}$ to define a norm. Unlike in the case of the conjugate gradient method, however, reduction of the residual in the preconditioned MINRES algorithm is in a norm that is dependent on the preconditioner. Thus one must be careful not to select a preconditioner that simply distorts this norm.

Writing the MINRES algorithm for (6.6) in terms of vectors associated with (6.3) leads to the following algorithm:

Algorithm 6.1: THE PRECONDITIONED MINRES METHOD
$\mathbf{v}^{(0)} = \mathbf{0}, \mathbf{w}^{(0)} = \mathbf{0}, \mathbf{w}^{(1)} = \mathbf{0}$
Choose $\mathbf{u}^{(0)}$, compute $\mathbf{v}^{(1)} = \mathbf{f} - A\mathbf{u}^{(0)}$
Solve $M\mathbf{z}^{(1)} = \mathbf{v}^{(1)}$, set $\gamma_1 = \sqrt{\langle \mathbf{z}^{(1)}, \mathbf{v}^{(1)} \rangle}$
Set $\eta = \gamma_1, s_0 = s_1 = 0, c_0 = c_1 = 1$
for $j = 1$ until convergence do
$\quad \mathbf{z}^{(j)} = \mathbf{z}^{(j)}/\gamma_j$
$\quad \delta_j = \langle A\mathbf{z}^{(j)}, \mathbf{z}^{(j)} \rangle$
$\quad \mathbf{v}^{(j+1)} = A\mathbf{z}^{(j)} - (\delta_j/\gamma_j)\mathbf{v}^{(j)} - (\gamma_j/\gamma_{j-1})\mathbf{v}^{(j-1)}$
$\quad$ solve $M\mathbf{z}^{(j+1)} = \mathbf{v}^{(j+1)}$
$\quad \gamma_{j+1} = \sqrt{\langle \mathbf{z}^{(j+1)}, \mathbf{v}^{(j+1)} \rangle}$
$\quad \alpha_0 = c_j\delta_j - c_{j-1}s_j\gamma_j$
$\quad \alpha_1 = \sqrt{\alpha_0^2 + \gamma_{j+1}^2}$
$\quad \alpha_2 = s_j\delta_j + c_{j-1}c_j\gamma_j$
$\quad \alpha_3 = s_{j-1}\gamma_j$
$\quad c_{j+1} = \alpha_0/\alpha_1; \; s_{j+1} = \gamma_{j+1}/\alpha_1$
$\quad \mathbf{w}^{(j+1)} = (\mathbf{z}^{(j)} - \alpha_3\mathbf{w}^{(j-1)} - \alpha_2\mathbf{w}^{(j)})/\alpha_1$
$\quad \mathbf{u}^{(j)} = \mathbf{u}^{(j-1)} + c_{j+1}\eta\mathbf{w}^{(j+1)}$
$\quad \eta = -s_{j+1}\eta$
$\quad$<Test for convergence>
enddo

A preconditioner M that requires $\mathcal{O}(n)$ work for the solution of $M\mathbf{z} = \mathbf{v}$ and for which the MINRES method yields the solution in a number of iterations independent of n will thus provide an optimal solver for an indefinite matrix system. This is analogous to the situation for positive-definite matrix systems and preconditioned CG.

6.2   Preconditioning

We now fix attention on the generic Stokes matrix system

$$\begin{bmatrix} \mathbf{A} & B^T \\ B & -C \end{bmatrix} \begin{bmatrix} \mathbf{u} \\ \mathbf{p} \end{bmatrix} = \begin{bmatrix} \mathbf{f} \\ \mathbf{g} \end{bmatrix}, \tag{6.8}$$

where the matrix C is zero in the case of inf–sup stable mixed approximation. The vector Laplacian matrix $\mathbf{A}$ is always positive definite (see (5.156)) and is of dimension $n_u \times n_u$.

Except where specifically mentioned, our spectral analysis applies for both stabilized ($C \neq 0$) *and* stable mixed approximation. Indeed, setting $C = 0$, the discrete stability conditions (5.184) and (5.161) can both be written in the generic form

$$\gamma^2 \leq \frac{\langle (B\mathbf{A}^{-1}B^T + C)\mathbf{q}, \mathbf{q} \rangle}{\langle Q\mathbf{q}, \mathbf{q} \rangle} \quad \text{for all } \mathbf{q} \in \mathbb{R}^{n_p} \text{ such that } \mathbf{q} \neq \mathbf{1}, \quad (6.9)$$

where γ is the (generalized) inf–sup constant, and Q is the positive-definite pressure mass matrix (5.154). In the case of stable mixed approximation, we also have the alternative characterization

$$\gamma^2 \leq \frac{\langle B^T Q^{-1} B \mathbf{v}, \mathbf{v} \rangle}{\langle \mathbf{A}\mathbf{v}, \mathbf{v} \rangle} \quad \text{for all } \mathbf{v} \in \mathbb{R}^{n_u} \text{ for which } \langle \mathbf{A}\mathbf{v}, \mathbf{u} \rangle = 0 \text{ for } \mathbf{u} \in \text{null}(B).$$
$$(6.10)$$

The upper bound on the eigenvalues of the scaled pressure Schur complement is given by (5.163) or (5.168):

$$\frac{\langle B\mathbf{A}^{-1}B^T \mathbf{q}, \mathbf{q} \rangle}{\langle Q\mathbf{q}, \mathbf{q} \rangle} \leq \Gamma^2 \quad \text{for all } \mathbf{q} \in \mathbb{R}^{n_p}, \quad (6.11)$$

where the boundedness constant $\Gamma < \sqrt{d}$ for a flow problem in d space dimensions. Using (5.159), we can also express this bound in the form

$$\frac{\langle B^T Q^{-1} B \mathbf{v}, \mathbf{v} \rangle}{\langle \mathbf{A}\mathbf{v}, \mathbf{v} \rangle} \leq \Gamma^2 \quad \text{for all } \mathbf{v} \in \mathbb{R}^{n_u}. \quad (6.12)$$

Note the presence of the hydrostatic pressure $\mathbf{q} = \mathbf{1}$ in the null space of the Schur complement for enclosed flow problems: the corresponding vector $[\mathbf{0}^T, \mathbf{1}^T]^T$ is associated with a zero eigenvalue of the Stokes coefficient matrix. In the simpler case of non-enclosed flow the restriction in (6.9) that $\mathbf{q} \neq \mathbf{1}$ can be dropped.

Remark 6.1. A different way to handle the system (6.8) is to multiply the second block equation by -1, producing the nonsymmetric linear system

$$\begin{bmatrix} \mathbf{A} & B^T \\ -B & C \end{bmatrix} \begin{bmatrix} \mathbf{u} \\ \mathbf{p} \end{bmatrix} = \begin{bmatrix} \mathbf{f} \\ -\mathbf{g} \end{bmatrix}. \quad (6.13)$$

This is clearly equivalent to (6.8). One possible advantage here is that the coefficient matrix in (6.13) has a positive semi-definite symmetric part, since

$$\begin{bmatrix} \mathbf{u}^T & \mathbf{p}^T \end{bmatrix} \begin{bmatrix} \mathbf{A} & B^T \\ -B & C \end{bmatrix} \begin{bmatrix} \mathbf{u} \\ \mathbf{p} \end{bmatrix} = \langle \mathbf{A}\mathbf{u}, \mathbf{u} \rangle + \langle C\mathbf{p}, \mathbf{p} \rangle.$$

The disadvantage is that iterative methods designed for symmetric matrices cannot be used in general. In particular, the option of using an optimal iterative method with fixed computational cost per iteration, such as MINRES, is lost. See Computational Exercise 6.7 for an exploration of these issues.

6.2.1   General strategies for preconditioning

For the Stokes problem, discretization error is measured in the energy norm for velocities and in the L_2 norm for pressure (see Section 5.4). Therefore, the natural matrix norm is $\|\mathbf{e}^{(k)}\|_E$ where

$$E = \begin{bmatrix} \mathbf{A} & 0 \\ 0 & Q \end{bmatrix} \tag{6.14}$$

(see (5.151), (5.152) and Problem 6.1). Since $K\mathbf{e}^{(k)} = \mathbf{r}^{(k)}$, in terms of the residual this is

$$\|\mathbf{e}^{(k)}\|_E^2 = \langle EK^{-1}\mathbf{r}^{(k)}, K^{-1}\mathbf{r}^{(k)} \rangle = \|\mathbf{r}^{(k)}\|_{K^{-1}EK^{-1}}^2 .$$

For the Stokes problem, with coefficient matrix (6.1), the relevant matrix is

$$K^{-1}EK^{-1} = (KE^{-1}K)^{-1} = \left(\begin{bmatrix} \mathbf{A} & B^T \\ B & -C \end{bmatrix} \begin{bmatrix} \mathbf{A}^{-1} & 0 \\ 0 & Q^{-1} \end{bmatrix} \begin{bmatrix} \mathbf{A} & B^T \\ B & -C \end{bmatrix} \right)^{-1}$$

$$= \begin{bmatrix} \mathbf{A} + B^TQ^{-1}B & B^T - B^TQ^{-1}C \\ B - CQ^{-1}B & B\mathbf{A}^{-1}B^T + CQ^{-1}C \end{bmatrix}^{-1} . \tag{6.15}$$

Since it is $\|\mathbf{r}^{(k)}\|_{M^{-1}}$ that is reduced by the MINRES method, it would appear that a good choice of preconditioner is the positive-definite matrix

$$M = \begin{bmatrix} \mathbf{A} + B^TQ^{-1}B & B^T - B^TQ^{-1}C \\ B - CQ^{-1}B & B\mathbf{A}^{-1}B^T + CQ^{-1}C \end{bmatrix} . \tag{6.16}$$

For uniformly stabilized approximation ($C = 0$), this has the form

$$M = \begin{bmatrix} \mathbf{A} + B^TQ^{-1}B & B^T \\ B & B\mathbf{A}^{-1}B^T \end{bmatrix} . \tag{6.17}$$

Here, the presence of the two relevant Schur complements for stability (see (6.9) and (6.10)) are evident. Notice that $B\mathbf{A}^{-1}B^T$ is a discrete operator representing $\nabla \cdot (\nabla^2)^{-1}\nabla$, and $B^TQ^{-1}B$ represents $\nabla(\nabla\cdot)$ on the vector of velocity components. It is clear, however, that these matrix operators are not suitable as components of a preconditioner for the Stokes system, since they do not satisfy the requirement concerning ease of solution of systems of the form $M\mathbf{z} = \mathbf{r}$.

We will now derive some effective and practical preconditioners in this setting. In Section 6.2.3 we will also show that under appropriate circumstances, the resulting strategies are in fact essentially as good as (6.16)–(6.17) with respect to the norm being minimized by MINRES.

The form of the Galerkin matrix (6.1) and the desired norm based on the matrix (6.14) suggests that it is important to take account of the block structure when preconditioning. We thus consider block diagonal preconditioning matrices of the form

$$M = \begin{bmatrix} \mathbf{P} & 0 \\ 0 & T \end{bmatrix}, \tag{6.18}$$

where both $\mathbf{P} \in \mathbb{R}^{n_u \times n_u}$ and $T \in \mathbb{R}^{n_p \times n_p}$ are symmetric and positive-definite. The convergence bound (6.7) then indicates that the speed of MINRES convergence depends on the eigenvalues λ of the generalized eigenvalue problem

$$\begin{bmatrix} \mathbf{A} & B^T \\ B & -C \end{bmatrix} \begin{bmatrix} \mathbf{u} \\ \mathbf{p} \end{bmatrix} = \lambda \begin{bmatrix} \mathbf{P} & 0 \\ 0 & T \end{bmatrix} \begin{bmatrix} \mathbf{u} \\ \mathbf{p} \end{bmatrix}. \tag{6.19}$$

It is readily seen that if $\mathbf{P} = \mathbf{A}$, then $\lambda = 1$ is an eigenvalue of multiplicity at least $n_u - n_p$ corresponding to any eigenvector $[\mathbf{u}^T, \mathbf{0}^T]^T$ with $B\mathbf{u} = 0$. The multiplicity comes simply from the size of the right null space of the rectangular matrix B; thus if B is of full rank, n_p, then the multiplicity is exactly $n_u - n_p$. In the uniformly stable case ($C = 0$), if also $T = B\mathbf{A}^{-1}B^T$, then the remaining eigenvalues satisfy,

$$(1 - \lambda)\mathbf{A}\mathbf{u} = -B^T\mathbf{p} \quad \text{and} \quad B\mathbf{u} = \lambda B\mathbf{A}^{-1}B^T\mathbf{p}$$

or by eliminating $\mathbf{u}$,

$$(\lambda^2 - \lambda - 1)B\mathbf{A}^{-1}B^T\mathbf{p} = 0.$$

Thus, since the assumed *inf–sup* stability in this case ensures that $B\mathbf{A}^{-1}B^T$ is positive-definite, we deduce that $\lambda = 1/2 \pm \sqrt{5}/2$ are the remaining eigenvalues, each with multiplicity n_p. This is an ideal situation from the point of view of convergence of MINRES — since the preconditioned matrix has only three distinct eigenvalues, there is a cubic polynomial with these three roots, and the convergence bound (6.7) will be zero for $k = 3$. That is, MINRES will terminate with the exact solution after three iterations irrespective of the size of the discrete problem.

This is an idealized situation, since the preconditioning operation with (6.18) requires the action of the inverses of $\mathbf{A}$ and of the Schur complement $B\mathbf{A}^{-1}B^T + C$. Three iterations require three such computations. The operation with the Schur complement is completely impractical since this is a full matrix. Note, moreover, that the congruence transform (6.2) would allow *direct solution* of this problem with two such operations with $\mathbf{A}$ and one with the Schur complement. However, this special choice of M suggests what is really needed, namely, a suitably chosen $\mathbf{P}$ that approximates $\mathbf{A}$, and a suitable T to approximate the Schur complement $B\mathbf{A}^{-1}B^T + C$.

We continue to consider the uniformly stable case for the moment, though our analysis in Section 6.2.2 covers also the stabilized case. The key to handling the Schur complement is provided by the stability condition (6.9) together with the boundedness condition (6.11): the sparse pressure mass matrix Q is spectrally equivalent to the dense matrix $B\mathbf{A}^{-1}B^T$ so that there is a lot to gain by selecting $T = Q$. The analysis with this choice of T is similar to that given above. Start with

$$\mathbf{Au} + B^T \mathbf{p} = \lambda \mathbf{Au}$$
$$B\mathbf{u} = \lambda Q\mathbf{p}.$$

The case $\lambda = 1$ arises with the same eigenvectors (and thus multiplicity) as above, and for $\lambda \neq 1$, eliminating $\mathbf{u}$ using the first of these equations gives

$$B\mathbf{A}^{-1}B^T \mathbf{p} = \mu Q\mathbf{p},$$

where $\mu = \lambda(\lambda - 1)$. For each eigenvalue μ, there is a pair of eigenvalues

$$\lambda = \tfrac{1}{2} - \tfrac{1}{2}\sqrt{1 + 4\mu} < 0 \quad \text{and} \quad \lambda = \tfrac{1}{2} + \tfrac{1}{2}\sqrt{1 + 4\mu} > 1$$

of the original problem. Now, since discrete *inf–sup* stability and boundedness imply $\gamma^2 \leq \mu \leq \Gamma^2$, every eigenvalue λ lies in

$$\left[\frac{1 - \sqrt{1 + 4\Gamma^2}}{2}, \frac{1 - \sqrt{1 + 4\gamma^2}}{2}\right] \cup \{1\} \cup \left[\frac{1 + \sqrt{1 + 4\gamma^2}}{2}, \frac{1 + \sqrt{1 + 4\Gamma^2}}{2}\right].$$

That is, the multiple eigenvalue $\lambda = 1$ is retained and the remaining eigenvalues are pairwise symmetric about $\tfrac{1}{2}$ and lie in small intervals that are uniformly bounded from $\pm\infty$ and uniformly bounded away from the origin. In this situation, MINRES will not terminate in three iterations, but nevertheless convergence will be fast and, crucially, the number of MINRES iterations needed will be independent of the size of the discrete problem. This observation holds simply because there is nothing depending on h in the convergence bound (6.7). For an unstable approximation and in the absence of stabilization, we have $\gamma = 0$ (or $\gamma \to 0$ under mesh refinement). This means that the negative eigenvalues would not be bounded away from the origin, leading to slower and slower convergence as h is reduced.

Before proceeding to more general theory, we motivate other approximations that will preserve the form of this "ideal" preconditioner but lead to a more practical preconditioner. It is apparent that block preconditioning as above requires at each MINRES iteration the solution of two systems of equations of size n_u and n_p, with coefficient matrices $\mathbf{P}$ and T respectively. The "ideal" choice $\mathbf{P} = \mathbf{A}$ requires an exact solution of a Poisson equation for each of the velocity components, since $\mathbf{A}$ is the vector Laplacian coming from approximation of the viscous terms (5.153) and (5.155). Use of a good preconditioner for the Laplacian (see Section 2.2), however, provides a practical and effective alternative approach. Moreover, it is not necessary to use an inner preconditioned conjugate gradient iteration to effect an exact solution (of the Poisson equation), but instead we are in a position to select $\mathbf{P}$ to be any convenient preconditioner for the Laplacian. $\mathbf{P}$ could, for example, be derived from domain decomposition or a multigrid cycle. That is, for a solution domain $\Omega \in \mathbb{R}^d$ we simply let $\mathbf{P}$ be a $d \times d$ block diagonal matrix each block of which is a preconditioner for the scalar Laplacian.

We will henceforth assume that

$$\delta \leq \frac{\langle \mathbf{Au}, \mathbf{u} \rangle}{\langle \mathbf{Pu}, \mathbf{u} \rangle} \leq \Delta \quad \text{for all } \mathbf{u} \in \mathbb{R}^{n_u}. \tag{6.20}$$

If we use an optimal preconditioner such as a multigrid V-cycle, then the h-independent convergence bound in Theorem 2.15 leads directly to (6.20) with δ and Δ both independent of h. This is shown in the following result.

Lemma 6.2. *If the iteration error for a simple iteration*

$$\mathbf{u}^{(i+1)} = (I - \mathbf{P}^{-1}\mathbf{A})\mathbf{u}^{(i)} + \mathbf{P}^{-1}\mathbf{f} \tag{6.21}$$

for the solution of the linear system $\mathbf{Au} = \mathbf{f}$ *satisfies*

$$\|\mathbf{u} - \mathbf{u}^{(i+1)}\|_{\mathbf{A}} \leq \rho \|\mathbf{u} - \mathbf{u}^{(i)}\|_{\mathbf{A}} \tag{6.22}$$

with the contraction factor satisfying $\rho < 1$, *then*

$$1 - \rho \leq \frac{\langle \mathbf{Av}, \mathbf{v} \rangle}{\langle \mathbf{Pv}, \mathbf{v} \rangle} \leq 1 + \rho \quad \text{for all } \mathbf{v}. \tag{6.23}$$

Proof Since the exact solution $\mathbf{u}$ certainly satisfies

$$\mathbf{u} = (I - \mathbf{P}^{-1}\mathbf{A})\mathbf{u} + \mathbf{P}^{-1}\mathbf{f}$$

(6.22) is the statement that

$$\langle \mathbf{A}(I - \mathbf{P}^{-1}\mathbf{A})\mathbf{e}, (I - \mathbf{P}^{-1}\mathbf{A})\mathbf{e} \rangle \leq \rho^2 \langle \mathbf{Ae}, \mathbf{e} \rangle$$

for any $\mathbf{e} = \mathbf{u} - \mathbf{u}^{(i)}$. Putting $\mathbf{v} = \mathbf{A}^{\frac{1}{2}}\mathbf{e}$, this can be written as

$$\langle (I - \mathbf{A}^{\frac{1}{2}}\mathbf{P}^{-1}\mathbf{A}^{\frac{1}{2}})\mathbf{v}, (I - \mathbf{A}^{\frac{1}{2}}\mathbf{P}^{-1}\mathbf{A}^{\frac{1}{2}})\mathbf{v} \rangle \leq \rho^2 \langle \mathbf{v}, \mathbf{v} \rangle$$

for all $\mathbf{v}$. Thus

$$-\rho\langle \mathbf{v}, \mathbf{v} \rangle \leq \langle (I - \mathbf{A}^{\frac{1}{2}}\mathbf{P}^{-1}\mathbf{A}^{\frac{1}{2}})\mathbf{v}, \mathbf{v} \rangle \leq \rho\langle \mathbf{v}, \mathbf{v} \rangle$$

from which it follows that

$$1 - \rho \leq \frac{\langle \mathbf{Av}, \mathbf{v} \rangle}{\langle \mathbf{Pv}, \mathbf{v} \rangle} \leq 1 + \rho$$

for all $\mathbf{v}$. $\square$

With a W-cycle preconditioner, Theorem 2.14 gives the necessary contraction bound (with $\rho = \eta_\ell$). For a V-cycle preconditioner, the numerical results in Tables 2.3 and 2.5 indicate a typical contraction rate of $\rho \approx 0.15$, so that $\delta \approx 0.85$ and $\Delta \approx 1.15$. The bottom line here is that with a single multigrid cycle to represent $\mathbf{P}$, only a few extra MINRES iterations will be required than if we were to make the more expensive choice $\mathbf{P} = \mathbf{A}$! By way of contrast, if we use a weak preconditioner such as diagonal scaling, for which $\delta = \mathcal{O}(h^2)$, $\Delta = \mathcal{O}(1)$ (see Theorem 1.119), then the number of required MINRES iterations will increase as h is reduced for any fixed convergence tolerance.

It is a much simpler matter to approximate the ideal choice of T without significantly affecting the convergence of MINRES; a simple approximation of the pressure mass matrix, Q, is all that is required. Discontinuous pressure approximation like $\mathbf{P}_0$ and $\mathbf{P}_{-1}$ is straightforward since the corresponding mass matrix is diagonal. On the other hand, for a continuous pressure approximation the choice $T = \mathtt{diag}(Q)$ still provides a uniform approximation of Q. Thus, let us assume the following bounds:

$$\theta^2 \le \frac{\langle Q\mathbf{q}, \mathbf{q}\rangle}{\langle T\mathbf{q}, \mathbf{q}\rangle} \le \Theta^2 \quad \text{for all } \mathbf{q} \in \mathbb{R}^{n_p}. \tag{6.24}$$

Notice that for the more naive choice $T = I$, Proposition 1.29 guarantees that the ratio Θ/θ will be independent of h for continuous $\mathbf{P}_1$ or $\mathbf{Q}_1$ pressure approximation. The additional benefit of choosing $T = \mathtt{diag}(Q)$ is that it is guaranteed that Θ and θ are independent of h individually. We summarize this fact in the following lemma.

Lemma 6.3. *If $\mathbf{P}_1$ pressure approximation is used on any subdivision in $\mathbb{R}^2$, and $T = \mathtt{diag}(Q)$, then (6.24) is satisfied with $\theta = 1/\sqrt{2}$ and $\Theta = \sqrt{2}$.*

Proof Direct calculation of the element mass matrix $Q^{(e)}$ for any $\mathbf{P}_1$ triangular element shows that

$$Q^{(e)} = \frac{1}{6} a^{(e)} \begin{bmatrix} 2 & 1 & 1 \\ 1 & 2 & 1 \\ 1 & 1 & 2 \end{bmatrix}, \tag{6.25}$$

where $a^{(e)}$ is the area of the triangular element.

Now, if n_{el} is the total number of elements, assembly of $\{Q^{(e)}, e = 1, \ldots, n_{el}\}$ gives the mass matrix Q. But also if $T^{(e)} := \mathtt{diag}(Q^{(e)})$ for each e, then assembly of $\{T^{(e)}, e = 1, \ldots, n_{el}\}$ gives $T = \mathtt{diag}(Q)$. This is because the diagonal entries of each $Q^{(e)}$ are precisely all of the elementwise contributions for each of the integrals $\int_\Omega \psi_i \psi_i$, $i = 1, \ldots, n_p$ where n_p is the number of nodes and $\{\psi_j, j = 1, \ldots, n_p\}$ are the nodal $\mathbf{P}_1$ basis functions. Correspondingly, all of the off-diagonal entries of each $Q^{(e)}$ are precisely all of the element contributions for each $\int_\Omega \psi_i \psi_j$, $i \ne j$! Thus

$$\min_e \frac{\langle Q^{(e)} p_e, p_e \rangle}{\langle T^{(e)} p_e, p_e \rangle} \le \frac{\langle Q\mathbf{p}, \mathbf{p} \rangle}{\langle T\mathbf{p}, \mathbf{p} \rangle} \le \max_e \frac{\langle Q^{(e)} p_e, p_e \rangle}{\langle T^{(e)} p_e, p_e \rangle}. \tag{6.26}$$

It follows from (6.25) that all diagonally scaled element mass matrices are equal to

$$
\begin{bmatrix}
1 & \frac{1}{2} & \frac{1}{2} \\
\frac{1}{2} & 1 & \frac{1}{2} \\
\frac{1}{2} & \frac{1}{2} & 1
\end{bmatrix}
\tag{6.27}
$$

independently of any geometric information about the mesh. Since the eigenvalues of (6.27) are $\frac{1}{2}$, $\frac{1}{2}$, and 2, it follows that

$$
\frac{1}{2} \le \frac{\langle Q\mathbf{p}, \mathbf{p}\rangle}{\langle T\mathbf{p}, \mathbf{p}\rangle} \le 2. \qquad\qquad \square
$$

Remark 6.4. The inequalities (6.26) hold generally for any conforming finite element approximation with the appropriate element mass matrices, see Wathen [210]. The choice $T = \text{diag}(Q)$ then leads to bounds (6.24) for any conforming pressure approximation: for P_1 (trilinear) approximation on a tetrahedral subdivision in $\mathbb{R}^3$ the bounds are $\theta = 1/\sqrt{2}$ and $\Theta = \sqrt{5/2}$; for rectangular Q_1 approximation we have that $\theta = 1/2$ and $\Theta = 3/2$; for Q_1 approximation on brick elements we have that $\theta = 1/\sqrt{8}$ and $\Theta = \sqrt{27/8}$; and for rectangular Q_2 elements we have $\theta = 1/2$ and $\Theta = 5/4$. These results will turn out to be important in developing preconditioners for the linearized Navier–Stokes equations (see Section 8.2).

6.2.2 Eigenvalue bounds

Our analysis proceeds with the assumptions (6.20), (6.24) together with the following bound on the eigenvalues of the stabilization matrix:

$$
\langle C\mathbf{q}, \mathbf{q}\rangle \le \Upsilon \langle Q\mathbf{q}, \mathbf{q}\rangle \quad \text{for all } \mathbf{q} \in \mathbb{R}^{n_p}.
\tag{6.28}
$$

Notice that $\Upsilon = 1$ in the case of *ideal* stabilization (see (5.185)). In the stable case, we have that $C = 0$ and so $\Upsilon = 0$.

Our goal is to determine upper and lower bounds on both the negative eigenvalues of (6.19) and the positive eigenvalues, for use in the MINRES convergence bound (6.7) As in the situation of using ideal (but impractical) preconditioning, it turns out that the stability of the underlying mixed approximation is vital in order to get bounds that are independent of the mesh parameter. Computations also demonstrate the necessity of ensuring stability if one is to get rapid iterative convergence. This is explored in Computational Exercise 6.3. The following lemma will be useful later.

Lemma 6.5. *If V and W are symmetric and positive-definite matrices of the same dimension, then $\langle V\mathbf{r}, \mathbf{r}\rangle \le \langle W\mathbf{r}, \mathbf{r}\rangle$ for all $\mathbf{r}$ if and only if $\langle V^{-1}\mathbf{r}, \mathbf{r}\rangle \ge \langle W^{-1}\mathbf{r}, \mathbf{r}\rangle$ for all $\mathbf{r}$.*

Proof See Problem 6.5. □

Theorem 6.6. *For the preconditioned Stokes system* (6.19), *let the inf–sup stability condition* (6.9) *and boundedness condition* (6.12) *hold, together with the spectral equivalences* (6.20) *and* (6.24) *and the ideal stabilization bound* (6.28). *Then all negative eigenvalues* λ *of* (6.19) *satisfy*

$$-\Theta^2(1 + \Upsilon) \le \lambda \tag{6.29}$$

and

$$\lambda \le \tfrac{1}{2}\left(\delta - \sqrt{\delta^2 + 4\delta\gamma^2\theta^2}\right), \tag{6.30}$$

and all positive eigenvalues λ *of* (6.19) *satisfy*

$$\delta \le \lambda \tag{6.31}$$

and

$$\lambda \le \Delta + \Gamma^2\Theta^2. \tag{6.32}$$

Proof If λ is an eigenvalue of (6.19), then there are vectors $\mathbf{u}$, $\mathbf{p}$ not both zero satisfying

$$\mathbf{Au} + B^T\mathbf{p} = \lambda\mathbf{Pu} \tag{6.33}$$
$$B\mathbf{u} - C\mathbf{p} = \lambda T\mathbf{p}. \tag{6.34}$$

First, note that if $\lambda > 0$ then $\mathbf{u} \ne \mathbf{0}$. Otherwise, if $\mathbf{u} = \mathbf{0}$, then (6.34) implies

$$-\langle C\mathbf{p}, \mathbf{p}\rangle = \lambda\langle T\mathbf{p}, \mathbf{p}\rangle.$$

But C is positive semi-definite and T is positive-definite, so this leads to the contradiction that $\mathbf{p} = \mathbf{0}$ also. Similarly, if $\lambda < 0$, then $\mathbf{p} \ne \mathbf{0}$ since otherwise (6.33) implies

$$\langle \mathbf{Au}, \mathbf{u}\rangle = \lambda\langle \mathbf{Pu}, \mathbf{u}\rangle.$$

It follows that $\mathbf{u} = \mathbf{0}$ since $\mathbf{A}$ and $\mathbf{P}$ are positive-definite.

Taking the scalar product of (6.33) with $\mathbf{u}$ and the scalar product of (6.34) with $\mathbf{p}$ and subtracting gives

$$\langle \mathbf{Au}, \mathbf{u}\rangle + \langle C\mathbf{p}, \mathbf{p}\rangle = \lambda\langle \mathbf{Pu}, \mathbf{u}\rangle - \langle T\mathbf{p}, \mathbf{p}\rangle,$$

since $\langle B^T\mathbf{p}, \mathbf{u}\rangle = \langle B\mathbf{u}, \mathbf{p}\rangle$. Now $\langle C\mathbf{p}, \mathbf{p}\rangle \ge 0$ and T is positive-definite, so

$$\langle \mathbf{Au}, \mathbf{u}\rangle - \lambda\langle \mathbf{Pu}, \mathbf{u}\rangle \le -\langle T\mathbf{p}, \mathbf{p}\rangle \le 0$$

and use of (6.20) implies

$$(\delta - \lambda)\langle \mathbf{P}\mathbf{u}, \mathbf{u}\rangle \leq 0.$$

This gives (6.31), since $\mathbf{P}$ is positive-definite and $\mathbf{u} \neq \mathbf{0}$.

Further, for $\lambda > 0$, $\lambda T + C$ must be positive-definite, so (6.34) gives

$$\mathbf{p} = (\lambda T + C)^{-1} B\mathbf{u} = \frac{1}{\lambda}\left(T + \frac{1}{\lambda}C\right)^{-1} B\mathbf{u}.$$

Substituting this in (6.33) gives

$$\mathbf{A}\mathbf{u} + \frac{1}{\lambda}B^{T}\left(T + \frac{1}{\lambda}C\right)^{-1} B\mathbf{u} = \lambda \mathbf{P}\mathbf{u}$$

and hence

$$\langle \mathbf{A}\mathbf{u}, \mathbf{u}\rangle + \frac{1}{\lambda}\left\langle \left(T + \frac{1}{\lambda}C\right)^{-1} B\mathbf{u}, B\mathbf{u}\right\rangle = \lambda\langle \mathbf{P}\mathbf{u}, \mathbf{u}\rangle. \qquad (6.35)$$

For $\lambda > 0$, since C is positive semi-definite, it follows that for all $\mathbf{r}$

$$\left\langle \left(T + \frac{1}{\lambda}C\right)\mathbf{r}, \mathbf{r}\right\rangle \geq \langle T\mathbf{r}, \mathbf{r}\rangle \geq \frac{1}{\Theta^{2}}\langle Q\mathbf{r}, \mathbf{r}\rangle,$$

where (6.24) has been used to obtain the last inequality. Thus using Lemma 6.5 we have

$$\left\langle \left(T + \frac{1}{\lambda}C\right)^{-1} B\mathbf{u}, B\mathbf{u}\right\rangle \leq \Theta^{2}\langle Q^{-1}B\mathbf{u}, B\mathbf{u}\rangle \leq \Theta^{2}\Gamma^{2}\langle \mathbf{A}\mathbf{u}, \mathbf{u}\rangle,$$

where the last inequality is a consequence of (6.12). Therefore from (6.35)

$$(1 + \Theta^{2}\Gamma^{2}/\lambda)\langle \mathbf{A}\mathbf{u}, \mathbf{u}\rangle \geq \lambda\langle \mathbf{P}\mathbf{u}, \mathbf{u}\rangle,$$

and application of (6.20) yields the quadratic inequality

$$0 \geq \lambda^{2} - \lambda\Delta - \Theta^{2}\Gamma^{2}\Delta, \qquad (6.36)$$

where we have multiplied through by λ (> 0) and used the fact that $\langle \mathbf{P}\mathbf{u}, \mathbf{u}\rangle > 0$. The inequality (6.36) implies that eigenvalues λ of (6.19) must lie in an interval between the roots of the corresponding quadratic equation. The smaller root gives a weaker bound than (6.31) and so is not useful. However, from the larger root we get

$$\lambda \leq \frac{\Delta}{2} + \frac{1}{2}\sqrt{\Delta^{2} + 4\Delta\Theta^{2}\Gamma^{2}}. \qquad (6.37)$$

This is a stronger bound than that quoted in the theorem; we can weaken it slightly by completing the square under the square root sign to give

$$\lambda \leq \frac{\Delta}{2} + \frac{1}{2}\sqrt{\Delta^2 + 4\Delta\Theta^2\Gamma^2 + 4\Theta^4\Gamma^4}$$

$$= \frac{\Delta}{2} + \frac{1}{2}\left(\Delta + 2\Theta^2\Gamma^2\right)$$

$$= \Delta + \Theta^2\Gamma^2,$$

which is the simpler bound (6.32).

For $\lambda < 0$, $\mathbf{A} - \lambda\mathbf{P}$ is certainly positive-definite and so (6.33) gives $\mathbf{u} = -(\mathbf{A} - \lambda\mathbf{P})^{-1}B^T\mathbf{p}$. Using this to eliminate $\mathbf{u}$ in (6.34) and taking the inner product with $\mathbf{p}$ leads to

$$\langle(\mathbf{A} - \lambda\mathbf{P})^{-1}B^T\mathbf{p}, B^T\mathbf{p}\rangle + \langle C\mathbf{p}, \mathbf{p}\rangle = -\lambda\langle T\mathbf{p}, \mathbf{p}\rangle. \tag{6.38}$$

Now $\langle(\mathbf{A} - \lambda\mathbf{P})\mathbf{r}, \mathbf{r}\rangle \geq \langle\mathbf{A}\mathbf{r}, \mathbf{r}\rangle$ for negative λ and any $\mathbf{r}$, so applying Lemma 6.5 to the first term in (6.38) gives

$$\langle\mathbf{A}^{-1}B^T\mathbf{p}, B^T\mathbf{p}\rangle + \langle C\mathbf{p}, \mathbf{p}\rangle \geq -\lambda\langle T\mathbf{p}, \mathbf{p}\rangle.$$

Further use of (6.9), (6.28) and (6.24) to express each contribution in terms of the mass matrix gives

$$\Gamma^2\langle Q\mathbf{p}, \mathbf{p}\rangle + \Upsilon\langle Q\mathbf{p}, \mathbf{p}\rangle \geq -\lambda\Theta^{-2}\langle Q\mathbf{p}, \mathbf{p}\rangle$$

from which (6.29) readily follows since $\langle Q\mathbf{p}, \mathbf{p}\rangle > 0$.

Finally, for the upper bound (6.30) on the negative eigenvalues, note that for any $\mathbf{r}$, (6.20) implies

$$\langle(\mathbf{A} - \lambda\mathbf{P})\mathbf{r}, \mathbf{r}\rangle = \langle\mathbf{A}\mathbf{r}, \mathbf{r}\rangle - \lambda\langle\mathbf{P}\mathbf{r}, \mathbf{r}\rangle$$

$$\leq \langle\mathbf{A}\mathbf{r}, \mathbf{r}\rangle - \frac{\lambda}{\delta}\langle\mathbf{A}\mathbf{r}, \mathbf{r}\rangle,$$

so that application of Lemma 6.5 gives for any $\mathbf{r}$

$$\langle(\mathbf{A} - \lambda\mathbf{P})^{-1}\mathbf{r}, \mathbf{r}\rangle \geq \frac{1}{(1 - \lambda/\delta)}\langle\mathbf{A}^{-1}\mathbf{r}, \mathbf{r}\rangle. \tag{6.39}$$

Use of (6.39) in (6.38) then gives

$$\frac{1}{(1 - \lambda/\delta)}\langle\mathbf{A}^{-1}B^T\mathbf{p}, B^T\mathbf{p}\rangle + \langle C\mathbf{p}, \mathbf{p}\rangle \leq -\lambda\langle T\mathbf{p}, \mathbf{p}\rangle. \tag{6.40}$$

Since $0 < (1 - \lambda/\delta)^{-1} < 1$, we can write

$$\frac{1}{(1 - \lambda/\delta)}\left(\langle B\mathbf{A}^{-1}B^T\mathbf{p}, \mathbf{p}\rangle + \langle C\mathbf{p}, \mathbf{p}\rangle\right) \leq -\lambda\langle T\mathbf{p}, \mathbf{p}\rangle \tag{6.41}$$

and so using the stability statement (6.9), we find that except for the hydrostatic pressure

$$\frac{1}{(1 - \lambda/\delta)}\gamma^2\langle Q\mathbf{p}, \mathbf{p}\rangle \leq -\lambda\langle T\mathbf{p}, \mathbf{p}\rangle \leq -\lambda\theta^{-2}\langle Q\mathbf{p}, \mathbf{p}\rangle \qquad (6.42)$$

which rearranges to the quadratic inequality

$$0 \leq \lambda^2 - \lambda\delta - \gamma^2\theta^2\delta. \qquad (6.43)$$

The useful upper bound (6.30) on the negative eigenvalues comes from considering the negative root of the corresponding quadratic equation. □

We emphasize that by simply taking $C = 0$ above so that $\Upsilon = 0$, the case of uniformly stable mixed finite element approximation is included.

Corollary 6.7. *For a uniformly stable mixed finite element approximation, under the conditions of the above theorem, the eigenvalues λ of (6.19) satisfy*

$$-\Theta^2 \leq \lambda,$$

as well as (6.30), (6.31) and (6.32).

The bounds (6.29), (6.30), (6.31) and (6.32) define a pair of inclusion intervals

$$\left[-\Theta^2(1 + \Upsilon), \tfrac{1}{2}\left(\delta - \sqrt{\delta^2 + 4\delta\gamma^2\theta^2}\right)\right] \cup [\delta, \Delta + \Gamma^2\Theta^2] \qquad (6.44)$$

for the eigenvalues of the preconditioned system (6.19). This in turn leads to an upper bound on the convergence factor of preconditioned MINRES through (6.7). Before considering the convergence factor in more detail, let us highlight the roles of the relevant quantities δ, Δ, γ, Γ, θ, Θ and Υ (which arises only in a stabilized formulation):

- δ, Δ: depend on how well the preconditioning block $\mathbf{P}$ approximates the discrete vector Laplacian $\mathbf{A}$. Note that a simple scaling of $\mathbf{P}$ can be used to fix one of these parameters, so it is the ratio Δ/δ that determines the effectiveness of $\mathbf{P}$ as a preconditioner for $\mathbf{A}$. This ratio is the same condition number, κ, that arises in the classical CG convergence bound of Theorem 2.4.
- γ, (generalized) inf–sup constant: bounded away from zero independently of the mesh.
- Γ, upper bound on the pressure Schur complement: $\Gamma = 1$ for enclosed flow problems, and otherwise $\Gamma \leq \sqrt{d}$ for any flow domain $\Omega \subset \mathbb{R}^d$.
- θ, Θ: constants that are independent of the mesh for the simple choice $T = \mathtt{diag}(Q)$.
- Υ: upper bound on the eigenvalues of the stabilization matrix C. For *ideal* stabilization $\Upsilon = 1$.

The stability parameter γ may depend on the geometry of the domain, *but it is only δ and Δ that can depend explicitly on the size of the discrete problem.* That is, if a suitable stable or stabilized discretization is employed, so that (6.9) and (6.28) hold, and assuming that a suitable approximation of the pressure mass matrix (such as its diagonal) is used so that (6.24) is satisfied, then the convergence of the preconditioned MINRES algorithm will be completely determined by the quality of the Laplacian preconditioner, $\mathbf{P}$. The only way that mesh-size dependence arises is through a dependence of δ or Δ on the representative mesh-size, h. Of course, if an unstable mixed element is used, it is clear that the bound (6.30) is zero (or approaches zero as $h \rightarrow 0$), so that the negative eigenvalues can approach zero. We reiterate these two important points. First, with proper discretization, the steady Stokes equations are essentially as difficult (or as easy) to solve as the Poisson equation. Second, stability of the discretization, in addition to affecting the quality of the finite element approximation, also impacts on the convergence properties of iterative methods for the Stokes equations.

Because of Lemma 6.2, the most favorable situation is where δ, Δ are independent of h as arises when a spectrally equivalent preconditioner for the Laplacian is used. An important practical case is when $\mathbf{P}$ represents a multigrid cycle (see Section 2.5). In this situation the eigenvalue inclusion set (6.44) is bounded both above and away from the origin independently of h, so the MINRES convergence bound (6.7) is also independent of h. Convergence to a fixed tolerance is therefore guaranteed in a number of iterations that is independent of h, that is, not dependent on the dimension of the discrete problem. To illustrate this, Table 6.1 shows the number of MINRES iterations required to solve discrete versions of the driven cavity flow problem in Example 5.1.3 to a fixed tolerance on a sequence of uniform square grids. We take T to be the diagonal of the mass matrix, and we use a multigrid V-cycle for the scalar Laplacian with damped

Table 6.1 *Number of* MINRES *iterations needed to satisfy* $\|\mathbf{r}^{(k)}\|/\|\mathbf{r}^{(0)}\| \leq 10^{-6}$ *for Example 5.1.3 with multigrid preconditioning for the Laplacian. ℓ is the grid refinement level; h is $2^{1-\ell}$ for $\mathbf{Q}_1$ velocity approximation, and $2^{2-\ell}$ for $\mathbf{Q}_2$ approximation. In brackets is the number of seconds of CPU time using* `matlab` *6.5 on a Sun sparcv9 502 MHz processor. The final column shows the CPU time for a sparse direct solve for $\mathbf{Q}_2$–$\mathbf{P}_0$ approximation.*

| | Mixed approximation | | | | | `matlab\` |
ℓ	Q_1–Q_1	Q_1–P_0	Q_2–Q_1	Q_2–P_1	Q_2–P_0	Q_2–P_0
4	32 (4.8)	36 (5.6)	45 (5.4)	29 (4.9)	25 (4.6)	(0.3)
5	33 (8.9)	38 (7.9)	50 (9.2)	31 (7.3)	25 (6.2)	(3.0)
6	32 (18.9)	38 (20.6)	50 (26.9)	31 (19.0)	27 (16.0)	(30.7)
7	31 (68.1)	37 (75.5)	49 (102.8)	29 (68.6)	27 (59.2)	(220.9)
8	30 (295)	36 (313)	47 (427)	29 (305)	27 (267)	(8961)

Jacobi smoothing and a single pre- and post-smoothing step in order to define $\mathbf{P}$. Looking down the columns in the table, it can be seen that the number of MINRES iterations is essentially constant.

The timings are only included to give an indication of the complexity of the computation. It can be seen that as h is halved (so that the number of degrees of freedom increases by a factor of four) the time also approximately increases by a factor of four. This indicates an optimal solver since the work is proportional to the discrete problem size as this gets larger. In contrast, the timings for the `matlab` sparse direct solver can be seen to grow at a rather faster rate,[2] though efficiency is very good on coarse grids.

For stretched grids, Jacobi smoothing is less effective. With other smoothers, however, the overall performance of preconditioned MINRES is largely unaffected by the increased element aspect ratio, see Computational Exercise 6.8.

The bounds of Theorem 6.6 are also useful when δ and/or Δ have some dependence on h. Without loss of generally (since we can simply scale $\mathbf{P}$ as discussed above), assume that Δ is an $\mathcal{O}(1)$ constant and that $\delta \to 0$ as $h \to 0$. For example, using a simple diagonal preconditioner for the Laplacian $\mathbf{P} = \mathrm{diag}(\mathbf{A})$, we have that $\delta = \mathcal{O}(h^2)$. In any such situation, the right-hand side of (6.30) can be written as

$$\tfrac{1}{2}\delta - \tfrac{1}{2}\sqrt{\delta^2 + 4\delta\gamma^2\theta^2} = \tfrac{1}{2}\delta - \sqrt{\delta}\,2\gamma\theta\left(1 + \delta/(4\gamma^2\theta^2)\right)^{1/2}$$
$$= -2\gamma\theta\,\sqrt{\delta} + \tfrac{1}{2}\delta + \mathcal{O}(\delta^{3/2}),$$

and we see that the leading term in the asymptotic expression is $\mathcal{O}(\sqrt{\delta})$. Moreover, since the lower bound on the positive eigenvalues (6.31) is δ, we deduce that the positive eigenvalues λ of the preconditioned Stokes matrix will approach the origin twice as fast as the negative eigenvalues as $h \to 0$. To show that these asymptotic bounds are indicative of what actually happens computationally, a sample set of extremal eigenvalues is presented in Table 6.2.

Table 6.2 *Extremal negative and positive eigenvalues of the diagonally preconditioned Stokes matrix for a sequence of square grids defined on $\Omega_\square = [-1, 1] \times [-1, 1]$ with $Q_2 - Q_1$ mixed approximation: ℓ is the grid refinement level and $h = 2^{2-\ell}$.*

ℓ	Negative eigenvalues		Positive eigenvalues	
4	-0.4382	-0.1142	0.0510	1.5366
5	-0.4434	-0.0673	0.0128	1.5483
6	-0.4435	-0.0348	0.0032	1.5513
7	-0.4442	-0.0178	0.0008	1.5520

---

[2] Ultimately it will grow by a factor of 16, since the bandwidth is inversely proportional to the grid size.

Remark 6.8. The *penalty method* approach for *unstable* mixed approximations, in which the role of the stabilization matrix C in (6.8) is replaced by ϵQ for some small value ϵ (the *penalty parameter*) is sometimes advocated in elasticity applications. Unfortunately, it leads to a loss of mass (which may be acceptable for very small values of ϵ), and it has the additional difficulty of causing severe ill-conditioning for the linear systems. Thus, rapidly convergent iterative methods are not possible. This can be readily seen from the above theory: replacing C by ϵQ leads to $\gamma^2 = \epsilon$ in (6.9) as $BA^{-1}B^T$ is singular for an unstable element. The upper bound on the negative eigenvalues of the preconditioned Stokes system (6.30) then behaves like $\mathcal{O}(\epsilon)$ regardless of the quality of the preconditioner. This bound is indicative of the behavior of the extreme eigenvalues seen for the unstable penalty method in computations — the negative eigenvalues are as close to zero as ϵ. One is then faced with the contradictory requirements of needing a small ϵ for accuracy and a large ϵ for acceptable convergence of the iterative solver!

It should be said that the penalty approach applied to a uniformly stable approximation of the Stokes problem has little effect on iterative convergence. Basically, the addition of the term $\epsilon\langle Q\mathbf{q}, \mathbf{q}\rangle$ in (6.9) can only increase the value of γ for the unpenalized uniformly stable approximation. The use of small values for ϵ is thus perfectly acceptable in the sense that the rate of iterative convergence is similar to that which would be obtained for the unpenalized mixed method.

6.2.3 Equivalent norms for MINRES

We now return to the discussion of norms at the start of this section. We observed earlier that if the MINRES algorithm is to minimize the "correct" norm of the error — that is, the norm (6.14) associated with the finite element approximation error — then it should be run with the specific preconditioner defined by (6.16). Our analysis also shows that an optimal preconditioning strategy which derives from the block diagonal preconditioning (6.18) provides a spectrally equivalent representation of the error whenever δ and Δ are both independent of h. This means that (6.16) and (6.18) determine the same quadratic form up to a bounded constant was long as δ and Δ are both independent of h. This attractive equivalence of norms is established in the form of a theorem below. We restrict attention to stable mixed approximation for simplicity.

Theorem 6.9. *Given* $\mathbf{e} \in \mathbb{R}^{n_u+n_p}$, *let* $\mathbf{r} = K\mathbf{e}$ *where* K *is the Stokes coefficient matrix* (6.1) *arising from an inf–sup stable approximation (i.e.* $C = 0$). *Let* E *be the block diagonal matrix* (6.14), *and let* M *be given by* (6.18) *with* $\mathbf{P}$ *and* T *satisfying* (6.20) *and* (6.24), *respectively. If* δ *and* Δ *in* (6.20) *are independent of* h, *then there exist positive constants* c *and* C *such that*

$$c\|\mathbf{e}\|_E \le \|\mathbf{r}\|_{M^{-1}} \le C\|\mathbf{e}\|_E. \tag{6.45}$$

Proof Write

$$\mathbf{e} = \begin{bmatrix} \mathbf{u} \\ \mathbf{p} \end{bmatrix} = \begin{bmatrix} \mathbf{v} \\ 0 \end{bmatrix} + \begin{bmatrix} \mathbf{w} \\ \mathbf{p} \end{bmatrix}, \tag{6.46}$$

where $\mathbf{v} \in \texttt{null}(B)$ and $\langle \mathbf{s}, A\mathbf{w} \rangle = 0$ for any $\mathbf{s} \in \texttt{null}(B)$. Thus

$$\|\mathbf{e}\|_E^2 = \langle A\mathbf{v}, \mathbf{v} \rangle + \langle A\mathbf{w}, \mathbf{w} \rangle + \langle Q\mathbf{p}, \mathbf{p} \rangle$$

and

$$\mathbf{r} = \begin{bmatrix} A\mathbf{u} + B^T\mathbf{p} \\ B\mathbf{u} \end{bmatrix}.$$

Using (6.20) and (6.24), we have

$$
\begin{aligned}
\|\mathbf{r}\|_{M^{-1}}^2 &= \langle \mathbf{P}^{-1}(A\mathbf{u} + B^T\mathbf{p}), (A\mathbf{u} + B^T\mathbf{p}) \rangle + \langle T^{-1}B\mathbf{u}, B\mathbf{u} \rangle \\
&\geq \delta \langle A^{-1}(A\mathbf{u} + B^T\mathbf{p}), (A\mathbf{u} + B^T\mathbf{p}) \rangle + \theta^2 \langle Q^{-1}B\mathbf{u}, B\mathbf{u} \rangle \\
&= \delta \langle A\mathbf{v}, \mathbf{v} \rangle + \delta \langle A\mathbf{w}, \mathbf{w} \rangle + \theta^2 \langle Q^{-1}B\mathbf{w}, B\mathbf{w} \rangle \\
&\quad + 2\delta \langle B\mathbf{w}, \mathbf{p} \rangle + \delta \langle BA^{-1}B^T\mathbf{p}, \mathbf{p} \rangle \\
&\geq \delta \left[\langle A\mathbf{v}, \mathbf{v} \rangle + (1 + \theta^2\gamma^2/\delta)\langle A\mathbf{w}, \mathbf{w} \rangle + 2\langle B\mathbf{w}, \mathbf{p} \rangle + \langle BA^{-1}B^T\mathbf{p}, \mathbf{p} \rangle \right],
\end{aligned}
$$

$$\tag{6.47}$$
$$\tag{6.48}$$

where (6.48) is derived from (6.47) by use of (6.10). We now make the specific choice $\alpha^2 = \frac{1}{2}\theta^2\gamma^2/\delta + \sqrt{1 + \frac{1}{4}\theta^4\gamma^4/\delta^2}$, and rearrange (6.48) to give the bound

$$
\begin{aligned}
\|\mathbf{r}\|_{M^{-1}}^2 \geq \ \delta \Bigg[&\langle A\mathbf{v}, \mathbf{v} \rangle + \left(1 + \frac{\theta^2\gamma^2}{\delta} - \alpha^2\right)\langle A\mathbf{w}, \mathbf{w} \rangle + \left(1 - \frac{1}{\alpha^2}\right)\langle BA^{-1}B^T\mathbf{p}, \mathbf{p} \rangle \\
&+ \alpha^2\langle A\mathbf{w}, \mathbf{w} \rangle + 2\langle B\mathbf{w}, \mathbf{p} \rangle + \frac{1}{\alpha^2}\langle BA^{-1}B^T\mathbf{p}, \mathbf{p} \rangle \Bigg].
\end{aligned}
$$

Then, noting that

$$
\begin{aligned}
&\alpha^2\langle A\mathbf{w}, \mathbf{w} \rangle + 2\langle B\mathbf{w}, \mathbf{p} \rangle + \frac{1}{\alpha^2}\langle BA^{-1}B^T\mathbf{p}, \mathbf{p} \rangle \\
&= \left\langle A^{-1}\left(\alpha A\mathbf{w} + \frac{1}{\alpha}B^T\mathbf{p}\right), \left(\alpha A\mathbf{w} + \frac{1}{\alpha}B^T\mathbf{p}\right) \right\rangle \geq 0.
\end{aligned}
$$

where the inequality holds because A^{-1} is positive-definite, we see that

$$\|\mathbf{r}\|_{M^{-1}}^2 \geq \delta \left[\langle A\mathbf{v}, \mathbf{v} \rangle + \left(1 + \frac{\theta^2\gamma^2}{\delta} - \alpha^2\right)\langle A\mathbf{w}, \mathbf{w} \rangle + \left(1 - \frac{1}{\alpha^2}\right)\langle BA^{-1}B^T\mathbf{p}, \mathbf{p} \rangle \right].$$

The choice of α makes $1 + \theta^2\gamma^2/\delta - \alpha^2 = 1 - 1/\alpha^2$ which is a positive constant since $\alpha > 1$. Using (6.9) then gives

$$\|\mathbf{r}\|_{M^{-1}}^2 \geq \delta\langle A\mathbf{v}, \mathbf{v} \rangle + \delta(1 - 1/\alpha^2)\langle A\mathbf{w}, \mathbf{w} \rangle + \delta(1 - 1/\alpha^2)\gamma^2\langle Q\mathbf{p}, \mathbf{p} \rangle. \tag{6.49}$$

This gives the required lower bound with

$$c = \delta(1 - 1/\alpha^2)\gamma^2 = \delta\gamma^2 \left(1 + \tfrac{1}{2}\theta^2\gamma^2/\delta - \sqrt{1 + \tfrac{1}{4}\theta^4\gamma^4/\delta^2}\right).$$

The upper bound is established in an analogous fashion:

$$\|\mathbf{r}\|_{M^{-1}}^2 = \langle \mathbf{P}^{-1}(A\mathbf{u} + B^T\mathbf{p}), (A\mathbf{u} + B^T\mathbf{p})\rangle + \langle T^{-1}B\mathbf{u}, B\mathbf{u}\rangle$$

$$\leq \Delta\langle A^{-1}(A\mathbf{u} + B^T\mathbf{p}), (A\mathbf{u} + B^T\mathbf{p})\rangle + \Theta^2\langle Q^{-1}B\mathbf{u}, B\mathbf{u}\rangle$$

$$= \Delta\left[\langle A\mathbf{v}, \mathbf{v}\rangle + \left(2 + \frac{\Theta^2\Gamma^2}{\Delta}\right)\langle A\mathbf{w}, \mathbf{w}\rangle + 2\langle BA^{-1}B^T\mathbf{p}, \mathbf{p}\rangle\right.$$

$$\left. - \langle A^{-1}(A\mathbf{w} - B^T\mathbf{p}), (A\mathbf{w} - B^T\mathbf{p})\rangle\right]$$

$$\leq \Delta\left[\langle A\mathbf{v}, \mathbf{v}\rangle + \left(2 + \frac{\Theta^2\Gamma^2}{\Delta}\right)\langle A\mathbf{w}, \mathbf{w}\rangle + 2\Gamma^2\langle Q\mathbf{p}, \mathbf{p}\rangle\right].$$

This establishes the desired bound with $C = \max\{2\Delta + \Theta^2\Gamma^2, 2\Delta\Gamma^2\}$. $\qquad\square$

This theorem shows that if $\mathbf{e} = \mathbf{e}^{(k)}$ is the error vector at the kth MINRES step, with residual $\mathbf{r} = \mathbf{r}^{(k)}$, then $\|\mathbf{r}^{(k)}\|_{M^{-1}}$, the quantity minimized by MINRES, is equivalent to the "ideal" measure $\|\mathbf{e}^{(k)}\|_E = \|\mathbf{r}^{(k)}\|_{K^{-1}EK^{-1}}$ up to a constant. Consequently, essentially nothing is lost by minimizing with respect to a norm that is easy to work with.

Corollary 6.10. *Let* $\mathbf{u}^{(k)}$ *and* $\mathbf{p}^{(k)}$ *denote the velocity and pressure iterates obtained at the kth step of the preconditioned* MINRES *iteration, where the preconditioner M of (6.18) is such that $\mathbf{P}$ is spectrally equivalent to the vector Laplacian and T is spectrally equivalent to the pressure mass matrix. Let $\vec{u}_h^{(k)}$ and $p_h^{(k)}$ be the associated finite element functions. Then*

$$\|\nabla(\vec{u} - \vec{u}_h^{(k)})\| + \|p - p_h^{(k)}\| \leq \|\nabla(\vec{u} - \vec{u}_h)\| + \|p - p_h\| + (\sqrt{2}/c)\|\mathbf{r}^{(k)}\|_{M^{-1}}.$$

Proof From the triangle inequality,

$$\|\nabla(\vec{u} - \vec{u}_h^{(k)})\| + \|p - p_h^{(k)}\|$$

$$\leq \|\nabla(\vec{u} - \vec{u}_h)\| + \|p - p_h\| + \|\nabla(\vec{u}_h - \vec{u}_h^{(k)})\| + \|p_h - p_h^{(k)}\|.$$

The assertion then follows from the inequalities:

$$\|\nabla(\vec{u}_h - \vec{u}_h^{(k)})\| + \|p_h - p_h^{(k)}\| \leq \sqrt{2}\|\mathbf{e}^{(k)}\|_E \leq (\sqrt{2}/c)\|\mathbf{r}^{(k)}\|_{M^{-1}}. \qquad\square$$

Remark 6.11. This property of optimally preconditioned MINRES is analogous to the norm equivalence property of CG for Galerkin approximation of the Poisson problem, see Remark 2.3.

Remark 6.12. For an enclosed flow the Galerkin matrix (6.1) will be singular with a one-dimensional null space associated with the hydrostatic pressures. Because the finite element pressure basis is a partition of unity, this is $\mathrm{span}\left\{ \begin{bmatrix} \mathbf{0} \\ \mathbf{1} \end{bmatrix} \right\}$. As discussed in Section 5.3, the consistency of the discrete Stokes system is equivalent to conservation of mass. The question arises how the rank deficiency affects iterative convergence to a solution of the consistent system. A similar issue arises for Navier–Stokes systems and we defer a detailed consideration of possible consequences to Section 8.3.4. For the preconditioned iterations we have described, the singularity causes no difficulty in obtaining convergence to a consistent solution of the Stokes system provided a non-singular preconditioner is used. In particular the rate of preconditioned MINRES convergence depends only on the nonzero part of the spectrum.

6.2.4 *MINRES convergence analysis*

Given the descriptive eigenvalue inclusion intervals (6.44), it remains to convert these spectral estimates to descriptive preconditioned MINRES convergence bounds through the polynomial approximation problem (6.7). As in Chapter 2 (in fact, exactly as in the replacement of (2.11) by (2.12)), the bound (6.7) based on the discrete eigenvalues λ_j of $M^{-1}K$

$$\frac{\|\mathbf{r}_k\|_{M^{-1}}}{\|\mathbf{r}_0\|_{M^{-1}}} \leq \min_{p_k \in \Pi_k, p_k(0)=1} \max_j |p_k(\lambda_j)| \qquad (6.50)$$

is weakened to a bound in terms of continuous intervals. That is, we overestimate the contraction factor via

$$\frac{\|\mathbf{r}_k\|_{M^{-1}}}{\|\mathbf{r}_0\|_{M^{-1}}} \leq \min_{p_k \in \Pi_k, p_k(0)=1} \max_{z \in [-a,-b] \cup [c,d]} |p_k(z)|, \qquad (6.51)$$

where the negative eigenvalues λ_j are contained in $[-a, -b]$, and the positive eigenvalues λ_j are contained in $[c, d]$. We note that if a, b, c, d are extreme eigenvalues, then little information is lost by weakening (6.50) to (6.51) as long as there are no significant gaps between consecutive eigenvalues in either interval.

We may now use the eigenvalue inclusion intervals (6.44),

$$a = \Theta^2(1+\Upsilon), \quad b = \tfrac{1}{2}\sqrt{\delta^2 + 4\delta\gamma^2\theta^2} - \tfrac{1}{2}\delta, \quad c = \delta, \quad d = \Delta + \Gamma^2\Theta^2$$

in (6.51) to give a convergence bound for preconditioned MINRES. As mentioned earlier, the solution of this approximation problem is not so accessible as in the case of solving a positive definite symmetric system using CG. Although the exact solution of (6.7) is expressible in general in terms of elliptic functions (see [74]), in certain situations more accessible bounds can be obtained using standard Chebyshev polynomials as is the case here.

We assume as before that, via scaling if necessary, Δ is an $\mathcal{O}(1)$ constant, so that a and d are constants independent of h. We may then extend the smaller of the intervals $[-a, -b]$ or $[c, d]$ by increasing a (to $\hat{a}$, say) or d (to $\hat{d}$) by a constant amount independent of h, so that either $[-\hat{a}, -b]$ and $[c, d]$ are of the same length or else $[-a, -b]$ and $[c, \hat{d}]$ are of the same length. To avoid confusion in the notation, let us henceforth assume that the two intervals $[-a, -b]$ and $[c, d]$ are of the same length with a and d being constants independent of h.

The key idea here is that we can construct a polynomial of degree $2k$

$$p_{2k}(t) = \tau_k(q(t))/\tau_k(q(0)), \tag{6.52}$$

which is symmetric about the midpoint $\frac{1}{2}(b + c)$, by taking $q(t)$ to be the quadratic polynomial

$$q(t) = 2\frac{(t + b)(t - c)}{bc - ad} + 1 \tag{6.53}$$

and τ_k to be the standard Chebyshev polynomial given by (2.14). This polynomial $p_{2k}(t)$ is a candidate polynomial for use in the bound (6.51) at the $2k$th iteration. In fact, (6.52) gives the best possible bound for the given equal length intervals.

Analysis of the Chebyshev polynomial in a manner exactly analogous to that for CG (see Section 2.1.1 and Problem 6.6) then gives the desired MINRES convergence bound.

Theorem 6.13. *After $2k$ steps of the minimum residual method, the iteration residual $\mathbf{r}^{(2k)} = \mathbf{b} - K\mathbf{u}^{(2k)}$ satisfies the bound*

$$\|\mathbf{r}^{(2k)}\|_{M^{-1}} \leq 2\left(\frac{\sqrt{ad} - \sqrt{bc}}{\sqrt{ad} + \sqrt{bc}}\right)^k \|\mathbf{r}^{(0)}\|_{M^{-1}}. \tag{6.54}$$

Note that $\|\mathbf{r}^{(2k+1)}\|_{M^{-1}} \leq \|\mathbf{r}^{(2k)}\|_{M^{-1}}$, because of the minimization property of MINRES. However, the possibility that no reduction in the residual norm occurs at every other step is not precluded. Indeed, "staircasing" where $\|\mathbf{r}^{(2k+1)}\|_{M^{-1}} = \|\mathbf{r}^{(2k)}\|_{M^{-1}}$ is often seen in computations. This behavior can also be predicted analytically when there is a certain exact symmetry of the eigenvalues about the midpoint $\frac{1}{2}(b + c)$ and of coefficients in the eigenvector expansion, and we will return to this issue in Chapter 8.

In cases where a, b, c and d are all independent of h, Theorem 6.13 merely confirms that convergence of MINRES will occur at a rate independent of h. That is, the number of iterations to achieve a given reduction in the residual norm will be essentially constant however the mesh size is varied. This is demonstrated by the results in Table 6.1. When a and d are independent of h, but $c = \mathcal{O}(\delta)$, $b = \mathcal{O}(\sqrt{\delta})$ with $\delta \to 0$ as $h \to 0$ as discussed in Section 6.2.2, then Theorem 6.13 shows that $\|\mathbf{r}^{(k)}\|_{M^{-1}}$ is guaranteed to decrease at every other step by a factor $1 - \mathcal{O}(\delta^{3/4})$. Correspondingly, for an estimate of the number of MINRES iterations required, $\|\mathbf{r}^{(2k)}\|_{M^{-1}}/\|\mathbf{r}^{(0)}\|_{M^{-1}} \leq \epsilon$ should be satisfied within

$$k \approx \tfrac{1}{2} |\log(\epsilon/2)| \, \delta^{-3/2} \qquad (6.55)$$

iterations, see Computational Exercise 6.9.

Discussion and bibliographical notes

Traditional methods for the solution of Stokes problems involve iterating between separate solution steps for pressure variables and velocity variables. Often a segregated solution step for velocity components is followed by a pressure update (or pressure correction) step. There are a number of variants, but the principal methods are named after Uzawa. The main reference is Fortin & Glowinski [78], but basic descriptions can be found in a number of places including Quarteroni & Valli [152]. An important issue with any of these methods is the selection of parameters.

6.1. MINRES was originally derived by Paige & Saunders [142]. In the same paper SYMMLQ, another Krylov subspace iterative method for indefinite linear systems based on the Lanczos procedure was introduced. SYMMLQ requires the same amount of computational work as MINRES per iteration, but does not have a minimization property with respect to Krylov subspaces. Instead, it has an optimality property over different spaces. Though computational rounding error can have more effect on MINRES than SYMMLQ, see [177], this is only relevant for ill-conditioned problems. The optimality property of MINRES and the desirable norm equivalence expressed in Theorem 6.9 makes optimally preconditioned MINRES the method of choice for Stokes and related problems.

Fischer [74] gives an attractive and thorough description (with `matlab` algorithms) of Krylov subspace methods for symmetric matrix systems.

6.2. Klawonn [123] analyzed the use of block triangular preconditioners for the Stokes problem similar to those described in Chapter 8. Bramble & Pasciak [21] describe a preconditioning method that leads to a positive-definite preconditioned matrix which is self-adjoint in a nonstandard inner product for the Stokes problem. Analysis of this approach and related segregated methods can be found in Zulehner [223]. For a careful comparison of the performance of several different iterative solution methods, all involving the use of multigrid, see Elman [56]. The notion of "black-box" preconditioning — using algebraic multigrid to effect the action of the inverse of the Laplacian operator in a block preconditioner — is described by Powell & Silvester [150].

Eigenvalue bounds for discrete saddle-point problems were established by Rusten & Winther [161], and independently for the Stokes problem by Wathen & Silvester [213]. The eigenvalue bounds presented here come from Silvester & Wathen [175]. Tighter estimates of bounds on the convergence of preconditioned MINRES in situations where extremal eigenvalues have an asymptotic dependence on the mesh parameter can be found in Wathen et al. [211, 212].

Problems

6.1. Observe that if $\mathbf{v}$ is the vector of coefficients for the finite element function $\vec{v}_h$ and $\mathbf{q}$ correspondingly for q_h in terms of the appropriate expansions (5.31), (5.32) in terms of basis functions, then

$$\|\nabla \vec{v}_h\|^2 + \|q_h\|^2 = \langle \mathbf{Av}, \mathbf{v} \rangle + \langle Q\mathbf{q}, \mathbf{q} \rangle = \langle E\mathbf{e}, \mathbf{e} \rangle,$$

where E is given by (6.14) and $\mathbf{e}^T = [\mathbf{v}^T, \mathbf{q}^T]$. Then by establishing that

$$\sqrt{a+b} \le \sqrt{a} + \sqrt{b} \le \sqrt{2}\sqrt{a+b}$$

for any positive values a, b, show that

$$\|\mathbf{e}\|_E \le \|\nabla \vec{v}_h\| + \|q_h\| \le \sqrt{2}\|\mathbf{e}\|_E.$$

6.2. In the case of the ideal preconditioner $\mathbf{P} = \mathbf{A}$, $T = Q$ show that the eigenvalue bounds (6.29), (6.30), (6.31) and (6.37) define the eigenvalue inclusion intervals

$$\lambda \in \left[-1, \frac{1 - \sqrt{1 + 4\gamma^2}}{2} \right] \cup \left[1, \frac{1 + \sqrt{1 + 4\Gamma^2}}{2} \right],$$

in the case $C = 0$.

6.3. If a multigrid cycle contracts the error at a rate 0.15 so that (6.23) is satisfied with $\rho = 0.15$, use Theorem 2.4 to establish that CG with this single multigrid cycle as a preconditioner will contract the error at a rate faster than 0.0755 at each iteration.

6.4. Verify that the element mass matrix for a Q_1 rectangle of side lengths h and l is

$$Q^{(k)} = \frac{hl}{36} \begin{bmatrix} 4 & 2 & 1 & 2 \\ 2 & 4 & 2 & 1 \\ 1 & 2 & 4 & 2 \\ 2 & 1 & 2 & 4 \end{bmatrix}$$

(see Section 1.3.2 for definition of the basis functions). Hence obtain the bounds for rectangular grids of Q_1 elements given in Remark 6.4.

6.5. By expanding in terms of an orthonormal eigenvector basis of eigenvectors of the symmetric and positive definite matrix V, show that

$$\langle V\mathbf{r}, \mathbf{r} \rangle \le \langle \mathbf{r}, \mathbf{r} \rangle$$

for all vectors $\mathbf{r}$, is equivalent to $\lambda_{\max}(V) \le 1$ where $\lambda_{\max}$ is the largest eigenvalue. Then, by writing $\mathbf{s} = W^{\frac{1}{2}}\mathbf{r}$, show that

$$\langle V\mathbf{r}, \mathbf{r} \rangle \le \langle W\mathbf{r}, \mathbf{r} \rangle$$

for all $\mathbf{r}$, is equivalent to $\lambda_{\max}(W^{-\frac{1}{2}}VW^{-\frac{1}{2}}) = \lambda_{\max}(W^{-1}V) \leq 1$. Deduce that $\langle V\mathbf{r}, \mathbf{r} \rangle \leq \langle W\mathbf{r}, \mathbf{r} \rangle$ for all $\mathbf{r}$ if and only if $\langle V^{-1}\mathbf{r}, \mathbf{r} \rangle \geq \langle W^{-1}\mathbf{r}, \mathbf{r} \rangle$ for all $\mathbf{r}$.

6.6. By proving that if $a - b = d - c$ then

$$2\frac{(t+b)(t-c)}{bc - ad} + 1 = 2\frac{(t+a)(t-d)}{bc - ad} - 1,$$

show that, under the same condition

$$q : [c, d] \to [-1, 1] \quad \text{and} \quad q : [-a, -b] \to [-1, 1],$$

where q is the quadratic polynomial given by (6.53). Deduce (6.54) from (6.51) employing the polynomial (6.52) and using (2.16) with argument $t = q(0) = (ad + bc)/(ad - bc)$.

6.7. This question concerns the convergence of (unpreconditioned) MINRES iteration for $K\mathbf{x} = \mathbf{b}$ in the situation that the eigenvalues of the symmetric and indefinite matrix K are included in $[-a, -b] \cup [c, d]$, with positive constants $a = d$ and $b = c$ so that the inclusion intervals are symmetrically placed on either side of the origin. In this situation the convergence bound (6.54) simplifies to

$$\frac{\|\mathbf{r}^{(2k)}\|}{\|\mathbf{r}^{(0)}\|} \leq 2 \left(\frac{d - c}{d + c} \right)^k.$$

Show that the eigenvalues of the symmetric and positive-definite matrix $K^T K = K^2$ lie in the interval $[c^2, d^2]$ and hence that the CG convergence bound (2.17) for $K^2 \mathbf{x} = K\mathbf{b}$ is given by

$$\frac{\|\mathbf{r}^{(k)}\|}{\|\mathbf{r}^{(0)}\|} = \frac{\|\mathbf{e}^{(k)}\|_{K^2}}{\|\mathbf{e}^{(0)}\|_{K^2}} \leq 2 \left(\frac{d - c}{d + c} \right)^k,$$

where $\{\mathbf{r}^{(k)}\}$ and $\{\mathbf{e}^{(k)}\}$ are the CG residual and error vectors, respectively.

Observe that these two convergence bounds are essentially the same. Notice that a single iteration of CG for $K^2 \mathbf{x} = K\mathbf{b}$ requires two matrix-vector multiplications involving K, so that k iterations of CG and $2k$ iterations of MINRES both require $2k$ matrix-vector multiplications. (The inclusion bounds for Stokes matrices derived in this chapter never have such perfect symmetry about the origin!)

Computational exercises

The driver `stokes_testproblem` can be used to set up any of the reference problems in Examples 5.1.1–5.1.4. The function `it_solve` uses the built-in `matlab` routine `minres` and offers the possibility of a diagonal, block diagonal or a multigrid V-cycle as the preconditioning block $\mathbf{P}$, whilst $T = \texttt{diag}(Q)$.

6.1. For Example 5.1.1, use `stokes_testproblem` to compute the blocks `Ast`, `Bst` and `C` for the Stokes matrix `K = [Ast,Bst';Bst,-(1/4)*C]` using stabilized Q_1–P_0 approximation on a 4×4 grid. Compute the eigenvalues of K using `eig(full(K))`, and check that the number of negative and zero eigenvalues agree with the dimension of the pressure space. Compare the number of unit and positive eigenvalues with the number of boundary velocity nodes and the number of interior velocity nodes, respectively.

Repeat the exercise for the problem in Example 5.1.3 using both stabilized approximation and unstabilized approximation and note any differences.

6.2. For the very simple (invertible!) indefinite matrix

$$K = \begin{bmatrix} 0 & 1 \\ 1 & 0 \end{bmatrix}$$

and any random right-hand side vector, run CG and MINRES via

$$[\text{X}, \text{FLAG}, \text{RELRES}, \text{ITER}, \text{RESVEC}] = \text{pcg}(\text{K}, \text{b});$$

$$[\text{X}, \text{FLAG}, \text{RELRES}, \text{ITER}, \text{RESVEC}] = \text{minres}(\text{K}, \text{b});$$

What do you observe? (Also, beware of the right-hand side vector `b=[1,1]'`; since this is an eigenvector of K, both methods converge immediately!).

6.3. Consider solving the regularized driven cavity flow problem in Example 5.1.3 using Q_1–P_0 approximation without stabilization. The effect on the computed pressure solution will be self-evident, but the effect on iterative convergence can be observed by then running `it_solve` using any of the preconditioner choices. Contrast the iteration counts with those obtained using the same preconditioner in the case of stabilized Q_1–P_0 approximation.

6.4. Use of `stokes_testproblem` automatically computes the pressure mass matrix Q. Select the Q_2–Q_1 or Q_1–Q_1 option for either a uniform or stretched grid and then compute and plot the eigenvalues of the diagonally preconditioned (scaled) Q_1 mass matrix and compare the results with the bounds in Remark 6.4.

6.5. Following Exercise 6.4, calculate the maximum eigenvalue of $Q^{-1}C$ for either of the stabilized elements and compare with (6.28). (Note that it is convenient to use `eig(full(C),full(Q))` to compute the generalized eigenvalues of $C - \lambda Q$ which are the same as the eigenvalues of $Q^{-1}C$).

6.6. Consider the enclosed flow problem in Example 5.1.3 discretized using stabilized Q_1–Q_1 approximation. Estimate γ by computing the eigenvalues of $BA^{-1}B^T + C - \lambda Q$ and using (6.9). Then, analytically calculate the eigenvalue inclusion intervals (6.44) for block diagonal preconditioning ($\mathbf{P} = \mathbf{A}, T = Q$) using Remark 6.4 and the estimate of Υ from Exercise 6.5. For a coarse enough grid (say 8×8) you should be able compute extemal eigenvalues and compare with the analytic bounds.

6.7. The Quasi-Minimum Residual method, QMR, is a Krylov subspace iterative method applicable to systems of linear equations with a nonsymmetric coefficient matrix. It is equivalent to MINRES for a symmetric matrix (see [80] for details). It is also a built-in `matlab` function (try typing `help qmr`). There is also an efficient symmetric implementation of QMR (described in [81]) that has the special property that a symmetric indefinite preconditioner can be used if the coefficient matrix is itself symmetric indefinite. (Note that MINRES does not converge in this situation!)

Use `stokes_testproblem` to compute the blocks `Ast`, `Bst` and `C` for the Stokes matrix `K = [Ast,Bst';Bst,-1/4*C]` using stabilized Q_1–P_0 approximation, together with the right-hand side vector `[fst;gst]`. Then, investigate the convergence of QMR with the symmetric positive definite and symmetric indefinite block diagonal preconditioners

$$\begin{bmatrix} \mathbf{A} & 0 \\ 0 & Q \end{bmatrix} \quad \text{and} \quad \begin{bmatrix} \mathbf{A} & 0 \\ 0 & -Q \end{bmatrix},$$

respectively, and compare iteration counts with MINRES with the definite preconditioner.

6.8. Consider solving the regularized driven cavity flow problem in Example 5.1.3 using stabilized Q_1–P_0 approximation. Select a stretched 32×32 grid and run `it_solve` with multigrid preconditioning and Jacobi smoothing using one pre- and one post-smoothing step. Compare the convergence profile with that obtained for a uniform 32×32 grid. Then repeat the exercise using ILU smoothing instead of Jacobi.

6.9. Run MINRES with diagonal preconditioning for a uniform sequence grids using an inf–sup stable approximation of the step flow problem in Example 5.1.2. Hence, compare the asymptotic trend in the number of iterations required to satisfy a fixed convergence tolerance with that expressed in the estimate (6.55). (Hint: recall that $\delta = \mathcal{O}(h^2)$ for diagonal preconditioning.)

7

THE NAVIER–STOKES EQUATIONS

Adding a convection term and a forcing term $\vec{f}$ to the Stokes system discussed in Chapter 5 leads to the steady-state Navier–Stokes equation system

$$-\nu\nabla^2\vec{u} + \vec{u}\cdot\nabla\vec{u} + \nabla p = \vec{f}, \tag{7.1}$$

$$\nabla\cdot\vec{u} = 0,$$

where $\nu > 0$ is a given constant called the *kinematic viscosity*. The Navier–Stokes system is the basis for computational modeling of the flow of an incompressible Newtonian fluid, such as air or water. Finding effective approximation methods for this system is at the heart of a broad range of engineering applications, from airplane design to nuclear reactor safety evaluation. It is also the basic model for weather prediction, so is firmly embedded in everyday life, at least in the British Isles! As in Chapter 5, the variable $\vec{u}$ represents the velocity of the fluid and p represents the pressure. The convection term $\vec{u}\cdot\nabla\vec{u}$ is simply the vector obtained by taking the convective derivative of each velocity component in turn, that is, $\vec{u}\cdot\nabla\vec{u} := (\vec{u}\cdot\nabla)\vec{u}$. The fact that this term is *nonlinear* is what makes life interesting — boundary value problems associated with the Navier–Stokes equations can have more than one stable solution! Solution non-uniqueness presents an additional challenge for the numerical analysis of approximations to the system (7.1) and will be a key feature of this chapter. The use of mixed approximation methods leads to nonlinear algebraic systems of equations which can only be solved using iteration. Two such strategies, namely Picard iteration and Newton iteration, are discussed later.[1] The solution of the linearized equations that arise at each level of nonlinear iteration will be considered in Chapter 8.

The boundary value problem that is considered is the system (7.1) posed on a two- or three-dimensional domain Ω, together with boundary conditions on $\partial\Omega = \partial\Omega_D \cup \partial\Omega_N$ given by

$$\vec{u} = \vec{w} \text{ on } \partial\Omega_D, \quad \nu\frac{\partial\vec{u}}{\partial n} - \vec{n}p = \vec{0} \text{ on } \partial\Omega_N, \tag{7.2}$$

where $\vec{n}$ denotes the outward-pointing normal to the boundary. This is completely analogous to the Stokes equations in Chapter 5. If the velocity is specified everywhere on the boundary, that is, if $\partial\Omega_D \equiv \partial\Omega$, then the pressure solution

---

[1]These two nonlinear solver strategies are implemented in the IFISS software.

to the Navier–Stokes problem (7.1) and (7.2) is only unique up to a hydrostatic constant.

The presence of the convection term in (7.1) means that, in contrast to what happens for Stokes flow, layers in the solution are likely to arise. The discussion of exponential boundary layers in Chapter 3 is thus relevant here. Indeed, if one wishes to model flow in a channel, then a "hard" Dirichlet outflow condition, that is $\vec{u} = \vec{w}$ in (7.2) with $\vec{w} \cdot \vec{n} > 0$, is likely to lead to difficulties in outflow regions. A Neumann outflow condition is much less intrusive.

As in Chapter 3, having a quantitative measure of the relative contributions of convection and viscous diffusion is very useful. This can be achieved by normalizing the system (7.1) with respect to the size of the domain and the magnitude of the velocity. To this end, let L denote a characteristic length scale for the domain Ω; for example, L can be the Poincaré constant of Lemma 1.2. If points in Ω are denoted by $\vec{x}$, then $\vec{\xi} = \vec{x}/L$ denotes points of a normalized domain. In addition, let the velocity $\vec{u}$ be defined so that $\vec{u} = U\vec{u}_*$ where U is a reference value — for example, the maximum magnitude of velocity on the inflow. If the pressure is scaled so that $p(L\vec{\xi}) = U^2 p_*(\vec{\xi})$ on the normalized domain, then the momentum equation in (7.1) can be rewritten as

$$-\frac{1}{\mathcal{R}}\nabla^2 \vec{u}_* + \vec{u}_* \cdot \nabla \vec{u}_* + \nabla p_* = \frac{L}{U^2}\vec{f}. \qquad (7.3)$$

Here, the relative contributions of convection and diffusion are defined by the *Reynolds number*

$$\mathcal{R} := \frac{UL}{\nu}. \qquad (7.4)$$

If L and U are suitably chosen, then the condition $\mathcal{R} \leq 1$ means that (7.3) is diffusion-dominated and the flow solution can be shown to be uniquely defined. Details are given later. In contrast, the flow problem is convection-dominated when $\mathcal{R} \gg 1$. Notice that taking the limit $\mathcal{R} \to \infty$ in (7.3) gives the reduced hyperbolic problem

$$\vec{u}_* \cdot \nabla \vec{u}_* + \nabla p_* = \vec{f}_*, \qquad (7.5)$$
$$\nabla \cdot \vec{u}_* = 0.$$

This system is known as the *incompressible Euler equations*, which defines a standard model for inviscid flow. The model is especially useful away from fixed walls — which are always "slippery" in (7.5) since a tangential velocity boundary condition cannot be imposed. The breakdown of the inviscid flow model induced by the inclusion of viscous diffusion is the origin of *shear layers* that arise in the solution of (7.1) and (7.2). We give an example in the next section.

A final point here is that the Stokes equations (5.1) cannot be recovered from the dimensionless system (7.3). If very viscous fluids are being modelled, then an alternative scaling of the pressure is likely to be more appropriate (see Problem 7.1).

7.1   Reference problems

The first three examples show the effect of adding convection in the context of the reference flow problems discussed in Chapter 5. In all cases the reference velocity U is unity, and the reference length L is of order unity. This means that the Reynolds number of the flow is typically given by $1/\nu$, so that a flow problem is convection-dominated whenever $\nu < 1$. The final example is a classical model of boundary layer flow. The forcing term $\vec{f}$ is zero in every case.

7.1.1 Example: Square domain $\Omega_\square = (-1, 1)^2$, parabolic inflow boundary condition, natural outflow boundary condition, analytic solution.

The Poiseuille channel flow solution

$$u_x = 1 - y^2; \quad u_y = 0; \quad p = -2\nu x; \tag{7.6}$$

illustrated in Figure 5.1 is also an analytic solution of the Navier–Stokes equations, since the convection term $\vec{u} \cdot \nabla \vec{u}$ is identically zero. It also satisfies the natural outflow condition

$$\begin{aligned} \nu \frac{\partial u_x}{\partial x} - p &= 0 \\ \frac{\partial u_y}{\partial x} &= 0. \end{aligned} \tag{7.7}$$

The only difference between (7.6) and the Stokes channel flow solution is that the pressure gradient is proportional to the viscosity parameter. This makes sense physically; if a fluid is not very viscous then only a small pressure difference is needed to maintain the flow. Notice also that in the extreme limit $\nu \to 0$, the parabolic velocity solution specified in (7.6) satisfies the Euler equations (7.5) together with a constant pressure solution.

It is clear from the discussion above that an effective iteration strategy for the nonlinear system of equations that arises from discretization of a Poiseuille flow problem is to use the corresponding discrete Stokes solution as the initial guess. Moreover, with the $\boldsymbol{Q}_2$–$\boldsymbol{Q}_1$ or the $\boldsymbol{Q}_2$–$\boldsymbol{P}_{-1}$ approximation method (discussed in Section 5.3.1), one would expect to compute the exact solution (7.6) independently of the grid. We will return to this point when discussing a priori error bounds in Section 7.4.1.

7.1.2 Example: L-shaped domain $\Omega_{\vdash}$, parabolic inflow boundary condition, natural outflow boundary condition.

This example is a fast flowing analogue of the Stokes flow over a step illustrated in Figure 5.2. The boundary conditions are unchanged; a Poiseuille flow profile is imposed on the inflow boundary ($x = -1; 0 \le y \le 1$), and a no-flow (zero velocity) condition is imposed on the walls. The Neumann condition (7.7) is applied at the outflow boundary ($x = 5; -1 < y < 1$) and automatically sets the mean outflow pressure to zero.

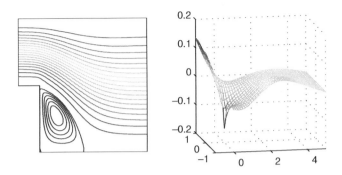

FIG. 7.1. Equally spaced streamline plot (left) and pressure plot (right) of a
    Q_2–Q_1 approximation of Example 7.1.2 with $\nu = 1/50$.

As can be readily seen by comparing Figures 7.1 and 5.2, there is a qualitative
difference between convection-dominated and diffusion-dominated flow. The
increased velocity caused by convection makes it harder for the fluid to flow
around the corner, and a slow-moving component of the fluid becomes entrained
behind the step. There are two sets of streamlines at equally spaced levels plotted in Figure 7.1; one set is associated with positive streamfunction values and
shows the path of particles introduced at the inflow. These pass over the step
and exit at the outflow. The second set of streamlines is associated with negative
values of the streamfunction. These streamlines show the path of particles in the
recirculation region near the step; they are much closer in value, reflecting the
fact that the recirculating flow is relatively slow-moving.

If L is taken to be the height of the outflow region, then the flow pattern shown
in Figure 7.1 corresponds to a Reynolds number of 100. If the viscosity parameter
were an order of magnitude smaller, then steady flow would be unstable — in a
laboratory experiment the corresponding flow would be turbulent — and thus
there is little sense in trying to compute steady solutions for $\nu < 1/1000$ in this
case. Notice also that a longer channel should really be used at higher Reynolds
numbers. If the flow at the exit is not well developed (i.e. if it does not have an
essentially parabolic profile), then the outflow boundary condition (7.7) is not
appropriate.

The singularity at the origin is an important feature of the flow even in the
convection-dominated case. The key point is that "close" to the corner the fluid
is slow moving and thus the Stokes equations are a good approximation to the
Navier–Stokes equations in this vicinity. We return to this point when discussing
a posteriori error estimation in Section 7.4.2.

7.1.3 Example: Square domain $\Omega_{\square}$, regularized cavity boundary condition.

This example is a fast-flowing analogue of the Stokes flow in a cavity illustrated in Figure 5.3. The solution shown in Figure 7.2 corresponds to a Reynolds

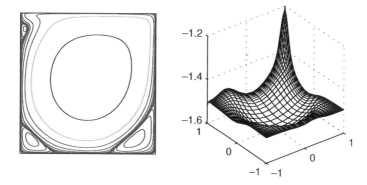

FIG. 7.2. Exponentially spaced streamline plot (left) and pressure plot (right)
   of a Q_2–Q_1 approximation of Example 7.1.3 with $\nu = 1/1000$.

number of 2000. Comparing the solution with that in Figure 5.3 reveals once
again that the convection-dominated flow is very different. The centre of the
primary recirculation is now close to the centre of the cavity, and particles in
the body of the fluid move in a circular trajectory. Indeed, at high Reynolds
number, classical analysis predicts that the core fluid will move as a solid body
with uniform vorticity, see Batchelor [11]. Also, two secondary recirculations
are generated in the bottom corners. These are stronger and more prominent
features than the little Moffatt eddies. The third recirculation near the upper
left corner is a feature of stable flows for Reynolds numbers of this order of
magnitude. The major qualitative difference in the pressure solution is that the
corner singularities are much less prominent in the convection-dominated case.
Notice also that the regions of constant pressure in Figure 7.2 correspond to the
secondary recirculations, where the fluid is relatively slow-moving.

   Steady flow in a two-dimensional cavity is not stable for Reynolds number
much greater than 10^4. At a critical Reynolds number (approximately 13,000)
the flow pattern develops into a time-periodic state with "waves" running around
the cavity walls. In a real-life practical setting, however, three-dimensional effects
are fundamentally important. In the case of a cube shaped cavity, flow generated
by moving one of the walls is unsteady at Reynolds number much less than 1000;
see Shankar & Deshpande [171].

7.1.4 Example: Non-Lipschitz domain $\Omega_{\boxminus} = (-1, 5) \times (-1, 1)$, parallel flow
   boundary condition, natural outflow boundary condition.

   This example, known as *Blasius flow*, models boundary layer flow over a flat
plate. Its analytic solution is a cornerstone of classical fluid mechanics and can
be found in any elementary text on fluid mechanics, see for example, Acheson
[1, pp. 266–275]. The problem is equivalent to that of computing the steady flow
over a flat plate moving at a constant speed through a fluid that is at rest.

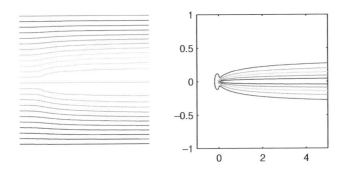

FIG. 7.3. Equally spaced streamline plot (left) and horizontal velocity contour plot (right) of a 64×64 geometrically stretched grid ($\alpha = 1.1$) solution of Example 7.1.4 with $\nu = 1/200$.

To model this flow, the "parallel flow" Dirichlet condition $u_x = 1$, $u_y = 0$ is imposed at the inflow boundary ($x = -1; -1 \leq y \leq 1$) and also on the top and bottom of the channel ($-1 \leq x \leq 5; y = \pm 1$), representing walls moving from left to right with speed unity. The plate is modelled by imposing a no-flow condition on the internal boundary ($0 \leq x \leq 5; y = 0$), and the usual Neumann condition (7.7) is applied at the outflow boundary ($x = 5; -1 < y < 1$). The flow Reynolds number is defined by $\mathcal{R} = \mathrm{U}L/\nu$, where $\mathrm{U} = 1$ is the free-stream velocity, and $\mathrm{L} = 5$ is the length of the plate.

The main reason why this flow problem is of such historical importance is that in the inviscid limit, $\mathcal{R} \to \infty$, the solution is simply $u_x = 1$, $u_y = 0$ and $p = \text{constant}$. That is, the plate is "invisible". In reality, there is a *shear boundary layer*, of width proportional to $\sqrt{\nu}$, within which the horizontal velocity increases rapidly from zero to unity. A typical finite Reynolds number solution is illustrated in Figure 7.3. The contours of the horizontal velocity are equally spaced between 0 and 0.95 and show the evolution of the boundary layer as the fluid passes from the leading edge of the plate to the outflow. The parabolic shape of the velocity contours is completely consistent with asymptotic theory; see Acheson [1]. This is explored further in Computational Exercise 7.2.

Since highly stretched grids are typically needed to compute the flow accurately at large Reynolds numbers, the problem represents a stiff test for any Navier–Stokes solver. The singularity in the pressure at the leading edge of the plate is another major source of difficulty — the asymptotic theory breaks down in its vicinity. Ultimately, for a critical Reynolds number of order 10^6, the flow becomes unstable and transition to a so-called *turbulent boundary layer* occurs.

7.2 Weak formulation and linearization

The weak formulation of the Navier–Stokes flow problem (7.1) and (7.2) is determined exactly as in Chapter 5. Indeed, if we define the same solution and test spaces

$$\mathbf{H}_E^1 := \{\vec{u} \in \mathcal{H}^1(\Omega)^d \mid \vec{u} = \vec{w} \text{ on } \partial\Omega_D\}, \tag{7.8}$$

$$\mathbf{H}_{E_0}^1 := \{\vec{v} \in \mathcal{H}^1(\Omega)^d \mid \vec{v} = \vec{0} \text{ on } \partial\Omega_D\}, \tag{7.9}$$

then the standard weak formulation is the following:

Find $\vec{u} \in \mathbf{H}_E^1$ and $p \in L_2(\Omega)$ such that

$$\nu \int_\Omega \nabla\vec{u} : \nabla\vec{v} + \int_\Omega (\vec{u} \cdot \nabla\vec{u}) \cdot \vec{v} - \int_\Omega p\,(\nabla \cdot \vec{v}) = \int_\Omega \vec{f} \cdot \vec{v} \quad \text{for all } \vec{v} \in \mathbf{H}_{E_0}^1$$

$$\tag{7.10}$$

$$\int_\Omega q\,(\nabla \cdot \vec{u}) = 0 \quad \text{for all } q \in L_2(\Omega). \tag{7.11}$$

From the identity (5.19) it is clear that any solution of (7.1) and (7.2) also satisfies (7.10)–(7.11).

The main difference between this formulation and the Stokes analogue (5.22)–(5.23) is the convection term. This can be identified with a trilinear form $c : H_{E_0}^1 \times H_{E_0}^1 \times H_{E_0}^1 \to \mathbb{R}$ defined as follows:

$$c(\vec{z}; \vec{u}, \vec{v}) := \int_\Omega (\vec{z} \cdot \nabla\vec{u}) \cdot \vec{v}. \tag{7.12}$$

The key to establishing the *existence* of a weak solution is the Lax–Milgram lemma used in Chapter 3. The other ingredient is a version of Brouwer's fixed point theorem; details are given in Quarteroni & Valli [152, pp. 341–342] or Girault & Raviart [85, pp. 284–287]. To see the connection with Chapter 3, let the subspace of divergence-free velocities be given by

$$\mathbf{V}_{E_0} := \{\vec{z} \in \mathbf{H}_{E_0}^1; \nabla \cdot \vec{z} = 0 \text{ in } \Omega\}, \tag{7.13}$$

and let us define the bilinear form $a_{\vec{z}}(\cdot, \cdot) : \mathbf{V}_{E_0} \times \mathbf{V}_{E_0} \to \mathbb{R}$, via

$$a_{\vec{z}}(\vec{u}, \vec{v}) := \nu \int_\Omega \nabla\vec{u} : \nabla\vec{v} + c(\vec{z}; \vec{u}, \vec{v}), \tag{7.14}$$

where $\vec{z}$ is a given function in $\mathbf{V}_{E_0}$. As in Chapter 3, the coercivity and continuity of the underlying bilinear form needs to be established with respect to the norm $\|\nabla\vec{v}\|$. The Dirichlet case $\partial\Omega = \partial\Omega_D$ is explored in Problem 7.2. The essential point is that the convection term is skew-symmetric: $c(\vec{z}; \vec{u}, \vec{v}) = -c(\vec{z}; \vec{v}, \vec{u})$

over $\mathbf{V}_{E_0}$. This means that $c(\vec{z}; \vec{u}, \vec{u}) = 0$ and so ellipticity is straightforward

$$a_{\vec{z}}(\vec{u}, \vec{u}) = \nu \, \|\nabla \vec{u}\|^2 \quad \text{for all } \vec{u} \in \mathbf{V}_{E_0}. \tag{7.15}$$

Establishing continuity, that is,

$$c(\vec{z}; \vec{u}, \vec{v}) \le \Gamma \, \|\nabla \vec{z}\| \, \|\nabla \vec{u}\| \, \|\nabla \vec{v}\| \tag{7.16}$$

is more technical, see Girault & Raviart [85, p. 284].

The nonlinearity of the Navier–Stokes operator complicates analysis of the well-posedness of the weak formulation (7.10)–(7.11). To see this, let us define $\vec{u} := \vec{u}_1 - \vec{u}_2$ and $p := p_1 - p_2$ where $(\vec{u}_1, p_1)$ and $(\vec{u}_2, p_2)$ represent distinct solutions. The fact that

$$c(\vec{u}_1; \vec{u}_1, \vec{v}) - c(\vec{u}_2; \vec{u}_2, \vec{v}) \ne c(\vec{u}_1 - \vec{u}_2; \vec{u}_1 - \vec{u}_2, \vec{v})$$

means that it is not possible to use a homogeneous problem analogous to (5.25)–(5.26) to establish uniqueness.

Instead, given the existence of at least one weak solution satisfying (7.10)–(7.11), it is possible to specify conditions on the forcing function $\vec{f}$ and boundary data $\vec{w}$ that ensure the uniqueness of such a solution (up to a constant pressure in the case of $\partial\Omega = \partial\Omega_D$). The case of enclosed flow $\vec{w} = \vec{0}$ is the easiest to analyze. Specifically, if we define

$$\|\vec{f}\|_* := \sup_{\vec{v} \in \mathbf{V}_{E_0}} \frac{(\vec{f}, \vec{v})}{\|\nabla \vec{v}\|},$$

then a well-known (sufficient) condition for uniqueness (see e.g. [85, Theorem 2.2]) is that the forcing function is small in the sense that

$$\|\vec{f}\|_* \le \frac{\nu^2}{\Gamma_*}, \tag{7.17}$$

where Γ_* is the best possible constant such that (7.16) holds. The more interesting case of nonzero boundary data can also be studied, see [85, pp. 287–293], with the same conclusion; a unique weak solution can be guaranteed only if the viscosity parameter is large enough.

The issues of non-existence and non-uniqueness of steady-state flow solutions is closely related to the classical concepts of *hydrodynamic stability* and *bifurcation*. We review some of the underlying ideas in the following section. This also introduces the notion of linearization, which is the subject of the subsequent section.

7.2.1 *Stability theory and bifurcation analysis*

Suppose there is a weak solution $\vec{u}(\vec{x})$ and $p(\vec{x})$ satisfying the steady-state formulation (7.10)–(7.11). We will refer to this solution as a "fixed point". The essential idea of stability theory — motivated by practical flow experiments — is to study the evolution of deviations from such a fixed point. If perturbations of a steady solution are amplified as time evolves, then the fixed point is unstable, and the evolution naturally drives the system away from the steady state. If, in contrast, perturbations decay with time, then the fixed point is stable and the computed steady solution is a candidate time-independent solution in a practical setting.

To investigate this, let us define a time-dependent flow solution such that

$$\vec{v}(\vec{x}, t) = \vec{u}(\vec{x}) + \delta\vec{u}(\vec{x}, t), \quad q(\vec{x}, t) = p(\vec{x}) + \delta p(\vec{x}, t). \tag{7.18}$$

We assume that the velocity $\vec{v}(\vec{x}, t)$ satisfies the essential boundary condition in (7.2) so that $\delta\vec{u}(\cdot, t)$ is zero on the Dirichlet boundary for all time. If this solution is substituted into the time-dependent analogue of the weak formulation, then the perturbations $\delta\vec{u}(\vec{x})$ and $\delta p(\vec{x})$ satisfy the following initial-boundary value problem: find $\delta\vec{u}(\vec{x}, t)$ and $\delta p(\vec{x}, t)$ such that

$$\int_\Omega \frac{\partial \delta\vec{u}}{\partial t} \cdot \vec{v} + D(\vec{u}, \delta\vec{u}, \vec{v}) + \nu \int_\Omega \nabla\delta\vec{u} : \nabla\vec{v} - \int_\Omega \delta p \, (\nabla \cdot \vec{v}) = \vec{0}, \tag{7.19}$$

$$\int_\Omega q \nabla \cdot \delta\vec{u} = 0, \tag{7.20}$$

for all $\vec{v} \in \mathbf{H}^1_{E_0}$ and $q \in L_2(\Omega)$, where $D(\vec{u}, \delta\vec{u}, \vec{v})$ is the difference in the nonlinear terms, that is,

$$\begin{aligned} D(\vec{u}, \delta\vec{u}, \vec{v}) &= \int_\Omega (\delta\vec{u} + \vec{u}) \cdot \nabla(\delta\vec{u} + \vec{u}) \cdot \vec{v} - \int_\Omega (\vec{u} \cdot \nabla\vec{u}) \cdot \vec{v} \\ &= \int_\Omega (\delta\vec{u} \cdot \nabla\delta\vec{u}) \cdot \vec{v} + \int_\Omega (\delta\vec{u} \cdot \nabla\vec{u}) \cdot \vec{v} + \int_\Omega (\vec{u} \cdot \nabla\delta\vec{u}) \cdot \vec{v} \\ &= c(\delta\vec{u}; \delta\vec{u}, \vec{v}) + c(\delta\vec{u}; \vec{u}, \vec{v}) + c(\vec{u}; \delta\vec{u}, \vec{v}). \end{aligned} \tag{7.21}$$

The goal of stability theory is to extract information about the evolution of the perturbations without solving the problem (7.19)–(7.21).

A standard "linear stability" analysis is obtained by dropping the quadratic term $c(\delta\vec{u}; \delta\vec{u}, \vec{v})$ in (7.21), and then assuming that the perturbations grow or decay exponentially in time; that is, we redefine $\delta\vec{u}(\vec{x}, t) \leftarrow e^{\lambda t}\delta\vec{u}(\vec{x})$ and $\delta p(\vec{x}, t) \leftarrow e^{\lambda t}\delta p(\vec{x})$. This leads to the following eigenproblem: find eigenvalues λ and corresponding eigenvectors $(\delta\vec{u}, \delta p)$ such that

$$c(\delta\vec{u}; \vec{u}, \vec{v}) + c(\vec{u}; \delta\vec{u}, \vec{v}) + \nu \int_\Omega \nabla\delta\vec{u} : \nabla\vec{v} - \int_\Omega \delta p \, (\nabla \cdot \vec{v}), = \lambda \int_\Omega \delta\vec{u} \cdot \vec{v}$$

$$\int_\Omega q \, (\nabla \cdot \delta\vec{u}) = 0. \tag{7.22}$$

If the real parts of all the eigenvalues are negative, then the original flow solution $(\vec{u}, p)$ is (linearly) stable. As an example, analysis of the Poiseuille flow problem in Example 7.1.1 establishes that the steady state flow is unstable if $\mathcal{R} = 1/\nu > 5772$. See [1, p. 324] for details.

Linear stability is only a necessary condition for full nonlinear stability. To derive a sufficient condition requires a so-called *energy analysis*, which is sketched below. We start with the nonlinear formulation (7.19)–(7.21) and assume a Dirichlet boundary condition, so that the test function $\vec{v}$ is zero everywhere on $\partial\Omega$. Let t be fixed. Given that $\delta\vec{u} \in \mathbf{H}^1_{E_0}$ we can take $\vec{v} = \delta\vec{u}$. If we also take $q = \delta p$ then the pressure term drops out of (7.19) to give

$$\int_\Omega \frac{\partial \delta\vec{u}}{\partial t} \cdot \delta\vec{u} + D(\vec{u}, \delta\vec{u}, \delta\vec{u}) + \nu \left\|\nabla\delta\vec{u}\right\|^2 = 0. \tag{7.23}$$

Notice that

$$\int_\Omega \frac{\partial \delta\vec{u}}{\partial t} \cdot \delta\vec{u} = \frac{\mathrm{d}}{\mathrm{d}t}\left(\frac{1}{2}\|\delta\vec{u}\|^2\right). \tag{7.24}$$

To simplify the second term of (7.23) we note that choosing $q = \nabla \cdot \delta\vec{u}$ in (7.20) implies that $\nabla \cdot \delta\vec{u} = 0$ in Ω. Then, using the skew-symmetry property in Problem 7.2, we see that $c(\delta\vec{u}; \delta\vec{u}, \delta\vec{u}) = 0$. Similarly, since $\nabla \cdot \vec{u} = 0$ in Ω we have that $c(\vec{u}; \delta\vec{u}, \delta\vec{u}) = 0$, and so, using the expansion (7.21), we deduce that $D(\vec{u}, \delta\vec{u}, \delta\vec{u}) = c(\delta\vec{u}; \vec{u}, \delta\vec{u}) = -c(\delta\vec{u}; \delta\vec{u}, \vec{u})$. Substituting into (7.23) and using (7.24) thus leads to the following estimate for the decay of the L_2 norm of the velocity perturbation:

$$\frac{\mathrm{d}}{\mathrm{d}t}\left(\frac{1}{2}\|\delta\vec{u}\|^2\right) = c(\delta\vec{u}; \delta\vec{u}, \vec{u}) - \nu \left\|\nabla\delta\vec{u}\right\|^2. \tag{7.25}$$

At this point it is convenient to rescale the problem by non-dimensionalizing the original nonlinear formulation (7.19)–(7.21). Specifically, consider the dimensionless variables

$$\vec{\xi} = \vec{x}/\mathrm{L}, \quad t_* = t\nu/\mathrm{L}^2, \quad \vec{u}_* = \vec{u}/\mathrm{U}, \quad p_* = p\mathrm{L}/\nu\mathrm{U},$$

where L is the Poincaré constant of Lemma 1.2 and U is the maximum magnitude of the velocity in Ω. This rescaling is analogous to that used in Problem 7.1. Rewriting the bound (7.25) in terms of the dimensionless variables then gives

$$\frac{\mathrm{d}}{\mathrm{d}t}\left(\frac{1}{2}\|\delta\vec{u}\|^2\right) = \mathcal{R}c(\delta\vec{u}; \delta\vec{u}, \vec{u}) - \left\|\nabla\delta\vec{u}\right\|^2, \tag{7.26}$$

where $\mathcal{R}$, the Reynolds number, is now the only parameter in the problem. The viscous term is the "good guy" here: if $\mathcal{R}$ is zero, then the right-hand side of (7.26) is negative and all perturbations monotonically decay in time.

The influence of the Reynolds number on the stability of the flow solution is now becoming apparent. Note that the dimensionless velocity is scaled so that $\|\vec{u}\|_\infty = 1$, and so if we bound the convection term as in (3.18)

$$c(\delta\vec{u}; \delta\vec{u}, \vec{u}) = \int_\Omega \delta\vec{u} \cdot \nabla\delta\vec{u} \cdot \vec{u} \leq \|\delta\vec{u}\|\|\nabla\delta\vec{u}\| \leq \frac{1}{2}\left(\mathcal{R}\|\delta\vec{u}\|^2 + \frac{1}{\mathcal{R}}\|\nabla\delta\vec{u}\|^2\right),$$

and then use the fact that rescaling with respect to L gives a Poincaré bound $\|\delta\vec{u}\| \leq \|\nabla\delta\vec{u}\|$, we find that

$$\frac{\mathrm{d}}{\mathrm{d}t}\left(\frac{1}{2}\|\delta\vec{u}\|^2\right) \leq \frac{\mathcal{R}^2}{2}\|\delta\vec{u}\|^2 - \frac{1}{2}\|\nabla\delta\vec{u}\|^2 \leq (\mathcal{R}^2 - 1)\frac{1}{2}\|\delta\vec{u}\|^2.$$

Thus, defining $K(t) := \|\delta\vec{u}(\cdot, t)\|^2$ to be the energy of the perturbation, we have a simple linear differential inequality

$$\frac{\mathrm{d}K}{\mathrm{d}t} + (1 - \mathcal{R}^2)K \leq 0. \tag{7.27}$$

To solve this, we introduce the positive integrating factor $e^{(1-\mathcal{R}^2)t}$ and integrate from time $t = 0$. This gives the energy decay bound

$$K(t) \leq K(0)e^{-(1-\mathcal{R}^2)t}, \tag{7.28}$$

and we see that a sufficient condition for the decay of all perturbations is that $\mathcal{R} < 1$. In this case the flow problem is said to be *nonlinearly stable*.

The discussion above is in complete agreement with laboratory fluid flow experiments. In reality, there are at least two ways that a steady flow can lose stability. A first example is the flow around a cylinder placed in the center of a channel. At low flow rates (i.e. small Reynolds numbers), steady "laminar" flow patterns are observed. If the flow rate is gradually increased, then a critical Reynolds number will be reached when the steady flow loses stability and a time-periodic vortex shedding flow is obtained. Mathematically, this behavior is realized by a *Hopf bifurcation* of the Navier–Stokes equations. This corresponds to the existence of a critical Reynolds number $\mathcal{R}^*$ at which a single conjugate pair of eigenvalues $\lambda = \pm\mu\mathbf{j}$ of the linearized stability problem (7.22) would cross the imaginary axis if the Reynolds number were further increased.

A second, very different, form of bifurcation occurs if linearized stability is lost at a critical Reynolds number $\mathcal{R}_*$, with a real eigenvalue moving from the left half to the right half-plane as the Reynolds number is increased. For $\mathcal{R} > \mathcal{R}_*$ the Navier–Stokes problem can have multiple steady solutions, some of which are stable and some which are not. An example of this behavior in a laboratory is that of flow in a symmetric smoothly expanding channel. This is an example of a symmetry-breaking *pitchfork bifurcation*. At low flow rates, the flow is symmetric about the mid-channel and two equally sized recirculating eddies exist downstream of the expansion. Above a critical flow rate the flow remains steady but one of the eddies is bigger than the other and the flow is no longer symmetric about the mid-plane. There are two such stable states: either the left-hand eddy is the bigger one, or else the right-hand eddy is bigger. The original symmetric flow is now an unstable state. More generally, if the channel is not symmetric but close to being so, then one would still expect to see singular behavior at a Reynolds number close to $\mathcal{R}_*$. The difference in this case is that the basic flow solution would not be symmetric in the first place. This flow would retain its stability when the real eigenvalue crossed from the left half to the right half-plane. At the critical Reynolds number, called a quadratic fold point, a secondary flow will also become stable, thus giving two branches of stable solutions exactly as in the case of symmetry-breaking bifurcation. Details of the computation of such a flow are given in the review by Cliffe, Spence & Taverner [46].

From the point of view of approximation, any bifurcation point presents a challenge. In the cases of pitchfork bifurcations and quadratic fold points, the linearized problem (7.22) has a zero eigenvalue when $\mathcal{R} = \mathcal{R}_*$. This also has an impact on the convergence of the nonlinear iteration methods that are discussed in the next section. A common assumption that enables theoretical analysis concerns the existence of a *local branch of non-singular solutions*, see for example Girault & Raviart [85, pp. 297–300], Gunzburger [95, p. 14]. This is a set of solutions $\{(\vec{u}(\mathcal{R}), p(\mathcal{R}))\}$ of (7.10)–(7.11), parameterized by the Reynolds number. The assumption is that the linearized problem underlying (7.22) is non-singular up to a constant pressure. That is, for any $\vec{u}$ on the branch, there exists unique $\vec{w} \in \mathbf{H}^1_{E_0}$ and $r \in L_2(\Omega)$ (unique up to a constant) such that

$$c(\vec{u}; \vec{w}, \vec{v}) + c(\vec{w}; \vec{u}, \vec{v}) + \frac{1}{\mathcal{R}} \int_\Omega \nabla \vec{w} : \nabla \vec{v} - \int_\Omega r \, (\nabla \cdot \vec{v}), = \int_\Omega \vec{f} \cdot \vec{v}$$

$$\int_\Omega q \, (\nabla \cdot \vec{w}) = 0 \tag{7.29}$$

for $\vec{v} \in \mathbf{H}^1_{E_0}$ and $q \in L_2(\Omega)$. Making this assumption is the key to analyzing convergence to multiple solutions — that is when $\mathcal{R}$ is bigger that the critical Reynolds number $\mathcal{R}_*$ in the example above. As long as the Reynolds number is not too close to a critical value, the approximation results mirror those obtained in the chapter on the Stokes problem. We return to this topic in Section 7.4.1.

7.2.2 *Nonlinear iteration*

Solving the Navier–Stokes equations requires nonlinear iteration with a linearized problem being solved at every step. Thus, given an "initial guess"[2] $(\vec{u}_0, p_0) \in \mathbf{H}^1_E \times L_2(\Omega)$, a sequence of iterates $(\vec{u}_0, p_0), (\vec{u}_1, p_1), (\vec{u}_2, p_2), \ldots \in \mathbf{H}^1_E \times L_2(\Omega)$ is computed, which (we hope) converges to the solution of the weak formulation. We

---

[2]In the IFISS software, $(\vec{u}_0, p_0)$ is the solution of the corresponding Stokes problem.

introduce two classical linearization procedures below. The mixed finite element approximation of the resulting linearized systems is discussed in subsequent sections.

Newton iteration turns out to be a very natural approach. Given the iterate $(\vec{u}_k, p_k)$, we start by computing the *nonlinear residual* associated with the weak formulation (7.10)–(7.11). This is the pair $R_k(\vec{v}), r_k(q)$ satisfying

$$R_k = \int_\Omega \vec{f} \cdot \vec{v} - c(\vec{u}_k; \vec{u}_k, \vec{v}) - \nu \int_\Omega \nabla \vec{u}_k : \nabla \vec{v} + \int_\Omega p_k \, (\nabla \cdot \vec{v}),$$

$$r_k = -\int_\Omega q \, (\nabla \cdot \vec{u}_k)$$

for any $\vec{v} \in \mathbf{H}^1_{E_0}$ and $q \in L_2(\Omega)$. With $\vec{u} = \vec{u}_k + \delta \vec{u}_k$ and $p = p_k + \delta p_k$, the solution of (7.10)–(7.11), it is easy to see that the corrections $\delta \vec{u}_k \in \mathbf{H}^1_{E_0}$, $\delta p_k \in L_2(\Omega)$ satisfy

$$D(\vec{u}_k, \delta \vec{u}_k, \vec{v}) + \nu \int_\Omega \nabla \delta \vec{u}_k : \nabla \vec{v} - \int_\Omega \delta p_k \, (\nabla \cdot \vec{v}) = R_k(\vec{v}) \qquad (7.30)$$

$$\int_\Omega q \, (\nabla \cdot \delta \vec{u}_k) = r_k(q) \qquad (7.31)$$

for all $\vec{v} \in \mathbf{H}^1_{E_0}$ and $q \in L_2(\Omega)$, where $D(\vec{u}_k, \delta \vec{u}_k, \vec{v})$ is the difference in the non-linear terms, as in (7.21). Expanding $D(\vec{u}_k, \delta \vec{u}_k, \vec{v})$ and dropping the quadratic term $c(\delta \vec{u}_k; \delta \vec{u}_k, \vec{v})$ exactly as in the derivation of (7.22) gives the linear problem: for all $\vec{v} \in \mathbf{H}^1_{E_0}$ and $q \in L_2(\Omega)$, find $\delta \vec{u}_k \in \mathbf{H}^1_{E_0}$ and $\delta p_k \in L_2(\Omega)$ satisfying

$$c(\delta \vec{u}_k; \vec{u}_k, \vec{v}) + c(\vec{u}_k; \delta \vec{u}_k, \vec{v}) + \nu \int_\Omega \nabla \delta \vec{u}_k : \nabla \vec{v} - \int_\Omega \delta p_k \, (\nabla \cdot \vec{v}) = R_k(\vec{v}),$$

$$\int_\Omega q \, (\nabla \cdot \delta \vec{u}_k) = r_k(q).$$

$$(7.32)$$

The solution of (7.32) is the so-called Newton correction. Updating the previous iterate via $\vec{u}_{k+1} = \vec{u}_k + \delta \vec{u}_k$, $p_{k+1} = p_k + \delta p_k$ defines the next iterate in the sequence.

Let us check the consistency of the iteration process, by letting $\vec{u}_k = \vec{u}$ and $p_k = p$. Substituting this into the update formula reveals that the right-hand side of (7.32) is zero. This is where the notion of a non-singular branch of solutions is useful. Recalling the definition (7.29), we see that this assumption ensures consistency—if the right-hand side is zero then the only solution of (7.32) is $\delta \vec{u}_k = \vec{0}$ and $\delta p_k = 0$.

We discuss the finite-dimensional analogue of the Newton iteration strategy in the next section. Looking ahead to this, it is well known that if the Jacobian matrix is non-singular at the fixed point, then locally, Newton iteration converges quadratically. That is, once the current iterate is close to the fixed point, the iteration error is essentially squared at every step. An analogous result holds in the above infinite-dimensional case. Indeed, it can be rigorously established that if the initial velocity estimate $\vec{u}_0$ is "sufficiently close" to a branch of non-singular solutions, then quadratic convergence to a uniquely defined fixed point is guaranteed. Note that the initial pressure estimate p_0 can be arbitrary in this case. Further details can be found in Girault & Raviart [85, pp. 362–366].

The second approach for linearization is Picard's method. In terms of the representation (7.30)–(7.31), the quadratic term $c(\delta\vec{u}_k; \delta\vec{u}_k, \vec{v})$ is dropped along with the linear term $c(\delta\vec{u}_k; \vec{u}_k, \vec{v})$. Thus, instead of (7.32), we have the following linear problem: for all $\vec{v} \in \mathbf{H}^1_{E_0}$ and $q \in L_2(\Omega)$, find $\delta\vec{u}_k \in \mathbf{H}^1_{E_0}$ and $\delta p_k \in L_2(\Omega)$ satisfying

$$c(\vec{u}_k; \delta\vec{u}_k, \vec{v}) + \nu \int_\Omega \nabla\delta\vec{u}_k : \nabla\vec{v} - \int_\Omega \delta p_k \, (\nabla \cdot \vec{v}) = R_k(\vec{v}),$$

$$\int_\Omega q \, (\nabla \cdot \delta\vec{u}_k) = r_k(q). \tag{7.33}$$

The solution of (7.33) is the Picard correction. Updating the previous iterate via $\vec{u}_{k+1} = \vec{u}_k + \delta\vec{u}_k$, $p_{k+1} = p_k + \delta p_k$ defines the next iterate in the sequence.

If we substitute $\delta\vec{u}_k = \vec{u}_{k+1} - \vec{u}_k$ and $\delta p_k = p_{k+1} - p_k$ into (7.33), then we obtain an explicit definition for the new iterate: for all $\vec{v} \in \mathbf{H}^1_{E_0}$ and $q \in L_2(\Omega)$, find $\vec{u}_{k+1} \in \mathbf{H}^1_E$ and $p_{k+1} \in L_2(\Omega)$ such that

$$c(\vec{u}_k; \vec{u}_{k+1}, \vec{v}) + \nu \int_\Omega \nabla\vec{u}_{k+1} : \nabla\vec{v} - \int_\Omega p_{k+1} \, (\nabla \cdot \vec{v}) = \int_\Omega \vec{f} \cdot \vec{v}, \tag{7.34}$$

$$\int_\Omega q \, (\nabla \cdot \vec{u}_{k+1}) = 0. \tag{7.35}$$

The formulation (7.34)–(7.35) is commonly referred to as the *Oseen system*. Comparing (7.34)–(7.35) with the weak formulation we see that the Picard iteration corresponds to a simple fixed point iteration strategy for solving (7.10)–(7.11), with the convection coefficient evaluated at the current velocity. As a result, the rate of convergence of Picard iteration is only linear in general. We present some numerical results in the next chapter. Note that an initial pressure p_0 is not needed if the iteration is coded in the alternative form (7.34)–(7.35).

The main drawback of Newton's method is that the radius of the ball of convergence is typically proportional to the viscosity parameter ν. Thus, as the Reynolds number is increased, better and better initial guesses are needed in order for the Newton iteration to converge. The advantage of the Picard iteration is that, relative to Newton iteration, it has a huge ball of convergence.

As an example of this, let us consider the case of enclosed flow, where $\vec{w} = \vec{0}$. If the standard uniqueness condition (7.17) is satisfied, then it can be shown that Picard iteration is *globally* convergent; that is, it will converge to the uniquely defined flow solution for any initial velocity $\vec{u}_0$. See Karakashian [116] or Roos et al. [159, p. 282] for a rigorous statement and proof.

7.3 Mixed finite element approximation

Discretization aspects are discussed in this section. The mixed approximation methods in Section 5.3 are the obvious starting point. It should be emphasized up front that we do not know how to generalize the streamline-diffusion methodology discussed in Section 3.3.2 to give a consistent stabilization of the convection operator in (7.10). As a result we only consider standard Galerkin discretization in the remainder of the book.[3] There are two other issues that arise in generalizing the material of Chapter 5 on the Stokes equations to the Navier–Stokes equations. The first concerns the need to linearize the convection term. We focus on this issue in this section. The second issue concerns the need to stabilize the approximation of the linearized convection–diffusion operator on coarse grids when multigrid is used to solve the discrete problem. This can be done using the SD stabilization discussed in Chapter 3, and we defer further discussion until the next chapter.

As in Chapter 5, a discrete weak formulation is defined using finite dimensional spaces $\mathbf{X}_0^h \subset \mathbf{H}_{E_0}^1$ and $M^h \subset L_2(\Omega)$. Specifically, given a velocity solution space $\mathbf{X}_E^h$, the discrete version of (7.10)–(7.11) is: find $\vec{u}_h \in \mathbf{X}_E^h$ and $p_h \in M^h$ such that

$$\nu \int_\Omega \nabla \vec{u}_h : \nabla \vec{v}_h + \int_\Omega (\vec{u}_h \cdot \nabla \vec{u}_h) \cdot \vec{v}_h - \int_\Omega p_h (\nabla \cdot \vec{v}_h)$$

$$= \int_\Omega \vec{f} \cdot \vec{v}_h \quad \text{for all } \vec{v}_h \in \mathbf{X}_0^h, \tag{7.36}$$

$$\int_\Omega q_h (\nabla \cdot \vec{u}_h) = 0 \quad \text{for all } q_h \in M^h. \tag{7.37}$$

Implementation entails defining appropriate bases for the finite element spaces, leading to a nonlinear system of algebraic equations. Linearization of this system using Newton iteration gives the finite-dimensional analogue of (7.32): find corrections $\delta \vec{u}_h \in \mathbf{X}_0^h$ and $\delta p_h \in M^h$ (dropping the subscript k to avoid notational

---

[3]This also means the IFISS software is *not* general purpose: it is tacitly assuming that grids are fine enough to ensure that any layers in the flow solution are resolved.

clutter) satisfying

$$c(\delta\vec{u}_h; \vec{u}_h, \vec{v}_h) + c(\vec{u}_h; \delta\vec{u}_h, \vec{v}_h) + \nu \int_\Omega \nabla\delta\vec{u}_h : \nabla\vec{v}_h - \int_\Omega \delta p_h(\nabla \cdot \vec{v}_h) = R_k(\vec{v}_h),$$

$$\int_\Omega q_h (\nabla \cdot \delta\vec{u}_h) = r_k(q_h) \qquad (7.38)$$

for all $\vec{v}_h \in \mathbf{X}_0^h$ and $q_h \in M^h$. Here, $R_k(\vec{v}_h)$ and $r_k(q_h)$ are the nonlinear residuals associated with the discrete formulation (7.36)–(7.37). Dropping the term $c(\delta\vec{u}_h; \vec{u}_h, \vec{v}_h)$ gives the discrete analogue of the Picard update (7.33).

To define the corresponding linear algebra problem, we use a set of vector-valued basis functions $\{\vec{\phi}_j\}$, so that

$$\vec{u}_h = \sum_{j=1}^{n_u} \mathbf{u}_j\vec{\phi}_j + \sum_{j=n_u+1}^{n_u+n_\partial} \mathbf{u}_j\vec{\phi}_j, \quad \delta\vec{u}_h = \sum_{j=1}^{n_u} \Delta\mathbf{u}_j\vec{\phi}_j, \qquad (7.39)$$

and we fix the coefficients $\mathbf{u}_j : j = n_u + 1, \ldots, n_u + n_\partial$, so that the second term interpolates the boundary data on $\partial\Omega_D$. We also introduce a set of pressure basis functions $\{\psi_k\}$ and set

$$p_h = \sum_{k=1}^{n_p} \mathbf{p}_k\,\psi_k, \quad \delta p_h = \sum_{k=1}^{n_p} \Delta\mathbf{p}_k\,\psi_k. \qquad (7.40)$$

Substituting into (7.38) then gives a system of linear equations

$$\begin{bmatrix} \nu\mathbf{A} + \mathbf{N} + \mathbf{W} & B^T \\ B & O \end{bmatrix} \begin{bmatrix} \Delta\mathbf{u} \\ \Delta\mathbf{p} \end{bmatrix} = \begin{bmatrix} \mathbf{f} \\ \mathbf{g} \end{bmatrix}. \qquad (7.41)$$

The matrix $\mathbf{A}$ is the vector-Laplacian matrix and the matrix B is the divergence matrix, defined in (5.34) and (5.35), respectively. The new matrices in (7.41) are the vector-convection matrix $\mathbf{N}$ and the Newton derivative matrix $\mathbf{W}$. These depend on the current estimate of the discrete velocity $\vec{u}_h$, and the entries are given by

$$\mathbf{N} = [\mathbf{n}_{ij}], \quad \mathbf{n}_{ij} = \int_\Omega (\vec{u}_h \cdot \nabla\vec{\phi}_j) \cdot \vec{\phi}_i, \qquad (7.42)$$

$$\mathbf{W} = [\mathbf{w}_{ij}], \quad \mathbf{w}_{ij} = \int_\Omega (\vec{\phi}_j \cdot \nabla\vec{u}_h) \cdot \vec{\phi}_i \qquad (7.43)$$

for i and $j = 1, \ldots, n_u$. Notice that the Newton derivative matrix is symmetric. The right-hand side vectors in (7.41) are the nonlinear residuals associated with the current discrete solution estimates $\vec{u}_h$ and p_h, expanded via (7.39) and (7.40):

$$\mathbf{f} = [\boldsymbol{f}_i], \quad \boldsymbol{f}_i = \int_\Omega \vec{f} \cdot \vec{\phi}_i - \int_\Omega \vec{u}_h \cdot \nabla \vec{u}_h \cdot \vec{\phi}_i$$

$$- \nu \int_\Omega \nabla \vec{u}_h : \nabla \vec{\phi}_i + \int_\Omega p_h \left(\nabla \cdot \vec{\phi}_i \right), \qquad (7.44)$$

$$\mathbf{g} = [\boldsymbol{g}_k], \quad \boldsymbol{g}_k = \int_\Omega \psi_k \left(\nabla \cdot \vec{u}_h \right). \qquad (7.45)$$

The system (7.41) is referred to as the *discrete Newton problem*. For Picard iteration, we omit the Newton derivative matrix to give the *discrete Oseen problem*:

$$\begin{bmatrix} \nu \mathbf{A} + \mathbf{N} & B^T \\ B & O \end{bmatrix} \begin{bmatrix} \boldsymbol{\Delta}\mathbf{u} \\ \boldsymbol{\Delta}\mathbf{p} \end{bmatrix} = \begin{bmatrix} \mathbf{f} \\ \mathbf{g} \end{bmatrix}. \qquad (7.46)$$

In general, the components of velocity are approximated using a single finite element space. Given the two-dimensional splitting in (5.38), it can be shown (see Problem 7.3) that the system (7.41) can be rewritten as

$$\begin{bmatrix} A + N + W_{xx} & W_{xy} & B_x^T \\ W_{yx} & A + N + W_{yy} & B_y^T \\ B_x & B_y & O \end{bmatrix} \begin{bmatrix} \boldsymbol{\Delta}\mathbf{u}_x \\ \boldsymbol{\Delta}\mathbf{u}_y \\ \boldsymbol{\Delta}\mathbf{p} \end{bmatrix} = \begin{bmatrix} \mathbf{f}_x \\ \mathbf{f}_y \\ \mathbf{g} \end{bmatrix}, \qquad (7.47)$$

where the matrix N is the $n \times n$ scalar convection matrix

$$N = [n_{ij}], \quad n_{ij} = \int_\Omega (\vec{u}_h \cdot \nabla \phi_j)\phi_i, \qquad (7.48)$$

which was introduced and studied in Section 3.5. The $n \times n$ matrices W_{xx}, W_{xy}, W_{yx} and W_{yy} represent weak derivatives of the current velocity in the x and y directions; for example, defining the x component of $\vec{u}_h$ by u_x we have that

$$W_{xy} = [w_{xy,ij}], \quad w_{xy,ij} = \int_\Omega \frac{\partial u_x}{\partial y}\phi_i\phi_j. \qquad (7.49)$$

As discussed in Section 5.3.2, the lowest order mixed approximations like $\boldsymbol{Q}_1$–$\boldsymbol{P}_0$ and $\boldsymbol{Q}_1$–$\boldsymbol{Q}_1$ are unstable. In this case, the discrete approximation is stabilized by replacing the zero block in the Newton system (7.41) and the Oseen system (7.46) with a stabilization matrix. For example, the stabilized analogue of the Oseen system (7.46) is given by

$$\begin{bmatrix} \nu \mathbf{A} + \mathbf{N} & B^T \\ B & -\frac{1}{\nu}C \end{bmatrix} \begin{bmatrix} \mathbf{u} \\ \mathbf{p} \end{bmatrix} = \begin{bmatrix} \mathbf{f} \\ \mathbf{g} \end{bmatrix}, \qquad (7.50)$$

where C is the stabilization matrix in (5.84) or (5.94). Note that the scaling of the stabilization matrix in (7.50) is inversely proportional to the viscosity parameter. This scaling is included to ensure that if the convection matrix $\mathbf{N}$ is set to zero then the computed velocity solution is independent of ν and solves the corresponding stabilized Stokes system.

A sufficient condition for these discrete Oseen and Newton problems to be well-posed is given by the following general linear algebra result.

Proposition 7.1. *Define the generic system matrix,*

$$K = \begin{bmatrix} \mathbf{F} & B^T \\ B & -C \end{bmatrix},$$

where C is a symmetric and positive semi-definite matrix. If $\langle \mathbf{F}\,\mathbf{u}, \mathbf{u} \rangle > 0$ for all $\mathbf{u} \neq \mathbf{0}$, then

$$\mathtt{null}(K) = \left\{ \begin{bmatrix} \mathbf{0} \\ \mathbf{p} \end{bmatrix} \,\middle|\, \mathbf{p} \in \mathtt{null}(B\mathbf{F}^{-1}B^T + C) \right\}.$$

Proof See Problem 7.4. □

7.4 Theory of errors

Our treatment of a priori error bounds is intentionally brief since the technical material needed for a rigorous treatment is beyond the scope of this book. Mathematically minded readers are referred to Girault & Raviart [85, pp. 316–323] for definitive results. The development of efficient a posteriori error estimation techniques for the Navier–Stokes problem is an active research topic. The books of Ainsworth & Oden [4, pp. 219–222] or Verfürth [204, pp. 84–99] provide the basis for the error estimation strategy that is described in Section 7.4.2.

Throughout this section we restrict attention to the simplest case of enclosed flow with $\partial\Omega = \partial\Omega_D$ and $\vec{w} = \vec{0}$. We also recall, from (5.100), the continuous bilinear forms $a : \mathbf{H}^1 \times \mathbf{H}^1 \to \mathbb{R}$, and $b : \mathbf{H}^1 \times L_2(\Omega) \to \mathbb{R}$

$$a(\vec{u}, \vec{v}) := \nu \int_\Omega \nabla\vec{u} : \nabla\vec{v}, \quad b(\vec{v}, q) := -\int_\Omega q\,(\nabla \cdot \vec{v}). \tag{7.51}$$

Given the continuous functional $\ell : \mathbf{H}^1 \to \mathbb{R}$

$$\ell(\vec{v}) := \int_{\partial\Omega} \vec{f} \cdot \vec{v}; \tag{7.52}$$

the underlying weak formulation (7.10)–(7.12) may be restated as

---

Find $\vec{u} \in \mathbf{H}^1_{E_0}$ and $p \in L_2(\Omega)$ such that

$$a(\vec{u}, \vec{v}) + c(\vec{u}; \vec{u}, \vec{v}) + b(\vec{v}, p) = \ell(\vec{v}) \quad \text{for all } \vec{v} \in \mathbf{H}^1_{E_0},$$
$$b(\vec{u}, q) = 0 \quad \text{for all } q \in L_2(\Omega). \tag{7.53}$$

With a conforming mixed approximation, the corresponding discrete problem (7.36)–(7.37) is given by

Find $\vec{u}_h \in \mathbf{X}_0^h$ and $p_h \in M^h$ such that

$$a(\vec{u}_h, \vec{v}_h) + c(\vec{u}_h; \vec{u}_h, \vec{v}_h) + b(\vec{v}_h, p_h) = \ell(\vec{v}_h) \quad \text{for all } \vec{v}_h \in \mathbf{X}_0^h,$$
$$b(\vec{u}_h, q_h) = 0 \quad \text{for all } q_h \in M^h. \tag{7.54}$$

Our aim is to bound $\|\vec{u} - \vec{u}_h\|_X$ and $\|p - p_h\|_M$ with respect to the energy norm for the velocity $\|\vec{v}\|_X := \|\nabla \vec{v}\|$ and the quotient norm for the pressure $\|p\|_M := \|p\|_{0,\Omega}$. The essential message of the next section is that the mixed methods for the Stokes equations can be used to approximate branches of non-singular solutions of the Navier–Stokes problem (7.53) with the same order of accuracy.

7.4.1   A priori error bounds

Given the ellipticity condition (7.15) and the continuity condition (7.16) associated with the existence of a weak solution satisfying (7.53), the key to an effective approximation turns out to be the *inf–sup* condition. We summarize this fact in the following theorem.

Theorem 7.2. *Let $\Lambda \subset \mathbb{R}_+$ and let $\{(\vec{u}(\mathcal{R}), p(\mathcal{R})); \mathcal{R} \in \Lambda\}$ denote a branch of non-singular solutions of (7.53) in the sense of (7.29) with Reynolds number $\mathcal{R} = 1/\nu$. Assume that there exists a constant $\gamma^* > 0$, independent of h, such that the mixed approximation spaces in (7.54) satisfy the inf–sup condition*

$$\max_{\vec{v}_h \in \mathbf{X}_0^h} \frac{b(\vec{v}_h, q_h)}{\|\vec{v}_h\|_X} \geq \gamma^* \|q_h\|_M \quad \text{for all } q_h \in M^h.$$

Then, for h sufficiently small, there exists a unique branch of non-singular solutions $\{(\vec{u}_h(\mathcal{R}), p_h(\mathcal{R})); \mathcal{R} \in \Lambda\}$ satisfying (7.54). Moreover, the discrete solution branch is continuously differentiable with respect to $\mathcal{R}$, and for any fixed $\mathcal{R} \in \Lambda$ there exists a constant $C > 0$, such that

$$\|\vec{u} - \vec{u}_h\|_X + \|p - p_h\|_M \leq C \left(\inf_{\vec{v}_h \in \mathbf{X}_0^h} \|\vec{u} - \vec{v}_h\|_X + \inf_{q_h \in M^h} \|p - q_h\|_M \right).$$

$$\tag{7.55}$$

Proof See Girault & Raviart [85, Theorem 4.1]. $\square$

Remark 7.3. The bound (7.55) can be generalized to cover finite element approximation of singular cases including limit points and symmetry breaking bifurcation points. See Cliffe, Spence and Taverner [46] for details.

The importance of the bound (7.55) is that satisfaction of the discrete inf-sup stability condition ensures that an optimal rate of convergence is achieved whenever the underlying Navier–Stokes problem is sufficiently regular.

Definition 7.4 ((Navier–Stokes $\mathcal{H}^k$ regularity)). The variational problem (7.53) is said to be $\mathcal{H}^k$-regular if $\mathcal{R} \to (\vec{u}(\mathcal{R}), p(\mathcal{R}))$ is a continuous mapping from the solution branch Λ into the space $\mathbf{H}^k(\Omega) \times \mathcal{H}^{k-1}(\Omega)$.

A specific example is given in the following theorem.

Theorem 7.5. *If an $\mathcal{H}^3$-regular variational problem* (7.53) *is solved using either $\mathbf{Q}_2$–$\mathbf{Q}_1$ approximation or $\mathbf{Q}_2$–$\mathbf{P}_{-1}$ approximation on a grid $\mathcal{T}_h$ of rectangular elements, and if the aspect ratio condition is satisfied (see Definition 1.18), then for all $\mathcal{R} \in \Lambda$, there exists a constant C_2 such that*

$$\|\nabla(\vec{u} - \vec{u}_h)\| + \|p - p_h\|_{0,\Omega} \leq C_2 \, h^2 \, (\|D^3\vec{u}\| + \|D^2 p\|), \qquad (7.56)$$

where h is the length of the longest edge in $\mathcal{T}_h$.

Proof The bound for $\mathbf{Q}_2$–$\mathbf{Q}_1$ approximation follows from combining (7.55) with the interpolation error bounds (5.118) and (5.119). □

Remark 7.6. In the special case of Poiseuille flow (7.6), the right-hand side of (7.56) is identically zero. This means that, independently of the grid that is used, computed $\mathbf{Q}_2$–$\mathbf{Q}_1$ and $\mathbf{Q}_2$–$\mathbf{P}_{-1}$ solutions of Example 7.1.1 should always be *exact*, see Computational Exercise 7.1.

Remark 7.7. As in Chapters 3 and 5, the statement of the bound (7.56) needs to be qualified in the presence of layers and singularities. In the case of the flow problem in Example 7.1.4, streamline derivatives are inversely proportional to $\sqrt{\nu}$, and so $\|D^3 u\|_{\square_k}$ blows up as $\nu \to 0$ in elements $\square_k$ that lie within such a layer. The flow problem in Example 7.1.2 is not even $\mathcal{H}^2$-regular. In this case, the rate of convergence is slower than $O(h)$ for a uniform grid sequence, independent of the order of the mixed approximation.

To complete the section, we state the analogue of Theorem 5.11.

Proposition 7.8. *Suppose an $\mathcal{H}^2$-regular problem* (7.53) *is solved using an ideally stabilized $\mathbf{Q}_1$–$\mathbf{P}_0$ approximation on a grid $\mathcal{T}_h$ of rectangular elements. That is, with $\beta = \beta^* = 1/4$ fixed, the discrete incompressibility constraint in* (7.54) *is replaced by*

$$b(\vec{u}_h, q_h) - \frac{\beta^*}{\nu} \sum_{M \in \mathcal{T}_M} |M| \sum_{e \in \Gamma_M} \langle [\![p_h]\!]_e, [\![q_h]\!]_e \rangle_{\bar{E}} = 0 \quad \text{for all } q_h \in M^h, \quad (7.57)$$

see (7.50), *where each common boundary between the macroelements $M \in \mathcal{T}_M$ has at least one interior velocity node. Then, if the usual aspect ratio condition* *is satisfied, for all $\mathcal{R} \in \Lambda$ there exists a constant C_1, such that*

$$\|\nabla(\vec{u} - \vec{u}_h)\| + \|p - p_h\|_{0,\Omega} \leq C_1 \, h \, \left(\|D^2\vec{u}\| + \|D^1 p\|\right), \qquad (7.58)$$

where h is the length of the longest edge in $\mathcal{T}_h$.

7.4.2 A posteriori error bounds

We have a specific objective in this section, namely, to explore the effectiveness of a particular strategy for a priori error estimation. (This technique is built into the IFISS software.) The estimator requires the solution of local Poisson problems and is the natural extension of the approach developed in Sections 1.5.2 and 5.4.2.

Recalling the "big" symmetric bilinear form

$$B((\vec{u}, p); (\vec{v}, q)) = a(\vec{u}, \vec{v}) + b(\vec{u}, q) + b(\vec{v}, p), \qquad (7.59)$$

associated with the Stokes operator in (7.53), and introducing the functional $F((\vec{v}, q)) = \ell(\vec{v})$ associated with the forcing term, we have that the errors $\vec{e} := \vec{u} - \vec{u}_h \in \mathbf{H}^1_{E_0}$ and $\epsilon := p - p_h \in L_2(\Omega)$ associated with (7.53) and (7.54) satisfy

$$\begin{aligned}
B((\vec{e}, \epsilon); (\vec{v}, q)) &= B((\vec{u} - \vec{u}_h, p - p_h); (\vec{v}, q)) \\
&= B((\vec{u}, p); (\vec{v}, q)) - B((\vec{u}_h, p_h); (\vec{v}, q)) \\
&= F((\vec{v}, q)) - c(\vec{u}; \vec{u}, \vec{v}) - B((\vec{u}_h, p_h); (\vec{v}, q)) \\
&= -c(\vec{u}; \vec{u}, \vec{v}) + \ell(\vec{v}) - a(\vec{u}_h, \vec{v}) - b(\vec{v}, p_h) - b(\vec{u}_h, q),
\end{aligned}$$
$$\qquad (7.60)$$

for all $(\vec{v}, q) \in \mathbf{H}^1_{E_0} \times L_2(\Omega)$. To simplify (7.60), we recall the difference operator $D(\cdot, \cdot, \cdot)$ from (7.21) and note that

$$D(\vec{u}_h, \vec{e}, \vec{v}) = c(\vec{e} + \vec{u}_h; \vec{e} + \vec{u}_h, \vec{v}) - c(\vec{u}_h; \vec{u}_h, \vec{v}) = c(\vec{u}; \vec{u}, \vec{v}) - c(\vec{u}_h; \vec{u}_h, \vec{v}).$$

If we then combine this with (7.60), we get

$$\begin{aligned}
B((\vec{e}, \epsilon); &(\vec{v}, q)) + D(\vec{u}_h, \vec{e}, \vec{v}) \\
&= -c(\vec{u}_h; \vec{u}_h, \vec{v}) + \ell(\vec{v}) - a(\vec{u}_h, \vec{v}) - b(\vec{v}, p_h) - b(\vec{u}_h, q). \qquad (7.61)
\end{aligned}$$

The error can be localized by breaking up the right-hand side of (7.61) and integrating by parts using (5.140). Following the path taken in Section 5.4.2, and defining the equidistributed *stress jump* operator

$$\vec{R}^*_E := \begin{cases} \frac{1}{2} \llbracket \nu \nabla \vec{u}_h - p_h \vec{\mathbf{I}} \rrbracket & E \in \mathcal{E}_{h,\Omega} \\ -\left(\nu \frac{\partial \vec{u}_h}{\partial n_{E,T}} - p_h \vec{n}_{E,T}\right) & E \in \mathcal{E}_{h,N} \\ 0 & E \in \mathcal{E}_{h,D}, \end{cases} \qquad (7.62)$$

and the interior residuals $\vec{R}_T := \{\vec{f} + \nu\nabla^2\vec{u}_h - \vec{u}_h \cdot \nabla\vec{u}_h - \nabla p_h\}|_T$ and $R_T := \{\nabla \cdot \vec{u}_h\}|_T$, we find that the errors $\vec{e} \in \mathbf{H}^1_{E_0}$ and $\epsilon \in L_2(\Omega)$ satisfy the nonlinear equation

$$B((\vec{e}, \epsilon); (\vec{v}, q)) + D(\vec{u}_h, \vec{e}, \vec{v}) = \sum_{T \in \mathcal{T}_h} \left[(\vec{R}_T, \vec{v})_T - \sum_{E \in \mathcal{E}(T)} \left\langle \vec{R}_E^*, \vec{v} \right\rangle_E + (R_T, q)_T \right] \quad (7.63)$$

for all $(\vec{v}, q) \in \mathbf{H}_{E_0}^1 \times L_2(\Omega)$.

The error characterization (7.63) is the starting point for a posteriori error analysis. The crucial question here is to determine the best way of handling the nonlinear term on the left-hand side of (7.63). To provide an answer, we follow the argument in Section 7.2.1. First, as in (7.21), we note that

$$D(\vec{u}_h, \vec{e}, \vec{v}) = c(\vec{e}; \vec{e}, \vec{v}) + c(\vec{e}; \vec{u}_h, \vec{v}) + c(\vec{u}_h; \vec{e}, \vec{v}).$$

Second, dropping the quadratic term, we see that the problem (7.63) can be approximated by the linear problem: find $\vec{e} \in \mathbf{H}_{E_0}^1$ and $\epsilon \in L_2(\Omega)$ such that

$$c(\vec{e}; \vec{u}_h, \vec{v}) + c(\vec{u}_h; \vec{e}, \vec{v}) + \nu \int_\Omega \nabla \vec{e} : \nabla \vec{v} - \int_\Omega \epsilon \, (\nabla \cdot \vec{v})$$

$$= \sum_{T \in \mathcal{T}_h} \left[(\vec{R}_T, \vec{v})_T - \sum_{E \in \mathcal{E}(T)} \left\langle \vec{R}_E^*, \vec{v} \right\rangle_E \right]$$

$$- \int_\Omega q \, (\nabla \cdot \vec{e}) = \sum_{T \in \mathcal{T}_h} (R_T, q)_T. \quad (7.64)$$

Notice the close connection between (7.64) and the Newton update problem (7.32). Basically, (7.64) corresponds to a Newton linearization about the discrete velocity solution $\vec{u}_h$. This connection reinforces the importance of the a priori error analysis. The conditions of the statement of Theorem 7.2 imply the existence a unique branch of non-singular solutions $\{(\vec{u}_h(\mathcal{R}), p_h(\mathcal{R})); \mathcal{R} \in \Lambda\}$ that is arbitrarily close to the branch $\{(\vec{u}(\mathcal{R}), p(\mathcal{R})); \mathcal{R} \in \Lambda\}$ solving (7.53). In this case the linearized problem (7.64) is guaranteed to be well posed.

We shall concentrate on the stabilized Q_1–P_0 or P_1–P_0 approximation methods in two dimensions for the remainder of the section. Notice that for either of these low order approximations, the divergence residual R_T is piecewise constant and the stress jump term $\vec{R}_E^*$ is piecewise linear. The other element residual is given by $\vec{R}_T = \{\vec{f} - \vec{u}_h \cdot \nabla \vec{u}_h\}|_T$.

A direct approximation of (7.64) suggests that a local linearized Navier–Stokes problem needs to be solved in order to compute a local error estimator. Following the strategy of Ainsworth and Oden [4, section 9.3]) however, we can omit the convective term on the left-hand side of (7.64) and solve local Poisson problems in exactly the same way as for the Stokes equations. To this end, recalling the Stokes characterization (5.143)–(5.144), we introduce the higher order approximation space $\vec{\mathcal{Q}}_T := (\mathcal{Q}_T)^2$ for the two velocity components and compute the element function $\vec{e}_T \in \vec{Q}_T$ satisfying the uncoupled Poisson problems

$$\nu(\nabla \vec{e}_T, \nabla \vec{v})_T = (\vec{R}_T, \vec{v})_T - \sum_{E \in \mathcal{E}(T)} \left\langle \vec{R}_E^*, \vec{v} \right\rangle_E \quad \text{for all } \vec{v} \in \vec{Q}_T. \quad (7.65)$$

With $\epsilon_T = \nabla \cdot \vec{u}_h$, the local error estimate is given by

$$\eta_T^2 := \|\nabla \vec{e}_T\|_T^2 + \|\epsilon_T\|_T^2 = \|\nabla \vec{e}_T\|_T^2 + \|\nabla \cdot \vec{u}_h\|_T^2, \quad (7.66)$$

and the global error estimator is $\eta := \left(\sum_{T \in \mathcal{T}_h} \eta_T^2 \right)^{1/2}$. The theoretical underpinning for this strategy is the following result.

Proposition 7.9. *Let the condition* (7.17) *be satisfied so that there exists a unique solution* $(\vec{u}, p)$ *to the variational problem* (7.53) *in the enclosed flow case* $\vec{u} = \vec{0}$ *on* $\partial\Omega$. *If the discrete problem* (7.54) *is solved using a grid of* $\boldsymbol{Q}_1$–$\boldsymbol{P}_0$ *rectangular elements with a pressure jump stabilization term* (5.130), *then the estimator* η_T *in* (7.66) *satisfies the upper bound*

$$\|\nabla(\vec{u} - \vec{u}_h)\| + \|p - p_h\|_{0,\Omega} \leq C \left(\sum_{T \in \mathcal{T}_h} \eta_T^2 \right)^{1/2}, \quad (7.67)$$

where C *depends only on the aspect ratio constant given in Definition* 1.18 *and the continuous B-stability constant* γ_Ω *in Theorem* 5.18.

The proof of (7.67) is technical and is omitted. A global upper bound for the associated residual estimator, that is

$$\|\nabla(\vec{u} - \vec{u}_h)\| + \|p - p_h\|_{0,\Omega}$$

$$\leq C(\gamma_\Omega) \left(\sum_{T \in \mathcal{T}_h} \left\{ h_T^2 \|\vec{R}_T\|_T^2 + \|R_T\|_T^2 + \sum_{E \in \mathcal{E}(T)} h_E \|\vec{R}_E^*\|_E^2 \right\} \right)^{1/2}$$

is established in Verfürth [204, Prop. 3.19]. The small data assumption (7.17) is needed in the statement of Proposition 7.9 in order to get a bound on the nonlinear term — we simply state a well-known result here.

Lemma 7.10. *If* (7.17) *is satisfied then* $\|\nabla \vec{u}\| \leq \theta \frac{\nu}{\Gamma_*}$ *for some fixed number* $\theta \in (0, 1]$, *and the nonlinear term in* (7.63) *can be bounded as follows:*

$$D(\vec{u}_h, \vec{e}, \vec{v}) \leq (2\nu + \Gamma_* \|\nabla \vec{e}\|) \|\nabla \vec{e}\| \|\nabla \vec{v}\| \quad \text{for all } \vec{v} \in \mathbf{H}_{E_0}^1. \quad (7.68)$$

Proof See Ainsworth and Oden [4, pp. 220–222].   □

Remark 7.11. The efficiency of the error estimator η_T in (7.66) is an open question. The bound on the estimator for the convection–diffusion equation in (3.66) and the bounds [204, Prop. 3.19] suggest that η_T will provide a local lower bound on the error as long as the mesh Peclet number is small in the regions of rapid variation of the streamline derivative of the solution $\vec{u}$.

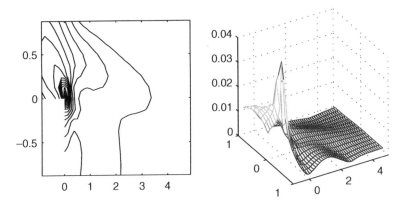

FIG. 7.4. Estimated error η_T associated with a 32×96 square grid ($\ell = 5$) $\boldsymbol{Q}_1$–$\boldsymbol{P}_0$ solution for the flow over a step, Example 7.1.2: $\nu = 1/50$, $\beta = \beta^*$.

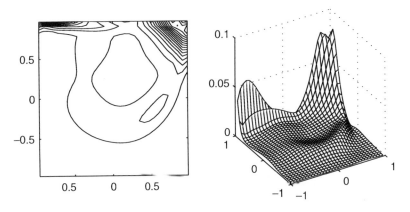

FIG. 7.5. Estimated error η_T associated with a 64×64 square grid ($\ell = 6$) $\boldsymbol{Q}_1$–$\boldsymbol{P}_0$ solution to the regularized driven cavity problem, Example 5.1.3: $\nu = 1/200$, $\beta = \beta^*$.

Some plots of the estimated error distribution associated with computed solutions to the problems in Examples 7.1.2 and 7.1.3 are presented in Figure 7.4 and Figure 7.5, respectively. The plot in Figure 7.4 corresponds to the solution illustrated in Figure 7.1 computed with $\boldsymbol{Q}_1$–$\boldsymbol{P}_0$ approximation instead of $\boldsymbol{Q}_2$–$\boldsymbol{Q}_1$. Comparing Figure 7.4 with the corresponding error plot for the Stokes solution in Figure 5.20, we can see that the effect of corner singularity is smeared out in the Navier–Stokes case, leading to a much smoother error distribution. The error plot for the driven cavity problem in Figure 7.5 should be compared with its Stokes flow analogue in Figure 5.21. The most important feature in the convection-dominated case is the shear layer, which originates at the lid

Table 7.1 *Estimated errors for regularized driven cavity flow using Q_1–P_0 approximation; $\nu = 1/200$, $\beta = \beta^*$, ℓ is the grid refinement level and $h = 2^{1-\ell}$.*

ℓ	$\|\nabla \cdot \vec{u}_h\|_\Omega$	η	n_u
4	5.288×10^{-2}	3.605×10^0	450
5	1.997×10^{-2}	1.972×10^0	1922
6	5.970×10^{-3}	8.845×10^{-1}	7938
7	1.593×10^{-3}	4.166×10^{-1}	32258

boundary and is swept around the cavity to give circular-shaped error contours. Some computed errors are given in Table 7.1 to show the effect of grid refinement in this case. These results, together with the bound (7.67), suggest that the Q_1–P_0 approximation is converging to the exact flow solution at close to optimal rate, that is $O(h)$. Notice that the error estimates for the Navier–Stokes cavity problem are much larger than the Stokes flow analogues given in Table 5.5, suggesting that much finer grids are needed to get the same level of accuracy. The divergence error is super-convergent however, mirroring the results for Stokes flow. The case of flow over a step is explored in Computational Exercise 7.3.

Discussion and bibliographical notes

There is an alarming number of alternative ways of formulating the Navier–Stokes equations. A complete list of possibilities can be found in Gresho & Sani [92, pp. 452–456]. In particular, the convection term $\vec{u} \cdot \nabla \vec{u}$ may also be expressed in a divergence form, rotational form or skew-symmetric form. Although these alternative representations are equivalent in the continuum, they typically lead to discretized models that are different. The *skew-symmetric* form is given by

$$-\nu\nabla^2\vec{u} + \vec{u} \cdot \nabla\vec{u} + \tfrac{1}{2}\vec{u}(\nabla \cdot \vec{u}) + \nabla p = \vec{f},$$

$$\nabla \cdot \vec{u} = 0.$$

This formulation of the Navier–Stokes equations is a convenient starting point for analysis (it was introduced for this reason by Temam, see [191]). For enclosed flow boundary conditions, the convection matrix in the corresponding discrete system is guaranteed to be skew-symmetric. This is true for the standard formulation (7.1) *only if* the discrete velocity $\vec{u}_h$ is divergence-free everywhere in the flow domain.

7.1. Channel flow over a step is a standard test problem. A detailed discussion of the stability of the steady-state flow can be found in Gresho et al. [90]. An important feature of driven-cavity flows is that they enable detailed comparisons to be made between results obtained by experiment, theory and computation. The results of Ghia et al. [84] are still used as a benchmark for comparison purposes. An accurate solution of the Blasius flow problem can be found in Shi et al. [172].

7.2. Further discussion of the theoretical aspects of the existence, uniqueness, and continuous dependence on the data for the steady-state Navier–Stokes equations can be found in the classic texts by Ladyzhenskaya [126] and Temam [191, 192]. The discussion of stability theory in Section 7.2.1 is taken from the excellent book of Doering & Gibbon [50, chap. 2].

In practice, the flow solution at a given Reynolds number may be of limited interest. For example, one might be more interested in predicting the critical Reynolds number corresponding to a Hopf bifurcation (signalling the onset of a time-periodic solution, see Cliffe et al. [46]), or in computing a stable flow solution beyond a fold point. In such cases, starting from a Stokes flow solution and using Picard or Newton iteration is unlikely to work since the solution is ill-conditioned (in terms of being sensitive to perturbations) at the bifurcation point. The only way forward in such cases is to employ a so-called *continuation* technique. The simplest idea — continuation in Reynolds number — is to solve a problem at a lower Reynolds number $\mathcal{R}_0$ and then to use the solution $(\vec{u}_h(\mathcal{R}_0), p_h(\mathcal{R}_0))$ as the starting guess for Newton's iteration at the desired Reynolds number $\mathcal{R}_* > \mathcal{R}_0$. Further details are given in Gunzburger [95, pp. 105–108]. Reynolds number continuation will not work however as a turning point is approached. Near such points a different continuation parameter must be used, such as the arc-length along the solution curve. This latter idea is known as *pseudo-arclength continuation* and was developed by Keller, see [121]. For an excellent introduction to the mathematics underlying bifurcation, see Spence & Graham [181].

7.3. We have restricted our attention to standard Galerkin approximations throughout the chapter. We can get away with this because we are restricting attention to steady flows, and these only occur at low Reynolds number. Solving flows at higher Reynolds numbers would almost certainly necessitate the introduction of stabilization methods of streamline-diffusion type. Such stabilization methods were introduced for incompressible fluid flow problems by Johnson & Saranen [113], and were developed by Tobiska & Lube, see [193]. Analysis of a more general class of so-called GLS (Galerkin Least Squares) methods can be found in Roos et al. [159, chap. IV].

The "local" a posteriori error estimation technique discussed in Section 7.4.1 is limited to low Reynolds number flows. In general, error estimation in the vicinity of bifurcation points is fraught with difficulty, see Johnson et al. [112]. In cases of dominant convection, the transport of the local error needs to be accurately measured if the estimator is to be used to drive an adaptive mesh refinement process. A practical strategy for doing this — solving a so-called *linearized dual problem* — has been pioneered by Rannacher [155].

Problems

7.1. Show that non-dimensionalizing the momentum equation using the pressure scaling $p(L\vec{\xi}) = (\nu U/L)\, p_*(\vec{\xi})$ gives the following alternative to (7.3)

$$-\nabla^2 \vec{u}_* + \mathcal{R}\vec{u}_* \cdot \nabla \vec{u}_* + \nabla p_* = \vec{f}_*.$$

This shows that taking the limit $\nu \to \infty$ gives the Stokes equations (5.1).

7.2. Use Green's theorem to show that

$$\int_\Omega (\vec{z} \cdot \nabla \vec{u}) \cdot \vec{v} = -\int_\Omega (\vec{z} \cdot \nabla \vec{v}) \cdot \vec{u} - \int_\Omega (\vec{u} \cdot \vec{v}) \, \nabla \cdot \vec{z} + \int_{\partial\Omega} (\vec{u} \cdot \vec{v}) \, \vec{z} \cdot \vec{n}.$$

Deduce that the form $c(\vec{z}; \vec{u}, \vec{v})$ is skew-symmetric if $\nabla \cdot \vec{z} = 0$ in Ω and either $\vec{z} \cdot \vec{n} = 0$ or $\vec{u} = \vec{0}$ or $\vec{v} = \vec{0}$ on $\partial\Omega$.

7.3. Verify that the specific choice of basis functions (5.38) applied to the discrete Newton problem (7.41) leads to the matrix partitioning (7.47).

7.4. Prove Proposition 7.1.

Computational exercises

7.1. Verify that the *exact* solution (7.6) can be computed using Q_2–Q_1 approximation and the Newton iteration strategy built into IFISS. Take a sequence of uniform square grids, and a range of viscosity parameters ν. What happens in the limit $\nu \to 0$? Repeat this experiment using the stabilized Q_1–P_0 method instead. Can you explain the difference in the observed behavior?

7.2. Consider solving the Blasius flow problem in Example 7.1.4 with values $\nu = 0.01$, 0.001, and 0.0001. Take a 64×64 stretched grid discretization with Q_2–Q_1 approximation and three distinct stretch ratios; 1, 1.1 and 1.2. Classical asymptotic theory predicts that, away from the leading edge, the boundary layer thickness (when $u_x = 0.99$, say) is proportional to $\sqrt{\nu x}$, where x is the distance to the tip of the plate. Do the numerical results agree with the theory?

7.3. Consider solving the expansion flow problem in Example 7.1.2 using the default stabilized Q_1–P_0 method. Tabulate the global error estimator η for a sequence of uniform square grids, and hence estimate the rate of convergence. Compare your results with those obtained in the Stokes flow case. Is there any difference in the rates of convergence observed in these two cases?

7.4. This is a fundamental modeling exercise. It explores the natural outflow condition in Example 7.1.2. Instead of making the default choice, generate a grid on a domain which extends to twice the original length. (The outflow is at $x = 10$ rather than $x = 5$.) Take the locally mass conserving Q_2–P_{-1} approximation method, and the default square grid. Then, use `flowvolume` to compare the computed horizontal velocity profiles at $x = 5$ for increasing Reynold numbers (specifically; $\nu = 1/50$, $\nu = 1/125$ and $\nu = 1/200$,) on the original and the long domain.

7.5. This exercise explores the length of the recirculating eddy in the step flow problem in Example 7.1.2. Take Q_2–P_{-1} approximation and a sequence of square grids and compute (to an accuracy of three digits) the point on the wall where the eddy "reattaches" in the case $\nu = 1/50$. Repeat this experiment using the nonconservative Q_2–Q_1 approximation method.

7.6. This exercise builds on Exercise 5.7. Consider the contracting channel flow featured in Exercise 1.5. Take the default grid, and using `flowvolume`, compare the computed horizontal velocity profiles at $x = 4$ using Q_2–P_{-1} approximation in the cases $\nu = 1$, $\nu = 1/100$ and $\nu = 1/1000$.

8

SOLUTION OF DISCRETE NAVIER–STOKES PROBLEMS

In this final chapter of the book, we bring together our ideas underlying fast solvers for convection–diffusion and Stokes problems to develop effective strategies for solving the discrete systems that arise in the solution of the Navier–Stokes equations.

The convection–diffusion equation studied in Chapters 3 and 4 and the Stokes equations studied in Chapters 5 and 6 both contain the Laplacian operator as an important component. However, both sets of equations model more complex physical processes than pure viscous diffusion. Like the Poisson equation, the convection–diffusion equation is of scalar form, but the presence of convection in the model leads to a nonsymmetric coefficient matrix, and the conjugate gradient method has to be replaced with variants designed to handle nonsymmetry. Similarly, although the vector Laplacian appearing in the Stokes operator is a self-adjoint coercive operator on the constraint space $\{\vec{u} \,|\, \nabla \cdot \vec{u} = 0\}$, the coupled velocity–pressure system leads to an indefinite matrix, and the MINRES method must be used to handle this. Equally important (indeed, perhaps more so), in order for these Krylov subspace methods to display rapid convergence, preconditioning or splitting strategies are needed that well represent the physical and mathematical properties of each of the models. Splittings that incorporate properties of the flows are important for efficient solution of convection-dominated systems and preconditioners directly tied to the saddle point form of the Stokes equations are needed for rapid convergence of MINRES.

As shown in Section 7.3, the algebraic systems that arise from the Navier–Stokes equations have a block structure like that of the Stokes equations, but now it is a discrete convection–diffusion operator that is the important subsidiary scalar component of the system. The solution algorithms developed here will generalize the ideas of both Chapters 4 and 6, borrowing from them in many ways and augmenting them to handle the structure of the new discrete problem. Moreover, the results of the earlier chapters, including Chapter 2, will be used to implement the ideas described here. In particular, efficient solvers for the Poisson equation and the convection–diffusion equation will be used as building blocks to produce new methods for the Navier–Stokes equations.

Our perspective in designing these solution strategies is that the algorithmic technology represented in the material of Chapters 2, 4 and 6 has reached a mature state. Taken collectively, the solvers discussed in these chapters can be used as a basis for the derivation of new efficient preconditioning algorithms for the Navier–Stokes equations. These new algorithms respect the coupling of velocities and pressures in the system and they are designed to accurately represent the relative contributions of convection and diffusion in the flow. Although their analytic convergence properties are not fully understood, as we will see, they offer the potential to avoid the slow convergence or limited applicability seen with traditional approaches for the Navier–Stokes equations such as SIMPLE iteration, pseudo-timestepping, or multigrid methods based on block Gauss–Seidel smoothers.

8.1 General strategies for preconditioning

Throughout this chapter, we will assume that the systems of equations to be solved have the generic form

$$
\begin{bmatrix} \mathbf{F} & B^T \\ B & -C \end{bmatrix} \begin{bmatrix} \mathbf{u} \\ \mathbf{p} \end{bmatrix} = \begin{bmatrix} \mathbf{f} \\ \mathbf{g} \end{bmatrix},
\tag{8.1}
$$

arising from Newton iteration (7.41) or Picard iteration (7.46) applied to the Navier–Stokes equations. The coefficient matrices are nonsymmetric, and the Krylov subspace methods discussed in Chapter 4, such as GMRES, can be used to solve these systems. Therefore, we can proceed directly to considerations of preconditioning.

Suppose first that the coefficient matrix has the form

$$
K = \begin{bmatrix} \mathbf{F} & B^T \\ B & 0 \end{bmatrix},
\tag{8.2}
$$

arising from uniformly stable ($C = 0$) mixed approximation of the Navier–Stokes equations. To simplify the derivation of preconditioning strategies, we will proceed under the assumption that K is non-singular. The case arising in enclosed flow, where the undetermined hydrostatic pressure causes B^T, and therefore K, to be rank-deficient by one, will be discussed in Section 8.3.4. In Chapter 6, we showed that preconditioners having a block diagonal structure lead to very efficient iterative solvers for the Stokes equations, and we will begin with this point of view. Consider a preconditioning operator of the form

$$
M = \begin{bmatrix} \mathbf{M_F} & 0 \\ 0 & M_S \end{bmatrix}.
\tag{8.3}
$$

The motivation for using this notation rather than $\mathbf{M_F} = \mathbf{P}$, $M_S = T$ as in Chapter 6 will be made evident below. For the moment, we will take $\mathbf{M_F} = \mathbf{F}$ and explore what is needed for M_S.

As in Chapter 6, insight can be obtained from the generalized eigenvalue problem

$$
\begin{bmatrix} \mathbf{F} & B^T \\ B & 0 \end{bmatrix} \begin{bmatrix} \mathbf{u} \\ \mathbf{p} \end{bmatrix} = \lambda \begin{bmatrix} \mathbf{F} & 0 \\ 0 & M_S \end{bmatrix} \begin{bmatrix} \mathbf{u} \\ \mathbf{p} \end{bmatrix}.
\tag{8.4}
$$

There are two possibilities: $\lambda = 1$ or $\lambda \neq 1$. The eigenvalue $\lambda = 1$ has both algebraic and geometric multiplicity equal to $n_u - n_p$, see Problem 8.1. Consider the case $\lambda \neq 1$. Manipulation of the first block equation leads to $\mathbf{u} = (1/(\lambda - 1))\mathbf{F}^{-1}B^T\mathbf{p}$ and then eliminating $\mathbf{u}$ from the second block equation gives

$$BF^{-1}B^T\mathbf{p} = \mu M_S\mathbf{p}, \tag{8.5}$$

where $\mu = \lambda(\lambda - 1)$. Thus, the optimal choice is $M_S = S$, where

$$S = B\mathbf{F}^{-1}B^T \tag{8.6}$$

is the *Schur complement operator*. This means $\mu = 1$ in (8.5) and there are precisely three eigenvalues of (8.4),

$$\lambda_1 = 1, \quad \lambda_{2,3} = \frac{1 \pm \sqrt{5}}{2}.$$

This preconditioning strategy would thus result in convergence to the solution of (8.2) in three GMRES steps, see Murphy et al. [135].

Using a preconditioner M within a Krylov subspace iteration entails applying the action of M^{-1} to a vector. As discussed in Section 6.2, it is not feasible to use the Schur complement operator for M_S in (8.3). Thus, we seek a matrix operator that represents a good approximation to S, and for which applying M_S^{-1} to a vector is inexpensive. To motivate our choice of M_S we first summarize the analysis in Chapter 6. If the system (8.2) comes from the Stokes equations, that is, if $\mathbf{F}$ consists of a set of discrete Laplacian operators, then we know that a good choice for M_S is the pressure mass matrix Q or even its diagonal. Indeed, for the specific choice $M_S = \texttt{diag}(Q)$, the eigenvalues $\{\mu\}$ of (8.5) are contained in an interval that does not depend on the discretization parameter. Moreover, the eigenvalues $\{\lambda\}$ of (8.4) corresponding to the preconditioned Stokes operator are either unity or lie in two such intervals, one to the right of unity and one to the left of zero. This leads to rapid convergence of the MINRES algorithm.

Suppose we also follow this line of reasoning here. That is, suppose that when (8.5) arises from the linearized Navier–Stokes equations, we have the goal of making the eigenvalues $\{\mu\}$ tightly clustered, say in a small region near $(1,0)$ in the complex plane. If this can be done, then under the mapping

$$\lambda \mapsto \frac{1 \pm \sqrt{1 + 4\mu}}{2},$$

the eigenvalues of (8.4) will also be tightly clustered, this time in two regions, one on each side of the imaginary axis.

Theorem 8.1. *Suppose the eigenvalues satisfying* (8.5) *lie inside a rectangle* $[a, b] \times [-c, c]$ *in the complex plane, where* a, b, *and* c *are positive constants. Then the eigenvalues of the associated problem* (8.4) *are either unity or else are* *contained in two rectangular regions given by,*

$$\left[\tfrac{1}{2}(1+\hat{a}), \tfrac{1}{2}(1+\hat{b})\right] \times [-\hat{c}, \hat{c}] \quad \text{and} \quad \left[\tfrac{1}{2}(1-\hat{b}), \tfrac{1}{2}(1-\hat{a})\right] \times [-\hat{c}, \hat{c}],$$

where $\hat{a} = (1+4a)^{1/2}$, $\hat{b} = \left[\tfrac{1}{2}\left(1+4b+\sqrt{1+8b+16(b^2+c^2)}\right)\right]^{1/2}$, *and* $\hat{c} = c/(1+4a)^{1/2}$.

Proof The proof is given by writing $\mu = \alpha + i\beta$ and $\sqrt{1+4\mu} = \sigma + i\tau$ and then deriving expressions for σ and τ in terms of α and β; the details can be found in Elman & Silvester [58]. □

Notice that the eigenvalue regions are symmetric with respect to the line $\Re(\lambda) = 1/2$. The convergence analysis of GMRES (see Section 4.1) can then be used to establish bounds on rates of convergence. In particular, Theorem 4.2 can be generalized by enclosing each of the two rectangular regions inside an ellipse.

Of course, all of this will depend on our finding an effective preconditioning operator M_S. Before addressing this crucial issue, we first look more closely at the structure of the preconditioning operator M. For the discrete Stokes equations in Chapter 6, there is a very good reason to use the block diagonal form (8.3): when M is symmetric and positive-definite, the preconditioned problem retains symmetry. This means that the MINRES method is applicable, and all the desirable features of that algorithm (short-term recurrences and optimality) remain in effect.

Here, however, we are starting with a nonsymmetric coefficient matrix, and this restricts our options with respect to Krylov subspace iteration. We must give up either short recurrences or optimality. With either choice, the advantages of block diagonal preconditioners are lost and it turns out that a slight variation on the structure of the preconditioner yields other improvements. In particular, consider as an alternative to (8.3) the block triangular matrix

$$M = \begin{bmatrix} \mathbf{M_F} & B^T \\ 0 & -M_S \end{bmatrix}. \tag{8.7}$$

Assuming that $\mathbf{M_F} = \mathbf{F}$ for the time being, the generalized eigenvalue problem analogous to (8.4) is

$$\begin{bmatrix} \mathbf{F} & B^T \\ B & 0 \end{bmatrix} \begin{bmatrix} \mathbf{u} \\ \mathbf{p} \end{bmatrix} = \lambda \begin{bmatrix} \mathbf{F} & B^T \\ 0 & -M_S \end{bmatrix} \begin{bmatrix} \mathbf{u} \\ \mathbf{p} \end{bmatrix}. \tag{8.8}$$

Once again, there are the two possibilities: $\lambda = 1$ and $\lambda \neq 1$. In this case, the first block equation yields

$$\mathbf{Fu} + B^T\mathbf{p} = 0 \quad \text{or} \quad \mathbf{u} = -\mathbf{F}^{-1}B^T\mathbf{p}.$$

Substitution into the second equation then gives (8.5) with $\mu = \lambda$. That is, if $\mathbf{M_F} = \mathbf{F}$, then the eigenvalues of the preconditioned operator derived from the block triangular preconditioner (8.7) consist of unity together with the eigenvalues of (8.5).

The advantage of this approach over the analogous block diagonal preconditioning is that all the eigenvalues of the preconditioned system lie on one side of the imaginary axis, and this leads to more rapid convergence. Indeed, for a carefully chosen starting vector, it can be shown that GMRES iteration with the block triangular preconditioner requires half as many steps as when the block diagonal preconditioner is used.

Theorem 8.2. *Consider right-preconditioned* GMRES *iteration applied to* (8.1) *in the unstabilized case* $C = 0$. *Given arbitrary* $\mathbf{p}^{(0)}$, *assume that* $\mathbf{u}^{(0)} = \mathbf{F}^{-1}(\mathbf{f} - B^T\mathbf{p}^{(0)})$. *Let* $[\mathbf{u}^{(k)}, \mathbf{p}^{(k)}]$ *denote the iterates generated using the block triangular preconditioner* (8.7) *with* $\mathbf{M_F} = \mathbf{F}$, *and let* $[\hat{\mathbf{u}}^{(k)}, \hat{\mathbf{p}}^{(k)}]$ *denote the iterates generated using the block diagonal preconditioner* (8.3), *also with* $\mathbf{M_F} = \mathbf{F}$. *Then*

$$\begin{bmatrix} \hat{\mathbf{u}}^{(2k)} \\ \hat{\mathbf{p}}^{(2k)} \end{bmatrix} = \begin{bmatrix} \hat{\mathbf{u}}^{(2k+1)} \\ \hat{\mathbf{p}}^{(2k+1)} \end{bmatrix} = \begin{bmatrix} \mathbf{u}^{(k)} \\ \mathbf{p}^{(k)} \end{bmatrix}.$$

Proof See Fischer et al. [75]. □

The special choice of $\mathbf{u}^{(0)}$ makes the initial residual vector $\mathbf{0}$ in the first component. Theorem 8.2 is also indicative of general trends. That is, for any choice of $\mathbf{u}^{(0)}$, Krylov subspace iteration with the block triangular preconditioner requires approximately half as many steps as with the block diagonal preconditioner, see Elman & Silvester [58]. Moreover, the costs per step of the two preconditioners are essentially the same. To see this, observe that applying the block triangular preconditioner entails solving the system of equations

$$\begin{bmatrix} \mathbf{M_F} & B^T \\ 0 & -M_S \end{bmatrix} \begin{bmatrix} \mathbf{w} \\ \mathbf{s} \end{bmatrix} = \begin{bmatrix} \mathbf{v} \\ \mathbf{q} \end{bmatrix},$$

where $\mathbf{v}$ and $\mathbf{q}$ are given. This requires solution of the individual systems

$$M_S\mathbf{s} = -\mathbf{q}, \quad \mathbf{M_F}\mathbf{w} = \mathbf{v} - \mathbf{B}^T\mathbf{s}. \tag{8.9}$$

The only cost not incurred by the block diagonal preconditioner is that of a (sparse) matrix-vector product with B^T, which, at $O(n_u)$ floating point operations, is negligible.

There is another way to understand the effect of (8.7) which we will discuss in the context of the more general form of the coefficient matrix,

$$K = \begin{bmatrix} \mathbf{F} & B^T \\ B & -C \end{bmatrix}. \tag{8.10}$$

This will identify what to look for in M_S when the stabilized discretizations discussed in Section 5.3 are used. Consider the block LU-decomposition $K = \mathcal{L}\mathcal{U}$

given by

$$\begin{bmatrix} \mathbf{F} & B^T \\ B & -C \end{bmatrix} = \begin{bmatrix} I & 0 \\ B\mathbf{F}^{-1} & I \end{bmatrix} \begin{bmatrix} \mathbf{F} & B^T \\ 0 & -(B\mathbf{F}^{-1}B^T + C) \end{bmatrix}.$$

If the upper triangular factor $\mathcal{U}$ is viewed as a preconditioner, then all the eigenvalues of the preconditioned matrix $\mathcal{L} = K\mathcal{U}^{-1}$ are identically one. The GMRES algorithm would need precisely two steps to compute the solution to this preconditioned problem, for any discretization mesh size h and Reynolds number $\mathcal{R}$, see Problem 8.2. As we have observed, it is not feasible to explicitly use the Schur complement as part of the preconditioning operator, and a practical preconditioning strategy is derived by making an approximation to $\mathcal{U}$. In particular, we seek $M_S \approx S$, where now

$$S = B\mathbf{F}^{-1}B^T + C \qquad (8.11)$$

is the Schur complement of the stabilized discrete operator. In this case, the algebraic multiplicity of the eigenvalue $\lambda = 1$ is at least n_u, and it may be higher if M_S is a good approximation to S, see Problem 8.3.

Up to this point we have assumed that $\mathbf{M_F} = \mathbf{F}$, which means preconditioning requires the application of the action of $\mathbf{F}^{-1}$ to a vector in the discrete velocity space. If K is a discrete Oseen operator, as would be the case for the Picard iteration (7.46), then $\mathbf{F}$ is a block diagonal matrix containing a discrete convection–diffusion operator in each of its nonzero blocks. Our preconditioning methodology thus requires a fast solver strategy for discrete convection–diffusion problems (which itself is the focus of Chapter 4). If we are to mimic the Stokes preconditioning strategy of Section 6.2 then the action of $\mathbf{M_F}^{-1}$ must be effected using a process like multigrid. We will explain how to do this in Section 8.3. For now, the main issue is to determine effective choices of the Schur complement preconditioner M_S. This is the subject of the next section.

8.2   Approximations to the Schur complement operator

In this section, we describe two ways to approximate the Schur complement operator, both of which can be used for M_S in the preconditioner (8.7). Before presenting them, we note again that a good choice for M_S for Stokes problems is the pressure mass matrix Q. This idea also has merit for discrete Navier–Stokes problems. With the choice $M_S = (1/\nu)Q$ (or with a simpler operator in place of Q, such as the diagonal of Q), the preconditioning methodologies described in the previous section also display mesh-independent convergence rates; see Elman & Silvester [58] and Klawonn & Starke [124]. This approach is highly competitive for low Reynolds numbers (say, of the order of $\mathcal{R} \leq 10$). However, it does not take into account the effects of convection on the Schur complement operator, and convergence rates deteriorate as the Reynolds number increases. Our aim here is to develop more faithful approximations to the Schur complement matrix that better reflect the balance of convection and diffusion in the problem and so lead to improved convergence rates at higher Reynolds numbers.

We will consider two ways to approximate the Schur complement matrix $BF^{-1}B^T$. Both of them can be motivated by starting with the Oseen operator (7.46) and observing that it contains as a component a discrete version of a convection–diffusion operator

$$\mathcal{L} = -\nu\nabla^2 + \vec{w}_h \cdot \nabla, \tag{8.12}$$

defined on the velocity space. Here $\vec{w}_h$ is the approximation to the discrete velocity computed at the most recent nonlinear iteration. Let us suppose that there is an analogous operator

$$\mathcal{L}_p = (-\nu\nabla^2 + \vec{w}_h \cdot \nabla)_p$$

defined on the pressure space. In doing this, we are overlooking the fact that the pressures are only taken to be in $L_2(\Omega)$ and simply assuming (or pretending) that it makes sense to consider this differential form. Let us also suppose that the commutator of the convection–diffusion operators with the gradient operator

$$\mathcal{E} = (-\nu\nabla^2 + \vec{w}_h \cdot \nabla)\nabla - \nabla(-\nu\nabla^2 + \vec{w}_h \cdot \nabla)_p \tag{8.13}$$

is small in some sense. The commutator would be zero if $\vec{w}_h$ were constant and the operators were defined on an unbounded domain. The use of the convection–diffusion operator on the pressure space can be viewed as a *model reduction* of the block version of the operator on the velocity space.

8.2.1 *The pressure convection–diffusion preconditioner*

For the first approach for approximating the Schur complement, assume for the moment that a C^0 approximation is used for the pressure. Let $\mathbf{Q}$ denote the velocity mass matrix,

$$\mathbf{Q} = [\mathbf{q}_{ij}], \quad \mathbf{q}_{ij} = \int_\Omega \vec{\phi}_j \cdot \vec{\phi}_i,$$

where $\{\vec{\phi}_i\}$ is the basis for the discrete velocity space. The matrix representation of the discrete gradient operator is $\mathbf{Q}^{-1}B^T$ and the representation of the discrete (negative) divergence operator is $Q^{-1}B$. (See Problem 8.4 for a derivation; the mass matrices are needed to give the correct scaling.) A similar derivation of discrete convection–diffusion operators on both the velocity space and the pressure space leads to the following discrete version of the commutator in terms of finite element matrices:

$$\mathcal{E}_h = (\mathbf{Q}^{-1}\mathbf{F})(\mathbf{Q}^{-1}B^T) - (\mathbf{Q}^{-1}B^T)(Q^{-1}F_p). \tag{8.14}$$

Note in particular that if $\{\psi_j\}$ is the basis for the C^0 pressure discretization, then

$$F_p = [f_{p,ij}], \quad f_{p,ij} = \nu \int_\Omega \nabla \psi_j \cdot \nabla \psi_i + \int_\Omega (\vec{w}_h \cdot \nabla \psi_j) \, \psi_i . \tag{8.15}$$

Let us transform (8.14) to isolate the Schur complement by premultiplying by $BF^{-1}Q$ and postmultiplying by $F_p^{-1}Q$. The assumption that the commutator is small identifies an approximation to the Schur complement matrix:

$$BF^{-1}B^T \approx BQ^{-1}B^T F_p^{-1}Q . \tag{8.16}$$

There are some details to work out before this derivation leads to a practical preconditioner. In general, the matrix on the right side of (8.16) is not a practical choice for M_S because $BQ^{-1}B^T$ will not be easy to work with. (In particular, it will be dense.) To see how to handle this, recall the observation from Chapter 5 (Section 5.5.1), that for certain classes of problems and discretizations, there is an alternative inf-sup condition that leads to spectral equivalence between $BQ^{-1}B^T$ and the (*sparse*) pressure Laplacian matrix A_p given by,

$$A_p = [a_{p,ij}], \quad a_{p,ij} = \int_\Omega \nabla \psi_j \cdot \nabla \psi_i . \tag{8.17}$$

(The details are in (5.172) and (5.173).) Although such an equivalence holds only for problems with enclosed flow, this observation suggests that in general a discrete Laplacian on the pressure space is what is needed in place of $BQ^{-1}B^T$. For enclosed flow boundary conditions, we can use (8.17).

This brings us to the more general question of the role boundary conditions should play in the definitions of F_p and A_p. Irrespective of the boundary conditions of the underlying flow problem (see (7.2)), the entries of the matrices of (8.15) and (8.17) are the same as those coming from discrete elliptic equations posed with Neumann boundary conditions (see (1.17) and (3.11)). In such cases, the solution (to the convection–diffusion problem or the Poisson problem) is unique only up to a constant. For enclosed flow, the pressure solution is also unique up to a constant, and it seems natural to simply use (8.15) and (8.17) to define F_p and A_p.[1] However, preconditioners built from these singular operators are not effective for problems with inflow and outflow boundaries. In the case of the step problem of Example 7.1.2, we use what we know about the convection–diffusion equation to modify the definition of F_p given in (8.15) when there are inflow boundaries present. As discussed in Chapter 3, for the reduced problem (3.3) occurring in the limiting case of pure convection, the solution is determined by specified Dirichlet boundary conditions on the inflow boundary, where $\vec{w} \cdot \vec{n} < 0$. This suggests using Dirichlet boundary conditions along inflow boundaries to define F_p. This means that the rows and columns of F_p corresponding to pressure nodes on an inflow boundary are treated as though they are

---

[1]This means that the preconditioner will use singular operators; the computational issues associated with this are discussed in Section 8.3.4.

associated with Dirichlet boundary conditions. At nodes on other (characteristic or outflow) components of $\partial\Omega$, the entries of F_p are defined by (8.15). A_p is defined in an analogous manner so that F_p and A_p are derived from consistent boundary conditions.

We note that this discussion only concerns the definition of the algebraic operators, F_p and A_p. That is, the only place where boundary conditions have any impact is on the *definition* of the preconditioning operator. In particular, there are no boundary conditions imposed on the discrete pressures, no values of Dirichlet conditions to determine, and there is no right-hand side that is affected by these boundary node modifications.

Substituting the operator A_p into the right-hand side of (8.16) gives the following Schur complement approximation:

$$M_S := A_p F_p^{-1} Q. \tag{8.18}$$

We refer to this generic strategy — using a block triangular matrix preconditioner (8.7) with (8.18) in the (2,2) block — as *pressure convection–diffusion preconditioning*. As can be seen from (8.9), the preconditioning strategy requires the action of

$$M_S^{-1} = Q^{-1} F_p A_p^{-1}.$$

Therefore, implementing the strategy requires the action of a Poisson solve, a mass matrix solve and a matrix-vector product with a specially constructed matrix F_p. Comparing the overall strategy with more conventional solution methods that are currently used in practice, the matrix F_p is the "magic ingredient". We return to this point in Section 8.3.5.

Pressure convection–diffusion preconditioning is also effective in conjunction with stabilized $\boldsymbol{Q}_1$–$\boldsymbol{Q}_1$ and $\boldsymbol{Q}_1$–$\boldsymbol{P}_0$ mixed approximation, that is, where the coefficient matrix is as in (8.10). To justify this statement, we note that the assumption of a small commutator $\mathcal{E}$ in (8.13) implies that

$$\nabla^2 (-\nu\nabla^2 + \vec{w}_h \cdot \nabla)_p^{-1} \approx \nabla \cdot (-\nu\nabla^2 + \vec{w}_h \cdot \nabla)^{-1} \nabla. \tag{8.19}$$

In the case of unstabilized mixed approximation the right-hand side of (8.19) can be identified with the discrete operator

$$(Q^{-1}B)(\mathbf{Q}^{-1}\mathbf{F})^{-1}(\mathbf{Q}^{-1}B^T) = Q^{-1}(B\,\mathbf{F}^{-1}B^T) = Q^{-1}S.$$

If, however, the mixed approximation is stabilized then we have a generalized Schur complement, $S := B\mathbf{F}^{-1}B^T + C$, and the discrete version of (8.19) is

$$(Q^{-1}A_p)(Q^{-1}F_p)^{-1} \approx Q^{-1}S = Q^{-1}(B\,\mathbf{F}^{-1}B^T + C). \tag{8.20}$$

This suggests that the preconditioning operator of (8.18) remains a suitable approximation

$$M_S := A_p F_p^{-1} Q \approx B\,\mathbf{F}^{-1}B^T + C, \tag{8.21}$$

provided appropriate choices are made for the matrix operators A_p and F_p.

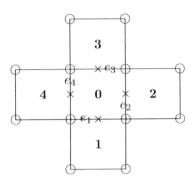

FIG. 8.1. Interior patch of five elements: Q_1–P_0 approximation.

For a C^0 pressure approximation, natural choices for A_p and F_p come from (8.15) and (8.17). As a result, pressure convection–diffusion preconditioning for stabilized Q_1–Q_1 mixed approximation is completely straightforward. The construction of A_p and F_p in the case of stabilized Q_1–P_0 is much more of a challenge — specifically, the issue of discontinuous pressure approximation needs to be addressed. The key idea here is to construct approximations using jump operators defined on inter-element edges/faces. To visualize such a construction, consider the (interior) patch of five rectangular elements illustrated in Figure 8.1, and suppose that we wish to construct a discrete approximation to the convective derivative $\vec{w}_h \cdot \nabla p_h$ in the central element, $\mathbf{0}$, where $\vec{w}_h$ is the Q_1 discrete velocity field defined by the vertex values.

The weak form of the derivative is the starting point. It is given by $\int_\Omega q_h \vec{w}_h \cdot \nabla p_h$ where q_h is the characteristic function, that is, q_h is equal to unity on some specific element T and equal to zero on all the other elements. Ignoring the fact that the weak derivative is in fact zero, we break the integral into element contributions and integrate by parts elementwise. That is, defining the edge set $\mathcal{E}(T)$ associated with element T and the outward-pointing normal $\vec{n}_{E,T}$, we have the element contribution

$$\int_T q_h \vec{w}_h \cdot \nabla p_h = -\int_T q_h p_h \nabla \cdot \vec{w}_h - \sum_{E \in \mathcal{E}(T)} \langle q_h \vec{w}_h \cdot \vec{n}_{E,T}, p_h \rangle_E$$

$$= -p|_T \int_T \nabla \cdot \vec{w}_h - \sum_{E \in \mathcal{E}(T)} p|_T \int_E \vec{w}_h \cdot \vec{n}_{E,T}.$$

The fact that p_h is discontinuous means that the first term on the right-hand side is zero, see (5.63). Thus, summing the element contributions gives

$$\int_\Omega q_h \vec{w}_h \cdot \nabla p_h = \sum_{T \in \mathcal{T}_h} \int_T q_h \vec{w}_h \cdot \nabla p_h = -\sum_{E \in \mathcal{E}(T)} p_h|_T \int_E \vec{w}_h \cdot \vec{n}_{E,T}. \qquad (8.22)$$

Given that the characteristic function in the neighboring element, S say, sharing edge E will generate a corresponding contribution, we introduce the *scalar jump operator*

$$[\![p_h]\!]_{E,T} := p_h|_T - p_h|_S, \tag{8.23}$$

where T is the interior element and S is the exterior element. We now employ the tactic used in deriving a posteriori error estimates and *equidistribute* the contributions in (8.22) to the elements adjoining edge E in equal proportion. Thus,

$$\int_\Omega q_h \vec{w}_h \cdot \nabla p_h = -\frac{1}{2} \sum_{E \in \mathcal{E}(T)} [\![p_h]\!]_{E,T} \int_E \vec{w}_h \cdot \vec{n}_{E,T}. \tag{8.24}$$

This gives a nonzero approximation to the area-weighted derivative,

$$
\begin{aligned}
|\Box_T|\, \vec{w}_h \cdot \nabla p_h|_T &\approx \int_T q_h \vec{w}_h \cdot \nabla p_h \\
&= -\frac{1}{2} \sum_{E \in \mathcal{E}(T)} [\![p_h]\!]_{E,T} \int_E \vec{w}_h \cdot \vec{n}_{E,T} \\
&= -\frac{1}{2} \sum_{E \in \mathcal{E}(T)} h_E\, [\![p_h]\!]_{E,T}\, \vec{w}_h \cdot \vec{n}_{E,T}(\vec{x}_\times), \tag{8.25}
\end{aligned}
$$

where h_E is the length of the edge and $\vec{x}_\times$ is the mid-point coordinate (see Figure 8.1).

For a uniform rectangular grid, the construction (8.25) gives the classical centered finite difference approximation to the pressure defined at the centroid of the elements. To see this, suppose that $\vec{w}_h = (1,0)$ and consider the central element in Figure 8.1,

$$|\Box_T|\, \vec{w}_h \cdot \nabla p_h|_T = h_x h_y \frac{\partial p_h}{\partial x}(\vec{x}_0) \approx -\frac{1}{2} h_y\, [\![p_h]\!]_{2,0} + \frac{1}{2} h_y\, [\![p_h]\!]_{4,0}. \tag{8.26}$$

Thus,

$$\frac{\partial p_h}{\partial x}(\vec{x}_0) \approx \frac{1}{2h_x}(-[\![p_h]\!]_{2,0} + [\![p_h]\!]_{4,0}) = \frac{1}{2}\left(\frac{p_2 - p_0}{h_x} + \frac{p_0 - p_4}{h_x}\right), \tag{8.27}$$

as expected.

A discrete pressure Laplacian $-\nabla^2 p_h$ can be constructed by differencing the first derivative operators. Specifically, the scaled jump terms $-[\![p_h]\!]_{2,0}/h_x$ and $[\![p_h]\!]_{4,0}/h_x$ in (8.27) can be identified with x-derivatives defined on the edges $E_{2,0}$ and $E_{4,0}$, respectively. Differencing these edge derivatives gives a centered approximation to the second derivative:

$$
\begin{aligned}
\frac{\partial^2 p_h}{\partial x^2}(\vec{x}_0) &\approx \frac{1}{h_x}\left(\frac{\partial p_h}{\partial x}(\times_{4,0}) - \frac{\partial p_h}{\partial x}(\times_{2,0}) \right) \\
&= -\frac{1}{h_x^2}\left([\![p_h]\!]_{4,0} + [\![p_h]\!]_{2,0} \right).
\end{aligned}
\tag{8.28}
$$

Combining (8.28) with the analogous difference approximation in the y-direction gives the following approximation to the scaled Laplacian at the middle element in Figure 8.1:

$$
|\Box_T|\,(-\nabla^2 p_h)|_T \approx \left(\frac{h_y}{h_x} \right) \left([\![p_h]\!]_{4,T} + [\![p_h]\!]_{2,T} \right) + \left(\frac{h_x}{h_y} \right) \left([\![p_h]\!]_{3,T} + [\![p_h]\!]_{1,T} \right).
\tag{8.29}
$$

The right-hand sides of (8.25) and (8.29) define the entries of the matrices N_p and A_p respectively, (with $F_p = \nu A_p + N_p$) in the case of P_0 pressure approximation.[2] In either case, jump terms corresponding to the boundary edges are omitted for elements adjoining the boundary of the domain. This is consistent with the definitions (8.15) and (8.17) since for enclosed flow problems it gives matrices A_p and F_p representing discretization of problems posed with Neumann boundary conditions. For non-enclosed flow problems, the matrices are further modified to represent Dirichlet boundary conditions for rows and columns corresponding to the elements that adjoin the inflow boundary. This mirrors the strategy used in the continuous pressure case.

The construction (8.29) leads to a particularly simple definition of the matrix A_p in the case of a uniform grid of square elements:

$$
A_p = [a_{p,ij}], \quad a_{p,ij} = \sum_{E \in \mathcal{E}_{h,\Omega}} \langle [\![\psi_j]\!], [\![\psi_i]\!] \rangle_{\bar{E}},
\tag{8.30}
$$

where $\{\psi_j\}$ is the basis of the P_0 approximation, $\mathcal{E}_{h,\Omega}$ is the set of interior edges, and $\langle p, q \rangle_{\bar{E}} = (1/|E|) \int_E pq$ as in Section 5.3.2. The spectral properties of the resulting discrete Laplacian A_p are explored in Computational Exercise 8.2.

Remark 8.3. Generalizing the construction of N_p and A_p embodied in (8.25) and (8.29) to Q_2–P_{-1} approximation is completely straightforward.[3] Moreover, the Q_2–P_{-1} convection–diffusion preconditioner is surprisingly effective, see Computational Exercise 8.3.

Remark 8.4. The pressure convection–diffusion strategy is motivated by assuming commutativity of the underlying differential operators. In general, however, the associated discrete commutator $\mathcal{E}_h$ will not necessarily be "small" in the sense that it has a small matrix norm, even though it may have few nonzero

---

[2] Encoded in the IFISS function fpsetup_q0.m.
[3] Encoded in the IFISS function fpsetup_q2p1.m.

entries. Fortunately for us, the effectiveness of the strategy is not affected by this. In the next section, by way of contrast, an alternative preconditioning strategy will be developed. This is designed to minimize a specific norm of the discrete commutator $\mathcal{E}_h$.

8.2.2   The least-squares commutator preconditioner

Our second approach for approximating the Schur complement operator is only applicable when the mixed approximation is uniformly stable with respect to the inf-sup condition (5.49). Rather than deriving an approximation to the Schur complement from the continuous version of the commutator (8.13), we will instead define an approximation to the matrix operator F_p that makes the discrete commutator (8.14) small. To set things up, note that for a C^0 pressure approximation the L_2-norm of the commutator, viewed as an operator defined on the pressure space M^h, is

$$\sup_{p_h \neq 0} \frac{\| [(-\nu\nabla^2 + \vec{w}_h \cdot \nabla)\nabla - \nabla(-\nu\nabla^2 + \vec{w}_h \cdot \nabla)_p] \, p_h \|}{\|p_h\|}$$

$$= \sup_{\mathbf{p} \neq 0} \frac{\| [(\mathbf{Q}^{-1}\mathbf{F})(\mathbf{Q}^{-1}B^T) - (\mathbf{Q}^{-1}B^T)(Q^{-1}F_p)] \, \mathbf{p} \|_\mathbf{Q}}{\|\mathbf{p}\|_\mathbf{Q}},$$

where $\|\mathbf{v}\|_\mathbf{Q} = \langle \mathbf{Q}\mathbf{v}, \mathbf{v} \rangle^{1/2}$. Suppose we now try to construct F_p so that this norm is small. One way to do this is to minimize the individual vector norms of the columns of the discrete commutator one by one, that is, by defining the jth column of F_p to solve the weighted least squares problem

$$\min \| [\mathbf{Q}^{-1}\mathbf{F}\mathbf{Q}^{-1}B^T]_j - \mathbf{Q}^{-1}B^T Q^{-1}[F_p]_j \|_\mathbf{Q}.$$

The associated normal equations are

$$Q^{-1}B\mathbf{Q}^{-1}B^T Q^{-1}[F_p]_j = [Q^{-1}B\mathbf{Q}^{-1}\mathbf{F}\mathbf{Q}^{-1}B^T]_j,$$

which leads to the following definition of F_p:

$$F_p = Q\,(B\mathbf{Q}^{-1}B^T)^{-1}(B\mathbf{Q}^{-1}\mathbf{F}\mathbf{Q}^{-1}B^T).$$

Substitution of this expression into (8.16) then gives an approximation to the Schur complement matrix:

$$BF^{-1}B^T \approx (B\mathbf{Q}^{-1}B^T)\,(B\mathbf{Q}^{-1}\mathbf{F}\mathbf{Q}^{-1}B^T)^{-1}(B\mathbf{Q}^{-1}B^T). \qquad (8.31)$$

In contrast to what is done for the pressure convection–diffusion preconditioner, this methodology does not require the explicit construction of the matrix F_p. What is still needed is an appropriate means of applying the action of the inverse of the preconditioner. For this, as in the discussion leading to (8.17), it is not practical to work with $\mathbf{Q}^{-1}$ in (8.31), since it is a dense matrix. A practical algorithm is obtained by replacing $\mathbf{Q}$ with $\hat{\mathbf{Q}} = \texttt{diag}(\mathbf{Q})$ (see Lemma 6.3 and Remark 6.4), from which the sparse (scaled) discrete Laplacian $B\hat{\mathbf{Q}}^{-1}B^T$ can be constructed (see also (5.173)). Replacing $\mathbf{Q}$ with $\hat{\mathbf{Q}}$ everywhere in (8.31) gives the following Schur complement approximation:

$$M_S := (B\hat{\mathbf{Q}}^{-1}B^T)\,(B\hat{\mathbf{Q}}^{-1}\mathbf{F}\hat{\mathbf{Q}}^{-1}B^T)^{-1}(B\hat{\mathbf{Q}}^{-1}B^T). \qquad (8.32)$$

We refer to this alternative strategy — using a block triangular matrix preconditioner (8.7) with (8.32) in the (2,2) block — as *least-squares commutator preconditioning*. Implementing the strategy requires the action of the inverse

$$M_S^{-1} = (B\hat{\mathbf{Q}}^{-1}B^T)^{-1}\,(B\hat{\mathbf{Q}}^{-1}\mathbf{F}\hat{\mathbf{Q}}^{-1}B^T)(B\hat{\mathbf{Q}}^{-1}B^T)^{-1}$$

and so involves two discrete Poisson solves and matrix-vector products with the matrices B, B^T, $\mathbf{F}$ and (the diagonal matrix) $\hat{\mathbf{Q}}^{-1}$.

We briefly summarize the characteristics of the two strategies at this point. The main advantage of the least-squares approach is that it is fully automated, that is, it is defined in terms of matrices such as $\mathbf{F}$ and B available in the statement of the problem, and it does not require the construction of the auxiliary operators F_p and A_p that are needed for the pressure convection–diffusion preconditioner. On the other hand, the least-squares preconditioner requires two Poisson solves at each step, rather than one. The pressure convection–diffusion preconditioner is a better strategy in the sense that is applicable to mixed approximation methods that require stabilization. Moreover, only this preconditioner does the "right thing" in the Stokes limit: the Schur complement tends to $(1/\nu)BA^{-1}B^T$ and the preconditioning matrix tends to $(1/\nu)Q$. A detailed assessment of the performance of both strategies is given in Section 8.3.

8.3   Performance and analysis

This section presents the results of some computational experiments and convergence analysis that enable us to assess the merits of the pressure convection–diffusion and least-squares commutator preconditioners.[4] Our specific goals are to demonstrate the sensitivity to the mesh size h and the Reynolds number $\mathcal{R}$. We also show that iterative methods can be used effectively to handle the Poisson and convection–diffusion equations that arise as subproblems; when multilevel methods are used here, the resulting Navier–Stokes solvers are of optimal complexity. The superiority of both preconditioning methods to more conventional approaches is demonstrated in Section 8.3.5. Additional results comparing the effectiveness of Picard and Newton iterations for the nonlinear systems are presented in Section 8.4.

We consider two benchmark flow problems: Example 7.1.2 (flow over a step) and Example 7.1.3 (driven cavity flow). The driven cavity flow problem has steady solutions for large Reynolds numbers of the order of 10^4 (see the discussion of Example 7.1.3), and we explore values of $\mathcal{R} = 2/\nu$ between 10 and 1000. For the step problem, it was observed in Chapter 7 that the natural outflow boundary

---

[4]The IFISS software provides an ideal test-bed for performance comparisons of these ideas.

condition (7.7) provides reasonable accuracy (that is, the flow is "settled" at the truncated boundary) only if the Reynolds number is small enough. Hence, for this problem we limit the maximum value to $\mathcal{R} = 200$. For both flow problems, discretization is done on a sequence of uniform square grids with edge length h.

The nonlinear problem arising after discretization of the weak formulation is given by (7.36)–(7.37). Expressed in terms of the matrices defined in Section 7.3, the nonlinear algebraic system in the more general case of stabilized mixed approximation can be written as,

$$\mathbf{F}(\mathbf{u})\mathbf{u} + B^T\mathbf{p} = \mathbf{f},$$
$$B\mathbf{u} - C\mathbf{p} = \mathbf{g}, \qquad (8.33)$$

where $\mathbf{F}(\mathbf{u})\mathbf{u}$ is the discretization of the nonlinear convection–diffusion operator $-\nu\nabla^2\vec{u} + (\vec{u} \cdot \nabla)\vec{u}$, and C is the stabilization matrix. Our modus operandi is to solve the discrete nonlinear system (8.33) using either Picard or Newton iteration as described in Section 7.2.2. The initial iterate $\left[\mathbf{u}^{(0)}, \mathbf{p}^{(0)}\right]$ is obtained by solving the corresponding discrete Stokes problem (that is dropping the convection term when setting up the left-hand side of (8.33)). The nonlinear iteration is terminated when the vector Euclidean norm of the nonlinear residual has a relative error of 10^{-5}, that is,

$$\left\| \begin{bmatrix} \mathbf{f} - \left(\mathbf{F}(\mathbf{u}^{(m)})\mathbf{u}^{(m)} + B^T\mathbf{p}^{(m)}\right) \\ \mathbf{g} - \left(B\mathbf{u}^{(m)} - C\mathbf{p}^{(m)}\right) \end{bmatrix} \right\| \leq 10^{-5} \left\| \begin{bmatrix} \mathbf{f} \\ \mathbf{g} \end{bmatrix} \right\|. \qquad (8.34)$$

Our comparison of preconditioner performance is based on the number of GMRES iterations required to solve the linear system arising at the *last* step of the nonlinear iteration. In all cases, the linearized Newton and Picard systems (see (7.50)) are solved to give a correction to the nonlinear iterate. The starting guess for the linearized iteration is zero, and the stopping criterion is

$$\|\mathbf{r}^{(k)}\| \leq 10^{-6} \|\mathbf{s}^{(m)}\|,$$

where $\mathbf{r}^{(k)}$ is the residual of the linear system and $\mathbf{s}^{(m)}$ is the left-hand side residual in (8.34) associated with the final nonlinear system. Our experience is that the GMRES iteration counts so obtained are typical of performance at all stages of the nonlinear iteration.

8.3.1 *Ideal versions of the preconditioners*

Recall that implementation of the pressure convection–diffusion and least squares commutator preconditioners requires solution of two subsidiary problems, one on the velocity space and one on the pressure space, see (8.9). We first consider so-called "ideal" versions of the preconditioners in which $\mathbf{M_F} = \mathbf{F}$ in (8.7) and the required actions of the inverses of A_p and Q (see (8.18)) or of $B\hat{Q}^{-1}B^T$ (see (8.32)) are computed to high precision. In these experiments we simply use the sparse direct method built into matlab (via the "backslash" operator) to

Table 8.1 GMRES *iterations needed to satisfy* $\|\mathbf{r}^{(k)}\|/\|\mathbf{s}^{(m)}\| < 10^{-6}$, *starting with zero initial iterate, for the step problem (Example 7.1.2), Reynolds number* $\mathcal{R} = 2/\nu$, *and "ideal" preconditioning. Here ℓ is the grid refinement level; h is $2^{1-\ell}$ for $\mathbf{Q}_2$–$\mathbf{Q}_1$ approximation, and $2^{2-\ell}$ for $\mathbf{Q}_1$–$\mathbf{P}_0$ approximation.*

		Pressure convection–diffusion						Least-squares commutator		
		$\mathbf{Q}_2$–$\mathbf{Q}_1$			$\mathbf{Q}_1$–$\mathbf{P}_0$			$\mathbf{Q}_2$–$\mathbf{Q}_1$		
	ℓ	$\mathcal{R}=10$	100	200	10	100	200	10	100	200
Picard system	4	22	31	42	28	39	57	11	18	30
	5	25	33	39	33	44	56	15	17	23
	6	30	42	47	35	58	66	19	21	22
Newton system	4	23	49	74	29	62	95	12	22	36
	5	27	53	71	34	66	95	15	19	27
	6	31	61	79	38	80	102	19	26	28

solve the subsidiary systems associated with matrices $\mathbf{F}$, A_p, Q and $B\hat{Q}^{-1}B^T$. (The issue of singular systems is dealt with in Section 8.3.4.) This strategy will not be practical, especially for three-dimensional problems, because of the costs of the sparse solves. Our primary goal here is to assess the raw potential of two preconditioning strategies.

Some performance results are summarized in Table 8.1 for the step problem, and in Table 8.2 for the cavity problem. The top of each table gives preconditioned GMRES iteration counts for the system obtained using Picard iteration, and the bottom of each table the corresponding counts for the system arising from Newton iteration.[5] These tables of iterations are supplemented by the graphical displays in Figures 8.2 and 8.3 showing the convergence behavior during the course of the iterative solution for some of these examples.

Let us highlight the main trends revealed by these experiments.

1. *Dependence on discretization parameter h.* By and large, the iteration counts obtained using the pressure convection–diffusion preconditioner are independent of the grid size. This is clearly true for the cavity flow problem for both types of discretization, with the exception of the case $\mathcal{R} = 1000$, where the iterations actually *decrease* as the grid is refined. (We attribute

---

[5] For the cavity problem with $\mathcal{R} = 1000$, Newton's method was not convergent when started with the Stokes solution, and for these results the Newton iteration was preceded by three Picard steps. Also, for this problem on a 16×16 grid, the Picard iteration was stopped after 15 steps without having satisfied the stopping criterion, and the iteration counts reported are for the system at the 15th Picard step.

Table 8.2 GMRES *iterations needed to satisfy* $\|\mathbf{r}^{(k)}\|/\|\mathbf{s}^{(m)}\| < 10^{-6}$, *starting with zero initial iterate, for the driven cavity problem (Example 7.1.3), Reynolds number* $\mathcal{R} = 2/\nu$, *and "ideal" preconditioning. Here* ℓ *is the grid refinement level;* h *is* $2^{1-\ell}$ *for* $\mathbf{Q}_2$–$\mathbf{Q}_1$ *approximation, and* $2^{2-\ell}$ *for* $\mathbf{Q}_1$–$\mathbf{P}_0$ *approximation.*

		Pressure convection–diffusion						Least-squares commutator		
		$\mathbf{Q}_2$–$\mathbf{Q}_1$			$\mathbf{Q}_1$–$\mathbf{P}_0$			$\mathbf{Q}_2$–$\mathbf{Q}_1$		
	ℓ	$\mathcal{R} = 10$	100	1000	10	100	1000	10	100	1000
Picard system	4	14	24	75	18	28	56	8	13	55
	5	15	25	76	18	28	80	11	16	62
	6	15	26	65	18	28	83	14	21	55
	7	15	26	55	18	28	66	18	27	45
Newton system	4	16	30	83	20	36	>200	8	18	78
	5	16	32	139	20	37	135	11	21	106
	6	17	34	141	20	37	164	15	27	107
	7	17	34	128	19	37	162	19	34	95

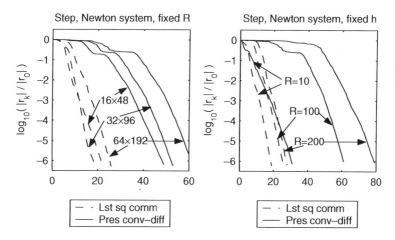

FIG. 8.2. GMRES iterations with "ideal" preconditioners, for linear systems arising from Newton iteration applied to the backward facing step problem with $\mathbf{Q}_2$–$\mathbf{Q}_1$ discretization. Left: $\mathcal{R} = 100$ and various grids. Right: 64×192 grid and various $\mathcal{R}$.

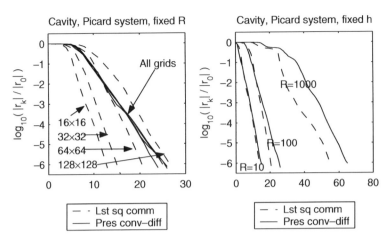

FIG. 8.3. GMRES iterations with "ideal" preconditioners, for linear systems arising from Picard iteration applied to the driven cavity problem with Q_2–Q_1 discretization. Left: $\mathcal{R} = 100$ and various grids. Right: 64×64 grid and various $\mathcal{R}$.

this to inaccuracy in the discretization on the coarse grids, in both the operator and the preconditioner.) The fact that performance does not depend on h is consistent with the bounds of Theorem 8.5. Although the trends are not quite as clear-cut for the step flow problem, our expectation is that the iteration counts will also settle to a constant value as the grid is further refined.

The effect of grid size on the least-squares commutator preconditioner is a bit harder to discern. For the smaller values of $\mathcal{R}$, the counts are clearly increasing as the grid is refined. Equally importantly, the slopes of the curves for this methodology become less steep in Figure 8.3 as the grid is refined, indicating a slower rate of convergence with grid refinement for the cavity flow problem. However, this dependence becomes less pronounced as the Reynolds number increases. We also note that the performance of this preconditioner is dramatically superior than for earlier versions of it developed by Elman in [60], which did not take the scaling associated with the velocity mass matrix into account. Results in [60] suggested that the iteration counts grow in proportion to $h^{-1/2}$, but we do not have rigorous analysis that predicts this behavior. It is also evident that for any given mesh, performance may be very competitive with, and often superior to, that of the pressure convection–diffusion preconditioner.

2. *Dependence on Reynolds number* $\mathcal{R}$. The iteration counts are mildly dependent on the Reynolds number. With the pressure convection–diffusion preconditioner, for Picard iteration and the cavity flow problem, a one-hundred-fold increase in $\mathcal{R}$ results in an approximate three-fold

increase in iterations. Similarly, for the step flow problem, a twenty-fold increase in $\mathcal{R}$ leads to a doubling of iteration counts.

3. *Asymptotic convergence factors.* From results in Chapter 4, we know that the residuals obtained by GMRES iteration satisfy $\log\left(\|\mathbf{r}^{(k)}\|/\|\mathbf{r}^{(0)}\|\right) \approx k \log \rho$, where ρ is the asymptotic convergence factor, see Section 4.1. It can be seen from Figures 8.2 and 8.3 that for the pressure convection–diffusion preconditioner, the asymptotic convergence factor is independent of both grid size and Reynolds number. This point is studied in [69], where it is shown that the increased latency seen with larger Reynolds number is caused by a small number of outlying eigenvalues. In contrast, the convergence factors for the least-squares commutator preconditioner depend on both h and $\mathcal{R}$.

4. *Impact of nonlinear iteration.* For both benchmark problems, the linear systems arising from Newton's method are somewhat more difficult to solve than those arising from Picard iteration. This is consistent with Theorem 8.5 given below. By referring directly to the Jacobian matrix, the least-squares commutator preconditioner offers some advantage for Newton's method, and for several large values of $\mathcal{R}$, GMRES with this preconditioner requires fewer iterations than when the pressure convection–diffusion preconditioner is used. (See, e.g. Figure 8.2.) This advantage disappears for fine enough meshes.

5. *Impact of discretization.* For the pressure convection–diffusion preconditioner, there is no significant difference in performance using the stable Q_2–Q_1 and stabilized Q_1–P_0 mixed approximations.

At the time of this writing, the relative merits of these two strategies for preconditioning are not completely settled. The advantages of the pressure convection–diffusion approach are: it requires less work per step (one vs. two Poisson solves); it is more generally applicable (to both stable and stabilized elements); and it has been subjected to superior convergence analysis (see Section 8.3.3). The advantage of the least squares commutator approach lies in its use of the Jacobian noted above. In addition, this method requires no intervention on the part of a user; once the linear system (8.1) is specified all the tools needed for the preconditioner are available. In contrast, the pressure convection–diffusion preconditioner needs the matrices F_p and A_p. (We note however that Elman et al. [57] are working on a strategy for automating the construction of F_p without user intervention.) A general summary of some of the main properties of the two preconditioners is given in Table 8.3.

8.3.2   *Use of iterative methods for subproblems*

If the preconditioning strategies in Table 8.3 are to be efficient in practice, then the sparse solvers used for the subsidiary problems must be replaced by optimal complexity approximations based on iterative methods. The ideas in Chapters 2

Table 8.3 *Summary of the key properties of Schur complement preconditioners: pressure convection–diffusion* (FP); *least-squares commutator* (BFBT).

	FP	BFBT
effective for Newton systems	yes	yes
applies to stabilized approximation	yes	no
fully automated construction	tricky	yes
h-independent convergence	usually	sometimes
sensitivity to Reynolds number	mild	mild

and 4 are thus relevant in this more general setting. A naive approach would be to use iterative procedures to accurately solve the subsidiary problems, that is, to apply the actions of $\mathbf{F}^{-1}$, A_p^{-1} or $(B\hat{\mathbf{Q}}^{-1}B^T)^{-1}$, to high precision. A smarter, more cost-effective approach is to use a fixed number of multigrid iterations to generate approximate solutions instead. (For example, adopting the strategy taken in Chapter 6, the action of the inverse of the pressure Poisson matrix A_p is spectrally equivalent in the sense of (6.20) to a single V-cycle of multigrid.) We emphasize our philosophy that, in principle, *any* effective iterative solver can be used as a "black-box" fast solver for these subsidiary problems.[6] We refer to the general strategy of using iterative solvers in this way as *iterated* versions of the preconditioners.

A set of experiments showing the effectiveness of an iterated version of the pressure convection–diffusion preconditioner will now be discussed. We take a specific mixed approximation method, Q_2–Q_1 and repeat some of the experiments of the previous section, see also Computational Exercise 8.4. Here, however, to "solve" the subsidiary problems, standard multigrid components are used: a single V-cycle with one presmoothing and one postsmoothing step, with a direct solve on the coarsest grid (level $\ell = 2$). We consider two standard choices of smoother.[7] For the step problem, we use incomplete factorizations IC(0) and ILU(0) as smoothers for the Laplacian A_p and the convection–diffusion operator, respectively. For the cavity problem, motivated by the results and discussion in Chapter 3, we use a four-directional line Gauss–Seidel smoother for the convection–diffusion operator and a unidirectional line Gauss–Seidel smoother for the Laplacian. Notice that the character of the flow in the cavity is similar to that of the recirculating flow of Example 3.1.4 in Chapter 3. We also emphasize

---

[6]We note, however, that "inner" iterations derived from Krylov subspace methods or from a variable number of iterations determine a nonlinear function of the input, and "flexible" versions of the outer Krylov subspace solver will be needed; see for example [163], [176]. A fixed number of stationary iterations (such as multigrid iterations) defines a linear preconditioner for use with standard outer iterations.

[7]These are built into the IFISS software.

the fact that performance of the iterated preconditioning strategy is not greatly dependent on multigrid components as long as the smoothing operators capture the underlying physics in an appropriate way.

Although our real aim is to define a "black-box" implementation of the iterated preconditioner, there are several nonstandard issues concerning the preconditioner $\mathbf{M_F}$ used for the velocity in (8.9):

- First, note the difference between the matrices arising from the Picard (7.46) and Newton (7.41) iterations. For Picard iteration, $\mathbf{F}$ is a set of uncoupled discrete convection–diffusion operators, and so for $\mathbf{M_F}^{-1}$ we can apply multigrid approximation to the inverse of each of these in turn. For Newton iteration, $\mathbf{F}$ has the more complicated structure given in (7.41). (In particular, the velocity components are no longer uncoupled.) To deal with this, we can simply let $\hat{\mathbf{F}}$ be the discrete convection–diffusion operator that would arise if Picard iteration were used to advance to the next step. Then $\mathbf{M_F}^{-1}$ can again be derived from multigrid approximation to $\hat{\mathbf{F}}^{-1}$. In other words the preconditioner for the Jacobian matrix is the same as that which would be used for the corresponding Oseen equations.

- Second, bilinear approximation can always be used to define the velocity component of the preconditioner. That is, for the preconditioner, all the nodes (vertex, edge and centroid) of a Q_2 velocity approximation are treated as though they are vertex nodes of a fine-grid Q_1 discrete convection–diffusion operator. This greatly simplifies the construction of coarse grid operators — only bilinear interpolation and simple restriction operators are needed!

- Third, as discussed at the beginning of Section 7.3, it is not known how to consistently stabilize the Navier–Stokes discretization to give robustness, and accuracy (outside of layers) with coarse grids at high Reynolds number. On the other hand, we know for sure (see Chapter 4) that streamline-diffusion stabilization is crucially important if multigrid is to be a robust solution strategy for the convection–diffusion equation. This issue needs to be addressed if we are to construct an effective preconditioner for $\mathbf{F}$. In the case of Picard iteration, if each block of $\mathbf{F}$ is a standard discrete convection–diffusion operator, then a suitable preconditioner will corres-pond to a multigrid V-cycle applied to an augmented coefficient matrix $\mathbf{F} + \mathbf{D}$, where $\mathbf{D}$ is the corresponding block-stabilized SD matrix.[8] In addi-tion, as discussed in Section 4.3, a carefully designed streamline diffusion stabilization term must be included for all coarse grid systems generated within the multigrid process. For the Newton iteration we can simply use $\hat{\mathbf{F}} + \mathbf{D}$ where $\hat{\mathbf{F}}$ is as specified above.

---

[8]In the IFISS software the matrix $\mathbf{D}$ is the block diagonal matrix obtained by applying "optimal" SD stabilization, as discussed in Section 3.3.2.

Table 8.4 GMRES *iterations needed to satisfy* $\|\mathbf{r}^{(k)}\|/\|\mathbf{s}^{(m)}\| < 10^{-6}$, *starting with zero initial iterate, using multi-grid iterated pressure convection–diffusion preconditioning, for Q_2–Q_1 mixed approximation of Examples 7.1.2 and 7.1.3. $\mathcal{R} = 2/\nu$ is the Reynolds number, ℓ is the grid refinement level and $h = 2^{1-\ell}$.*

		Picard			Newton		
	ℓ	$\mathcal{R} = 10$	100	200	10	100	200
Backward step	4	35	57	97	38	74	131
Example 7.1.2	5	38	51	70	41	71	106
	6	43	58	67	46	79	101
	ℓ	$\mathcal{R} = 10$	100	1000	10	100	1000
Driven cavity	4	27	40	>200	31	54	>200
Example 7.1.3	5	29	38	181	31	52	>200
	6	30	36	138	32	51	>200
	7	31	37	88	34	52	166

We also remark that *algebraic multigrid methods* may be applied to these subsidiary (scalar) problems instead of the geometric methods described here. These offer the possibility of automatic construction of this component of the preconditioner from the coefficient matrix, without any knowledge of the underlying grid. This methodology can be extremely effective for these subsidiary problems, but it is much less effective for saddle point problems and problems of the form of the discrete linearized Navier–Stokes equations. For descriptions of algebraic multigrid, see for example Brezina et al. [29] or Ruge & Stüben [160].

With the above choice of $\mathbf{M_F}$ and the choice $M_S = \texttt{diag}(Q)$, all the operations required by the preconditioning are inexpensive, with costs proportional to the size of the discrete problem. Performance of the iterated pressure convection–diffusion preconditioner is typified by the results presented in Table 8.4. These iteration counts should be compared with the analogous data in Tables 8.1 and 8.2. Note that for large $\mathcal{R}$, as the grid is refined the improved quality of $\mathbf{M_F}$ leads to significant improvements in performance. In most cases fewer than twice as many iterations are needed (and often many fewer than that) than with the ideal preconditioner, to solve the complete system (8.1).

A comparison of convergence curves is presented in Figures 8.4 and 8.5. These clearly show that although there is an increase in the asymptotic contraction factor using the iterated version, it is still independent of the grid size.

To give the reader an idea of the effectiveness in terms of runtime of the iterated pressure convection–diffusion preconditioned GMRES method, Table 8.5

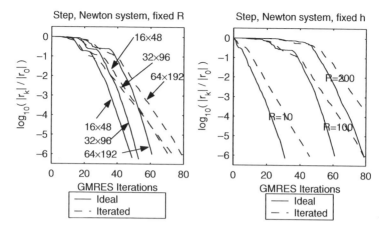

FIG. 8.4. GMRES iterations with "ideal" and iterated pressure convection–diffusion preconditioners, for linear systems arising from Newton iteration applied to the backward facing step problem with Q_2–Q_1 discretization. Left: $\mathcal{R} = 100$ and various grids. Right: 64×192 grid and various $\mathcal{R}$.

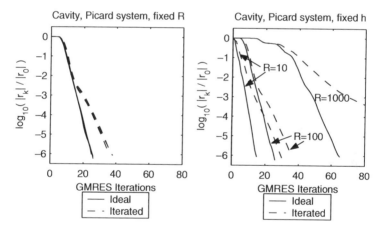

FIG. 8.5. GMRES iterations with "ideal" and iterated pressure convection–diffusion preconditioners, for linear systems arising from Picard iteration applied to the driven cavity problem with Q_2–Q_1 discretization. Left: $\mathcal{R} = 100$ and various grids (32×32, 64×64, 128×128). Right: 64×64 grid and various $\mathcal{R}$.

shows numerical results obtained when all the subsidiary problems are approximated by efficiently implemented multigrid procedures from the femlab 3.0 software library (see http://www.comsol.com). Specifically, for these results $\mathbf{M_F}$ is one V-cycle of the algebraic multigrid solver amgsol from femlab applied to $\mathbf{F}$,

Table 8.5 GMRES *iterations and (in parentheses) CPU time needed to satisfy* $\|\mathbf{r}^{(k)}\|/\|\mathbf{s}^{(m)}\| < 10^{-6}$, *starting with zero initial iterate using algebraic multigrid iterated pressure convection–diffusion preconditioning. The problem comes from* $\mathbf{Q}_2$–$\mathbf{Q}_1$ *mixed approximation of Example 7.1.3 and Picard iteration.* $\mathcal{R} = 2/\nu$ *is the Reynolds number,* ℓ *is the grid refinement level and* $h = 2^{1-\ell}$.

ℓ	$\mathcal{R} = 10$	100	200
4	23 (0.4)	29 (0.6)	58 (1.2)
5	24 (1.5)	31 (1.9)	41 (2.8)
6	23 (5.7)	31 (7.9)	34 (9.2)
7	21 (21.6)	28 (29.6)	31 (35.3)
8	21 (86.8)	27 (113.6)	31 (168.0)

using default settings; A_p is also one V-cycle of amgsol applied to the $\mathbf{Q}_1$ pressure Laplacian, also with default settings; and Q is approximated by diag(Q). The results shown in Table 8.5 are GMRES iteration counts and in brackets the number of seconds of CPU time running matlab 6.5 on a Sun sparcv9 502 MHz processor, to solve the final Oseen problem from a nonlinear Picard iteration for the driven cavity problem Example 7.1.3. Mesh-independence of the GMRES convergence rate is again apparent as is the mild increase in iterations required as the Reynolds number is increased. Most importantly here, the CPU time required for solution is seen to increase by a factor of four when the number of discrete variables is increased by four — that is with "black-box" algebraic multigrid for both the discrete convection–diffusion solve and the discrete Laplacian solve, the preconditioned Oseen solver is indeed optimal with respect to mesh size and has only mild Reynolds number dependence. Notice how convergence is poorer for the coarsest grid for the Reynolds number 200 problem: the grid is not sufficiently fine to well capture the features of the solution at this Reynolds number.

Section 8.3.5 shows numerical results using a preconditioner derived from the commonly used SIMPLE iteration, which can be compared to those shown here.

8.3.3 Convergence analysis

It is generally difficult to derive descriptive convergence analysis for nonsymmetric systems of equations. Thus, it is not surprising that there is no complete analysis of the convergence characteristics of the preconditioning methods described in this chapter at the time of writing. Nevertheless, there are some results that provide insight. In this section, we highlight one analytic result due to Loghin & Wathen [127] concerning the pressure convection–diffusion preconditioner.

Theorem 8.5. *Given a stable mixed approximation and a linearized system with coefficient matrix of the form* (8.2). *Assume that the condition*

$$\langle \mathbf{F}\mathbf{v}, \mathbf{v} \rangle \geq \eta \langle \mathbf{A}\mathbf{v}, \mathbf{v} \rangle \tag{8.35}$$

holds for some $\eta > 0$. Let $M_S = A_p F_p^{-1} Q$ be a pressure convection–diffusion approximation to $B\mathbf{F}^{-1}B^T$. Then the eigenvalues $\{\lambda\}$ of

$$\begin{bmatrix} \mathbf{F} & B^T \\ B & 0 \end{bmatrix} \begin{bmatrix} \mathbf{u} \\ \mathbf{p} \end{bmatrix} = \lambda \begin{bmatrix} \mathbf{F} & B^T \\ 0 & -M_S \end{bmatrix} \begin{bmatrix} \mathbf{u} \\ \mathbf{p} \end{bmatrix},$$

associated with block triangular preconditioning, satisfy $c_1 \leq |\lambda| \leq c_2$. Here $c_1 > 0$ and $c_2 > 0$ depend on η but are otherwise independent of the mesh parameter h.

Proof See Loghin & Wathen [127]. A summary of the proof can also be found in Elman, Loghin & Wathen [66]. □

This result implies that the eigenvalues of the pressure convection–diffusion preconditioned system KM^{-1} are bounded independent of the discretization mesh size h. Combined with the analysis of the GMRES algorithm presented in Chapter 4, the bounds imply that the asymptotic convergence factor of the preconditioned GMRES iteration is also independent of h.

Concerning Reynolds number dependence, the constant c_1 in the above theorem depends monotonically on η, so that a smaller value of η in (8.35) might be reasonably associated with slower GMRES convergence. This observation gives some indication of the relative merit of pressure convection–diffusion preconditioning in the cases of the Picard and Newton linearized systems. For the Oseen problems arising with Picard iteration, $\eta = \nu$ is possible under conditions on the wind that render the vector-advection matrix $\mathbf{N}$ (7.42) skew-symmetric. For Newton's method, however, the additional Newton derivative matrix $\mathbf{W}$ of (7.43) must also be taken into account and this necessarily leads to values of η smaller than ν.

This suggests that the performance of the preconditioned iterative solvers will be more sensitive to Reynolds numbers for Newton's method than for Picard iteration. This question will be explored via experimental results to be discussed in Section 8.4.

8.3.4 *Enclosed flow: singular systems are not a problem*

As observed in Chapters 5 (Section 5.3) and 7, when the Navier–Stokes equations are posed with enclosed flow boundary conditions ($\vec{u} = \vec{w}$ everywhere on $\partial\Omega$) the coefficient matrix of (8.1) is rank-deficient by one. Example 7.1.3, the driven cavity problem, is a problem of this type. For the system to be consistent, the right-hand side vector $\begin{bmatrix} \mathbf{f} \\ \mathbf{g} \end{bmatrix}$ must be orthogonal to $\begin{bmatrix} \mathbf{0} \\ \mathbf{1} \end{bmatrix}$, the null vector of the transpose of the coefficient matrix. In this section, we show that the GMRES algorithm, used in combination with the preconditioning strategies discussed above, will compute a solution to such consistent systems.

When GMRES with right-oriented preconditioning is used to solve the system (8.1), the construction of the Krylov subspace formally entails a set of matrix-vector products by the preconditioned operator KM^{-1} (see Algorithm 4.1). Here,

$$
\begin{aligned}
KM^{-1} &= \begin{bmatrix} \mathbf{F} & B^T \\ B & -C \end{bmatrix} \begin{bmatrix} \mathbf{M}_{\mathbf{F}}^{-1} & \mathbf{M}_{\mathbf{F}}^{-1} B^T M_S^{-1} \\ 0 & -M_S^{-1} \end{bmatrix} \\
&= \begin{bmatrix} \mathbf{FM}_{\mathbf{F}}^{-1} & (\mathbf{FM}_{\mathbf{F}}^{-1} - I) B^T M_S^{-1} \\ B\mathbf{M}_{\mathbf{F}}^{-1} & (B\mathbf{M}_{\mathbf{F}}^{-1} B^T + C) M_S^{-1} \end{bmatrix},
\end{aligned}
\tag{8.36}
$$

where K is the coefficient matrix of (8.1) and M is the block triangular preconditioning operator given in (8.7). As before, the Schur complement operator is $S = B\mathbf{F}^{-1} B^T + C$ and one may simply take $C = 0$ for uniformly stable approximation — the discussion here applies in general to both stable and stabilized elements. In both scenarios, S is singular with the single null vector $\mathbf{1}$. In the following discussion, we will use $\mathcal{S}^{\perp}$ to represent the orthogonal complement of $\mathbf{1}$ in $\mathbb{R}^{n_p}$. Note that S is a nonsingular operator from $\mathcal{S}^{\perp}$ to itself, see Problem 8.5.

Given an arbitrary starting vector $\begin{bmatrix} \mathbf{u}^{(0)}, \mathbf{p}^{(0)} \end{bmatrix}$, the initial residual is

$$
\mathbf{r}^{(0)} = \begin{bmatrix} \mathbf{f} \\ \mathbf{g} \end{bmatrix} - \begin{bmatrix} \mathbf{F} & B^T \\ B & -C \end{bmatrix} \begin{bmatrix} \mathbf{u}^{(0)} \\ \mathbf{p}^{(0)} \end{bmatrix}.
$$

Since both $\mathbf{range}(B)$ and $\mathbf{range}(C)$ are orthogonal to $\mathbf{1}$, $\mathbf{r}^{(0)}$ is orthogonal to $\begin{bmatrix} \mathbf{0} \\ \mathbf{1} \end{bmatrix}$. The $(k-1)$st step of the Arnoldi computation performed by GMRES produces the basis vector,

$$
\begin{bmatrix} \mathbf{w}^{(k)} \\ \mathbf{s}^{(k)} \end{bmatrix} \in \text{span} \left\{ \mathbf{r}^{(0)}, KM^{-1}\mathbf{r}^{(0)}, \ldots, (KM^{-1})^{k-1}\mathbf{r}^{(0)} \right\}
\tag{8.37}
$$

of the Krylov space. It follows from the form of the preconditioned operator that all members of this space are orthogonal to $\begin{bmatrix} \mathbf{0} \\ \mathbf{1} \end{bmatrix}$ or, equivalently, that $\mathbf{s}^{(k)}$ is always in $\mathcal{S}^{\perp}$.

The concern here is whether singularity causes any difficulty for the iteration, in particular, whether the solver "breaks down" before a solution is obtained. The fact that it does not is a consequence of the following result.

Theorem 8.6. *Suppose $\mathcal{F}$ is a rank-deficient square matrix and $\mathcal{F}\mathbf{x} = \mathbf{b}$ is a consistent linear system of equations. If $\text{null}(\mathcal{F}) \cap \text{range}(\mathcal{F}) = \{\mathbf{0}\}$, then GMRES constructs a solution without breaking down.*

Proof See Brown & Walker [38, Theorem 2.6]. □

In the present setting, $\mathcal{F} = KM^{-1}$ and we wish to understand the null space and range space of this matrix. Firstly we need to pay some attention to singularity in the various scalar components in preconditioners.

For the least-squares commutator preconditioner, where M_S is given by (8.32), this definition must be adapted to account for the singularity because $B^T \mathbf{1} = \mathbf{0}$. To do this, we will use the fact that both M_S and M_S^{-1} can be viewed as well-defined non-singular operators from the orthogonal complement space $\mathcal{S}^\perp$ to itself, and that $\mathbf{1}$ is a null vector for both M_S and M_S^{-1}. (Again, see Problem 8.5.) Similarly, for the pressure convection–diffusion preconditioner (8.18), it is natural in the setting of enclosed flow for the matrices A_p and F_p to be derived from Neumann boundary conditions, so that both matrices have $\mathbf{1}$ as a null vector. In this case, $\texttt{range}(M_S^{-1})$ will not in general be orthogonal to $\mathbf{1}$, since $\texttt{range}(F_p)$ will contain components in $\texttt{span}\{\mathbf{1}\}$ and Q is non-singular. However, as observed above, the second components of all members of the Krylov subspace generated by GMRES, those corresponding to the pressure space, lie in $\mathcal{S}^\perp$. Moreover, it follows from (8.36) that in the sequence of steps used to perform the preconditioned matrix-vector product, the action of M_S^{-1} is followed in the computation by multiplication by B^T and C, both of which map $\mathbf{1}$ to $\mathbf{0}$. This means that the effect of M_S^{-1} in the preconditioned matrix-vector product is the same as the effect of $P_1 M_S^{-1} P_1$, where

$$P_1 = I - \frac{\mathbf{1}\,\mathbf{1}^T}{\|\mathbf{1}\|^2} \tag{8.38}$$

is the orthogonal projection into $\mathcal{S}^\perp$, see Problem 8.6. Therefore, without loss of generality, for both preconditioning strategies M_S^{-1} can be treated as an operator that maps $\mathcal{S}^\perp$ to itself and $\mathbf{1}$ to $\mathbf{0}$.

We now return to Theorem 8.6 and examine the null space and range space of KM^{-1}. Suppose $\begin{bmatrix} \mathbf{v} \\ \mathbf{q} \end{bmatrix} \in \texttt{null}(KM^{-1})$. Use of (8.36) gives

$$\mathbf{v} = (\mathbf{M_F F}^{-1} - I)B^T M_S^{-1} \mathbf{q},$$

and then

$$S\,M_S^{-1}\mathbf{q} = \mathbf{0}. \tag{8.39}$$

There are two ways this condition can hold:

$$\text{either} \quad M_S^{-1}\mathbf{q} = \mathbf{0} \quad \text{or} \quad M_S^{-1}\mathbf{q} = \mathbf{1}. \tag{8.40}$$

If $\mathbf{q}$ satisfies either of these, then $B^T M_S^{-1}\mathbf{q} = \mathbf{0}$ and this implies that $\mathbf{v} = \mathbf{0}$. That is,

$$\texttt{null}(KM^{-1}) = \texttt{span}\left\{ \begin{bmatrix} \mathbf{0} \\ \mathbf{q} \end{bmatrix} \right\},$$

where $\mathbf{q}$ satisfies one of the conditions of (8.40). There is a nontrivial intersection with $\texttt{range}(KM^{-1})$ only if

$$\mathbf{FM_F^{-1}w} + (\mathbf{FM_F^{-1}} - I)B^T M_S^{-1}\mathbf{s} = \mathbf{0}$$
$$BM_F^{-1}\mathbf{w} + (BM_F^{-1}B^T + C)M_S^{-1}\mathbf{s} = \mathbf{q}$$

for some $\mathbf{w}$, $\mathbf{s}$. This leads to

$$S\,M_S^{-1}\mathbf{s} = \mathbf{q}. \qquad (8.41)$$

This condition can hold, and $\texttt{null}(KM^{-1})$ and $\texttt{range}(KM^{-1})$ can have a nontrivial intersection, only if $\mathbf{q} \in \texttt{range}(S) = \mathcal{S}^{\perp}$.

Let us explore whether this is possible. Since we know that M_S^{-1} maps $\mathcal{S}^{\perp}$ to itself and $\mathbf{1}$ to $\mathbf{0}$, only the first condition of (8.40) can hold, with $\mathbf{q} = \mathbf{1}$. If (8.41) were also valid, then this would imply that $\mathbf{1} \in \mathcal{S}^{\perp}$. But this contradicts the definition of $\mathcal{S}^{\perp}$, and it follows that the two spaces have no nontrivial intersection. In conclusion, Theorem 8.6 is applicable for either preconditioner and both strategies will compute a solution to (8.1) without breaking down.

In practice, use of either version of M_S^{-1} requires the application of the inverse of a singular matrix, A_p for the pressure convection–diffusion preconditioner, and $B\hat{Q}^{-1}B^T$ for the least-squares commutator preconditioner. Iterative solution with these symmetric positive semi-definite matrices causes no difficulty — see Section 2.3.

Our conclusion is that the preconditioning strategies discussed in previous sections can be implemented in a straightforward manner to cover the case of enclosed flow problems. In practice, GMRES convergence will not be affected by the rank deficiency. See Computational Exercise 8.5 for numerical verification on particular problems.

8.3.5   *Relation to SIMPLE iteration*

In an attempt to place the pressure convection–diffusion and least-squares commutator preconditioners into some comparative context, this section describes a widely used iterative method for solving the systems of equations arising from the steady-state Navier–Stokes equations is known as SIMPLE, the "Semi-Implicit Method for Pressure-Linked Equations." This method was originally published in the often-cited paper by Patankar & Spalding [148] and many variants of this approach have also been proposed, see for example, [110, 147, 200]. The pressure convection–diffusion and least-squares commutator preconditioners are related to SIMPLE — more precisely, as will be shown here, SIMPLE is essentialy a block iteration with a "poor man's" approximation to the Schur complement.

We follow Wesseling [215, pp. 296ff] for an algebraic description of the SIMPLE methodology. SIMPLE has been used primarily for stable finite-difference discretizations, where $C = 0$ in (8.1). It uses a block factorization

$$\begin{bmatrix} \mathbf{F} & B^T \\ B & 0 \end{bmatrix} \approx \begin{bmatrix} \mathbf{M_F} & 0 \\ B & -B\hat{\mathbf{F}}^{-1}B^T \end{bmatrix} \begin{bmatrix} I & \hat{\mathbf{F}}^{-1}B^T \\ 0 & I \end{bmatrix}, \tag{8.42}$$

where both $\mathbf{M_F}$ and $\hat{\mathbf{F}}$ are approximations to $\mathbf{F}$. The first operator $\mathbf{M_F}$ plays the same role as that discussed in Section 8.2, determined by an iteration to approximate the action of the inverse of $\mathbf{F}$. More details of this general issue are given in Section 8.3.

The second approximation $\hat{\mathbf{F}}$ is chosen so that the operator $B\hat{\mathbf{F}}^{-1}B^T$, which represents an approximation to the Schur complement $B\mathbf{F}^{-1}B^T$, can be used explicitly. The standard implementation of SIMPLE uses the diagonal of $\mathbf{F}$ for $\hat{\mathbf{F}}$ (and variants choose other diagonal approximations). This means that the approximate Schur complement $B\hat{\mathbf{F}}^{-1}B^T$ resembles a discrete Laplacian operator.

The SIMPLE solver for (8.1) is a stationary iteration derived from (8.42):

$$\begin{bmatrix} \mathbf{u}^{(k)} \\ \mathbf{p}^{(k)} \end{bmatrix} = \begin{bmatrix} \mathbf{u}^{(k-1)} \\ \mathbf{p}^{(k-1)} \end{bmatrix} + \begin{bmatrix} \omega_u\,\delta^{(k-1)} \\ \omega_p\,\eta^{(k-1)} \end{bmatrix}.$$

Here

$$\begin{bmatrix} \delta^{(k-1)} \\ \eta^{(k-1)} \end{bmatrix} = \begin{bmatrix} I & \hat{\mathbf{F}}^{-1}B^T \\ 0 & I \end{bmatrix}^{-1} \begin{bmatrix} \mathbf{M_F} & 0 \\ B & -B\hat{\mathbf{F}}^{-1}B^T \end{bmatrix}^{-1} \left(\begin{bmatrix} \mathbf{f} \\ \mathbf{g} \end{bmatrix} - \begin{bmatrix} \mathbf{F} & B^T \\ B & 0 \end{bmatrix} \begin{bmatrix} \mathbf{u}^{(k-1)} \\ \mathbf{p}^{(k-1)} \end{bmatrix} \right),$$

and ω_u and ω_p are parameters that can be tuned to accelerate convergence. That is, this is an iteration of the form (4.17) where the splitting operator is

$$M = \begin{bmatrix} \mathbf{M_F} & 0 \\ B & -B\hat{\mathbf{F}}^{-1}B^T \end{bmatrix} \begin{bmatrix} I & \hat{\mathbf{F}}^{-1}B^T \\ 0 & I \end{bmatrix} \begin{bmatrix} \omega_u^{-1}I & 0 \\ 0 & \omega_p^{-1}I \end{bmatrix}.$$

This iteration can easily be accelerated to produce a preconditioned Krylov subspace method. Indeed, in this case we see that SIMPLE represents a block preconditioner with a "poor man's" approximation $B\hat{\mathbf{F}}^{-1}B^T$ to the Schur complement. The attraction of SIMPLE is that it is easy to implement. However its effectiveness diminishes when the spatial mesh size becomes small or for flows that are convection-dominated. For further details, see [215].

To give some idea of the performance of a SIMPLE-like approach here, we present in Table 8.6 the results of using GMRES with a block triangular preconditioner of the form (8.7) with $\mathbf{M_F}$ defined as a single V-cycle of amgsol for the vector convection–diffusion operator with default settings and M_S defined as one V-cycle of amgsol for the Q_1 pressure Laplacian. Note that the computational work per preconditioning step is therefore essentially the same as for the pressure convection–diffusion preconditioner described in Section 8.2.1. These results can be directly compared to those in Table 8.5 where the same efficient "black-box" components are used with the pressure convection–diffusion preconditioner.

Table 8.6 *Number of* GMRES *iterations needed to satisfy* $\|\mathbf{r}^{(k)}\|/\|\mathbf{s}^{(m)}\| < 10^{-6}$, *starting with zero initial iterate, using iterated* SIMPLE-*like preconditioning and* $\mathbf{Q}_2$–$\mathbf{Q}_1$ *mixed approximation for the final Oseen solve in the Picard iteration for Example 7.1.3. The vector convection–diffusion operator and the pressure Laplacian are each approximated by one cycle of algebraic multigrid. In brackets is the CPU time required in seconds. The Reynolds number is* $\mathcal{R} = 2/\nu$, ℓ *is the grid refinement level and* $h = 2^{1-\ell}$.

ℓ	$\mathcal{R} = 10$	100	200
4	58 (2.3)	63 (2.5)	107 (5.2)
5	91 (12.3)	111 (16.4)	151 (25.6)
6	142 (108.2)	165 (124.2)	204 (173.9)
7	211 (766.4)	227 (865.9)	284 (1244.5)

8.4 Nonlinear iteration

In this chapter and the previous one, we have discussed two strategies, Newton's method and a fixed-point Picard iteration, for solving the nonlinear algebraic systems associated with the discrete Navier–Stokes equations. We conclude our discussion by looking more closely at the performance of these nonlinear strategies in the context of preconditioned iterative solution of the discrete Navier–Stokes equations for flows at Reynolds numbers commensurate with the existence of unique steady solutions.

As is well known, Newton's method is usually quadratically convergent for smooth problems, whereas Picard iteration has only a linear convergence rate. On the other hand, Newton's method is convergent only if the initial iterate is close enough to the solution. In contrast, Picard iteration is accepted as being globally convergent for flow problems that have a unique steady-state solution. It is also known (and shown in the previous section) that the costs of solving the linear problems from the nonlinear iterations tend to be higher when Newton's method is used than for Picard iteration. Moreover, as can be seen in the results in Tables 8.1–8.4, this cost difference becomes more pronounced as the Reynolds number increases.

Basically, there are two fundamental questions to address:

1. Will Newton's method produce a convergent sequence of iterates?
2. Will the faster convergence of Newton's method compensate for its greater cost per step?

In cases where the initial iterate (in our case, the solution to the Stokes equations) causes Newton's method to diverge, it is always possible to perform a few steps of Picard iteration to improve the quality of the approximate nonlinear solution, and then to switch to Newton's method to get faster convergence; the second question is relevant under these circumstances as well.

Since we are using iterative methods to solve the linear systems, an extra issue arises because if the nonlinear iterate is far from the solution, then it may not be necessary to solve the linearized system to high accuracy. This observation has led to the development of "inexact" methods for nonlinear equations, where the accuracy of the approximate linear system solution is tied to the accuracy of the nonlinear iterate.

To see what is involved, we briefly outline the analysis of inexact Newton methods, following Kelley [122]. Consider a "generic" system of nonlinear algebraic equations $\mathcal{F}(\mathbf{x}) = \mathbf{0}$, where $\mathcal{F}$ is an n-valued function defined on a subset of $\mathbb{R}^n$ that has a zero $\mathbf{x}^*$. Assume that the Jacobian matrix $\mathcal{F}'(\mathbf{x})$ is non-singular at $\mathbf{x}^*$ and that it is Lipschitz continuous near $\mathbf{x}^*$. (See [122] for more precise statements and definitions.) An inexact Newton method constructs a sequence of iterates

$$\mathbf{x}_{k+1} = \mathbf{x}_k + \mathbf{d}_k,$$

where $\mathbf{d}_k$ is computed to satisfy

$$\|\mathcal{F}'(\mathbf{x}_k)\mathbf{d}_k + \mathcal{F}(\mathbf{x}_k)\| \leq \eta_k \|\mathcal{F}(\mathbf{x}_k)\|. \tag{8.43}$$

The "forcing parameter" η_k serves as a stopping tolerance to be used for an iterative solution of the Jacobian system. If $\eta_k = 0$ then this is (exact) Newton's method.

The idea behind the inexact method is that in the early stages of the iteration, when the nonlinear residual solution is inaccurate, it may not be advantageous to solve the Jacobian system accurately. This is made precise in the following result.

Theorem 8.7. Let $\mathbf{e}_k = \mathbf{x}^* - \mathbf{x}_k$ denote the error at the kth step of inexact Newton iteration. If $\mathbf{x}_0$ is sufficiently close to $\mathbf{x}^*$, then there exist constants c_1 and c_2 such that

$$\|\mathbf{e}_{k+1}\| \leq c_1 \|\mathbf{e}_k\|^2 + c_2 \eta_k \|\mathbf{e}_k\|. \tag{8.44}$$

Proof See Kelley [122], Theorem 6.1.2.   □

The first term of this bound corresponds to the error associated with the exact Newton iteration, and the second term corresponds to an additional error resulting from approximate solution of the Jacobian system.

When iterative methods are used to solve the Jacobian system, there is a cost associated with using small η_k, since more iterations will be needed to meet a more stringent stopping criterion. This result shows that the extra work does not lead to improved accuracy of the nonlinear iterate for any η_k significantly smaller than the current error $\mathbf{e}_k$. It also suggests that a good strategy is to choose η_k so that the two terms are roughly equal in size. It can be shown that when $\mathbf{x}_k$ is near the solution, $\|\mathbf{e}_k\|$ is proportional to $\|\mathcal{F}(\mathbf{x}_k)\|$, which means that for the choice

$$\eta_k = \tau \, \|\mathcal{F}(\mathbf{x}_k)\|, \tag{8.45}$$

the two terms are in balance and the quadratic convergence rate of Newton's method is retained. (More generally, for $\eta_k = \tau \, \|\mathcal{F}(\mathbf{x}_k)\|^\alpha$, convergence is super-linear with rate $1 + \alpha$.) Here τ is a "safety factor" commonly set to a value around 0.1. Thus, with this strategy, the Jacobian system is solved with increased accuracy as the error becomes smaller, leading to a saving in the average cost per step with no degradation in the asymptotic convergence rate of the nonlinear iteration.

For the discrete Navier–Stokes equations with a uniformly stable discretization, the Jacobian system (7.41) arising from Newton's method is

$$\begin{bmatrix} \mathbf{F}_k & B^T \\ B & O \end{bmatrix} \begin{bmatrix} \Delta\mathbf{u}_k \\ \Delta\mathbf{p}_k \end{bmatrix} = \begin{bmatrix} \mathbf{f}_k \\ \mathbf{g}_k \end{bmatrix}, \tag{8.46}$$

where k is the iteration counter for the Newton iteration and $\mathbf{F}_k = \nu\mathbf{A} + \mathbf{W}_k + \mathbf{N}_k$, as shown Section 7.3. The right-hand side is the residual of the nonlinear system at step k of the nonlinear iteration,

$$\begin{bmatrix} \mathbf{f}_k \\ \mathbf{g}_k \end{bmatrix} = \begin{bmatrix} \mathbf{f} \\ \mathbf{g} \end{bmatrix} - \begin{bmatrix} \mathbf{F}_k & B^T \\ B & 0 \end{bmatrix} \begin{bmatrix} \mathbf{u}_k \\ \mathbf{p}_k \end{bmatrix}. \tag{8.47}$$

When an iterative method is used to solve this system, (8.43) takes the form

$$\left\| \begin{bmatrix} \mathbf{f}_k \\ \mathbf{g}_k \end{bmatrix} - \begin{bmatrix} \mathbf{F}_k & B^T \\ B & O \end{bmatrix} \begin{bmatrix} \Delta\mathbf{u}_k^{(m)} \\ \Delta\mathbf{p}_k^{(m)} \end{bmatrix} \right\| \leq \eta_k \left\| \begin{bmatrix} \mathbf{f}_k \\ \mathbf{g}_k \end{bmatrix} \right\|, \tag{8.48}$$

where $\{\Delta\mathbf{u}_k^{(m)}\}$, $\{\Delta\mathbf{p}_k^{(m)}\}$ are the mth iterates for the velocity and pressure updates obtained by the linear solver.

We will compare the performance of an exact Newton iteration with that of an inexact strategy. We use a tolerance of 10^{-5} in our stopping test for the Newton iteration, see (8.34). In light of the discussion above, any small enough fixed η_k will lead to a nonlinear solver that behaves like the exact Newton iteration. We have found $\eta_k = 10^{-6}$ to be small enough, in the sense that smaller values do not change the results in any significant way. (Generated solutions agree to within approximately four significant digits.) In the experiments described below, we refer to the strategy that uses this tolerance as the "exact" Newton method. We also note that results in Syamsudhuha & Silvester [189] indicate that values as large as $\eta_k = 10^{-3}$ lead to quadratic convergence. For the inexact method, we use

$$\eta_k = 10^{-2} \left\| \begin{bmatrix} \mathbf{f}_k \\ \mathbf{g}_k \end{bmatrix} \right\|, \tag{8.49}$$

that is, $\tau = 10^{-2}$ in (8.45).

The update for the Picard iteration also entails solving a system of the form (8.46), where in this case $\mathbf{F}_k = \nu\mathbf{A} + \mathbf{N}_k$. The Picard iteration is linearly convergent, that is the errors satisfy an inequality of the form $\|\mathbf{e}_{k+1}\| \leq c\|\mathbf{e}_k\|$ with $c < 1$. Consequently, for an inexact version of this strategy, a bound analogous to (8.44) has the form

$$\|\mathbf{e}_{k+1}\| \leq c_1 \|\mathbf{e}_k\| + c_2 \eta_k \|\mathbf{e}_k\|. \tag{8.50}$$

For small enough η_k, the second term of (8.50) is negligible in comparison to the first, and as above, in our experiments we simulate an exact Picard iteration using the tolerance $\eta_k = 10^{-6}$. Use of an inexact solver would have no impact on the asymptotic convergence behavior of the Picard iteration, but it is still possible to use a larger, fixed value of η_k for which the second term of (8.50) is slightly smaller than or comparable in size to the first term. Exactly as for Newton's method, this offers the possibility of eliminating unnecessary work from iterative solution of the linear equations. In the experiments described below, we use the fixed value $\eta_k = 10^{-2}$ for the inexact Picard iteration.

Thus, we have four sets of options:

$$\left. \begin{matrix} \text{Newton's method} \\ \text{or} \\ \text{Picard iteration} \end{matrix} \right\} \text{ coupled with } \left\{ \begin{matrix} \text{Exact linear solution} \\ \text{or} \\ \text{Inexact linear solution} \end{matrix} \right.$$

We explore these various combinations in a set of numerical experiments with the two benchmark problems considered in the previous section. GMRES with ideal pressure convection–diffusion preconditioning was used for the linear iteration. The results for the backward facing step flow, Example 7.1.2, using a 64×192 uniform square grid ($\ell = 5$) are shown in Table 8.7. Results for the driven cavity flow, Example 7.1.3, using a 128×128 uniform square grid ($\ell = 6$) are shown in Table 8.8.

In these experiments, the initial nonlinear iterate was obtained from the solution of Stokes equations. The tables report the number of nonlinear steps required to meet the stopping criterion (8.34), together with the total number of linear iterations and the average linear iterations per nonlinear step. The inexact Newton iteration used (8.49), so that a quadratic convergence rate is obtained. (If the expression of (8.49) was smaller than 10^{-6}, then the latter quantity was used for the tolerance.) Newton's method was convergent in all cases except when $\mathcal{R} = 1000$ for the cavity problem; for this example, three steps of Picard iteration were performed prior to using Newton's method.

We can evaluate these results by focusing on the total linear iterations. This reveals the superiority of the inexact versions of the nonlinear solvers, which

Table 8.7 *Performance of nonlinear solvers for the step flow problem, Example 7.1.2, using ideal pressure convection–diffusion preconditioning.* Q_2–Q_1 *mixed approximation: grid level $\ell = 5$.*

	Exact			Inexact		
	$\mathcal{R} = 10$	100	200	10	100	200
Newton steps	2	3	3	2	3	3
Total linear iterations	63	169	224	45	148	169
Average linear iterations	31.5	56.3	74.7	22.5	49.3	56.3
Picard steps	3	6	9	3	6	9
Total linear iterations	91	249	414	48	181	311
Average linear iterations	30.3	41.5	46.0	16.0	30.2	34.6

Table 8.8 *Performance of nonlinear solvers for the driven cavity problem, Example 7.1.3, using ideal pressure convection–diffusion preconditioning.* Q_2–Q_1 *mixed approximation: grid level $\ell = 6$.*

	Exact			Inexact		
	$\mathcal{R} = 10$	100	1000	10	100	1000
Newton steps	1	2	4 (3+1)	2	2	4 (3+1)
Total linear iterations	17	65	293	24	49	221
Average linear iterations	17.0	32.5	73.3	12.0	24.2	55.3
Picard steps	2	3	4	3	3	5
Total linear iterations	31	78	220	20	39	172
Average linear iterations	15.3	26.0	55.0	6.7	13.0	34.4

require fewer linear steps in five of the six examples. (The only exception is where one extra Newton step was needed for the cavity problem with $\mathcal{R} = 10$.) Thus, even for the simple variant of this idea used for the Picard iteration, it is advantageous to compute an approximate solution to the linear systems. Comparing the performance of the two nonlinear algorithms, we find a slight advantage for Newton's method: the Jacobian systems tend to require more linear iterations than the Oseen systems, but the smaller number of Newton steps compensates for this, and Newton's method "wins" in eight of the twelve examples. (However, Newton's method needs help from Picard iteration in the example with $\mathcal{R} = 1000$.) This conclusion applies to situations where there exist unique stable solutions. Our experience for larger Reynolds numbers, especially in cases where stability of solutions starts to break down, is that Newton's method is the only robust option.

These results together with those of Section 8.3 serve to highlight the real power of the methodology described in this chapter. At the heart of an effective nonlinear solver for the incompressible Navier–Stokes equations are preconditioned iterative linear solvers for the required sequence of linearized problems. It is the efficiency of the pressure convection–diffusion and least-squares commutator preconditioners that enables the Navier–Stokes solution strategy developed here.

Discussion and bibliographical notes

The convergence and stability properties of finite element discretization of methods of steady incompressible flow are summarized in the classic book by Girault & Raviart [85], and are well understood. The state of affairs concerning algorithms for solving the algebraic systems that arise from such discretizations is far less well-developed. The material presented in this chapter represents a summary of research undertaken by the authors and their collaborators in which this problem is addressed in a systematic way by taking advantage of the structure of the algebraic systems.

The methods presented build on strategies used for the Stokes equations, which are described in Chapter 6. Having the optimal preconditioned solvers for the Stokes problem, Golub & Wathen [86] analyzed their utility as preconditioners for the Oseen problem. Elman & Silvester [58] took essentially the same approach for the Oseen problem as that used successfully for the Stokes problem, where the Schur complement is approximated by the pressure mass matrix. Both of these preconditioning strategies showed mesh independence in the convergence rate, but this degraded like $\mathcal{O}(\nu^{-2})$ with the viscosity/Reynolds number. These were followed by various efforts to improve on the Elman & Silvester idea for problems where convection plays a significant role. In particular, a version of the least-squares commutator preconditioner was derived in Elman [60], and the pressure convection–diffusion method was derived in Kay, Loghin & Wathen [117] and Silvester et al. [174].

The least-squares commutator preconditioner presented in Section 8.2.2 was derived in Elman et al. [57], where it was shown that scaling using the velocity mass matrix dramatically improves performance. Development of methods of this type is currently an active area of research; other recent papers include Elman et al. [65], Gauthier, Saleri & Veneziani [82]. Olshanskii in [139] develops preconditioners for the Schur complement operator associated with the rotation form of the Navier–Stokes equations. Related work, with similar emphasis on the Schur complement matrix, is described in the recent book by Turek [197]. Although we have chosen the commutator (8.13) as the device for motivating these methods, we also point out that Kay, Loghin & Wathen [117] derived the pressure convection–diffusion preconditioner in a different way, using a fundamental solution tensor of the continuous form of the linearized Navier–Stokes operator. An alternative strategy is considered by Benzi & Golub [13]. This uses a perturbation of the skew-symmetric part of the coefficient matrix as a preconditioner for the Oseen operator.

To the best of our knowledge, these are the first systematic efforts to build solution algorithms for steady incompressible flow problems (other than multigrid methods, see below) based on preconditioners derived from the structure of the system. There is, however, a large literature that describes time-stepping methods for solving the evolutionary Navier–Stokes equations in terms of approximate block factorizations of matrices, see for example, Henriksen & Holmen [103] and Quarteroni, Saleri & Veneziani [151]. The flavor of this approach is as follows. For typical time-stepping schemes and stable spatial discretizations, the advancement in time takes the form

$$
\begin{bmatrix} \mathbf{F} & B^T \\ B & 0 \end{bmatrix} \begin{bmatrix} \mathbf{u}^{(m+1)} \\ \mathbf{p}^{(m+1)} \end{bmatrix} = \begin{bmatrix} \mathbf{f}^{(m)} \\ \mathbf{g}^{(m)} \end{bmatrix}, \tag{8.51}
$$

where m is a time-stepping index and both $\mathbf{F}$ and the right-hand side data depend on the time discretization. Approximate factorizations of the coefficient matrix can be expressed in the form

$$
\begin{bmatrix} \mathbf{F} & B^T \\ B & 0 \end{bmatrix} \approx \begin{bmatrix} \mathbf{F} & 0 \\ B & -B\mathbf{H}_1 B^T \end{bmatrix} \begin{bmatrix} I & \mathbf{H}_2 B^T \\ 0 & I \end{bmatrix} = \begin{bmatrix} \mathbf{F} & \mathbf{F}\mathbf{H}_2 B^T \\ B & B(\mathbf{H}_2 - \mathbf{H}_1)B^T \end{bmatrix},
$$

where $\mathbf{H}_1$ and $\mathbf{H}_2$ must be determined. In a standard example, if diffusion is treated implicitly, convection is treated explicitly, and a first order time differencing is used, then $\mathbf{F} = (1/\Delta t)\mathbf{Q} + \nu\mathbf{A}$. In this case, as shown by Perot [149], the choice $\mathbf{H}_1 = \mathbf{H}_2 = \Delta t\mathbf{Q}^{-1}$ corresponds exactly to the classic projection methods of Chorin [41, 42] & Temam [190].[9] Alternative choices of $\mathbf{H}_1$ and $\mathbf{H}_2$ are considered in [103] and [151]. The philosophy underlying these choices is for the approximate factorization not to lead to a larger discretization error than is inherited from the time discretization. Thus, these ideas have been restricted to evolutionary problems, although we note the similarity between them and SIMPLE iteration (8.42).

In contrast to the situation for algebraic preconditioners, there have been significant developments over a long period of time in multigrid algorithms that take advantage of the block structure of the discrete Navier–Stokes equations. The main requirement for multigrid methods in this setting is to develop smoothing operators for the block systems. Grid transfers between fine and coarse grids can be handled using standard approaches for both velocities and pressures. Since the problem as posed is indefinite, it is difficult to develop effective smoothers. A way around this was presented by Brandt & Dinar [26]. Our description follows Hackbusch [97, Section 11.3] and Wesseling [214, Section 9.7], [215, Section 7.6].

---

[9]We note that the importance of the commutator was observed by Chorin in [42, p. 352].

The idea is to make a change of variables

$$
\begin{bmatrix} \mathbf{F} & B^T \\ B & 0 \end{bmatrix} \begin{bmatrix} \mathbf{I} & B^T \\ 0 & -F_p \end{bmatrix} \begin{bmatrix} \hat{\mathbf{u}} \\ \hat{\mathbf{p}} \end{bmatrix} = \begin{bmatrix} \mathbf{f} \\ \mathbf{0} \end{bmatrix}, \quad \begin{bmatrix} \mathbf{u} \\ \mathbf{p} \end{bmatrix} = \begin{bmatrix} \mathbf{I} & B^T \\ 0 & -F_p \end{bmatrix} \begin{bmatrix} \hat{\mathbf{u}} \\ \hat{\mathbf{p}} \end{bmatrix}. \tag{8.52}
$$

The coefficient matrix for the transformed variables is

$$
\tilde{K} = \begin{bmatrix} \mathbf{F} & E \\ B & G \end{bmatrix},
$$

where $E = \mathbf{F}B^T - B^T F_p$ and $G = BB^T$. For Picard iteration, the $(1,1)$ entry of $\tilde{K}$ is the discrete convection–diffusion operator, and if F_p is defined as in Section 8.2.1, then the entry E in the $(1,2)$ block is a commutator. Using Taylor series arguments, it can be shown that this is small in cases where the discrete velocity and pressure share a common grid, and it follows that $\tilde{K}$ is essentially of block triangular form. The $(2,2)$ entry G is a scaled discrete Laplacian. In the *distributive relaxation* method developed by Brandt & Dinar [26], the smoothing strategy is determined using Gauss–Seidel smoothers for the two component operators $\mathbf{F}$ and G. This approach can also be used with smoothers generated by incomplete factorization, see Wittum [216, 217]. An alternative, developed by Vanka [201], is to build a smoother directly for the untransformed problem by grouping local velocities and pressures together. Extensive discussion of these ideas as well as many other references can be found in the recent book of Trottenberg et al. [196, Chapter 8].

It is evident that multigrid methods that use the change of variables (8.52) share many characteristics with the preconditioning methods considered in this chapter. A comparison of multigrid strategies and the preconditioned MINRES strategy described in Chapter 6 is given in Elman [59]. This study showed that the fastest variant of multigrid is somewhat more efficient than optimally preconditioned MINRES, but that there is no difference in the asymptotic behavior with respect to mesh size of the two methodologies. We know of no direct comparison for more general discrete Navier–Stokes systems. One drawback of the multigrid approach, however, concerns the commutator. In our experience, methods of this class are effective only if the velocities and pressure are defined on a common grid, so that $\tilde{K}$ is close to being block triangular. As discussed in Section 8.2.1, the commutator is not small for many stable approximation strategies, including Q_2–Q_1. This means that the algebraic preconditioning methods of this chapter are much more generally applicable in a mixed finite element approximation setting.

Our objective in developing the solution algorithms presented in this chapter is for them to be effective and adaptable to a variety of circumstances. Although this book is exclusively concerned with stationary problems, our ideas can also be adapted to handle evolutionary flow problems in a completely straightforward manner. Using fully implicit time-stepping leads to systems of the generic form (8.51), but with an operator $\mathbf{F}$ having a multiple of $(1/\Delta t)\mathbf{Q}$ added to the block diagonal. This makes the discrete convection–diffusion equations required by the preconditioners dramatically *easier* to solve. As a result, implicit treatment of the momentum equation — often viewed as leading to "difficult" algebraic systems — is in fact straightforward. Indeed, our preconditioning strategies open up the possibility of adaptive timestepping determined solely by the physics of the flow problem. Examples showing the effectiveness of our preconditioning strategy in a time-stepping context can be found in Elman [61]. With a physically realistic time step it can be seen that performance is essentially insensitive to the Reynolds number.

The preconditioning ideas presented here offer the possibility of being extended to more general systems such as those that include temperature or kinetics in the model. Indeed, although this book is about fluid flow problems, we believe that the ideas and approaches developed are of general relevance to the world of scientific computing.

Problems

8.1. Show that the generalized eigenvalue value problem (8.4) has an eigenvalue $\lambda = 1$ of algebraic and geometric multiplicity $n_u - n_p$. Show that if in addition the problem (8.5) has a complete set of linearly independent eigenvectors, then (8.4) has $2n_p$ additional eigenvalues and a complete set of eigenvectors.

8.2. Show that if GMRES is applied to a system of equations $Ax = b$ where A has the form

$$\begin{bmatrix} I & 0 \\ X & I \end{bmatrix}$$

with $X \neq 0$, then the exact solution is obtained in precisely two steps. (Hint: show that the minimal polynomial of A is $(z - 1)^2$.)

8.3. Show that the generalized eigenvalue value problem

$$\begin{bmatrix} \mathbf{F} & B^T \\ B & -C \end{bmatrix} \begin{bmatrix} \mathbf{u} \\ \mathbf{p} \end{bmatrix} = \lambda \begin{bmatrix} \mathbf{F} & B^T \\ 0 & M_S \end{bmatrix} \begin{bmatrix} \mathbf{u} \\ \mathbf{p} \end{bmatrix}$$

has an eigenvalue $\lambda = 1$ of algebraic multiplicity $n_u + m$ and geometric multiplicity n_u, where m is the number of eigenvalues of the problem

$$(B\mathbf{F}^{-1}B^T + C)p = \mu M_S p$$

that have the value unity.

8.4. Let $\mathcal{D}_h$ denote the discrete negative divergence operator defined on $\mathbf{X}_0^h$ by

$$(\mathcal{D}_h u_h, q_h) = (-\nabla \cdot u_h, q_h) \quad \text{for all } q_h \in M^h.$$

For u_h given and $p_h = \mathcal{D}_h u_h$, let

$$u_h = \sum_{i=1}^{n_u} \mathbf{u}_i \vec{\phi}_i, \quad p_h = \sum_{j=1}^{n_p} \mathbf{p}_j \psi_j$$

be representations in terms of bases $\{\vec{\phi}_i\}$ for $\mathbf{X}_0^h$ and $\{\psi_j\}$ for M^h. Show that the coefficient vectors satisfy

$$B\mathbf{u} = Q\mathbf{p}.$$

That is, $\mathcal{D}_h$ has the matrix representation $Q^{-1}B$.
Similarly, show that if $\mathcal{G}_h$ denotes the gradient operator defined on M^h by

$$(\mathcal{G}_h p_h, v_h) = -(p_h, \nabla \cdot v_h) \quad \text{for all } v_h \in \mathbf{X}_0^h,$$

then its matrix representation is $\mathbf{Q}^{-1}B^T$.

8.5. Suppose $\mathbf{F}$ and B of (8.2) come from a stable discretization of a problem with enclosed flow boundary conditions, so that B has rank $n_p - 1$ with $B^T\mathbf{1} = \mathbf{0}$.
a. Show that the Schur complement operator $B\mathbf{F}^{-1}B^T$ represents a non-singular operator from $\mathcal{S}^\perp$ to itself, where $\mathcal{S}^\perp$ is the orthogonal complement in $\mathbb{R}^{n_p}$ of $\mathbf{1}$. (Hint: use the singular value decomposition of B.)
b. Show that the least-squares commutator preconditioning matrix M_S of (8.32) is also a non-singular operator from $\mathcal{S}^\perp$ to itself.
c. Suppose instead that $\mathbf{F}$, B and C derive from a *stabilized* discretization of a problem with enclosed flow boundary conditions. Show that, again, the Schur complement $S = B\mathbf{F}^{-1}B^T + C$ is a non-singular operator from $\mathcal{S}^\perp$ to itself.

8.6. Let P_1 of (8.38) be the orthogonal projection of $\mathbb{R}^{n_p}$ onto $\mathcal{S}^\perp$. Work out the details of the use of P_1 for systems arising from enclosed flow. Show that in a right-preconditioned GMRES iteration, the result of the preconditioned matrix-vector product using the pressure convection–diffusion preconditioner is the same whether or not P_1 is used.

Computational exercises

The driver `navier_testproblem` can be used to set up any of the reference problems in Examples 7.1.1–7.1.4. Having solved one of these flow problems, the function `it_solve` allows the user to explore ideal or multgrid iterated preconditioning for the Oseen system associated with the latest flow solution.

8.1. Demonstrate the result in Theorem 8.2. That is, for a flow problem of your choice (the result is general!), show that GMRES combined with the "ideal" version of the block triangular preconditioner requires precisely half as many iterations to solve (8.1) as when the analogous block diagonal preconditioner is used. In order to do this, you will need to construct the block diagonal preconditioner. This can be done by modifying the block triangular preconditioning function solvers/m_fp.m.

8.2. Using the matlab eig function, compute the eigenvalues of the matrix A_p generated by the IFISS software when solving the driven cavity flow problem in Example 7.1.3 using Q_1–P_0 approximation on an 8×8 grid (grid level $\ell = 3$). Can you construct an analytic expression for the eigenvalues analogous to that in (1.122)? Repeat the exercise using Q_2–P_{-1} approximation on the same grid level (and be prepared for a surprise!).

8.3. Compare the effectiveness of the pressure convection–diffusion and least-squares commutator preconditioners for solving systems the step flow problem in Example 7.1.2 with the stable Q_2–P_{-1} mixed approximation method. Issues to consider include: whether the extra Poisson solve required by the least-squares commutator method represents a significant extra cost; how the stopping tolerance affects the relative merits of the two approaches; how performance is affected by the grid size and Reynolds number and the impact of irregular (stretched) grids on your conclusions.

8.4. Compare iteration counts and CPU timings for the ideal and iterated pressure convection–diffusion preconditioners by solving the driven cavity problem in Example 7.1.3 on a sequence of uniform grids. Compute results for Q_2–Q_1 and Q_1–P_0 mixed approximation and compare results with those in Table 8.2.

8.5. Use the IFISS software to explore the impact of singularity caused by enclosed flow boundary conditions on the performance of the pressure convection–diffusion and least squares commutator preconditioners. Use the driven cavity problem, Example 7.1.3.
Comments: this exercise entails "getting one's hands dirty" by going into the IFISS code and seeing what happens to intermediate quantities. For example, you can modify functions solvers/m_fp.m or solvers/m_xbfbt.m and examine whether application of the (pseudo)inverse of the discrete Laplacian makes some quantities contain a component of the null vector **1**. You can also check to see whether explicit use of the projection operator (8.38) has any impact on performance.

8.6. Explore the effect of the stopping tolerance for the linear equations on the performance of the inexact Newton and Picard algorithms. Determine "best" choices of the parameters for one of the benchmark problems. Again, these choices may be affected by grid size and Reynolds number.

8.7. Demonstrate that both Newton's method and Picard iteration exhibit the convergence behavior expected of them when they are used to solve the nonlinear algebraic equations arising from discretization of the Navier–Stokes equations. The errors for the Picard iteration should decrease linearly,

$$\|\mathbf{e}_{k+1}\| \leq c \|\mathbf{e}_k\|,$$

for some $c < 1$, and the errors for Newton's method,

$$\frac{\|\mathbf{e}_{k+1}\|}{\|\mathbf{e}_k\|^2},$$

should tend to a constant as k becomes large enough. Repeat this latter experiment for the inexact Newton algorithm. (Hint: it will be sufficient to compute convergence rates based on the residual error in (8.47) rather than the algebraic error.)

BIBLIOGRAPHY

Numbers in brackets following a reference give page numbers where the reference is cited.

[1] D. ACHESON, *Elementary Fluid Dynamics*, Oxford University Press, Oxford, 1990. [1, 278, 317, 318, 322]

[2] M. AINSWORTH AND P. COGGINS, *The stability of mixed hp-finite element methods for Stokes flow on high aspect ratio elements*, SIAM J. Numer. Anal., 38 (2000), pp. 1721–1761. [279]

[3] M. AINSWORTH AND J. ODEN, *A posteriori error estimates for Stokes' and Oseen's equations*, SIAM J. Numer. Anal., 34 (1997), pp. 228–245. [262]

[4] M. AINSWORTH AND J. ODEN, *A Posteriori Error Estimation in Finite Element Analysis*, Wiley, New York, 2000. [264, 266, 330, 334, 335]

[5] O. AXELSSON, *Conjugate gradient type methods for nonsymmetric and inconsistent systems of linear equations*, Linear Algebra Appl., 29 (1980), pp. 1–16. [208]

[6] O. AXELSSON, *Iterative Solution Methods*, Cambridge University Press, Cambridge, 1994. [88, 190, 209]

[7] I. BABUŠKA, *The finite element method with Lagrangian multipliers*, Numer. Math., 20 (1973), pp. 179–192. [279]

[8] R. E. BANK, *A comparison of two multilevel iterative methods for non-symmetric and indefinite elliptic finite element equations*, SIAM J. Numer. Anal., 18 (1981), pp. 724–723. [207]

[9] R. E. BANK AND A. WEISER, *Some a posteriori error estimators for elliptic partial differential equations*, Math. Comp., 44 (1985), pp. 283–301. [51]

[10] R. E. BANK AND B. WELFERT, *A posteriori error estimates for the Stokes problem*, SIAM J. Numer. Anal., 28 (1991), pp. 591–623. [279]

[11] G. BATCHELOR, *On steady laminar flow with closed streamlines at large reynolds numbers*, J. Fluid Mech., 1 (1956), pp. 77–90. [317]

[12] G. BATCHELOR, *An Inroduction to Fluid Dynamics*, Cambridge University Press, Cambridge, 2000. [1]

[13] M. BENZI AND G. H. GOLUB, *A preconditioner for generalized saddle point problems*, SIAM J. Matrix Anal. Appl., 26 (2004), pp. 20–41. [375]

[14] C. BERNARDI AND Y. MADAY, *Uniform inf-sup conditions for the spectral discretization of the Stokes problem*, Math. Methods Appl. Sci., 9 (1999), pp. 395–414. [278]

[15] J. BEY AND G. WITTUM, *Downwind numbering: robust multigrid for convection–diffusion problems*, Appl. Numer. Math., 23 (1997), pp. 177–192. [209, 210]

[16] P. BOCHEV, C. DOHRMANN, AND M. GUNZBURGER, *Stabilization of low-order mixed finite elements method for the Stokes equations*, SIAM J. Numer. Anal., 44 (2006), pp. 82–101. [261]

[17] J. BOLAND AND R. NICOLAIDES, *Stability of finite elements under divergence constraints*, SIAM J. Numer. Anal., 20 (1983), pp. 722–731. [229]

[18] J. BOLAND AND R. NICOLAIDES, *On the stability of bilinear-constant velocity pressure finite elements*, Numer. Math., 44 (1984), pp. 219–222. [279]

[19] D. BRAESS, *Finite Elements*, Cambridge University Press, London, 1997. [14, 20, 22, 25, 37, 39, 41, 43, 45, 78, 106, 207, 224]

[20] D. BRAESS AND W. HACKBUSCH, *A new convergence proof for the multigrid method including the V-cycle*, SIAM J. Numer. Anal., 20 (1983), pp. 967–975. [104]

[21] J. BRAMBLE AND J. PASCIAK, *A preconditioning technique for indefinite systems resulting from mixed approximations of elliptic problems*, Math. Comput., 50 (1988), pp. 1–17. [308]

[22] J. BRAMBLE AND J. PASCIAK, *New estimates for multilevel algorithms including the V-cycle*, Math. Comp., 60 (1993), pp. 447–471. [106]

[23] J. BRAMBLE, J. PASCIAK, J. WANG, AND J. XU, *Convergence estimates for multigrid algorithms without regularity assumptions*, Math. Comp., 57 (1991), pp. 23–45. [106]

[24] J. H. BRAMBLE, J. E. PASCIAK, AND J. XU, *The analysis of multigrid algorithms for nonsymmetric and indefinite elliptic problems*, Math. Comp., 51 (1988), pp. 389–414. [207]

[25] A. BRANDT, *Multi-level adaptive solutions to boundary-value problems*, Math. Comp., 31 (1977), pp. 333–390. [210]

[26] A. BRANDT AND N. DINAR, *Multigrid solutions to elliptic flow problems*, in Numerical Methods for Partial Differential Equations, S. V. Parter, ed., Academic Press, New York, 1979, pp. 53–147. [376, 377]

[27] A. BRANDT AND I. YAVNEH, *On multigrid solution of high-reynolds incompressible entering flows*, J. Comput. Phys., 101 (1992), pp. 151–164. [209, 210]

[28] S. C. BRENNER AND L. R. SCOTT, *The Mathematical Theory of Finite Element Methods*, Springer-Verlag, New York, 1994. [37, 38, 39, 43, 52, 53, 104, 106, 122, 224]

[29] M. BREZINA, R. FALGOUT, S. MACLACHLAN, T. MANTEUFFEL, S. MCCORMICK, AND J. RUGE, *Adaptive smoothed aggregation (αSA)*, SIAM J. Sci. Comput., 25 (2004), pp. 1896–1920. [362]

[30] F. BREZZI, *On the existence, uniqueness and approximation of saddle point problems arising from Lagrangian multipliers*, R.A.I.R.O. Anal. Numér., 8 (1974), pp. 129–154. [279]

[31] F. BREZZI, *Interacting with the subgrid world*, in Numerical Analysis 1999, D. F. Griffiths and G. A. Watson, eds., Chapman & Hall/CRC, Boca Raton, 2000. [162]

[32] F. BREZZI, *Stability of saddle-points in finite dimensions*, in Frontiers in Numerical Analysis, J. Blowey, A. Craig, and T. Shardlow, eds., Springer, Berlin, 2002, pp. 17–61. [280]

[33] F. BREZZI AND M. FORTIN, *Mixed and Hybrid Finite Element Methods*, Springer-Verlag, New York, 1991. [223, 249, 261]

[34] F. BREZZI, T. J. R. HUGHES, L. D. MARINI, A. RUSSO, AND E. SÜLI, *A priori error analysis of residual-free bubbles for advection-diffusion problems*, SIAM J. Numer. Anal., 36 (1999), pp. 1933–1948. [162]

[35] F. BREZZI, D. MARINI, AND A. RUSSO, *Applications of the pseudo residual-free bubbles to the stabilization of convection–diffusion problems*, Comput. Methods Appl. Mech. Eng., 166 (1998), pp. 51–63. [126, 162]

[36] F. BREZZI, D. MARINI, AND E. SÜLI, *Residual-free bubbles for advection-diffusion problems: the general error analysis*, Numer. Math., 85 (2000), pp. 31–47. [162]

[37] W. BRIGGS, V. HENSON, AND S. McCORMICK, *A Multigrid Tutorial*, SIAM, Philadelphia, 2000. [88]

[38] P. N. BROWN AND H. F. WALKER, *GMRES on (nearly) singular systems*, SIAM J. Matrix Anal. Appl., 18 (1997), pp. 37–51. [366]

[39] T. F. CHAN AND T. P. MATHEW, *Domain Decomposition Algorithms*, in Acta Numerica 1994, Cambridge University Press, 1994, pp. 61–143. [82]

[40] R. C. Y. CHIN AND T. A. MANTEUFFEL, *An analysis of block successive overrelaxation for a class of matrices with complex spectra*, SIAM J. Numer. Anal., 25 (1988), pp. 564–585. [209]

[41] A. J. CHORIN, *Numerical solution of the Navier–Stokes equations*, Math. Comp, 22 (1968), pp. 745–762. [376]

[42] A. J. CHORIN, *On the convergence of discrete approximations to the Navier–Stokes equations*, Math. Comp, 23 (1969), pp. 341–353. [376]

[43] I. CHRISTIE, D. F. GRIFFITHS, A. R. MITCHELL, AND O. C. ZIENKIEWICZ, *Finite element methods for second order differential equations with significant first derivatives*, Int. J. Numer. Methods Eng., 10 (1976), pp. 1389–1396. [162]

[44] P. G. CIARLET, *The Finite Element Method for Elliptic Problems*, North-Holland, Amsterdam, 1978. [35, 53]

[45] P. CLÉMENT, *Approximation by finite element functions using local regularization*, R.A.I.R.O. Anal. Numér., 2 (1975), pp. 77–84. [52]

[46] K. CLIFFE, A. SPENCE, AND S. TAVERNER, *The numerical analysis of bifurcation problem with application to fluid mechanics*, in Acta Numerica 2000, A. Iserles, ed., Cambridge University Press, Cambridge, 2000. [324, 331, 338]

[47] P. M. DE ZEEUW, *Matrix-dependent prolongations and restrictions in a blackbox multigrid solver*, J. Comput. Appl. Math., 3 (1990), pp. 1–27. [210]

[48] P. M. DE ZEEUW AND E. J. VAN ASSELT, *The convergence rate of multi-level algorithms applied to the convection–diffusion equation*, SIAM J. Sci. Comput., 2 (1985), pp. 492–503. [210]

[49] J. DENDY, *Black box multigrid for nonsymmetric problems*, Appl. Mat. Comp., 13 (1983), pp. 261–284. [210]

[50] C. R. DOERING AND J. GIBBON, *Applied Analysis of the Navier–Stokes Equations*, Cambridge University Press, Cambridge, 1995. [338]

[51] C. DOHRMANN AND P. BOCHEV, *A stabilized finite element method for the Stokes problem based on polynomial pressure projections*, Int. J. Numer. Methods Fluids, 46 (2004), pp. 183–201. [243]

[52] I. DUFF, A. ERISMAN, AND J. REID, *Direct Methods for Sparse Matrices*, Oxford University Press, Oxford, 1986. [68]

[53] W. ECKHAUS, *Asymptotic Analysis of Singular Perturbations*, North-Holland, Amsterdam, 1979. [117]

[54] M. EIERMANN AND O. ERNST, *Geometric aspects of the theory of Krylov subspace methods*, in Acta Numerica, Cambridge University Press, Cambridge, 2001, pp. 251–312. [209]

[55] S. C. EISENSTAT, H. C. ELMAN, AND M. H. SCHULTZ, *Variational iterative methods for nonsymmetric systems of linear equations*, SIAM J. Numer. Anal., 20 (1983), pp. 345–357. [208]

[56] H. ELMAN, *Multigrid and Krylov subspace methods for the discrete Stokes equations*, Int. J. Numer. Methods Fluids, 22 (1996), pp. 755–770. [308]

[57] H. ELMAN, V. E. HOWLE, J. SHADID, R. SHUTTLEWORTH, AND R. TUMINARO, *Block preconditioners based on approximate commutators*, SIAM J. Sci. Comput., 27 (2005), pp. 1651–1055. [359, 375]

[58] H. ELMAN AND D. SILVESTER, *Fast nonsymmetric iterations and preconditioning for Navier–Stokes equations*, SIAM J. Sci. Comput., 17 (1996), pp. 33–46. [344, 345, 346, 375]

[59] H. C. ELMAN, *Multigrid and Krylov subspace methods for the discrete Stokes equations*, Int. J. Numer. Methods Fluids, 227 (1996), pp. 755–770. [377]

[60] H. C. ELMAN, *Preconditioning for the steady-state Navier–Stokes equations with low viscosity*, SIAM J. Sci. Comput., 20 (1999), pp. 1299–1316. [358, 375]

[61] H. C. ELMAN, *Preconditioning Strategies for Models of Incompressible Flow*, J. Sci. Comput., 25 (2005), pp. 347–366. [378]

[62] H. C. ELMAN AND M. P. CHERNESKY, *Ordering effects on relaxation methods applied to the discrete one-dimensional convection–diffusion equation,* SIAM J. Numer. Anal., 30 (1993), pp. 1268–1290. [181, 209]

[63] H. C. ELMAN AND M. P. CHERNESKY, *Ordering effects on relaxation methods applied to the discrete convection–diffusion equation*, in Recent Advances in Iterative Methods, A. G. G. H. Golub and M. Luskin, eds., Springer-Verlag, New York, 1994, pp. 45–57. [209]

[64] H. C. ELMAN AND G. H. GOLUB, *Line iterative methods for cyclically reduced convection–diffusion problems*, SIAM J. Sci. Stat. Comput., 13 (1992), pp. 339–363. [209]

[65] H. C. ELMAN, V. E. HOWLE, J. SHADID, AND R. TUMINARO, *A parallel block multi-level preconditioner for the 3D incompressible Navier–Stokes equations*, J. Comput. Phys., 187 (2003), pp. 505–523. [375]

[66] H. C. ELMAN, D. LOGHIN, AND A. J. WATHEN, *Preconditioning techniques for Newton's method for the incompressible Navier–Stokes equations*, BIT, 43 (2003), pp. 961–974. [365]

[67] H. C. ELMAN AND A. RAMAGE, *An analysis of smoothing effects of upwinding strategies for the convection–diffusion equation*, SIAM J. Numer. Anal., 40 (2002), pp. 254–281. [157]

[68] H. C. ELMAN AND A. RAMAGE, *A characterisation of oscillations in the discrete two-dimensional convection–diffusion equation*, Math. Comp., 72 (2002), pp. 263–288. [157]

[69] H. C. ELMAN, D. J. SILVESTER, AND A. J. WATHEN, *Performance and analysis of saddle point preconditioners for the discrete steady-state Navier–Stokes equations*, Numer. Math., 90 (2002), pp. 665–688. [359]

[70] K. ERIKSSON, D. ESTEP, P. HANSBO, AND C. JOHNSON, *Computational Differential Equations*, Cambridge University Press, New York, 1996. [133, 161]

[71] V. FABER AND T. A. MANTEUFFEL, *Necessary and sufficient conditions for the existence of a conjugate gradient method*, SIAM J. Numer. Anal., 21 (1984), pp. 352–362. [209]

[72] V. FABER AND T. A. MANTEUFFEL, *Orthogonal error methods*, SIAM J. Numer. Anal., 24 (1987), pp. 170–187. [209]

[73] P. A. FARRELL, *Flow-conforming iterative methods or convectiondominated flows*, in IMACS Annals on Computing and Applied Mathematics, Basel, J. C. Baltzer, AG Scientific Co., 1989, pp. 681–686. [209, 210]

[74] B. FISCHER, *Polynomial Based Iteration Methods for Symmetric Linear Systems*, Wiley-Teubner, Chichester-Stuttgart, 1996. [87, 306, 308]

[75] B. FISCHER, A. RAMAGE, D. SILVESTER, AND A. J. WATHEN, *Minimum residual methods for augmented systems*, BIT, 38 (1998), pp. 527–543. [345]

[76] R. FLETCHER, *Conjugate gradient methods for indefinite systems*, in Numerical Methods Dundee 1975, G. A. Watson, ed., Springer-Verlag, New York, 1976, pp. 73–89. [173, 209]

[77] M. FORTIN, *An analysis of the convergence of mixed finite element methods*, R.A.I.R.O. Anal. Numér., 11 (1977), pp. 341–354. [279]

[78] M. FORTIN AND R. GLOWINSKI, *Augmented Lagrangian methods: applications to the numerical solutions of boundary-value problems*, North-Holland, Amsterdam-Oxford, 1983. [308]

[79] R. FREUND, G. H. GOLUB, AND N. M. NACHTIGAL, *Iterative solution of linear systems*, in Acta Numerica, Cambridge University Press, Cambridge, 1992, pp. 57–100. [209]

[80] R. FREUND AND N. M. NACHTIGAL, *QMR: a quasi-minimal residual method for non-Hermitian linear systems*, Numer. Math., 60 (1991), pp. 315–339. [176, 209, 312]

[81] R. W. FREUND AND N. M. NACHTIGAL, *A new Krylov-subspace method for symmetric indefinite linear systems*, in Proceedings of the 14th IMACS World Congress on Computational and Applied Mathematics, W. F. Ames, ed., IMACS, 1994, pp. 1253–1256. [312]

[82] A. GAUTHIER, F. SALERI, AND A. VENEZIANI, *A fast preconditioner for the incompressible Navier–Stokes equations*, Comput. Visual. Sci., 6 (2004), pp. 105–112. [375]

[83] A. GEORGE AND J. W. LIU, *Computer Solution of Large Sparse Positive Definite Systems*, Prentice-Hall, New Jersey, 1981. [68]

[84] U. GHIA, K. GHIA, AND C. SHIN, *High Re solution for incompressible viscous flow using the Navier–Stokes equations and a multigrid method*, J. Comput. Phys., 48 (1982), pp. 387–395. [338]

[85] V. GIRAULT AND P.-A. RAVIART, *Finite Element Methods for Navier–Stokes Equations*, Springer-Verlag, Berlin, 1986. [319, 320, 324, 326, 330, 331, 375]

[86] G. GOLUB AND A. WATHEN, *An iteration for indefinite systems and its application to the Navier–Stokes equations*, SIAM J. Sci. Comput., 19 (1998), pp. 530–539. [375]

[87] G. H. GOLUB AND C. F. V. LOAN, *Matrix Computations*, The Johns Hopkins University Press, Baltimore, third ed., 1996. [68, 285]

[88] G. H. GOLUB AND D. P. O'LEARY, *Some history of the conjugate gradient and Lanczos algorithms: 1948–1976*, SIAM Rev., 31 (1989), pp. 50–102. [84]

[89] A. GREENBAUM, *Iterative Methods for Solving Linear Systems*, SIAM, Philadelphia, 1997. [88, 209]

[90] P. GRESHO, D. GARTLING, J. TORCZYNSKI, K. CLIFFE, K. WINTERS, T. GARRATT, A. SPENCE, AND J. GOODRICH, *Is the steady viscous incompressible 2d flow over a backward facing step at Re = 800 stable?*, Int. J. Numer. Methods Fluids, 17 (1993), pp. 501–541. [337]

[91] P. GRESHO AND R. SANI, *Incompressible Flow and the Finite Element Method: Volume 1: Advection–Diffusion*, John Wiley, Chichester, 1998. [162]

[92] P. GRESHO AND R. SANI, *Incompressible Flow and the Finite Element Method: Volume 2: Isothermal Laminar flow*, John Wiley, Chichester, 1998. [217, 237, 273, 277, 279, 337]

[93] P. M. GRESHO AND R. L. LEE, *Don't supress the wiggles — they're telling you something*, Computers and Fluids, 9 (1981), pp. 223–253. [162]

[94] M. GUNZBURGER, *Navier–Stokes equations for incompressible flows: finite-element methods*, in Handbook of Computational Fluid Dynamics, R. Peyret, ed., Academic Press, London, 1996. [217]

[95] M. D. GUNZBURGER, *Finite Element Methods for Viscous Incompressible Flows*, Academic Press, San Diego, 1989. [324, 338]

[96] I. GUSTAFSSON, *A class of first order factorization methods*, BIT, 18 (1978), pp. 142–156. [82]

[97] W. HACKBUSCH, *Multi-Grid Methods and Applications*, Springer-Verlag, Berlin, 1985. [95, 107, 209, 210, 376]

[98] W. HACKBUSCH, *Iterative Solution of Large Sparse Systems of Equations*, Springer-Verlag, Berlin, 1994. [88, 209]

[99] W. HACKBUSCH AND T. PROBST, *Downwind Gauss–Seidel smoothing for convection dominated problems*, Numer. Linear Algebra Appl., 4 (1997), pp. 85–102. [209, 210]

[100] H. HAN, V. P. IL'IN, R. B. KELLOGG, AND W. YUAN, *Analysis of flow directed iterations*, J. Comp. Math., 10 (1992), pp. 57–76. [209, 210]

[101] F. HARLOW AND J. WELCH, *Numerical calculation of time-dependent viscous incompressible flow of fluid with free surface*, Phys. Fluids, 8 (1965), pp. 2182–2189. [278]

[102] G. W. HEDSTROM AND A. OSTERHELD, *The effect of cell Reynolds number on the computation of a boundary layer*, J. Comput. Phys., 37 (1980), pp. 399–421. [162]

[103] M. O. HENRIKSEN AND J. HOLMEN, *Algebraic splitting for incompressible Navier–Stokes equations*, J. Comput. Phys., 175 (2002), pp. 438–453. [376]

[104] M. R. HESTENES AND E. STIEFEL, *Methods of conjugate gradients for solving linear systems*, Journal of Research of the National Bureau of Standards, 49 (1952), pp. 409–435. [69]

[105] N. J. HIGHAM, *Accuracy and Stability of Numerical Algorithms*, SIAM, Philadelphia, 1996. [59]

[106] T. HUGHES, , W. LIU, AND A. BROOKS, *Finite element analysis of incompressible viscous flows by the penalty function formulation*, J. Comput. Phys., 30 (1979), pp. 1–60. [278]

[107] T. HUGHES AND A. BROOKS, *A multi-dimensional upwind scheme with no crosswind diffusion*, in Finite Element Methods for Convection Dominated Flows, T. Hughes, ed., AMD–vol 34, ASME, New York, 1979. [126, 162]

[108] T. J. R. HUGHES, *The Finite Element Method*, Prentice-Hall, New Jersey, 1987. [28, 32, 33, 278]

[109] B. IRONS, *A frontal solution program for finite element analysis*, Internat. J. Numer. Methods in Eng., 2 (1970), pp. 5–32. [68]

[110] R. I. ISSA, *Solution of the implicitly discretised fluid flow equations by operator splitting*, J. Comput. Phys., 62 (1986), pp. 40–65. [368]

[111] C. JOHNSON, *Numerical Solution of Partial Differential Equations by the Finite Element Method*, Cambridge University Press, New York, 1987. [35, 47, 162]

[112] C. JOHNSON, R. RANNACHER, AND M. BOMAN, *Numerics and hydrodynamic stability: towards error control in computational fluid dynamics*, SIAM J. Numer. Anal., 32 (1995), pp. 1058–1079. [338]

[113] C. JOHNSON AND J. SARANEN, *Streamline-diffusion methods for the incompressible Euler and Navier–Stokes equations*, Math. Comp., 47 (1986), pp. 1–18. [338]

[114] C. JOHNSON, A. H. SCHATZ, AND L. B. WAHLBIN, *Crosswind smear and pointwise error in streamlined diffusion finite element methods*, Math. Comp., 54 (1990), pp. 107–130. [162]

[115] D. JOSEPH, *Fluid Dynamics of Viscoelastic Liquids*, Springer-Verlag, New York, 1990. [7]

[116] O. KARAKASHIAN, *On a Galerkin-Lagrange multiplier method for the stationary Navier–Stokes equations*, SIAM J. Numer. Anal., 19 (1982), pp. 909–923. [327]

[117] D. KAY, D. LOGHIN, AND A. WATHEN, *A preconditioner for the steady-state Navier–Stokes equations*, SIAM J. Sci. Comput., 24 (2002), pp. 237–256. [375]

[118] D. KAY AND D. SILVESTER, *A posteriori error estimation for stabilized mixed approximations of the Stokes equations*, SIAM J. Sci. Comput., 21 (1999), pp. 1321–1336. [265, 266]

[119] D. KAY AND D. SILVESTER, *The reliability of local error estimators for convection–diffusion equations*, IMA J. Numer. Anal., 21 (2001), pp. 107–122. [144]

[120] N. KECHKAR AND D. SILVESTER, *Analysis of locally stabilized mixed finite element methods for the Stokes problem*, Math. Comp., 58 (1992), pp. 1–10. [259, 260]

[121] H. KELLER, *Numerical Methods in Bifurcation Problems*, Springer-Verlag, 1987. [338]

[122] T. KELLEY, *Iterative Methods for Linear and Nonlinear Equations*, SIAM, Philadelphia, 1995. [371]

[123] A. KLAWONN, *Block-triangular preconditioners for saddle point problems with a penalty term*, SIAM J. Sci. Comput., 19 (1998), pp. 172–184. [308]

[124] A. KLAWONN AND G. STARKE, *Block triangular preconditioners for nonsymmetric saddle point problems: field-of-values analysis*, Numer. Math., 81 (1999), pp. 577–594. [346]

[125] M. KŘÍŽEK, *On semiregular families of triangulations and linear interpolation*, Appl. Math., 36 (1991), pp. 223–232. [44]

[126] O. LADYZHENSKAYA, *The Mathematical Theory of Viscous Incompressible Flow*, Gordon & Breach, New York, 1969. [338]

[127] D. LOGHIN AND A. J. WATHEN, *Analysis of preconditioners for saddle-point problems*, SIAM J. Sci. Comput., 25 (2004), pp. 2029–2049. [365]

[128] J. MANDEL, *Multigrid convergence for nonsymmetric, indefinite variational problems and one smoothing step*, Appl. Math. Comput., 19 (1986), pp. 201–216. [207]

[129] K. MARDAL, X.-C. TAI, AND R. WINTHER, *A robust finite element method for Darcy–Stokes flow*, SIAM J. Numer. Anal., 40 (2002), pp. 1605–1631. [279]

[130] J. A. MEIJERINK AND H. A. VAN DER VORST, *An iterative solution method for linear systems of which the coefficient matrix is a symmetric m-matrix*, Math. Comp., 31 (1977), pp. 148–162. [81]

[131] G. MEURANT, *Computer Solution of Large Linear Systems*, North-Holland, Amsterdam, 1999. [82, 88, 209]

[132] J. J. H. MILLER, E. O'RIORDAN, AND G. I. SHISHKIN, *Fitted Numerical Methods for Singularly Perturbed Problems — Error Estimates in the Maximum Norm for Linear Problems in One and Two Dimensions*, World Scientific, Singapore, 1995. [140, 161, 162]

[133] H. MOFFATT, *Viscous and resistive eddies near a sharp corner*, J. Fluid Mech., 18 (1964), pp. 1–18. [278]

[134] K. W. MORTON, *Numerical Solution of Convection–Diffusion Problems*, Chapman & Hall, London, 1996. [152, 161, 162]

[135] M. F. MURPHY, G. H. GOLUB, AND A. J. WATHEN, *A note on preconditioning for indefinite linear systems*, SIAM J. Sci. Comput., 21 (2000), pp. 1969–1972. [343]

[136] R. A. NICOLAIDES, *Existence, uniqueness and approximation for generalized saddle point problems*, SIAM J. Numer. Anal., 19 (1982), pp. 349–357. [279]

[137] R. A. NICOLAIDES, *Analysis and convergence of the MAC scheme I*, SIAM J. Numer. Anal., 29 (1992), pp. 1579–1591. [278]

[138] J. OCKENDON, S. HOWISON, A. LACEY, AND A. MOVCHAN, *Applied Partial Differential Equations*, Oxford University Press, Oxford, 1999. [10]

[139] M. OLSHANSKII, *An iterative solver for the Oseen problem and numerical solution of incompressible Navier–Stokes equations*, Num. Lin. Alg. with Appl., 6 (1999), pp. 353–378. [375]

[140] M. A. OLSHANSKII AND A. REUSKEN, *Convergence analysis of a multigrid method for a convection-dominated model problem*, SIAM J. Numer. Anal., 42 (2004), pp. 1261–1291. [202]

[141] J. E. OSBORN, *Spectral approximation for compact operators*, Math. Comp., 29 (1975), pp. 712–725. [188]

[142] C. PAIGE AND M. SAUNDERS, *Solution of sparse indefinite systems of linear equations*, SIAM J. Numer. Anal., 12 (1975), pp. 617–629. [86, 308]

[143] S. V. PARTER, *On estimating the "rates of convergence" of iterative methods for elliptic difference operators*, Trans. Amer. Math. Soc., 114 (1965), pp. 320–354. [185]

[144] S. V. PARTER, *On the eigenvalues of second order elliptic difference operators*, SIAM J. Numer. Anal., 19 (1982), pp. 518–530. [188]

[145] S. V. PARTER AND M. STEUERWALT, *Block iterative methods for elliptic and parabolic difference equations*, SIAM J. Numer. Anal., 19 (1982), pp. 1173–1195. [185]

[146] S. V. PARTER AND M. STEUERWALT, *Block iterative methods for elliptic finite element equations*, SIAM J. Numer. Anal., 22 (1985), pp. 146–179. [185, 188, 190]

[147] S. V. PATANKAR, *Numerical Heat Transfer and Fluid Flow*, New York, MacGraw-Hill, 1980. [368]

[148] S. V. PATANKAR AND D. A. SPALDING, *A calculation procedure for heat, mass and momentum transfer in three dimensional parabolic flows*, Int. J. Heat and Mass Trans., 15 (1972), pp. 1787–1806. [368]

[149] J. B. PEROT, *An analysis of the fractional step method*, J. Comput. Phys., 108 (1993), pp. 51–58. [376]

[150] C. POWELL AND D. SILVESTER, *Black-box preconditioning for mixed formulation of self-adjoint elliptic PDEs*, in Challenges in Scientific Computing–CISC 2002, E. Bänsch, ed., Springer, Berlin, 2003. [308]

[151] A. QUARTERONI, F. SALERI, AND A. VENEZIANI, *Factorization methods for the numerical approximation of Navier–Stokes equations*, Comput. Methods Appl. Mech. Eng., 188 (2000), pp. 505–526. [376]

[152] A. QUARTERONI AND A. VALLI, *Numerical Approximation of Partial Differential Equations*, Springer-Verlag, Berlin, 1997. [122, 162, 308, 319]

[153] A. QUARTERONI AND A. VALLI, *Domain Decomposition Methods for Partial Differential Equations*, Oxford University Press, Oxford, 1999. [82]

[154] A. RAMAGE, *A multigrid preconditioner for stabilised discretisations of advection–diffusion problems*, J. Comp. Appl. Math., 110 (1999), pp. 187–203. [210]

[155] R. RANNACHER, *Adaptive Galerkin finite element methods for partial differential equations*, J. Comput. Appl. Math., 128 (2001), pp. 205–233. [338]

[156] M. RENARDY AND R. ROGERS, *An Introduction to Partial Differential Equations*, Springer-Verlag, Berlin, 1993. [14, 122]

[157] A. REUSKEN, *Convergence analysis of a multigrid method for convection–diffusion equations*, Numer. Math., 91 (2002), pp. 323–349. [208]

[158] T. J. RIVLIN, *An Introduction to the Approximation of Functions*, Blaisdell Publishing Company, Walham, Massachusetts, 1969. [74]

[159] H.-G. ROOS, M. STYNES, AND L. TOBISKA, *Numerical Methods for Singularly Perturbed Differential Equations*, Springer-Verlag, Berlin, 1996. [117, 123, 136, 141, 161, 162, 206, 327, 338]

[160] J. W. RUGE AND K. STÜBEN, *Algebraic multigrid (AMG)*, in Multigrid Methods, Frontiers in Applied Mathematics, S. F. McCormick, ed., SIAM, Philadelphia, 1987, pp. 73–130. [362]

[161] T. RUSTEN AND R. WINTHER, *A preconditioned iterative method for saddle-point problems*, SIAM J. Matrix Anal. Appl., 13 (1992), pp. 887–904. [308]

[162] Y. SAAD, *Krylov subspace methods for solving large unsymmetric systems of linear equations*, Math. Comp., 37 (1981), pp. 105–126. [208]

[163] Y. SAAD, *A flexible inner–outer preconditioned GMRES algorithm*, SIAM J. Sci. Comput., 14 (1993), pp. 461–469. [360]

[164] Y. SAAD, *Iterative Methods for Sparse Linear Systems*, PWS Publishing, Boston, 1996. Second Edition, SIAM, Philadelphia, 2003. [88, 171, 209]

[165] Y. SAAD AND M. H. SCHULTZ, *GMRES: A generalized minimal residual algorithm for solving nonsymmetric linear systems*, SIAM J. Sci. Stat. Comput., 7 (1986), pp. 856–869. [167, 208]

[166] R. SANI, P. GRESHO, R. LEE, AND D. GRIFFITHS, *On the cause and cure(?) of the spurious pressures generated by certain FEM solutions of the incompressible Navier–Stokes equations*, Int. J. Numer. Methods Fluids, 1 (1981), pp. 17–43 and 171–204. [279]

[167] D. SCHÖTZAU AND C. SCHWAB, *Mixed hp-finite element methods on aniso-tropic meshes*, Math. Methods Appl. Sci., 8 (1998), pp. 787–820. [279]

[168] D. SCHÖTZAU, C. SCHWAB, AND R. STENBERG, *Mixed hp-FEM on aniso-tropic meshes II: Hanging nodes and tensor products of boundary layer meshes*, Numer. Math., 83 (1999), pp. 667–697. [280]

[169] C. SCHWAB, *p- and hp-Finite Element Methods*, Oxford University Press, Oxford, 1998. [280]

[170] A. SEGAL, *Aspects of numerical methods for elliptic singular perturbation problems*, SIAM J. Sci. Stat. Comput., 3 (1982), pp. 327–349. [162]

[171] P. SHANKAR AND M. DESHPANDE, *Fluid mechanics in the driven cavity*, Annu. Rev. Fluid Mech., 32 (2000), pp. 93–136. [278, 317]

[172] J.-M. SHI, M. BREUER, AND F. DURST, *A combined analytical–numerical method for treating corner singularities in viscous flow predictions*, Int. J. Numer. Meth. Fluids, 45 (2004), pp. 659–688. [338]

[173] G. I. SHISHKIN, *Methods of constructing grid approximations for singularly perturbed boundary-value problems, condensing grid methods.*, Russ. J. Numer. Anal. Math. Modelling, 7 (1992), pp. 537–562. [162]

[174] D. SILVESTER, H. ELMAN, D. KAY, AND A. WATHEN, *Efficient precon-ditioning of the linearized Navier–Stokes equations for incompressible flow*, J. Comp. Appl. Math., 128 (2001), pp. 261–279. [375]

[175] D. SILVESTER AND A. WATHEN, *Fast iterative solution of stabilised Stokes systems. Part II: Using general block preconditioners*, SIAM J. Numer. Anal., 31 (1994), pp. 1352–1367. [308]

[176] V. SIMONCINI AND D. SZYLD, *Flexible inner-outer Krylov subspace methods*, SIAM J. Numer. Anal., 40 (2003), pp. 2219–2239. [360]

[177] G. SLEIJPEN, H. VAN DER VORST, AND J. MODERSITZKI, *Differences in the effects of rounding errors in Krylov solvers for symmetric indefinite linear systems*, SIAM J. Matrix Anal. Appl., 22 (2000), pp. 726–751. [308]

[178] G. L. G. SLEIJPEN AND D. R. FOKKEMA, *BICGSTAB(L) for linear equa-tions involving unsymmetric matrices with complex spectrum*, Elec. Trans. Numer. Math., 1 (1993), pp. 11–32. [176, 209]

[179] B. SMITH, P. BJORSTAD, AND W. GROPP, *Domain Decomposition*, Cambridge University Press, Cambridge, 1996. [82]

[180] P. SONNEVELD, *CGS, a fast Lanczos-type solver for nonsymmetric linear systems*, SIAM J. Sci. Stat. Comput., 10 (1989), pp. 36–52. [209]

[181] A. SPENCE AND I. GRAHAM, *Numerical methods for bifurcation problems*, in The Graduate Student's Guide to Numerical Analysis, M. Ainsworth, J. Levesley, and M. Marletta, eds., Springer, Berlin, 1999. [338]

[182] M. R. SPIEGEL, *Mathematical Handbook of Formulas and Tables*, Schaum's Outline Series, McGraw-Hill, New York, 1990. [159]

[183] R. STENBERG, *Analysis of mixed finite element methods for the Stokes problem: A unified approach*, Math. Comp., 42 (1984), pp. 9–23. [229, 256, 257]

[184] R. STENBERG AND M. SURI, *Mixed hp finite element methods for problems in elasticity and Stokes flow*, Numer. Math., 72 (1996), pp. 367–389. [279]

[185] G. STOYAN, *Towards discrete Velte decompositions and narrow bounds for inf-sup constants*, Comp. Math. Appl., 38 (1999), pp. 243–261. [280]

[186] G. STOYAN, $-\Delta = -\text{grad div} + \text{rot rot}$ *for matrices, with application to the finite element solution of the Stokes problem*, East–West J. Numer. Math., 8 (2000), pp. 323–340. [280]

[187] G. STRANG AND G. J. FIX, *An Analysis of the Finite Element Method*, Prentice-Hall, Englewood Cliffs, NJ, 1973. [12]

[188] M. STYNES AND L. TOBISKA, *Necessary l^2-uniform convergence conditions for difference schemes for two-dimensional convection–diffusion problems*, Comput. Math. Appl., 29 (1995), pp. 45–53. [162]

[189] SYAMSUDHUHA AND D. SILVESTER, *Efficient solution of the steady-state Navier–Stokes equations using a multigrid preconditioned Newton-Krylov method*, Int. J. Numer. Methods Fluids, 43 (2003), pp. 1407–1427. [372]

[190] R. TEMAM, *Sur l'approximation de la solution des équations de Navier–Stokes par la méthod des pas fractionnaires (II)*, Arch. Rational Mech. Anal., 33 (1969), pp. 377–385. [376]

[191] R. TEMAM, *Navier–Stokes Equations*, North-Holland, Amsterdam, 1979. [337, 338]

[192] R. TEMAM, *Navier–Stokes Equations and Nonlinear Functional Analysis*, SIAM, Philadelphia, 1983. [338]

[193] L. TOBISKA AND G. LUBE, *A modified streamline-diffusion method for solving the stationary Navier–Stokes equations*, Numer. Math., 59 (1991), pp. 13–29. [338]

[194] A. TOSELLI AND C. SCHWAB, *Mixed HP-finite element approximations on geometric edge and boundary layer meshes in three dimensions*, Numer. Math., 94 (2002), pp. 771–801. [280]

[195] L. N. TREFETHEN, *Pseudospectra of matrices*, in Numerical Analysis 1991, D. F. Griffiths and G. A. Watson, eds., Harlow, UK, 1992, Longman Scientific and Techincal, pp. 234–266. [209]

[196] U. TROTTENBERG, C. OOSTERLEE, AND A. SCHÜLLER, *Multigrid*, Academic Press, London, 2001. [88, 98, 196, 210, 377]

[197] S. TUREK, *Efficient Solvers for Incompressible Flow Problems*, Springer-Verlag, Berlin, 1999. [375]

[198] H. A. VAN DER VORST, *BI-CGSTAB: A fast and smoothly converging variant of BI-CG for the solution of nonsymmetric linear systems*, SIAM J. Sci. Stat. Comput., 10 (1992), pp. 631–644. [174, 209]

[199] H. A. VAN DER VORST, *Iterative Krylov Methods for Large Linear Systems*, Cambridge University Press, New York, 2003. [84, 88, 176, 209]

[200] J. P. VAN DOORMAAL AND G. D. RAITHBY, *Enhancements of the simple method for predicting incompressible fluid flows*, Numer. Heat Transfer, 7 (1984), pp. 147–163. [368]

[201] S. P. VANKA, *Block-implicit multigrid solution of Navier–Stokes equations in primitive variables*, J. Comput. Phys., 65 (1986), pp. 138–158. [377]

[202] R. S. VARGA, *Matrix Iterative Analysis*, Prentice-Hall, Englewood Cliffs, New Jersey, 1962. Second Edition, Springer-Verlag, New York, 2000. [190, 209]

[203] R. VERFÜRTH, *A posteriori error estimators for the Stokes equations*, Numer. Math., 55 (1989), pp. 309–325. [266, 279]

[204] R. VERFÜRTH, *A Review of A Posteriori Error Estimation and Adaptive Mesh-Refinement Techniques*, Wiley–Teubner, Chichester, 1996. [52, 54, 330, 335]

[205] R. VERFÜRTH, *A posteriori error estimators for convection–diffusion equations*, Numer. Math., 80 (1998), pp. 641–663. [144, 162]

[206] P. K. W. VINSOME, *Orthomin, an iterative methods for solving sparse sets of simultaneous equations*, in Proceedings of the 4th Symposium on Reservoir Simulation, Society of Petroleum Engineers of AIME, 1976, pp. 149–159. [208]

[207] V. V. VOEVODIN, *The question of non-self-adjoint extension of the conjugate gradients method is closed*, U.S.S.R. Comput. Maths. Math. Phys., 23 (1983), pp. 143–144. [209]

[208] F. WANG AND J. XU, *A block iterative methods for convection-dominated problems*, SIAM J. Sci. Comput., 21 (1993), pp. 620–645. [209, 210]

[209] J. WANG, *A convergence estimate for multigrid methods for nonselfadjoint and indefinite elliptic problems*, SIAM J. Numer. Anal., 30 (1993), pp. 275–285. [207]

[210] A. WATHEN, *Realistic eigenvalue bounds for the Galerkin mass matrix*, IMA J. Numer. Anal., 7 (1987), pp. 449–457. [296]

[211] A. WATHEN, B. FISCHER, AND D. SILVESTER, *The convergence rate of the minimum residual method for the Stokes problem*, Numer. Math., 71 (1995), pp. 121–134. [308]

[212] A. WATHEN, B. FISCHER, AND D. SILVESTER, *The convergence of iterative solution methods for symmetric and indefinite linear systems*, in Numerical Analysis 1997, D. Griffiths, D. Higham, and G. Watson, eds., Longman, 1998, pp. 230–243. [308]

[213] A. WATHEN AND D. SILVESTER, *Fast iterative solution of stabilised Stokes systems. Part I: Using simple diagonal preconditioners*, SIAM J. Numer. Anal., 30 (1993), pp. 630–649. [308]

[214] P. WESSELING, *An Introduction to Multigrid Methods*, John Wiley & Sons, New York, 1992. [107, 196, 209, 210, 376]

[215] P. WESSELING, *Principles of Computational Fluid Dynamics*, Springer-Verlag, Berlin, 2001. [368, 369, 376]

[216] G. WITTUM, *Multi-grid methods for the Stokes and Navier–Stokes equations*, Numer. Math., 54 (1989), pp. 543–564. [377]

[217] G. WITTUM, *On the convergence of multi-grid methods with transforming smoothers*, Numer. Math., 57 (1990), pp. 15–38. [377]

[218] C.-T. WU AND H. C. ELMAN, *Analysis and Comparison of Geometric and Algebraic Multigrid for Convection–Diffusion Operators*, tech. rep., Institute for Advanced Computer Studies, University of Maryland, 2004. [198, 199]

[219] I. YAVNEH, C. H. VENNER, AND A. BRANDT, *Fast multigrid solution of the advection problem with closed characteristics*, SIAM J. Sci. Comput., 19 (1998), pp. 111–125. [210]

[220] D. M. YOUNG, *Iterative Solution of Large Linear Systems*, Academic Press, New York, 1970. [182, 209]

[221] D. M. YOUNG AND K. C. JEA, *Generalized conjugate gradient acceleration of nonsymmetrizable iterative methods*, Linear Algebra Appl., 34 (1980), pp. 159–194. [208]

[222] G. ZHOU AND R. RANNACHER, *Pointwise superconvergence of the streamline diffusion finite element method*, Numer. Methods Partial Diff. Equations, 12 (1996), pp. 123–145. [162]

[223] W. ZULEHNER, *Analysis of iterative methods for saddle point problems: A unified approach*, Math. Comput., 71 (2001), pp. 479–505. [308]

INDEX

Made in the USA
Lexington, KY
10 December 2012